AF360700

Land Degradation Assessment with Earth Observation

Land Degradation Assessment with Earth Observation

Editor

Elias Symeonakis

MDPI • Basel • Beijing • Wuhan • Barcelona • Belgrade • Manchester • Tokyo • Cluj • Tianjin

Editor
Elias Symeonakis
Manchester Metropolitan University
UK

Editorial Office
MDPI
St. Alban-Anlage 66
4052 Basel, Switzerland

This is a reprint of articles from the Special Issue published online in the open access journal *Remote Sensing* (ISSN 2072-4292) (available at: https://www.mdpi.com/journal/remotesensing/special_issues/land_degradation_assessment_earth_observation).

For citation purposes, cite each article independently as indicated on the article page online and as indicated below:

LastName, A.A.; LastName, B.B.; LastName, C.C. Article Title. *Journal Name* **Year**, *Volume Number*, Page Range.

ISBN 978-3-0365-4227-0 (Hbk)
ISBN 978-3-0365-4228-7 (PDF)

Cover image courtesy of Elias Symeonakis

Contents

About the Editor

Elias Symeonakis

Elias Symeonakis is an Associate Professor of Earth Observation and GIS and the leader of the Remote Sensing group at Manchester Metropolitan University (MMU; https://www.remote-sensing-mmu.org/). His research is related to the assessment of land degradation in drylands, mainly in sub-Saharan Africa and the Mediterranean regions. Dr Symeonakis is one of the contributing authors of the First Mediterranean Assessment Report (https://www.medecc.org/first-mediterranean-assessment-report-mar1/) prepared by the Mediterranean Experts on Climate and environmental Change (MedECC; https://www.medecc.org/). For this work, MedECC was awarded the 2021 North–South Prize of the Council of Europe.

Editorial

Land Degradation Assessment with Earth Observation

Elias Symeonakis

Department of Natural Sciences, Manchester Metropolitan University, Manchester M1 5GD, UK;
e.symeonakis@mmu.ac.uk

Citation: Symeonakis, E. Land Degradation Assessment with Earth Observation. *Remote Sens.* 2022, 14, 1776. https://doi.org/10.3390/rs14081776

Received: 3 March 2022
Accepted: 2 April 2022
Published: 7 April 2022

Publisher's Note: MDPI stays neutral with regard to jurisdictional claims in published maps and institutional affiliations.

For decades now, land degradation has been identified as one of the most pressing problems facing the planet. Alarming estimates are often published by the academic community and intergovernmental organisations claiming that a third of the Earth's land surface is undergoing various degradation processes and almost half of the world's population is already residing in degraded lands. Moreover, as land degradation directly affects vegetation biophysical processes and leads to changes in ecosystem functioning, it has a knock-on effect on habitats and, therefore, on numerous species of flora and fauna that become endangered or/and extinct.

By far the most widely used approach in assessing land degradation has been to employ Earth observation (EO) data. Especially during the last decade, with technological advancements and the computational capacity of computers on the one hand, together with the availability of open access, remotely sensed data archives on the other, numerous studies dedicated in the study of the various aspects of land degradation have been undertaken. The spectral, spatial and temporal resolution of these studies varies considerably, and multiscale, multitemporal and multisensor approaches have also evolved.

This Special Issue (SI) on "Land Degradation Assessment with Earth Observation" provides 17 original research papers with a focus on land degradation in arid, semiarid and dry-subhumid areas (i.e., desertification) but also temperate rangelands, grasslands, woodlands and the humid tropics. The studies cover different spatial, spectral and temporal scales and employ a wealth of different optical, as well as radar sensors: from the finest spatial scale of an Unoccupied Aerial Vehicle (UAV), to PlanetScope, Sentinel-1 and -2, Gaofen, Landsat, MODIS, PROBA-V, SPOT VGT and AVHRR. Some of the ancillary datasets included in the methodological framework of a number of the papers are also derived from remotely sensed imagery, e.g., the SRTM digital elevation model or the Climate Hazards group Infrared Precipitation with Stations (CHIRPS) precipitation estimates. Many studies incorporate time-series analysis techniques that assess the general trend of vegetation or the timing and duration of the reduction in biological productivity brought about by land degradation (e.g., Mann—Kenndall test, Theil–Sen's slope, BFAST, TSS-RESTREND, LandTrendR). A number of papers employ statistical approaches in their analyses (e.g., principal components analysis, ordinary least squares or geographically weighted regression) or machine learning classification/regression techniques (e.g., Random Forests, Support Vector Machines). As anticipated from the latest trend in EO literature, some studies utilise the cloud computing infrastructure of Google Earth Engine to deal with the unprecedented volume of data that current methodological approaches entail.

Geographically, the papers of this SI are mostly related with areas within Africa (9 papers), which is unsurprising, as the African continent is the most severely affected by land degradation. The Asian region is also well represented with seven papers, and one paper is focused on an area in North America. In terms of the processes addressed, both abrupt and more salient changes and degradation processes are covered, with the most studied theme being the different aspects of vegetation degradation:

- Tomaszewska and Henebry [1] investigate pasture degradation in the Kyrgyz Republic, using spatiotemporal phenometrics with MODIS land surface temperature (LST) and Landsat Normalized Difference Vegetation Index (NDVI) data;

- Meng et al. [2] study grassland degradation in the Tibetan Plateau (China) with UAV and Gaofen data;
- Gedefaw et al. [3] look at rangeland degradation in New Mexico (USA) through a time-series analysis of the Global Inventory Modeling and Mapping Studies (GIMMS) NDVI and the Parameter elevation Regressions on Independent Slopes Model (PRISM) precipitation data;
- Wanyama et al. [4] study vegetation condition in the Mount Elgon ecosystem (Kenya and Uganda) with MODIS NDVI and CHIRPS precipitation data and a combination of trend and breakpoint analysis methods;
- Barvels and Fensholt [5] also look at vegetation condition. They use Landsat NDVI time-series and CHIRPS rainfall estimates with trend analysis techniques to assess greening and browning trends in the highlands of the Ethiopian Plateau;
- Adenle and Speranza [6] investigate degradation in the Nigerian Guinea savannah by combining MODIS-derived land degradation status estimates from a previous study with spatial data on different drivers of land degradation to identify socio-ecological archetypes of land degradation;
- The paper by Urban et al. [7] focuses on monitoring shrub encroachment in the Free State Province (South Africa) by incorporating a dense time-series of both radar (Sentinel-1) and optical (Sentinel-2) data;
- Li et al. [8] look at the spatial differences of vegetation response and associated land degradation due to multiple mining activities in northwestern China. They use Landsat imagery, monitor vegetation change using time-series analysis techniques and estimate the spatial heterogeneity of the change related specifically to mines.

Another common theme of the Special Issue is land degradation related to drought:

- Verhoeve et al. [9] study vegetation resilience under increasing drought conditions in two districts of Nothern Tanzania. They employ the National Oceanic and Atmospheric Administration (NOAA) Climate Data Record (CDR) AVHRR NDVI data together with Climate Research Unit (CRU) temperature and a combination of Cen-Trends and CHIRPS precipitation estimates;
- Kimura and Moriyama [10] look into drought conditions in Mongolia. They examine the trends in AVHRR- and MODIS-derived NDVI, as well as in an aridity index calculated using surface reflectance and LST data from MODIS, and propose a method to monitor land-surface dryness;
- Akinyemi [11] investigates the relationship between drought severity and land use/cover change in 17 constituencies in Botswana. She employs NDVI data from SPOT VGT and PROBA-V and land cover information from the European Space Agency's (ESA) Climate Change Initiative (CCI) and the Copernicus Climate Change Service (C3S-LC).

Soil erosion also appears as one of the land degradation processes of interest in the Special Issue:

- The study by Phinzi et al. [12] over an area of South Africa compares different classification algorithms and resampling methods to identify the optimal combination for the mapping of complex gully erosion systems, using PlanetScope data from the wet and dry seasons;
- Wang et al. [13] bring together MODIS NDVI and Land Aerosol Optical Depth data, climate assimilation and ancillary spatial data to develop a Google Earth Engine-based model for the delineation of the wind erosion potential of the entire Central Asian region (i.e., Kazakhstan, Uzbekistan, Turkmenistan, Kyrgyzstan and Tajikistan).

Land degradation related with the salinisation of the soil is also addressed in this Special Issue:

- Yu et al. [14] use Landsat data and integrate the salinization index, albedo, NDVI and the land surface soil moisture index to establish the salinized land degradation index (SDI) and apply their approach in an area that runs through Turkmenistan and Uzbekistan;

- Moussa et al. [15] compare a salinity index to an approach that employs Sentinel-2-derived NDVI time-series for detecting salt-affected soils in irrigated systems in an area of Niger.

One of the papers of the Special Issue deals with the humid tropics. In their study, Liu et al. [16] propose a framework for the improved accounting of reference levels (RLs) for the United Nations' Reducing Emissions from Deforestation and Forest Degradation (REDD+) programme. They combine the Intergovernmental Panel on Climate Change's (IPCC) Good Practice Guidance on Land Use, Land Use Change and Forestry with a land use change modelling approach and apply this to an area in southern China. Finally, the paper by Reith et al. (2021) focuses on the issue of land degradation monitoring and the methodology suggested by the United Nations Charter to Combat Desertification (UNCCD) to inform the sustainable development goal (SDG) 15.3.1 (i.e., "Proportion of degraded land over total land area"). Aiming to optimise the land degradation neutrality (LDN) efforts of the UN, Reith et al. [17] compare the land degradation assessments for an area in Central Tanzania derived using the coarser spatial-resolution MODIS and the finer Landsat data.

A clear message stemming from this SI is the ever-increasing relevance of Earth Observation technologies when it comes to assessing and monitoring land degradation. With the recently published IPCC AR6 Working Group II Report (https://www.ipcc.ch/report/ar6/wg2/downloads/report/IPCC_AR6_WGII_FinalDraft_FullReport.pdf, accessed on 3 March 2022), informing us of the severe impacts and risks to terrestrial and freshwater ecosystems and the ecosystem services they provide, the EO scientific community has a clear obligation to step up its efforts to address any remaining gaps—some of which have been identified in this SI—in order to produce highly accurate and relevant land degradation assessment and monitoring tools.

Funding: Elias Symeonakis is partly funded by a LEVERHULME TRUST INTERNATIONAL ACADEMIC FELLOWSHIP (Contract Number IF-2021–040).

Conflicts of Interest: The authors declare no conflict of interest.

References

1. Tomaszewska, M.A.; Henebry, G.M. Remote sensing of pasture degradation in the highlands of the kyrgyz republic: Finer-scale analysis reveals complicating factors. *Remote Sens.* **2021**, *13*, 3449. [CrossRef]
2. Meng, B.; Yang, Z.; Yu, H.; Qin, Y.; Sun, Y.; Zhang, J.; Chen, J.; Wang, Z.; Zhang, W.; Li, M.; et al. Mapping of kobresia pygmaea community based on umanned aerial vehicle technology and gaofen remote sensing data in alpine meadow grassland: A case study in eastern of Qinghai–Tibetan Plateau. *Remote Sens.* **2021**, *13*, 2483. [CrossRef]
3. Gedefaw, M.; Geli, H.; Abera, T. Assessment of rangeland degradation in New Mexico using time series segmentation and residual trend analysis (tss-restrend). *Remote Sens.* **2021**, *13*, 1618. [CrossRef]
4. Wanyama, D.; Moore, N.; Dahlin, K. Persistent vegetation greening and browning trends related to natural and human activities in the mount elgon ecosystem. *Remote Sens.* **2020**, *12*, 2113. [CrossRef]
5. Barvels, E.; Fensholt, R. Earth observation-based detectability of the effects of land management programmes to counter land degradation: A case study from the highlands of the Ethiopian Plateau. *Remote Sens.* **2021**, *13*, 1297. [CrossRef]
6. Adenle, A.; Speranza, C.I. Social-ecological archetypes of land degradation in the nigerian guinea savannah: Insights for sustainable land management. *Remote Sens.* **2020**, *13*, 32. [CrossRef]
7. Urban, M.; Schellenberg, K.; Morgenthal, T.; Dubois, C.; Hirner, A.; Gessner, U.; Mogonong, B.; Zhang, Z.; Baade, J.; Collett, A.; et al. Using sentinel-1 and sentinel-2 time series for slangbos mapping in the free state province, South Africa. *Remote Sens.* **2021**, *13*, 3342. [CrossRef]
8. Li, H.; Xie, M.; Wang, H.; Li, S.; Xu, M. Spatial heterogeneity of vegetation response to mining activities in resource regions of northwestern China. *Remote Sens.* **2020**, *12*, 3247. [CrossRef]
9. Verhoeve, S.L.; Keijzer, T.; Kaitila, R.; Wickama, J.; Sterk, G. Vegetation resilience under increasing drought conditions in northern Tanzania. *Remote Sens.* **2021**, *13*, 4592. [CrossRef]
10. Kimura, R.; Moriyama, M. Use of a modis Satellite-based aridity index to monitor drought conditions in Mongolia from 2001 to 2013. *Remote Sens.* **2021**, *13*, 2561. [CrossRef]
11. Akinyemi, F. Vegetation trends, drought severity and land use-land cover change during the growing season in semi-arid contexts. *Remote Sens.* **2021**, *13*, 836. [CrossRef]
12. Phinzi, K.; Abriha, D.; Szabó, S. Classification efficacy using k-fold cross-validation and bootstrapping resampling techniques on the example of mapping complex gully systems. *Remote Sens.* **2021**, *13*, 2980. [CrossRef]

13. Wang, W.; Samat, A.; Ge, Y.; Ma, L.; Tuheti, A.; Zou, S.; Abuduwaili, J. Quantitative soil wind erosion potential mapping for central Asia using the google earth engine platform. *Remote Sens.* **2020**, *12*, 3430. [CrossRef]
14. Yu, T.; Jiapaer, G.; Bao, A.; Zheng, G.; Jiang, L.; Yuan, Y.; Huang, X. Using synthetic remote sensing indicators to monitor the land degradation in a salinized area. *Remote Sens.* **2021**, *13*, 2851. [CrossRef]
15. Moussa, I.; Walter, C.; Michot, D.; Boukary, I.A.; Nicolas, H.; Pichelin, P.; Guéro, Y. Soil Salinity assessment in irrigated paddy fields of the niger valley using a four-year time series of sentinel-2 satellite images. *Remote Sens.* **2020**, *12*, 3399. [CrossRef]
16. Liu, G.; Feng, Y.; Xia, M.; Lu, H.; Guan, R.; Harada, K.; Zhang, C. Framework for accounting reference levels for redd+ in tropical forests: Case study from Xishuangbanna, China. *Remote Sens.* **2021**, *13*, 416. [CrossRef]
17. Reith, J.; Ghazaryan, G.; Muthoni, F.; Dubovyk, O. Assessment of land degradation in semiarid Tanzania—using multiscale remote sensing datasets to support sustainable development goal 15.3. *Remote Sens.* **2021**, *13*, 1754. [CrossRef]

remote sensing

Article

Remote Sensing of Pasture Degradation in the Highlands of the Kyrgyz Republic: Finer-Scale Analysis Reveals Complicating Factors

Monika A. Tomaszewska [1] and Geoffrey M. Henebry [1,2,*]

[1] Center for Global Change and Earth Observations, Michigan State University, East Lansing, MI 48823, USA; monikat@msu.edu

[2] Department of Geography, Environment, and Spatial Sciences, Michigan State University, East Lansing, MI 48824, USA

* Correspondence: henebryg@msu.edu; Tel.: +1-517-355-3899

Abstract: Degradation in the highland pastures of the Kyrgyz Republic, a small country in Central Asia, has been reported in several studies relying on coarse spatial resolution imagery, primarily MODIS. We used the results of land surface phenology modeling at higher spatial resolution to characterize spatial and temporal patterns of phenometrics indicative of the seasonal peak in herbaceous vegetation. In particular, we explored whether proximity to villages was associated with substantial decreases in the seasonal peak values. We found that terrain features—elevation and aspect—modulated the strength of the influence of village proximity on the phenometrics. Moreover, using contrasting hotter/drier and cooler/wetter years, we discovered that the growing season weather can interact with aspect to attenuate the negative influences of dry conditions on seasonal peak values. As these multiple contingent and interactive factors that shape the land surface phenology of the highland pastures may be blurred and obscured in coarser spatial resolution imagery, we discuss some limitations with prior and recent studies of pasture degradation.

Keywords: Kyrgyzstan; pastures; Landsat; MODIS; land surface phenology

Citation: Tomaszewska, M.A.; Henebry, G.M. Remote Sensing of Pasture Degradation in the Highlands of the Kyrgyz Republic: Finer-Scale Analysis Reveals Complicating Factors. *Remote Sens.* **2021**, *13*, 3449. https://doi.org/10.3390/rs13173449

Academic Editor: Elias Symeonakis

Received: 1 July 2021
Accepted: 26 August 2021
Published: 31 August 2021

Publisher's Note: MDPI stays neutral with regard to jurisdictional claims in published maps and institutional affiliations.

1. Introduction

The Kyrgyz Republic (Kyrgyzstan) is a small, highly mountainous nation of ~6.5 million in 2019, where more than 60% live in rural areas and where agropastoralism is the predominant land use [1]. Degradation of pasture resources in highly mountainous Kyrgyz Republic has been both widely reported [2–8] and disputed [9,10]. Many studies have attempted to detect land degradation in Central Asia by using remote sensing products based on finer temporal but coarser spatial resolution imagery [4,7,8,11–16]. While the finer temporal resolution products are better able to obtain clear views of the land surface, coarser spatial resolution products blend together heterogeneous surfaces. This blending or mixing is of particular concern in mountainous terrain, where differential insolation regimes arising from the interaction of aspect and slope generates microclimates that can accommodate distinct vegetation communities exhibiting different phenologies [3,17–19].

Untangling the differential influences of climate and human activity is complicated by coarser spatial resolution [7,14,20]. Pastoralism in Kyrgyzstan is characterized by transhumance, the seasonal movement of livestock to fresh pastures, and vertical transhumance in particular, where the herds are moved from low-elevation winter pastures near villages through transitional pastures in spring (and again in fall) to higher elevation summer pastures distant from settlements. This combination of spatial heterogeneity, terrain effects, and seasonality of pasture use presents challenges for detecting pasture degradation from coarser resolution remote sensing imagery. A further complication is presented by the high interannual variation in weather arising from the land-locked location of the country, its relatively high elevation, and regional climate change [12,21–24].

Here, we explore the pasture degradation question across rural Kyrgyz Republic using phenological metrics (phenometrics) derived from finer spatial resolution, but less frequent imagery (Landsat 30 m) linked to more frequent, but coarser spatial resolution products (MODIS 1 km) across 17 years (2001–2017). The phenometrics that we use here indicate the first seasonal peak in NDVI and the amount of thermal time (measured as accumulated growing degree-days calculated from land surface temperature time series) required to reach that peak NDVI. These phenometrics can be calculated for each pixel in each year for which there are sufficient, high-quality data and for which the joint time series exhibit sufficient seasonality to enable the modeling of the land surface phenology (LSP). Pasture degradation can be evaluated with phenometrics in a variety of ways, including trends, variation, extremes, and abrupt spatial and/or temporal shifts [12,20,25,26]. However, we are interested here in three questions about patterns that can complicate interpretation of patterns and trends: (1) What are the spatial patterns of the phenometrics as a function of distance from village center? (2) How do these patterns change as a function of elevation? and (3) How do these patterns change as a function of aspect?

Although we focus on the patterns of the temporal mean phenometrics during the study period, we also examine the patterns during two years with contrasting weather: hotter, drier 2007 versus cooler, wetter 2009. This study leverages the findings from prior studies [25,26] using the same approach to modeling land surface phenology and advances our understanding of the spatio-temporal dynamics of vegetation in the socio-ecological landscapes of montane Central Asia, specifically in the Kyrgyz Republic.

2. Materials and Methods

2.1. Study Area

The study area focuses on pasture lands within the territory of the Kyrgyz Republic that neighbors Uzbekistan (west), Kazakhstan (north), China (east and southeast), and Tajikistan (southwest) (Figure 1). The total area of the country is shy of 200,000 km^2 (96% in land), and the 2019 population was about 6.5 million, according to the World Bank (https://data.worldbank.org/country/kyrgyz-republic; accessed on 20 June 2021). It is a highly mountainous country, where more than 56% of the territory lies above 2500 m and where the mountain ranges of the Tien Shan, Pamir, and Alatau cover more than 90% of the total land area [27]. Pastoral rangelands constitute 87% [1] of the agricultural lands in the Kyrgyz Republic. Less than 10% of the land is used for crops, while forests cover only about 5%. Our study period extends from 2001 through 2017. The Kyrgyz Republic is divided into seven provinces (oblasts)—Talas, Chuy (including the capital city, Bishkek), and Issyk-Kul in the northern part as well as Jalal-Abad, Naryn, Osh, and Batken in the southern part—and 40 districts (rayons).

2.2. Geospatial Data

The land surface phenology metrics (or phenometrics) and terrain information used for this study were supplied from our previous work [25], where a detailed description of data, its processing, and phenometrics calculation methodology can be found. We provide an overview here.

To calculate LSP metrics, we used two products: MODIS land surface temperature (LST) and Landsat surface reflectance NDVI.

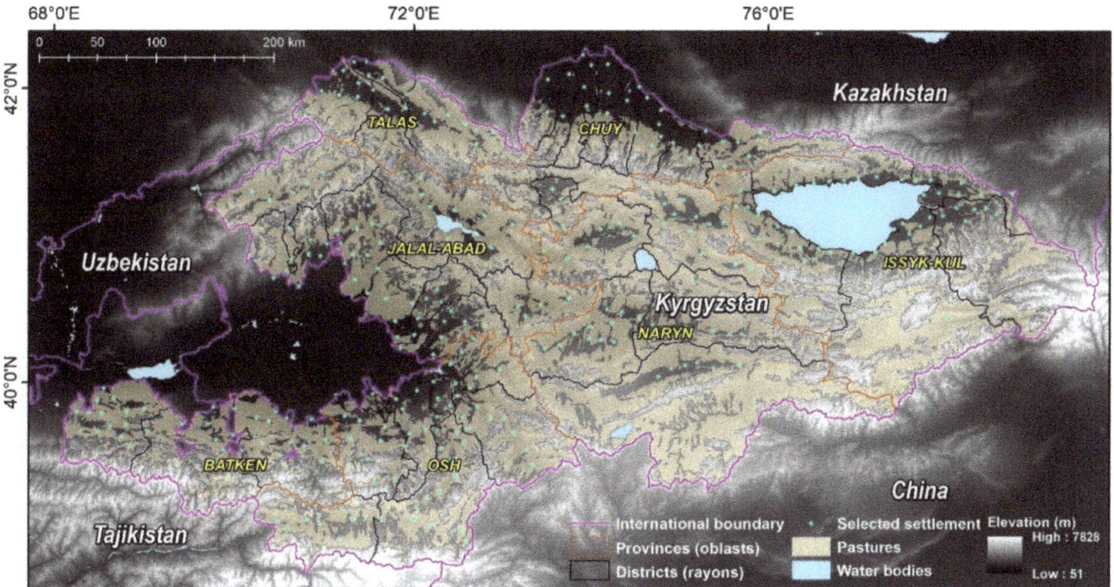

Figure 1. Pasture land use area (light tan) and selected settlement points (blue-green) in the Kyrgyz Republic (from [28,29]) draped over the SRTM 30 m DEM [30] (Projected coordinate system: Albers Conic Equal Area). Province (oblast) names appear in yellow.

We downloaded two tiles (h23v04 and h23v05) of 8-day MODIS Terra and MODIS Aqua Land Surface Temperature (MOD11A2/MYD11A2 V006) products at 1 km spatial resolution [31] from 2001 (MODIS/Terra) and from 2002 (MODIS/Aqua) up to the end of 2017. We merged tiles, removed poor quality pixels, converted units from Kelvin to °C, and reprojected data to Albers Conic Equal Area at 30 m spatial resolution using bilinear resampling.

The surface reflectance NDVI dataset from Landsat Collection 1 Tier 1 Level-1 Precision and Terrain (L1TP) of Landsat 5 Thematic Mapper (TM), Landsat 7 Enhanced Thematic Mapper Plus (ETM+), and Landsat 8 Operational Land Imager (OLI) was acquired for years 2001 to the end of 2017 from the USGS Earth Resources Observation and Science (EROS) Center Science Processing Architecture (ESPA) On-Demand Interface (https:// espa.cr.usgs.gov/). We downloaded 13,285 images across 33 unique tiles (WRS-2 Paths 147–155 and Rows 30–33) which were already projected into Albers Conic Equal Area. We masked poor-quality pixels and applied an inter-calibration equation to adjust Landsat 5 TM surface NDVI and Landsat 7 ETM+ surface NDVI to the surface Landsat 8 OLI NDVI, which on average was shown to have higher values (cf., Table 3 in [32], Surface NDVI from OLI = 0.0235 + 0.9723 ETM+). Because of the small differences between the Landsat 5 TM and Landsat 7 ETM+ data [33,34], we used the same equation for both datasets.

For the analyses conducted in this study, we additionally used three other geospatial datasets: (1) digital elevation model, (2) pasture land-use mask, and (3) point coverage of settlements. We downloaded 133 tiles of SRTMGL1, the NASA Shuttle Radar Topography Mission Global 1 arc second (~30 m) V003 elevation product [30] from USGS Earth Explorer (https://earthexplorer.usgs.gov/). Tiles were merged and reprojected into the Albers Conic Equal Area at 30 m spatial resolution using bilinear resampling. We then generated aspect and slope layers. For this study, we masked out pixels from all layers where slope was greater than 30 degrees. Additionally, we created sets of three terrain layers where we left only pixels on the contrasting aspects (northern: NW-N-NE-E, southern: SE-S-SW-W) at four elevation ranges: 1800–2400 m, 2400–2900 m, 2900–3400 m, and 3400–4000 m, and elevation−aspect interactions.

The pasture land-use class (122,405 km^2) was obtained from a Soviet-era land use map that was updated in 2008 using Landsat 7 ETM+ and MODIS datasets for the CACILM project [28,29]. We used those data to mask out non-pasture pixels in the terrain and LSP layers to focus only on the pasture land-use pixels.

The settlement point layer was also obtained from the CACILM project where the dataset was collected and consolidated by ECONET WWF Project [35] national teams. Over Kyrgyzstan, the layer has 703 points representing detailed administration levels, including small villages. We conducted a quality check on this dataset and eliminated points that were clearly mislocated, yielding 617 point locations (Table 1) for analysis.

Table 1. Number of settlement points after dataset revision and filtering per province (oblast).

Province (Oblast)	Total Points	Points after 10 km Filter
Batken	63	36
Chuy	67	33
Issyk-Kul	100	45
Jalal-Abad	113	58
Naryn	96	44
Osh	137	56
Talas	41	21
TOTAL	617	293

2.3. Methods

2.3.1. Land Surface Phenology

We used a downward-arching convex quadratic (CxQ) function to characterize LSP [36–38]. The model uses a vegetation index—here the NDVI from a Landsat surface reflectance time-series—as proxy for active green vegetation, and thermal time—here accumulated growing degree-days (AGDD) from MODIS LST—as proxy for insolation.

To obtain AGDD, we first transformed two diurnal and nocturnal observations from the MODIS on Terra and on Aqua into a mean LST using the following Equation (1):

$$\text{mean LSTt} = [\max(\text{LSTt}_{\text{TERRA1030}}, \text{LSTt}_{\text{AQUA1330}}) + \min(\text{LSTt}_{\text{TERRA2230}}, \text{LSTt}_{\text{AQUA0130}})]/2 \quad (1)$$

where $\text{LSTt}_{\text{TERRA1030}}$ is the LST for period t at the Terra daytime overpass, $\text{LSTt}_{\text{AQUA1330}}$ is the LSTt at the Aqua daytime overpass, $\text{LSTt}_{\text{TERRA2230}}$ is the LSTt at the Terra nighttime overpass, and $\text{LSTt}_{\text{AQUA0130}}$ is the LSTt at the Aqua nighttime overpass.

We filled gaps that resulted from missing or filtering excluded pixels using the Seasonally Decomposed Missing Value Imputation method [39], replaced all negative values with 0 °C, and calculated growing degree-days GDD (2) at compositing period t as the maximum of mean LST and T_{base}, where T_{base} was set to 0 °C [37,40].

$$\text{GDD}_t = \max(\text{mean LSTt} - T_{\text{base}}, 0) \quad (2)$$

For each year, we produced 46 GDD composites that were multiplied by 8 to account for the 8-day MODIS product composite period and accumulated each year into time series of AGDD (3), with an annual reset in January to 0 °C:

$$\text{AGDD}_t = \text{AGDD}_{t-1} + (\text{GDD}_t \times 8) \quad (3)$$

Prior the CxQ LSP model fitting, it was necessary to further clean and filter NDVI and AGDD datasets to reduce noise and spurious data. Therefore, we filtered out observations with NDVI < 0.1 and AGDD < 100 to avoid non-vegetated or snow-covered pixels. To account for cloud contamination that may have slipped through the masking process, we looked for unusual abrupt dips in the NDVI time-series. We first calculated the simple average of NDVI observations on either side of the focal observation. We then calculated the percentage difference between the average NDVI and the focal observation and excluded observations that were $\geq 15\%$ than the average of the two neighboring observations [25,26]. Having NDVI and AGDD datasets prepared for each pixel and each year (from 2001 to 2017), we applied the CxQ LSP model shown in (4):

$$NDVI = \alpha + \beta \times AGDD + \gamma \times AGDD^2 \qquad (4)$$

Using the fitted coefficients (4) for the intercept (α), slope (β), and quadratic (γ) parameters, we calculated two LSP phenometrics: Peak Height [PH=$\alpha - (\beta^2/4 \times \gamma)$], which is the maximum modeled NDVI; and Thermal Time to Peak [TTP $= -\beta/2 \times \gamma$], which is the quantity of AGDD required to reach PH and corresponds to the duration of green-up phase. To control CxQ LSP model fitting performance, we used a suite of six quality criteria: (i) the fitted quadratic parameter coefficient was less than zero ($\gamma < 0$); (ii) TTP greater than the AGDD at the first observation; (iii) adjusted R^2 greater than 0.7; (iv) Root Mean Square Difference (RMSD) less than 0.08; (v) at least seven observations in the time series to be fit, where at least three observations were distributed before and at least three after the PH; and (vi) the PH less than or equal to 1.0.

If any criterion was not fulfilled during the fitting process, then the last observation in the time series was removed and the model fitting procedure was rerun over the reduced time series. We repeated this fitting procedure until either the fitted model passed all criteria or the length of the time series was fewer than seven observations. In the latter case, the model fit for that pixel was labeled as failed and no phenometrics were calculated for that pixel.

Next, for each pixel where model fit was successful and phenometrics were obtained, we calculated the mean values of PH and TTP across 17 years. We also highlighted PH and TTP from the years 2007 and 2009, which were drier and wetter weather conditions, respectively. As mentioned in Section 2.2, we used the pasture land-use layer to mask data.

2.3.2. Settlement Ring Buffer Analyses

From the settlement point coverage, we randomly selected (ArcMap Software 10.6.0.8321, Random Generator Type: default ACM599, seed:0) points with the spatial constraint of 10 km from each other, which resulted in 293 focal points for analysis (Table 1).

Then, we created 10 ring buffers around each settlement focal point from 500 m to 5000 m distance in 500 m intervals (0–500 m, 500–1000 m, 1000–1500 m, 1500–2000 m, 2000–2500 m, 2500–3000 m, 3000–3500 m, 3500–4000 m, 4000–4500, and 4500–5000 m). The spatial constraint of 10 km ensured that the ring buffers would not intersect. Finally, we prepared nine datasets (raster stacks) at four elevation classes—(1) 1800–2400 m, (2) 2400–2900 m, (3) 2900–3400 m, and (4) 3400–4000 m—yielding 36 in total, as follows:

1. Elevation classes {1–4}, all aspects, slopes $< 30°$, PH_{mean}, TTP_{mean};
2. Elevation classes {1–4}, all aspects, slopes $< 30°$, PH_{2007}, TTP_{2007};
3. Elevation classes {1–4}, all aspects, slopes $< 30°$, PH_{2009}, TTP_{2009};
4. Elevation classes {1–4}, northern aspects, slopes $< 30°$, PH_{mean}, TTP_{mean};
5. Elevation classes {1–4}, northern aspects, slopes $< 30°$, PH_{2007}, TTP_{2007};
6. Elevation classes {1–4}, northern aspects, slopes $< 30°$, PH_{2009}, TTP_{2009};
7. Elevation classes {1–4}, southern aspects, slopes $< 30°$, PH_{mean}, TTP_{mean};
8. Elevation classes {1–4}, southern aspects, slopes $< 30°$, PH_{2007}, TTP_{2007};
9. Elevation classes {1–4}, southern aspects, slopes $< 30°$, PH_{2009}, TTP_{2009}.

Once these data were prepared, we extracted pixels from each set of settlement point ring buffers across all 36 datasets. During the extraction process, we ensured that only those pixels whose center point was within the ring buffer polygon were included so the same pixel would not spill into the neighboring ring buffers. Because of the prior pixel filtering by pasture land-use map, elevation classes, slope threshold value, and the fact that LSP CxQ modeling did not always succeed (i.e., no phenometrics calculated), the number of extracted pixels varied within each ring buffer for each dataset, and in some instances, there were no pixels to extract due to the multiple filtering steps. When there were extracted pixels, we calculated the mean values for each ring buffer, which were used for the complementary analyses of the influence of elevation, aspect, growing season weather, and distance from village on the two phenometrics.

3. Results

3.1. Buffer Mean Values of the Peak NDVI as a Function of Distance and Elevation

Figure 2 displays the spatial mean ±2SE values of the fitted NDVI Peak Height within each ring buffer calculated from the mean PH values across all years. Not surprisingly, there is considerable variation, but some patterns are evident. First, the PH decreased with increasing elevation. Second, the variation in the PH increased with elevation. Third, PH increased with distance from villages at 2400–2900 m and 2900–3400 m.

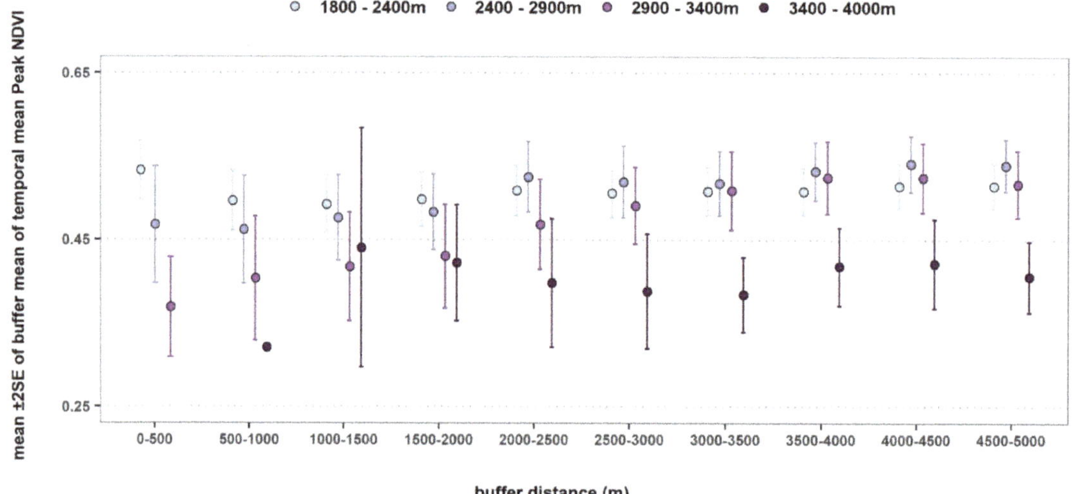

Figure 2. Elevational gradients in modeled peak NDVI: mean ±2SE of the ten ring buffer mean values calculated from the temporal mean Peak Height. Sequential color scheme starting from left represents four classes of elevation: 1800–2400 m, 2400–2900 m, 2900–3400 m, and 3400–4000 m.

3.2. Contrasting Mean Values of Phenometrics Nearby and Far from Villages

Figure 3 subsets the distributions to focus on either end of the spatial series: the 0–500 m ring buffer captured those pasture areas closest to villages and the 4500–5000 m ring buffer captured pastures that were far from the focal village and not intruding on the ring buffer of another village. For PH (Figure 3, left panel), there was no difference in the lowest elevation class (1800–2400 m), but the differences between PH values nearby and far from villages were strong at both 2400–2900 m and 2900–3400 m, where the PH values in the distant ring buffer were higher than in the nearby ring buffer. However, the distributions of the mean PH values were significantly different between the 0–500 m and 4500–5000 m buffers only at the 2900–3400 m elevation range, according to the Kolmogorov–Smirnov two-sample test with the Dunn-Šidák correction for post-hoc multiple comparisons [41]. Note that in the highest elevation class (3400–4000 m), there were no pasture areas in the ring buffer nearest (0–500 m) to villages. It is also evident that the variation of PH in the nearby ring buffer appeared larger at higher elevations relative to that in the lowest elevation class, but this difference may result from fewer samples at high elevations.

The Thermal Time to Peak phenometric (Figure 3, right panel) decreased with elevation, as expected, and the distant TTP ring buffer means were consistently—but not significantly—lower than the nearby values.

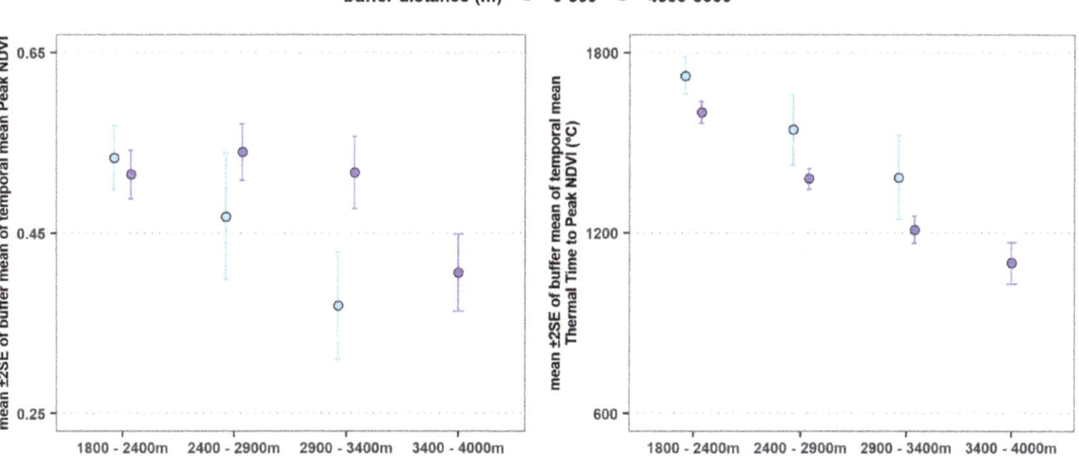

Figure 3. Contrasting values from pasture areas nearby and far from villages: mean ±2SE of the two (0–500 m in light blue, 4500–5000 m in purple) buffer mean values of temporal mean Peak Height (**left**) and mean Thermal Time to Peak (**right**) for the four elevation classes.

3.3. Influence of Weather on Phenometrics

Figure 4 shows the PH (left) and TTP (right) means for the ring buffers nearby versus far from the villages. For PH, the contrast between years and distance from villages was significant only for the 2900–3400 m elevation class (Figure 4, left). At the lowest elevation class, the nearby ring buffer means were not significantly different, but the distant means were. TTP was clearly lower during 2009, the wetter year, at all elevations, and lower at higher elevations, as expected. Evaluation by the Kolmogorov-Smirnov two-sample test (with the Dunn-Šidák correction) confirms what Figure 4 (right panel) shows: the TTP distributions were significantly different between 2007 and 2009 at every elevation class, but the difference between 0–500 m and 4500–5000 m ring buffer distributions was significant only at the 2900–3400 m elevation class and only in the drier year of 2007.

3.4. Influence of Aspect on Phenometrics

We tested for differences in the distributions of buffer means as a function of aspect (NW-N-NE-E vs. SE-S-SW-W) by distance from settlement point and elevation class using the Kolmogorov-Smirnov two-sample test with the Dunn-Šidák correction for post-hoc multiple comparisons. Northerly aspects consistently showed higher Peak Height means than in southerly aspects, but these differences were not statistically different at every distance or elevation class (Figure 5). In the lowest elevation class (1800–2400 m), the effect of aspect on PH was significant starting at the 2500–3000 m ring buffer; in the 2400–2900 m elevation class, the difference in PH between aspects appeared significant 500 m closer to the settlement point (i.e., at the 2000–2500 m ring buffer). In the two higher elevation classes, the aspect differences were not significant, but the variation about the means appears lower in the more distant ring buffers.

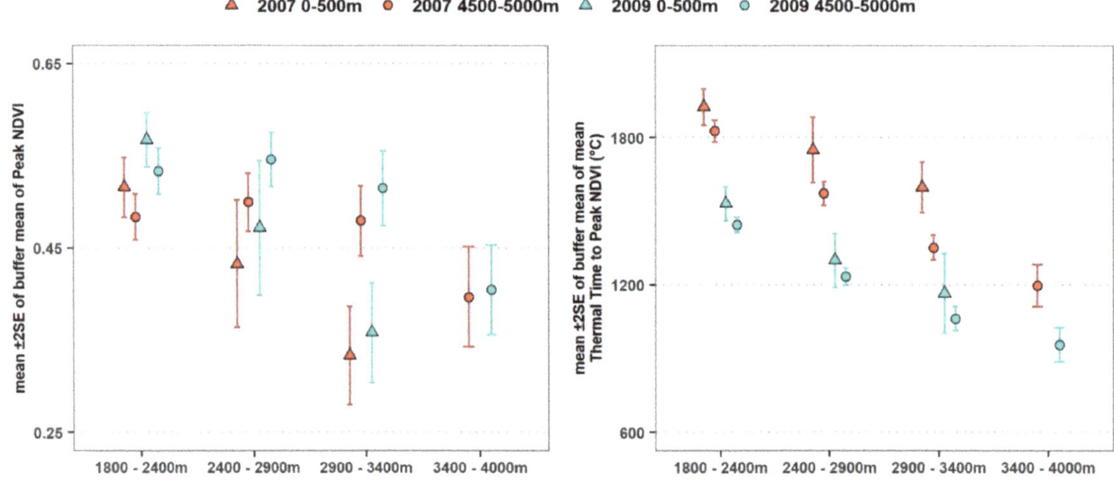

Figure 4. Differences in phenometric values (Peak Height at **left**, and Thermal Time to Peak at **right**) arising from distance and weather: mean ±2SE of the two ring buffer (0–500 m as a triangle, 4500–5000 m as a circle) means of 2007 (in red) and 2009 (in dark cyan) for the four elevation classes.

3.5. Interaction of Weather, Aspect, and Distance on Phenometrics

Figure 6 shows the distributions of phenometrics for northerly and southerly aspect slopes from pasture areas nearby (0–500 m) and far from (4500–5000 m) villages in the drier year of 2007 and the wetter year of 2009. The Kolmogorov-Smirnov two-sample test with the Dunn-Šidák correction reveals four patterns of interest in the Peak Height distributions: (1) the aspect effect on PH is significant at the two lower elevation classes only far from villages; (2) at the far ring buffer at each elevation class between 1800 m and 3400 m, the distributions of PH means from southerly aspects in the drier year of 2007 are significantly different from the northerly aspect PH means in the wetter year of 2009; (3) at each distance and elevation, distributions of PH means are not significantly different by year for the same aspect, i.e., the distributions of southerly (or northerly) aspect PH means are not different between years; and (4) the distributions of northerly aspect PH means in the drier year of 2007 are not significantly different from the southerly aspect PH means in the wetter year of 2009. This last pattern is clearer in Figure 7 (top panels) and illustrates how the northerly aspect slopes moderate the impact of dry years on pasture LSP.

The patterns in the TTP distributions were not surprising (Figure 6, right): (1) decreasing TTP with increasing elevation; (2) lower TTP values in the cooler, wetter year of 2009; and (3) no apparent aspect effect in the TTP buffer mean distributions. The bottom panels in Figure 7 show more clearly how strongly the TTP mean distributions diverge between years.

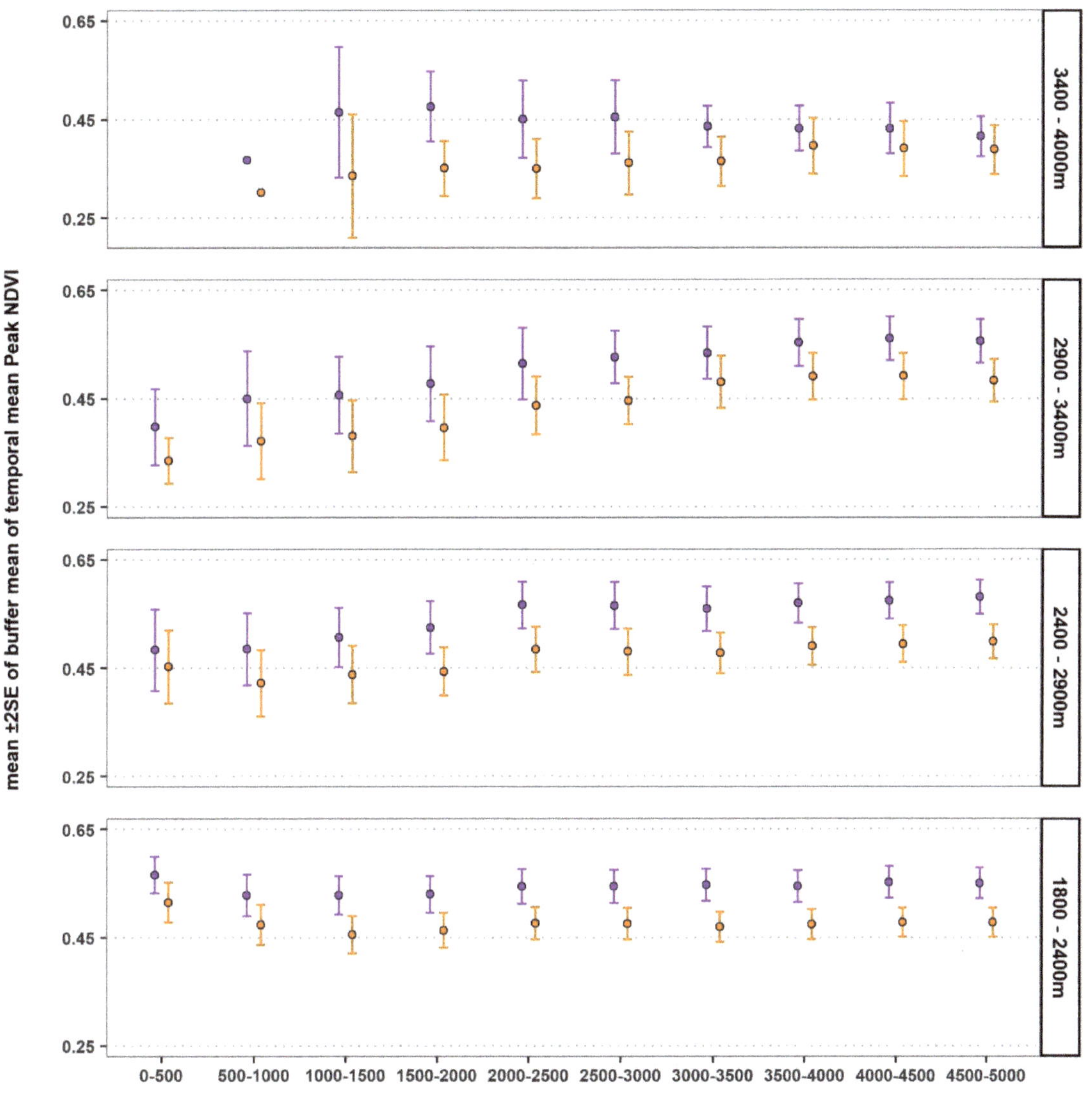

Figure 5. Contrasting values from pasture aspect: mean ±2SE of the ten ring buffer mean values of temporal mean Peak Height at two contrasting aspects (northern aspects in dark purple; southern aspects in dark orange) for the four elevation classes.

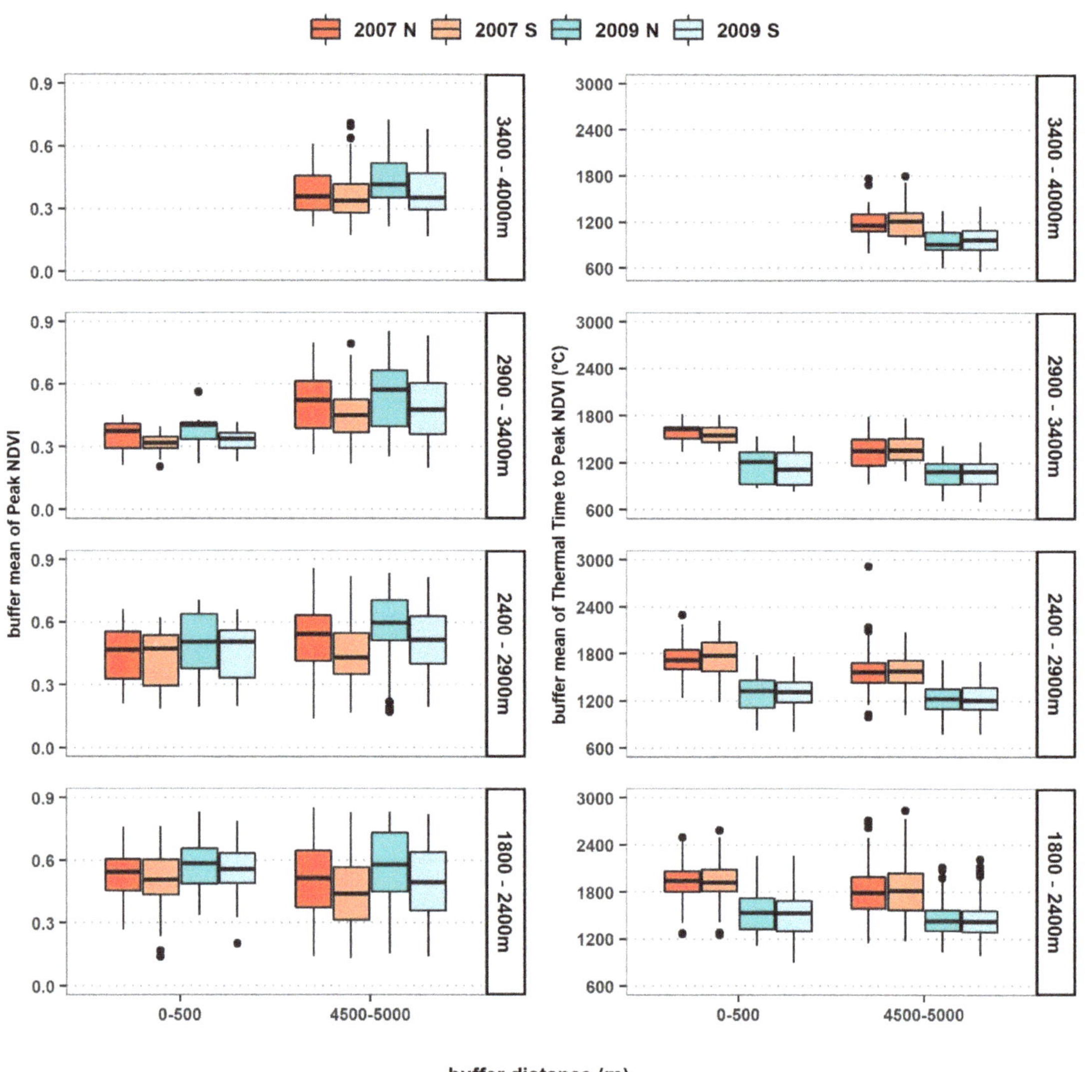

Figure 6. Interaction of weather, aspect, and distance: distributions of the phenometrics (Peak Height (**left**) and Thermal Time to Peak (**right**)) from two nearby (0–500 m) and distant (4500–5000 m) ring buffer mean values calculated from the hotter, drier year of 2007 (northern aspects [N] in red, southern aspects [S] in orange) and the cooler, wetter year of 2009 (northern aspects [N] in a dark cyan, southern aspects [S] in a light blue)) at the four elevation classes.

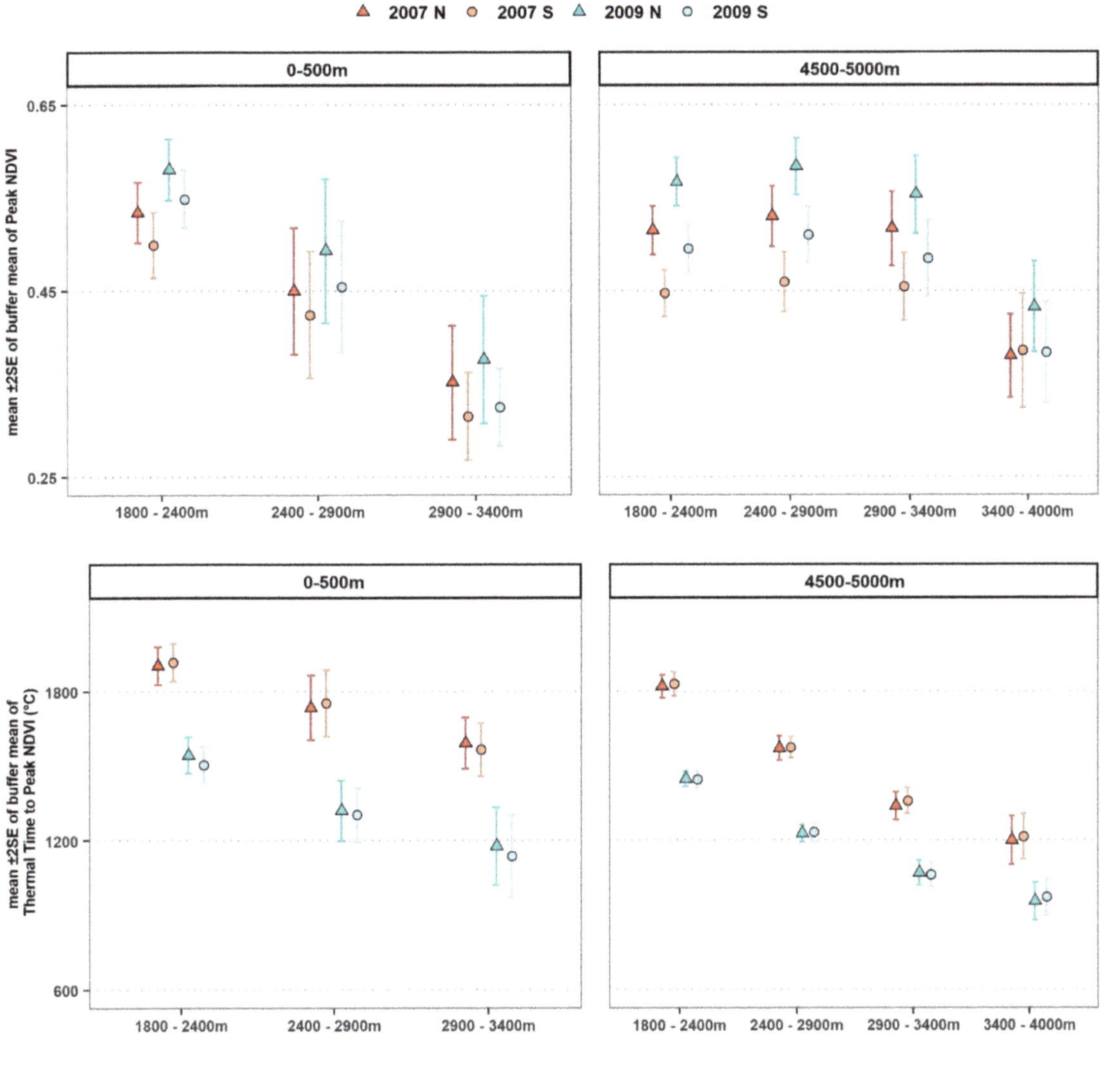

Figure 7. Differences in phenometric values (Peak Height on top and Thermal Time to Peak at bottom) arising from contrasts in weather and aspect: mean ±2SE of the two ring buffers at 0–500 m (**left**) and 4500–5000 m (**right**) from villages for hotter, drier 2007 (northern aspects [N] in a red triangle, southern aspects [S] in an orange circle) and cooler, wetter 2009 (northern aspects [N] in a dark cyan triangle, southern aspects [S] in a light blue circle) for the four elevation classes.

4. Discussion

The results of these linked analyses address the questions posed earlier. The Peak Height phenometric tends to increase in pasture areas as distance from the village increases. One interpretation of this pattern is the pasture areas near villages tend to be degraded. Winter pastures are closer to villages than either summer or transitional pastures and have been shown to exhibit significant differences in biogeochemistry and vegetation composition [5]. However, this general pattern can be modulated by the terrain, both elevation and aspect. Of the four elevation classes, significantly larger PH differences were seen between nearby and distant buffer rings in the 2900–3400 m elevation class (Figure 3, left). TTP decreased with elevation class but showed no clear difference between nearby and distant values (Figure 3, right), due in part to the much coarser spatial resolution of the LST data.

For the same reason, there was no clear influence of aspect on TTP (Figure 6, right panels; Figure 7, bottom panels). In contrast, the influence of aspect on PH was strongest in the lower elevation classes (Figure 5, bottom panels). When distant from a settlement, the PH on a northern aspect slope during a dry year appeared very similar to the PH on a southern aspect slope during a wet year (Figure 7, top right). However, in pasture areas closest to settlements, this aspect influence vanishes (Figure 7, top left).

The advantages of our study are several. First, we integrated long time series of two independent remote sensing products through a simple biometeorological model of land surface phenology that responds to the progress of growing season temperature rather than to the mere passing of days [37,38]. Second, the CxQ model has been shown to capture the initial seasonal peak well in grassland LSP [25,37,38]. Third, the finer spatial resolution of our analysis captured the influence of terrain features on vegetation growth and development as captured by the phenometrics [25].

Four limitations of our study are key. First, even the 30 m Landsat data can miss important landscape features influencing LSP [42]. Second, the 1 km spatial resolution of the MODIS LST product is coarser than would be optimal for use in rugged terrain. However, there are no effective alternatives at this point. Third, the CxQ model captures the initial seasonal peak but not necessarily the post-peak decay, particularly if there is heavy grazing post-peak [25]. This limitation is not as serious as it may first appear: we expect clear evidence of pasture degradation to appear only after several years of overgrazing, particularly if coupled with a severe drought. Fourth, while the pasture land-use mask was critical for omitting non-pastoral land-uses, there is substantial uncertainty associated with its development and accuracy. We have no accuracy assessment associated with it, and we expect that commission error is more likely than omission error, which would have the effect of increasing variance and decreasing the significance of the patterns. Finally, a key caveat for interpretation of these results should be noted: the 5 km extent of analysis was not meant to capture the summer pastures associated with villages. Summer pastures can be 10–50 km or more from a given village, and the spatial allocation and arrangement of summer pastures can be quite complicated.

Given our findings, to what extent do they raise questions about previous remote sensing studies on widespread degradation of pasture resources? The relationship between the scale of observation and the scale of the phenomenon of interest is crucial to the understanding and interpretability of the remote sensing data [43]. It is clear from these results that terrain effects can be obscured by coarser spatial resolution data, yet there is another scale to consider as well. Prior studies have demonstrated significant effects of climate oscillation modes on LSP in Central Asia [12,44,45]. However, more recent work at higher spatial resolution was not able to discern a significant influence from climate modes because the influence on LSP was overridden by local landscape structure [26].

Another facet of the degradation monitoring problem relates to the impact of spatial heterogeneity on the characterization of LSP. An important empirical study [46] explored how spatial heterogeneity in the land surface phenology observed at finer spatial resolution influences the timing of phenophase detection when observed at coarser spatial resolution. They found, when the land cover was homogeneous, that greater than 60% of Landsat 8 OLI 30 m pixels exhibiting start of season (SOS) detections were within 1 day of the SOS detection within a VIIRS pixel of 500 m spatial resolution. In contrast, less than 20% of the 30 m SOS detections were within 1 day if the land cover within the 500 m pixel was heterogeneous. They further reported that the SOS detection timing at the coarser spatial resolution was controlled by the timing when roughly 30% of the 30 m pixels within the 500 m pixel transitioned to SOS [46]. Thus, phenophase detections at coarser spatial resolutions are biased toward the earlier components in the vegetated land surface if the target landscape exhibits spatial heterogeneity. The study shows that the scaling effect on LSP is not resolvable with simple averaging. The mountain pastures of the Kyrgyz Republic and elsewhere in Central Asia exhibit a higher spatial heterogeneity than the croplands of central Iowa examined in [46]. Using a nonstandard MODIS product at 250 m spatial resolution, a study of pasture degradation in western Kyrgyz Republic found that pastures with a higher abundance of non-palatable vegetation exhibited later timing of peak NDVI during dry years in sub-alpine and mountain steppe ecozones but negligible impact in normal years [8]. The study is notable in that it included an extensive field component and demonstrated that higher NDVI values can accompany pasture degradation.

Several prior studies that have used MODIS data to detect trends in browning or greening in Kyrgyzstan and elsewhere in montane Central Asia may have been biased towards detecting browning trends and interpreting them as evidence of degradation due to using Collection 5 (C5) of the MODIS products. The differences between C5 and Collection 6 (C6 aka V006) are pronounced due to the loss of sensitivity of the red channel in the Terra MODIS [47,48]. Trend comparisons between the two collections have shown that significant negative trends in C5 were no longer significant in C6, and many nonsignificant trends in C5 appeared now as significant positive trends in C6 [49,50]. Two earlier studies [4,14] finding degradation in Kyrgyz pastures clearly used C5 products. Another study finding degradation [8] used data that was unlikely to have included the C6 corrections. A recent study [7] does not state which collection was used, but as the analysis period extended only until 2014, it is likely that it used C5 instead of C6 as well. The key point here is that as remote sensing products undergo upgrades that significantly change observed patterns, it is important to revisit those studies relying on earlier product versions to re-evaluate their findings in light of the new information.

A further concern is the use of simple linear slopes to detect significant trends. Applying ordinary least squares regression to a time series to fit a slope and declare the slope to be the trend has serious limitations from a statistical standpoint [51,52]. Positive autocorrelation reduces residual variance and inflates significance, increasing the risk of a Type I inferential error. Alternatives exist, such as the nonparametric Seasonal Kendall trend test, but it is not resistant to strong interannual autocorrelation. Fitting time series with autoregressive models has not been common in remote sensing studies.

Our findings suggest potential degradation in pasture areas as indicated by lower PHs closer to villages, but our analysis cannot rule out functional degradation arising from an increase in the coverage of nonpalatable species that can enhance NDVI without providing accessible forage [8,13,19]. Furthermore, what constitutes degradation in a particular socio-ecological system is frequently more subjective and culture-bound than is typically acknowledged [13,53–55].

5. Conclusions

We analyzed the landscape patterns of phenometrics based on fitted parameter coefficients of the land surface phenology model applied to 17 years of Landsat and MODIS seasonal time series across the montane pastures of the Kyrgyz Republic. We found that the Peak Height of NDVI generally decreased closer to villages, but the patterns were modulated—sometimes strongly—by elevation, aspect, and growing season weather. These findings raise questions about reports of pasture degradation based on coarser spatial resolution image time series. Our goal here was not to differentiate the relative contributions of these factors, because they are not susceptible to linear "unmixing". Rather, we sought to recognize their potential influence on the mixed signal observed at coarser spatial resolutions and to urge caution in interpreting, for instance, declines in NDVI trends with pasture degradation and increases with pasture remediation. The situation is more complicated. Due to the spatial heterogeneous distribution of pastures and pasture usage in mountainous landscapes, contextual information should be used to interpret remotely sensed patterns and trends in an appropriately nuanced manner.

Author Contributions: Conceptualization, G.M.H. and M.A.T.; methodology, G.M.H. and M.A.T.; software, M.A.T.; resources, G.M.H.; data curation, M.A.T.; writing—original draft preparation, G.M.H. and M.A.T.; writing—review, editing, and revision, G.M.H. and M.A.T.; visualization, M.A.T.; supervision, G.M.H.; project administration, G.M.H.; funding acquisition, G.M.H. Both authors have read and agreed to the published version of the manuscript.

Funding: This research was funded by the NASA LCLUC program, grant number 80NSSC20K0411. The APC was funded by the same grant.

Data Availability Statement: Part of the data used in this work were supplied from our previous work [25] and can be found in the NASA Land Use/Land Cover project repository: https://lcluc.umd.edu/metadatafiles/LCLUC-2015-PI-Henebry/RSE/, accessed on 20 June 2021. The remaining data presented in this study are available on request from the corresponding author (henebryg@msu.edu).

Acknowledgments: We appreciate the assistance and data sharing from Kamilya Kelgenbaeva. We also appreciate discussions with Munavar Zhumanova about pasture degradation. We would like to thank all reviewers and editors for their time and effort to improve the clarity, precision, and relevance of this publication.

Conflicts of Interest: The authors declare no conflict of interest.

References

1. FAOStats: Statistics page of the Food and Agriculture Organization (FAO). Available online: http://www.fao.org/faostat/en/#data/RL (accessed on 20 June 2021).
2. Akimaliev, D.A.; Zaurov, D.E.; Eisenman, S.W. The geography, climate and vegetation of Kyrgyzstan. In *Medicinal Plants of Central Asia: Uzbekistan and Kyrgyzstan*; Springer: New York, NY, USA, 2013; pp. 1–4.
3. Borchardt, P.; Schickhoff, U.; Scheitweiler, S.; Kulikov, M. Mountain pastures and grasslands in the SW Tien Shan, Kyrgyzstan—Floristic patterns, environmental gradients, phytogeography, and grazing impact. *J. Mt. Sci.* **2011**, *8*, 363–373. [CrossRef]
4. Dubovyk, O.; Landmann, T.; Dietz, A.; Menz, G. Quantifying the impacts of environmental factors on vegetation dynamics over climatic and management gradients of Central Asia. *Remote Sens.* **2016**, *8*, 600. [CrossRef]
5. Hoppe, F.; Kyzy, T.Z.; Usupbaev, A.; Schickhoff, U. Rangeland degradation assessment in Kyrgyzstan: Vegetation and soils as indicators of grazing pressure in Naryn Oblast. *J. Mt. Sci.* **2016**, *13*, 1567–1583. [CrossRef]
6. Kulikov, M.; Schickhoff, U.; Borchardt, P. Spatial and seasonal dynamics of soil loss ratio in mountain rangelands of south-western Kyrgyzstan. *J. Mt. Sci.* **2016**, *13*, 316–329. [CrossRef]
7. Wang, Y.; Yue, H.; Peng, Q.; He, C.; Hong, S.; Bryan, B.A. Recent responses of grassland net primary productivity to climatic and anthropogenic factors in Kyrgyzstan. *Land Degrad. Dev.* **2020**, *31*, 2490–2506. [CrossRef]
8. Zhumanova, M.; Mönnig, C.; Hergarten, C.; Darr, D.; Wrage-Mönnig, N. Assessment of vegetation degradation in mountainous pastures of the Western Tien-Shan, Kyrgyzstan, using eMODIS NDVI. *Ecol. Indic.* **2018**, *95*, 527–543. [CrossRef]
9. Kerven, C.; Steimann, B.; Dear, C.; Ashley, L. Researching the future of pastoralism in Central Asia's mountains: Examining development orthodoxies. *Mt. Res. Dev.* **2012**, *32*, 368–377. [CrossRef]
10. Robinson, S. Land degradation in Central Asia: Evidence, Perception and Policy. In *The End of Desertification?* Behnke, R., Mortimore, M., Eds.; Springer: Heidelberg, Germany, 2016; pp. 451–490. [CrossRef]
11. de Beurs, K.M.; Henebry, G.; Owsley, B.C.; Sokolik, I. Using multiple remote sensing perspectives to identify and attribute land surface dynamics in Central Asia 2001–2013. *Remote Sens. Environ.* **2015**, *170*, 48–61. [CrossRef]
12. de Beurs, K.M.; Henebry, G.; Owsley, B.C.; Sokolik, I.N. Large scale climate oscillation impacts on temperature, precipitation and land surface phenology in Central Asia. *Environ. Res. Lett.* **2018**, *13*, 065018. [CrossRef]
13. Eddy, I.M.; Gergel, S.E.; Coops, N.C.; Henebry, G.; Levine, J.; Zerriffi, H.; Shibkov, E. Integrating remote sensing and local ecological knowledge to monitor rangeland dynamics. *Ecol. Indic.* **2017**, *82*, 106–116. [CrossRef]
14. Kulikov, M.; Schickhoff, U. Vegetation and climate interaction patterns in Kyrgyzstan: Spatial discretization based on time series analysis. *Erdkunde* **2017**, *71*, 143–165. [CrossRef]
15. Xu, H.-J.; Wang, X.-P.; Zhang, X.-X. Decreased vegetation growth in response to summer drought in Central Asia from 2000 to 2012. *Int. J. Appl. Earth Obs. Geoinf.* **2016**, *52*, 390–402. [CrossRef]
16. Xu, H.-J.; Wang, X.-P.; Yang, T.-B. Trend shifts in satellite-derived vegetation growth in Central Eurasia, 1982–2013. *Sci. Total Environ.* **2017**, *579*, 1658–1674. [CrossRef] [PubMed]
17. Imanberdieva, N. Flora and plant formations distributed in At-Bashy Valleys–Internal Tien Shan in Kyrgyzstan and interactions with climate. In *Climate Change Impacts on High-Altitude Ecosystems*; Öztürk, M., Hakeem, K.R., Faridah-Hanum, I., Efe, R., Eds.; Springer: New York, NY, USA, 2015; pp. 569–590, ISBN 9783319128597.
18. Nowak, A.; Świerszcz, S.; Nowak, S.; Nobis, M. Classification of tall-forb vegetation in the Pamir-Alai and western Tian Shan Mountains (Tajikistan and Kyrgyzstan, Middle Asia). *Veg. Classif. Surv.* **2020**, *1*, 191–217. [CrossRef]

19. Zhumanova, M.; Wrage-Mönnig, N.; Jurasinski, G. Long-term vegetation change in the Western Tien-Shan Mountain pastures, Central Asia, driven by a combination of changing precipitation patterns and grazing pressure. *Sci. Total Environ.* **2021**, *781*, 146720. [CrossRef] [PubMed]
20. de Beurs, K.M.; Wright, C.K.; Henebry, G. Dual scale trend analysis for evaluating climatic and anthropogenic effects on the vegetated land surface in Russia and Kazakhstan. *Environ. Res. Lett.* **2009**, *4*, 045012. [CrossRef]
21. Groisman, P.; Bulygina, O.; Henebry, G.; Speranskaya, N.; Shiklomanov, A.; Chen, Y.; Tchebakova, N.; Parfenova, E.; Tilinina, N.; Zolina, O.; et al. Dryland belt of Northern Eurasia: Contemporary environmental changes and their consequences. *Environ. Res. Lett.* **2018**, *13*, 115008. [CrossRef]
22. Li, Z.; Chen, Y.; Li, W.; Deng, H.; Fang, G. Potential impacts of climate change on vegetation dynamics in Central Asia. *J. Geophys. Res. Atmos.* **2015**, *120*, 12345–12356. [CrossRef]
23. Tomaszewska, M.A.; Henebry, G. Changing snow seasonality in the highlands of Kyrgyzstan. *Environ. Res. Lett.* **2018**, *13*, 065006. [CrossRef]
24. Xenarios, S.; Gafurov, A.; Schmidt-Vogt, D.; Sehring, J.; Manandhar, S.; Hergarten, C.; Shigaeva, J.; Foggin, M. Climate change and adaptation of mountain societies in Central Asia: Uncertainties, knowledge gaps, and data constraints. *Reg. Environ. Chang.* **2019**, *19*, 1339–1352. [CrossRef]
25. Tomaszewska, M.A.; Nguyen, L.H.; Henebry, G. Land surface phenology in the highland pastures of montane Central Asia: Interactions with snow cover seasonality and terrain characteristics. *Remote Sens. Environ.* **2020**, *240*, 111675. [CrossRef]
26. Tomaszewska, M.A.; Henebry, G.M. How much variation in land surface phenology can climate oscillation modes explain at the scale of mountain pastures in Kyrgyzstan? *Int. J. Appl. Earth Obs. Geoinf.* **2020**, *87*, 102053. [CrossRef]
27. Azykova, E.K. Geographyical and landscape characteristics of mountain territories. In *Mountains of Kyrgyzstan*; Aidaraliev, A.A., Ed.; Technology: Bishkek, Kyrgyzstan, 2002.
28. Asian Development Bank. *Central Asia Atlas of Natural Resource*; Central Asian Countries Initiative for Land Management and Asian Development Bank: Manila, Philippines, 2010.
29. Asian Development Bank. Central Asian Countries Initiative for Land Management (CACILM) Multicountry Partnership Framework Support Project. Available online: https://www.adb.org/projects/38464-012/main (accessed on 20 June 2021).
30. NASA JPL. NASA Shuttle Radar Topography Mission Global 1 Arc Second. 2013. Available online: https://lpdaac.usgs.gov/products/srtmgl1v003/ (accessed on 20 June 2021). [CrossRef]
31. Wan, Z.; Hook, S.; Hulley, G. *MOD11A2 MODIS/Terra Land Surface Temperature/Emissivity 8-Day L3 Global 1 km SIN Grid V006*; NASA EOSDIS Land Processes DAAC: Sioux Falls, SD, USA, 2015.
32. Roy, D.; Kovalskyy, V.; Zhang, H.; Vermote, E.; Yan, L.; Kumar, S.S.; Egorov, A. Characterization of Landsat-7 to Landsat-8 reflective wavelength and normalized difference vegetation index continuity. *Remote Sens. Environ.* **2016**, *185*, 57–70. [CrossRef]
33. Fisher, J.I.; Mustard, J.F.; Vadeboncoeur, M. Green leaf phenology at Landsat resolution: Scaling from the field to the satellite. *Remote Sens. Environ.* **2006**, *100*, 265–279. [CrossRef]
34. Melaas, E.; Friedl, M.A.; Zhu, Z. Detecting interannual variation in deciduous broadleaf forest phenology using Landsat TM/ETM+ data. *Remote Sens. Environ.* **2013**, *132*, 176–185. [CrossRef]
35. Pereladova, O.; Krever, V.; Shestakov, A. *ECONET—Web for Life*; Central Asia: Moscow, Russia, 2006.
36. de Beurs, K.M.; Henebry, G. Land surface phenology, climatic variation, and institutional change: Analyzing agricultural land cover change in Kazakhstan. *Remote Sens. Environ.* **2004**, *89*, 497–509. [CrossRef]
37. Henebry, G.M.; de Beurs, K.M. Remote sensing of land surface phenology: A prospectus. In *Phenology: An Integrative Environmental Science*; Schwartz, M.D., Ed.; Springer Netherlands: Dordrecht, The Netherlands, 2013; Volume 9789400769, pp. 385–411, ISBN 9789400769250.
38. de Beurs, K.M.; Henebry, G.M. Spatio-temporal statistical methods for modelling land surface phenology. In *Phenological Research: Methods for Environmental and Climate Change Analysis*; Hudson, I.L., Keatley, M.R., Eds.; Springer: Dordrecht, The Netherlands, 2010; pp. 177–208, ISBN 978-90-481-3334-5.
39. Moritz, S.; Bartz-Beielstein, T. imputeTS: Time Series Missing Value Imputation in R. *R J.* **2017**, *9*, 207. [CrossRef]
40. Goodin, D.G.; Henebry, G. A technique for monitoring ecological disturbance in tallgrass prairie using seasonal NDVI trajectories and a discriminant function mixture model. *Remote Sens. Environ.* **1997**, *61*, 270–278. [CrossRef]
41. Šidák, Z. Rectangular confidence regions for the means of multivariate normal distributions. *J. Am. Stat. Assoc.* **1967**, *62*, 626–633. [CrossRef]
42. Cheng, Y.; Vrieling, A.; Fava, F.; Meroni, M.; Marshall, M.; Gachoki, S. Phenology of short vegetation cycles in a Kenyan rangeland from PlanetScope and Sentinel-2. *Remote Sens. Environ.* **2020**, *248*, 112004. [CrossRef]
43. Strahler, A.H.; Woodcock, C.E.; Smith, J.A. On the nature of models in remote sensing. *Remote Sens. Environ.* **1986**, *20*, 121–139. [CrossRef]
44. de Beurs, K.M.; Henebry, G. Northern annular mode effects on the land surface phenologies of Northern Eurasia. *J. Clim.* **2008**, *21*, 4257–4279. [CrossRef]
45. Wright, C.K.; de Beurs, K.M.; Henebry, G. Land surface anomalies preceding the 2010 Russian heat wave and a link to the North Atlantic oscillation. *Environ. Res. Lett.* **2014**, *9*, 124015. [CrossRef]
46. Zhang, X.; Wang, J.; Gao, F.; Liu, Y.; Schaaf, C.; Friedl, M.; Yu, Y.; Jayavelu, S.; Gray, J.; Liu, L.; et al. Exploration of scaling effects on coarse resolution land surface phenology. *Remote Sens. Environ.* **2017**, *190*, 318–330. [CrossRef]

47. Wang, D.; Morton, D.; Masek, J.; Wu, A.; Nagol, J.; Xiong, X.; Levy, R.; Vermote, E.; Wolfe, R. Impact of sensor degradation on the MODIS NDVI time series. *Remote Sens. Environ.* **2012**, *119*, 55–61. [CrossRef]
48. Lyapustin, A.; Wang, Y.; Xiong, X.; Meister, G.; Platnick, S.; Levy, R.; Franz, B.; Korkin, S.; Hilker, T.; Tucker, J.; et al. Scientific impact of MODIS C5 calibration degradation and C6+ improvements. *Atmos. Meas. Tech.* **2014**, *7*, 4353–4365. [CrossRef]
49. Zhang, Y.; Song, C.; Band, L.E.; Sun, G.; Li, J. Reanalysis of global terrestrial vegetation trends from MODIS products: Browning or greening? *Remote Sens. Environ.* **2017**, *191*, 145–155. [CrossRef]
50. Heck, E.; de Beurs, K.M.; Owsley, B.C.; Henebry, G.M. Evaluation of the MODIS collections 5 and 6 for change analysis of vegetation and land surface temperature dynamics in North and South America. *ISPRS J. Photogramm. Remote Sens.* **2019**, *156*, 121–134. [CrossRef]
51. de Beurs, K.; Henebry, G. Trend analysis of the Pathfinder AVHRR Land (PAL) NDVI data for the deserts of Central Asia. *IEEE Geosci. Remote Sens. Lett.* **2004**, *1*, 282–286. [CrossRef]
52. de Beurs, K.M.; Henebry, G. A statistical framework for the analysis of long image time series. *Int. J. Remote Sens.* **2005**, *26*, 1551–1573. [CrossRef]
53. Levine, J.; Isaeva, A.; Eddy, I.; Foggin, M.; Gergel, S.; Hagerman, S.; Zerriffi, H. A cognitive approach to the post-Soviet Central Asian pasture puzzle: New data from Kyrgyzstan. *Reg. Environ. Chang.* **2017**, *17*, 941–947. [CrossRef]
54. Levine, J.; Isaeva, A.; Zerriffi, H.; Eddy, I.M.S.; Foggin, M.; Gergel, S.E.; Hagerman, S.M. Testing for consensus on Kyrgyz rangelands: Local perceptions in Naryn oblast. *Ecol. Soc.* **2019**, *24*, 36. [CrossRef]
55. Liechti, K. The meanings of pasture in resource degradation negotiations: Evidence from post-socialist rural Kyrgyzstan. *Mt. Res. Dev.* **2012**, *32*, 304–312. [CrossRef]

remote sensing

Article

Mapping of *Kobresia pygmaea* Community Based on Umanned Aerial Vehicle Technology and Gaofen Remote Sensing Data in Alpine Meadow Grassland: A Case Study in Eastern of Qinghai–Tibetan Plateau

Baoping Meng [1,2], Zhigui Yang [1,2], Hongyan Yu [3], Yu Qin [4], Yi Sun [1,2], Jianguo Zhang [1,2], Jianjun Chen [5], Zhiwei Wang [6], Wei Zhang [4], Meng Li [1,2], Yanyan Lv [1,2] and Shuhua Yi [1,2,*]

[1] Institute of Fragile Eco-Environment, Nantong University, Nantong 226007, China; mengbp09@lzu.edu.cn (B.M.); 1822021025@stmail.ntu.edu.cn (Z.Y.); sunyi@ntu.edu.cn (Y.S.); sezjg@ntu.edu.cn (J.Z.); limeng@ntu.edu.cn (M.L.); lvyy09@lzu.edu.cn (Y.L.)

[2] School of Geographic Science, Nantong University, Nantong 226007, China

[3] Qinghai Service and Guarantee Center of Qilian Mountain National Park, Xining 810001, China; 18909718038@189.cn

[4] State Key Laboratory of Cryospheric Sciences, Northwest Institute of Eco-Environment and Resources, Chinese Academy of Sciences, 320 Donggang West Road, Lanzhou 730000, China; qiny@lzb.ac.cn (Y.Q.); zhangwei2015@lzb.ac.cn (W.Z.)

[5] College of Geomatics and Geoinformation, Guilin University of Technology, 12 Jiangan Road, Guilin 541004, China; chenjj@lzb.ac.cn

[6] Guizhou Institute of Prataculture, Guizhou Academy of Agricultural Sciences, Guiyang 550006, China; wzw1206@lzb.ac.cn

* Correspondence: yis@ntu.edu.cn

Citation: Meng, B.; Yang, Z.; Yu, H.; Qin, Y.; Sun, Y.; Zhang, J.; Chen, J.; Wang, Z.; Zhang, W.; Li, M.; et al. Mapping of *Kobresia pygmaea* Community Based on Umanned Aerial Vehicle Technology and Gaofen Remote Sensing Data in Alpine Meadow Grassland: A Case Study in Eastern of Qinghai–Tibetan Plateau. *Remote Sens.* **2021**, *13*, 2483. https://doi.org/10.3390/rs13132483

Academic Editor: Elias Symeonakis

Received: 21 May 2021
Accepted: 22 June 2021
Published: 25 June 2021

Abstract: The *Kobresia pygmaea* (KP) community is a key succession stage of alpine meadow degradation on the Qinghai–Tibet Plateau (QTP). However, most of the grassland classification and mapping studies have been performed at the grassland type level. The spatial distribution and impact factors of KP on the QTP are still unclear. In this study, field measurements of the grassland vegetation community in the eastern part of the QTP (Counties of Zeku, Henan and Maqu) from 2015 to 2019 were acquired using unmanned aerial vehicle (UAV) technology. The machine learning algorithms for grassland vegetation community classification were constructed by combining Gaofen satellite images and topographic indices. Then, the spatial distribution of KP community was mapped. The results showed that: (1) For all field observed sites, the alpine meadow vegetation communities demonstrated a considerable spatial heterogeneity. The traditional classification methods can hardly distinguish those communities due to the high similarity of their spectral characteristics. (2) The random forest method based on the combination of satellite vegetation indices, texture feature and topographic indices exhibited the best performance in three counties, with overall accuracy and Kappa coefficient ranged from 74.06% to 83.92% and 0.65 to 0.80, respectively. (3) As a whole, the area of KP community reached 1434.07 km^2, and accounted for 7.20% of the study area. We concluded that the combination of satellite remote sensing, UAV surveying and machine learning can be used for KP classification and mapping at community level.

Keywords: *Kobresia pygmaea* community; unmanned aerial vehicle; Gaofen satellite; spatial distribution

1. Introduction

Alpine meadow is the major vegetation type on the Qinghai–Tibet Plateau (QTP), China. It is important for animal husbandry, water conservation and biodiversity conservation [1,2]. Since the 1980s, due to the dual effects of climate change and human activities, alpine meadow grassland has experienced different extents of degradation, especially in the source region of Yellow River, which is on the eastern part of the QTP [3]. The degradation has restricted the sustainable development of animal husbandry and seriously threatened local ecological security [4]. The degradation succession stages of the alpine meadow grassland community include Poaceae, *Kobresia humilis* (KH), *Kobresia pygmaea* (KP) and black soil type (BS). The KP community is the key stage for the management of degraded grassland [5]. In the first two stages, the original community structure and function can be quickly restored under the grazing prohibition and artificial measures [6]. However, further degradation of KP community will cause irreversible degradation until the severest stage of black soil type [7]. Therefore, it is vital to map the current distribution of KP community grassland for mitigation and adaptation measures. However, previous grassland classifications have been performed at the vegetation type level, and few at the community level [8,9]. At present, the spatial distribution and impact factors of KP community on the QTP are still unclear [1,2,10]. Therefore, it is urgent to develop a method for mapping alpine meadow at community level.

Traditional grassland vegetation community samples are mainly obtained with the few field investigation, expert knowledge and literature reviews. Due to the complex distribution and dynamic of grassland vegetation communities, the field investigation cannot meet the accuracy requirement of classification [11–15]. In addition, remote sensing (RS) vegetation indices have been commonly used as classification variables, "the same object with different spectrum" or "the different object with same spectrum" have occurred frequently [16–18]. Successful classifications at the community level requires: (1) the RS images with proper temporal-spatial resolution, coverage, sensitive spectrum band; (2) massive field observations; and (3) effective classification methods.

Compared with traditional multi-spectral remote sensing (e.g., MODIS, Landsat, HJ-1A/1B), the Gao Fen 1 (GF1) and Gao Fen 6 (GF6) satellites have significant advantages in grassland resource monitoring [19]. Each of these satellites has a high resolution of 16 m (wide field view images, WFV), a relatively large detection width of 800 km and a short revisit period of two days (four days for each, two days for combination) [20]. Additionally, GF6 satellite adds the red edge band, which is beneficial to vegetation classification. Thus, it is easier to collect high quality remote sensing images at a regional scale [21].

The massive field observation is the basis of RS classification of grassland communities. However, the resolution of satellite images is insufficient to identify grassland communities, and traditional methods require large amounts of time, labor, cost and resources. In recent years, with the development of unmanned aerial vehicle (UAV) technology, the shortcoming of satellite and traditional methods in grassland resource monitoring are supplied [22–24]. On the one hand, the aerial photographs provided by UAV have high resolution, which can be used to identify the grassland vegetation community effectively [25]. On the other hand, UAV has a large observation range, which can save time and effort. Yi et al. (2017) [26] also developed a set of UAV aerial photography system with fixed-point, multi-site, collaborative observation, which can realize massive observation over large regions [27].

With the development of classification methods, machine learning algorithm has obvious advantages in RS image classification [28,29]. Based on neural network (NN), support vector machine (SVM), random forest (RF) and other machine learning algorithms, the satellite vegetation index, phrenological characteristics, image texture and topography are considered to improve the accuracy of RS classification [24,30–32]. However, RS classification in grassland mainly includes the land use type (e.g., grassland, non-grassland, woodland, etc.) [33], different biophysics characteristics (such as grassland with high, medium and low coverage) [34] and types in different climatic zones (such as class, group and type of grassland) [35].

In this study, we aimed to map the KP community over the eastern of QTP by using the combination of UAV aerial photographing, GF WFV images and machine learning algorithms. We hope this study can be helpful for guiding further mapping of the KP community over the whole QTP and provide scientific basics for restoration and management activities.

2. Data and Methods

2.1. Study Area

The study area is located at the eastern of the source region of the Yellow River, including Zeku County and Henan County of Qinghai province, and Maqu County of Gansu province (Figure 1). It is one of the most important animal husbandry basis on the QTP and also an important water source conservation area in China. The study area is located at 33°03′~35°33′N, 100°33′~102°33′E, with elevation ranging from 2871 to 4850 m (Figure 1c). The mean annual precipitation ranges from 400~600 mm, mean annual temperature is between −2.4~2.1 °C It belongs to the continental plateau temperate monsoon climate. Alpine meadow is one of the main alpine grassland types, accounting for 79.67% of the whole study area. Other than alpine meadow, mountain meadow, swamp meadow and alpine steppe account for 13.22%, 1.78%, and 1.69%, respectively (Figure 1b). The growth period of grassland plants is relatively short, only about 150 days, mainly from May to September. The grasslands are mainly used for yak and sheep grazing.

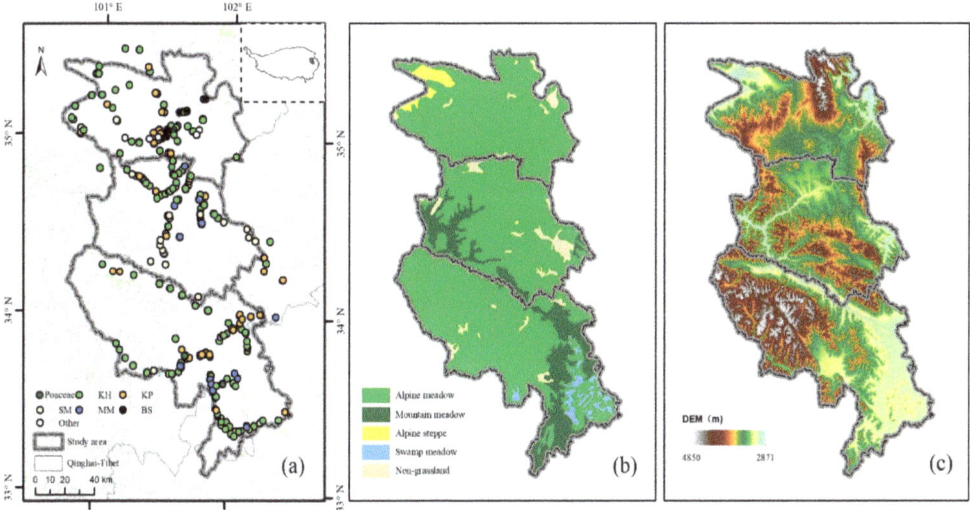

Figure 1. Location, grassland type and topography of study area. (**a**) Location of the study area in Qinghai–Tibetan Plateau and the observation sites; dots of different color represent the vegetation communities. Poaceae: *Elymus nutans + Stipa silena + Festuca ovina* community, KH: *Kobresia humilis* community, KP: *Kobresia pygmaea* community, BS: black soil type, MM: marsh meadow, SM: shrub meadow. (**b**) Grassland type of study area; (**c**) topography of study area.

2.2. Data and Preprocessing

2.2.1. Field Observation and Preprocess of Aerial Photographs

We carried out the field monitoring for vegetation communities of alpine meadow based on aerial photographs by Phantom 3 professional and Mavic 2 zoom Quad-Rotor intelligent UAVs (manufactured by DJI Innovation Industries; http://www.dji.com (accessed on 1 June 2018). According to grassland growth status and spatial representativeness, an area in the range of 250 × 250 m was selected as an observation site, and four flight routes were designed in each site, including one GRID flight way (200 × 200 m) and three BELT flight ways (40 × 40 m) (Figure 2a). The flight way of UAVs was designed by FragMAP [22], Phantom 3 professional was used to perform the GRID flight way at a height of 20 m (red dot in Figure 2a,b), Mavic 2 zoom was used to perform the BELT flight way at a height of 2 m (green dot in Figure 2a,c). The positional accuracy of two UAVs was ±1.5 m horizontally and ±0.5 m vertically. A photograph was then taken vertically downward at each way point automatically, the photograph resolutions of GRID and BELT were 1 and 0.09 cm, and the ground coverages were 26 × 35 m and 2.57 × 3.43 m, respectively.

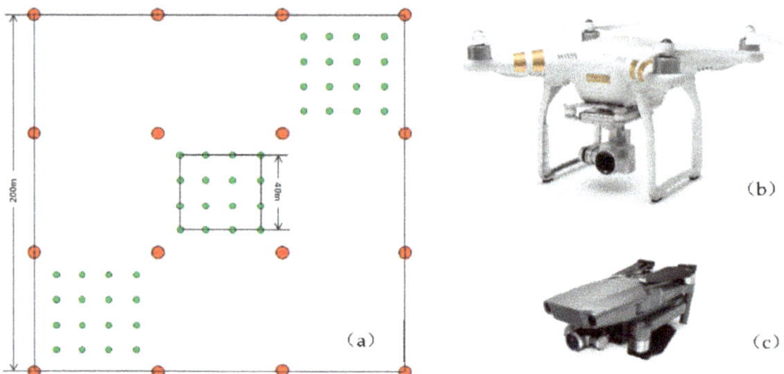

Figure 2. Strategy of field observation and data collection: (**a**) strategy of observation site; (**b**) Phantom 3 professional and (**c**) Mavic 2 zoom Quad-Rotor intelligent UAVs.

To better identify the vegetation species, about 9~15 aerial photographs were collected randomly by operating Mavic 2 zoom manually at a height of 0.5 m in each sample site. The number of photographs was determined by the uniformity of community growth status. These aerial photographs could clearly identify plant species, which was corresponding to the traditional ground observation quadrat (Supplementary Figure S1).

According to the dominant species of grass vegetation, grassland coverage, texture features and plateau pika (*Ochotona curzoniae*, hereafter pika) activities, the aerial photographs were divided into six types, including four alpine meadow vegetation communities of Poaceae, KH, KP and BS (Figure 3 and Table 1), two land covers of shrub meadow (SM) and marsh meadow (MM). Additionally, the forest and others (bare land, construction use and waters) were acquired based on the Google Earth images and GF WFV images. Field observation was carried out at the peak time of grassland growth, and 751 sample sites were observed from 2015 to 2019 in total (Figure 1c). About 30 sample sites were acquired for forest and others.

2.2.2. Region of Interest Construction

According to the GPS information recorded in FragMAP and stored in aerial photographs property files, the names of photographs were renamed by the number of 1 to 16 by the DJI Locator software [22] in each site. Then the region of interest was built based on photograph location information in the same observation site in ArcGIS and ENVI software (Figure 3d,h,l,p). Additionally, about 30 samples (region of interest, ROI) for forest and others (the water, bare land, and construction land) were selected in ENVI software, according to the GF WFV images and Google Earth images.

Figure 3. Classification criteria for aerial photographs of alpine meadow vegetation communities. (**a,e,i,m**) were aerial photographs taken at the height of 20 m by Phantom 3. (**b,c,f,g,j,k,n,o**) were aerial photographs taken at a height of 2 m by Mavic 2. (**c,g,k,o**) acquired with 2× wide-angle zoom lenses; (**d,h,l,p**) were Gaofen images of four types of vegetation communities corresponded; Poaceae, KH, KP and BS represented communities of *Elymus nutans* + *Stipa silena* + *Festuca ovina*, *Kobresia humilis*, *Kobresia pygmaea* and black soil type, respectively.

Table 1. Characteristics of vegetation communities in alpine meadow.

Community	Dominant Species	Coverage	Other Features
Poaceae	*Elymus nutans, Stipa silena, Festuca ovina*	More than 90%	Tall grassland height (20–50 cm in height), grassland was flat without any traces of pika activity
KH	*Kobresia humilis;* sub-dominant: *Elymus nutans* and *Festuca rubra*	More than 90%	Grassland was flat with low height (<10 cm in height) and high coverage, and small number of pika appeared
KP	*Kobresia pygmaea*	Between 30~80%	Grassland had a unique morphology and textural characteristics, with closed and monospecific builds (2~3 cm in height), polygonal crack patterns and a felty root mat, pika and poisonous weeds are invaded frequently
BS	Weeds	Less than 20%	Pika was rampant and weeds was overgrown

Poaceae, KH, KP and BS represented *Elymus nutans* + *Stipa silena* + *Festuca ovina*, Kobresia humilis, *Kobresia pygmaea* and black soil type, respectively.

2.2.3. Acquisition and Preprocessing of Remote Sensing Data

The remote sensing data, including GF1 and GF6 WFV imager images, were downloaded from the China Centre for Resources Satellite Data and Application (http://www.cresda.com/EN/ (accessed on 20 September 2019)). The WFV imager was carried by GF1 and GF2 satellites, with four multi-spectral bands (800 km of swath width) and eight multi-spectral bands (850 km of swath width), respectively. The resolution of WFV image was 16 m, and the revisit period for each satellite was 4 days. Together, the revisit period could be reached up to 2 days (Table 2). Three scenes of WFV images with no cloud cover in Zeku, Henan and Maqu, during the peak of grassland growth of 2019 and 2020 were downloaded (Table 3). The GF WFV data were preprocessed using ENVI 5.3 software, and the Radiometric Calibration module, FLAASH Atmospheric Correction module and RPC (Rational Polynomial Coefficient) Orthorectification module was used for converting the original DN value to atmospheric surface reflectance, atmospheric correction and precise geometric correction of WFV images, respectively. Then, the Band Math module was used to calculate the vegetation indices of NDVI, NDWI and NDMI. The Co-occurrence measures module was used to extract image texture features of WFV images based on a sliding window with 3 × 3 pixels, and the texture indices mainly included Mean, Variance, Homogeneity, Contrast, Dissimilarity, Entropy, Second Moment and Correlation.

Table 2. characterization of Gao Fen (GF) wide field view (WFV) cameras.

Satellite	Band	Spectral Range (μm)	Band Type	Spatial Resolution (m)	Swath Width (km)	Revisit Period (day)	Orbit Altitude (km)
GF-1	1	0.45–0.52	Blue	16	800	4	675
	2	0.52–0.59	Green				
	3	0.63–0.69	Red				
	4	0.77–0.89	NIR				
GF-6	1	0.45–0.52	Blue	16	800	4	645
	2	0.52–0.59	Green				
	3	0.63–0.69	Red				
	4	0.77–0.89	NIR				
	5	0.69–0.73	Red edge 1				
	6	0.73–0.77	Red edge 2				
	7	0.40–0.45	Purple				
	8	0.59–0.63	Yellow				

Table 3. List of GF1/6 WFV images used in this study.

County	Data of Satellite Images	Satellite	Path	Row	Central Latitude and Longitude	Cloud Percent
Zeku	2019.06.03	GF1	23	98	E 101.9, N 34.7	4%
Henan	2019.08.15	GF6	30	72	E 98.1, N 35.8	1%
Maqu	2020.08.25	GF6	18	72	E 104.7, N33.6	1%

The DEM data were 90 m shuttle radar topography mission (SRTM) images (version V004) (http://srtm.csi.cgiar.org/ (accessed on 1 June 2018) in Geo-TIFF format. The Slope, topographic position index (TPI) and aspect were calculated based on the DEM. Then, all indices above mentioned were uniformly projected as UTM_Zone_47N (same as GF WFV).

2.3. Vegetation Community Classification and Accuracy Evaluation

2.3.1. Classification Method

The maximum likelihood estimate (MLE), NN, SVM and RF classification methods were employed. MLE assuming each statistic of different types in every band was normally distributed, the likelihood of each pixel belonging to a certain training sample was calculated. Finally, the type of pixel was determined based on the highest likelihood [36]. NN (also called artificial neural network, ANN) referred to a multi-layer network structure, the Levenberg–Marquardt function algorithm was selected for NN training. The number of neurons and hidden layers were determined based on a trial-and-error process [37]. SVM was constructed by a set of hyperplanes in high- or infinite-dimensional space, the higher the functional margin, the lower the generalization error of the classifier. The radial basis function (RBF) was used as the kernel function, and the optimal cost and gamma values were obtained for final classification [38,39]. The RF algorithm was constructed by the classification tree, which applied a set of decision trees to improve prediction accuracy. The bootstrap sample was employed to construct a decision tree. The training samples were constantly selected to minimize the sum of the squared residuals until a complete tree was formed. Multiple decision trees were formed, and voting was used to obtain the final prediction [40,41]. MLE, NN and SVM methods were performed in ENVI supervised classification toolboxes of Maximum Likelihood Classification, Neural Net Classification and Support Vector Machine Classification, respectively. RF method was performed in ENVI Extensions toolbox of Random Forest Classification [42].

2.3.2. Classification and Accuracy Evaluation

Given the classification accuracy and efficiency, three input datasets were used: (1) GF1/6 WFV spectral band (band1 to band8); (2) vegetation and texture indices (NDVI, NDWI, SAVI, Contrast, Correlation, Dissimilarity, Entropy, Homogeneity, Mean, Second moment and Variance); (3) vegetation, texture, and topography indices (DEM, Slope, Asp and TPI). About 70% of observation sites were selected randomly as a training set, and the rest were used to validate classification accuracy in each county. The standard confusion matrix was employed to evaluate the classification accuracy of images, and the overall accuracy (OA), Kappa coefficient (Kappa), user's accuracy (UA) and producer' accuracy (PA) based on the validation datasets were used to assess the precision of classification results.

3. Results

3.1. Characteristics of Field Observation and Its Corresponding Multi-Indices

The distribution of observed sites was shown in Figure 1a. The vegetation communities of alpine meadow showed a considerable spatial heterogeneity. Among the 751 observed sites, the proportion of KH community is highest, with 56.32% of all observed sites. Followed by KP community (17.04% of all observed sites), the number of KP community observation sites were 68, 37 and 22 for Maqu, Zeku and Henan, respectively. The proportion of SM, MM, BS and Poaceae only accounted for 3.33~9.85% of all observed sites.

For the four types of alpine meadow grass communities and four types of land cover, the statistics of GF1/GF6 WFV image bands, vegetation indices, topography indices and texture indices were calculated in the study area (in Supplementary Materials). The result showed that the characteristics of multi-indices in alpine meadow vegetation communities were very similar, and it was difficult to distinguish with commonly used indices (Figure 4a,b,e). Even though eight land covers could be coarsely distinguished between each other in red edge bands (band5 and band6 of WFV image) and DEM, there was relatively large error in classification (with little difference in mean values and wide range in variation) (Figure 4c,d,f).

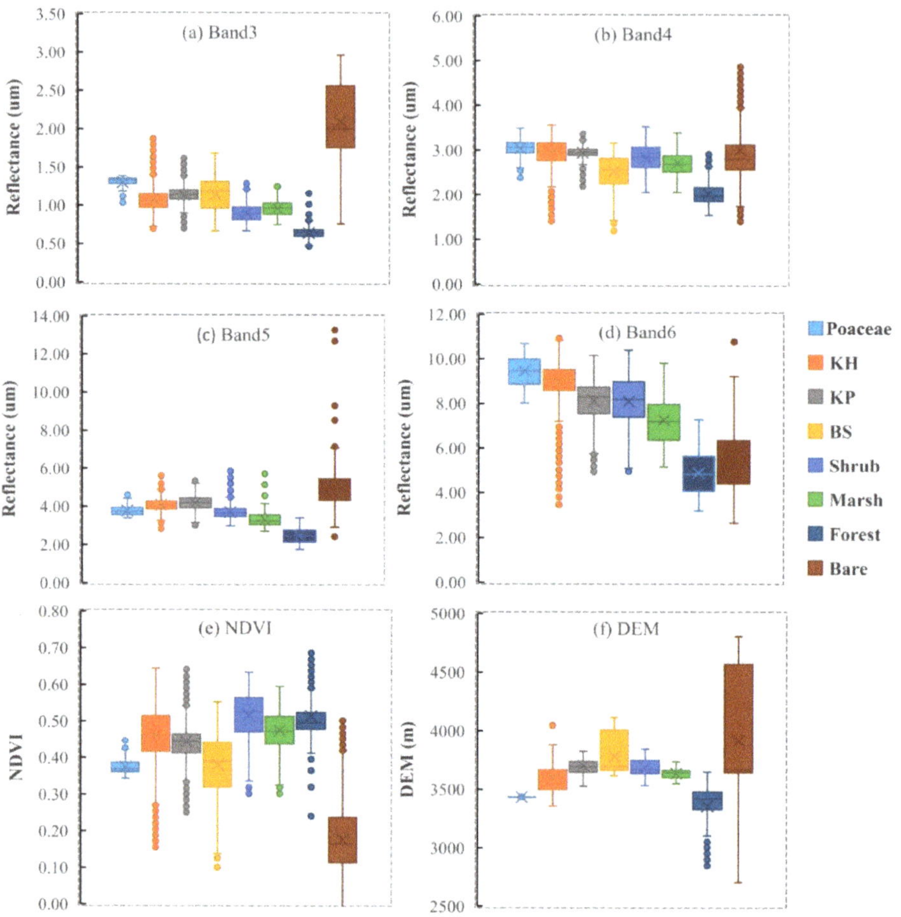

Figure 4. Statistical analysis results of band3 to band6 of GF1/GF6 images (**a–d**), NDVI (**e**) and DEM (**f**), respectively; Poaceae, KH, KP and BS represent *Elymus nutans + Stipa silena + Festuca ovina*, *Kobresia humilis*, *Kobresia pygmaea* and black soil type, respectively.

3.2. Accuracy Evaluation of the Different Classification Methods

Accuracy assessment of classification was performed with the validation samples listed in Table 4. Among the four classification methods, the RF method performed best, with the highest overall accuracy and Kappa coefficient ranged from 74.06% to 83.92% and from 0.65 to 0.80 in three counties, respectively. This was followed by the SVM method, with an overall accuracy that ranged from 69.39% to 78.53% and Kappa coefficient that ranged from 0.60 to 0.73. The accuracies of the NN and MLE method were the worst (overall accuracy ranged from 40.78% to 73.89%; Kappa coefficient ranged from 0.24 to 0.67). Among the three classifications input, in general, the MLE, SVM and RF methods based on the input data set of vegetation indices + texture + topography exhibited the best performance, followed by the spectrum and vegetation indices + texture. However, the performance of the NN method based on the above input data set showed contrary results to the MLE, SVM and RF methods.

Table 4. Overall accuracy and Kappa coefficient of eight land covers based on maximum likelihood estimate (MLE), neural network (NN), support vector machine (SVM) and random forest (RF) and difference input data set in County of Zeku, Henan and Maqu.

County	Input	Accuracy	Methods			
			MLE	NN	SVM	RF
Zeku	Spectrum	OA (%)	57.36	71.22	72.13	82.24
		Kappa	0.50	0.64	0.65	0.78
	Vegetation indices + texture	OA (%)	-	63.36	69.39	79.87
		Kappa	-	0.52	0.61	0.75
	Vegetation indices + texture + topography	OA (%)	63.75	40.78	78.53	83.92
		Kappa	0.57	0.24	0.73	0.80
Henan	Spectrum	OA (%)	68.03	75.14	74.96	81.32
		Kappa	0.60	0.67	0.66	0.76
	Vegetation indices + texture	OA (%)	72.04	65.89	73.31	80.39
		Kappa	0.64	0.54	0.65	0.75
	Vegetation indices + texture + topography	OA (%)	73.89	49.86	73.89	78.86
		Kappa	0.66	0.34	0.66	0.73
Maqu	Spectrum	OA (%)	65.67	70.04	70.28	75.96
		Kappa	0.56	0.60	0.60	0.68
	Vegetation indices + texture	OA (%)	51.38	67.12	73.78	74.06
		Kappa	0.40	0.55	0.65	0.65
	Vegetation indices + texture + topography	OA (%)	65.19	61.89	74.09	82.75
		Kappa	0.56	0.46	0.65	0.77

Note: overall accuracy, OA; maximum likelihood estimate, MLE; neural network, NN; support vector machine, SVM; random forest, RF.

The results of the standard confusion matrix were shown in Table 5. The PA and UA based on the RF method were highest in three counties, with 60.84% to 97.23% and 60.73% to 78.09%, respectively. The PA and UA based on other methods showed lower value, and the classification results of the KP community were easily confused with other grass communities and land cover types.

Table 5. Producer's accuracy and user's accuracy of *Kobresia pygmaea* community based on maximum likelihood estimate (MLE), neural network (NN), support vector machine (SVM) and random forest (RF) in Zeku, Henan and Maqu County.

County	Input	Accuracy (%)	Method			
			MLE	NN	SVM	RF
Zeku	Spectrum	PA	50.99	26.06	39.83	96.37
		UA	49.01	47.96	56.89	74.07
	Vegetation indices + texture	PA	-	38.83	19.61	86.94
		UA	-	46.85	46.49	69.94
	Vegetation indices + texture + topography	PA	69.30	64.93	84.82	97.23
		UA	46.45	43.34	57.38	65.68
Henan	Spectrum	PA	59.61	5.71	5.26	67.57
		UA	29.83	80.85	57.38	60.81
	Vegetation indices +texture	PA	26.58	10.06	26.88	76.73
		UA	41.75	44.97	54.43	67.59
	Vegetation indices + texture + topography	PA	35.83	-	32.83	68.67
		UA	41.86	-	46.40	67.65
Maqu	Spectrum	PA	70.68	64.83	52.74	67.23
		UA	43.68	49.41	50.67	59.29
	Vegetation indices + texture	PA	59.45	13.68	61.32	60.84
		UA	50.00	90.48	61.21	60.73
	Vegetation indices + texture + topography	PA	70.48	35.83	64.22	73.96
		UA	45.53	52.32	57.42	78.09

Note: producer's accuracy, PA; user's accuracy, UA; maximum likelihood estimate, MLE; neural network, NN; support vector machine, SVM; random forest, RF.

3.3. Distribution and Area of KP Community

According to the vegetation community distribution map acquired by the RF method, the spatial distribution of the KP community was fragmented with large spatial heterogeneity and small area (Figure 5). Among the three counties, the distribution of the KP community was mainly located in: the north, east and around the county urban area of Zeku County (around the town of Zequ, Qiakeri and Xipusha), with an area of 445.60 km^2 (6.82% of Zeku County); the northeast and central part of Henan County (east of county urban area, towns of Tuoyema and Duosun, and north of Saierlong), with an area of 176.76 km^2 (4.48% of Henan County); the part of county urban area, towns of Oulaxiuma, Muxihe and Awancang in Maqu County, with an area of 811.70 km^2 (8.59% of Maqu County). As a whole, the area of KP community reached 1434.07 km^2, and accounted for 7.20% of the study area.

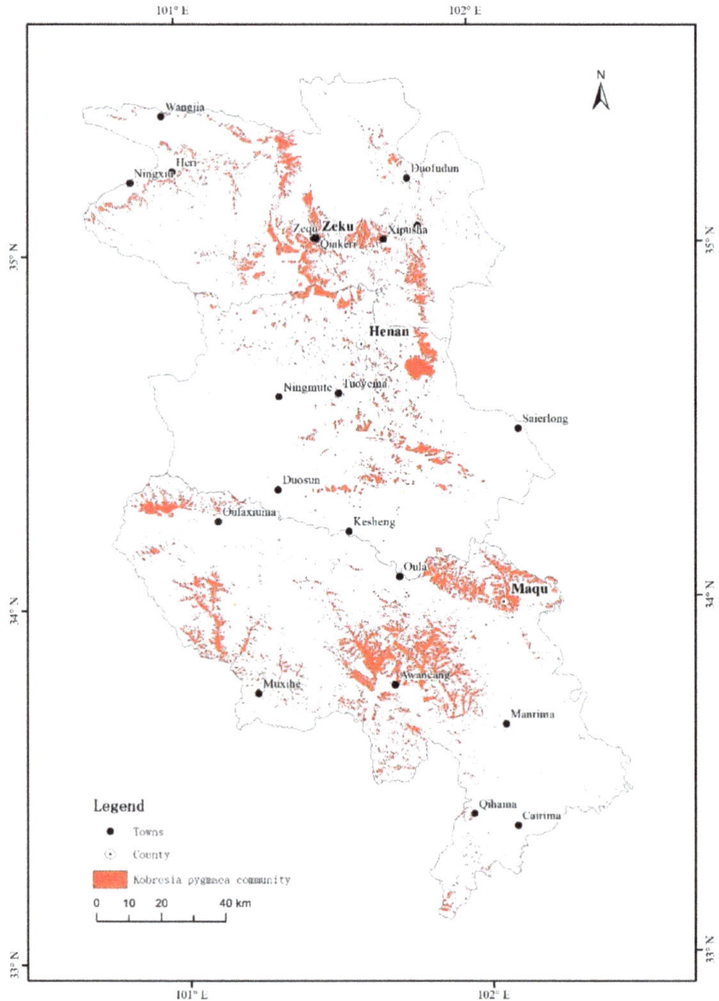

Figure 5. Distribution of *Kobresia pygmaea* community in Counties of Zeku, Henan and Maqu.

4. Discussion

4.1. Influence Factors of KP Community in the Qinghai–Tibet Plateau

Generally, the KP community builds almost closed, non-specific, golf-course like the lawn with a felty root mat. This characteristic mat not only protects soil against intensive trampling by herbivores, but also helps to cope with nutrient limitations enabling medium-term nutrient storage and increasing productivity and competitive ability of roots against leaching and other losses [43–45]. However, with browning (patchwise dieback of lawns), crack, collapse, fragmentation of KP community turf, the water budget [46], carbon cycle [47,48] and soil nutrition [44,45,49] have been significantly changed [10].

Pastoralism may have promoted the dominance of KP community and is a major driver for felty root mat formation [10]. However, the degradation of KP grassland may be caused by both human activities and climate change [9]. The mean annual precipitation in the northern and western parts of the QTP (the elevations ranged from 4400–4800 m) was less than 450 mm, with an increase of inter-annual variability towards the west [2,10]. Grassland suffered from co-limitation of summer rainfall and nutrient shortage [10,50–53]. The types of grassland were diverse, but the species richness was low [10,15]. Hence, the ten distinct plant communities were described in this area [2]. The grassland is dominated by KP community in closed lawns with covers of 98%, and companion species less than 10 [10,43].

Our study area is located at the eastern edge of the QTP (including three counties), the mean elevation is 3758 m (Figure 1c) and mean annual precipitation $\geq$ 450 mm. The alpine meadow in study area consists of four types of vegetation communities, including (a) the Poaceae community (*Elymus nutans* + *Stipa silena* + *Festuca ovina*), (b) the *Kobresia humilis* community, (c) the *Kobresia pygmaea* community (KP), and (d) the denuded black soil ecosystem. Those communities consist of more than 40 species, with mosaics of KP community patches and grasses, other sedges and perennial forbs growing as rosettes and cushions [54,55]. Overgrazing is the main inducing factor for grassland vegetation community variation [5,10,56], but effect of climate still cannot be eliminated. Although we have mapped the distribution of KP community, the relative contributions from climatic and anthropogenic forces require further investigation. The main effect factor can be distinguished by combining the potential distribution based on the ecological niche model [57] and realistic distribution based on remote sensing, which is very important for alpine meadow protection.

4.2. Challenges and Prospects for Alpine Meadow Grass Communities Classification

4.2.1. Field Observation

KP community plays a vital role in alpine meadow degradation succession in QTP. However, its spatial distribution is difficult to map: on the one hand, the field observation data is lacking; on the other hand, the distribution of the KP community is under a dynamic variation with different disturbances [5,43]. The massive field observation is the basis of RS classification for grassland community. Traditional grassland vegetation community samples were obtained with the few field investigation, expert knowledge and literature reviews [11–14,58]. Field observation is mainly carried out at quadrat, plot and belt transection scales [15,59,60]. Due to the complex distribution of grassland vegetation communities, the field investigation is difficult and time-consuming. Meanwhile, the expert knowledge and literature reviews cannot meet the accuracy requirement of classification, because of the subjective bias, the dynamic climate and anthropogenic activities [15,61].

In this study, the field observation was performed by UAV based on FragMAP [14]. The resolution of each aerial photo is ~0.87 cm and covers ~35 × 26 m of ground at the height of 20 m, which is close to the traditional ground observation plot [25,62]. Moreover, the UAV is efficient and easy to operate (about 15 min to finish each observation site), which provides the possibility for rapid observation in large regions [24]. Most importantly, the waypoints, once established, can be repeatedly used (the error of two flights of the same waypoint is 1–2 m, and two photos on the same waypoint from two different flights are almost overlapped). It is suitable to monitor the dynamic variation of grassland communities in a long-term period [25,62].

Limited by the UAV control range and battery life, the size of ROI was only 250 × 250 m, and the proportion of image raster used for training classification is relatively small. Besides, most of field observation sites were located in the flat area, which was near major traffic roads. Therefore, the spatial distribution of KP community still had some uncertainty in other regions of the study area. Moreover, the vegetation communities were distinguished by manual visual interpretation, and it requires good knowledge of plant taxonomy and time-consuming. Hence, the automatic identification of vegetation community based on aerial photograph and deep learning algorithm requires further exploration.

4.2.2. Classification Variables

NDVI, NDWI and SAVI have been commonly used as the classification variables for grassland classification [20,33,59]. The vertical variation of grassland vegetation is significantly changed with topographic features in the QTP [63], hence, topographical factor is an important classification basis in alpine vegetation communities classification [64]. Additionally, texture features are also essential variables in object-based classification, which usually reflect local spatial information relating to the change of image tone [16,17]. The common method in texture feature extraction is the grey level co-occurrence matrix (GLCM). The texture metric includes angular second moment, contrast correlation, entropy, homogeneity, difference, average and standard degrees [18]. Incorporating texture feature information usually enhances the recognition of "the same object with different spectrum" or "the different object with same spectrum" [16–18].

Our results showed that, the threshold range of these RS indices for identifying the alpine meadow communities are commonly confused during extraction and identification. According to the descriptive statistical value of those RS indices corresponding to the four alpine meadow grass communities, the threshold range of KH was close to KP, and that of Poaceae was close to BS among the NDVI, NDWI and SAVI (Figure 6a–c). Although four grass communities could be distinguished in topography and texture metrics, there were relatively few differences and large errors (with little difference in mean values and wide range in variation) (Figure 6d–i). Therefore, it was difficult to distinguish the alpine meadow grass communities based on single variable and simple combinations [33–35]. RS classification accuracy can be improved by combining the RS, topographic and texture indices (Tables 3 and 4).

Due to large errors in spatial quantification of some variables (such as texture indices), the classification still has some limitations and uncertainties [29]. Hence, we consider using high spatiotemporal resolution images in future research, such as the Sentinel- 2A/B satellite images, to reduce the effects of spatial heterogeneity on spectral reflectance and acquire more detailed texture features. Secondly, screening and reconstructing the remote sensing vegetation index: combining existing vegetation index, screening out indices that are more suitable for alpine meadow vegetation community classification.

4.2.3. Classification Method

Limited by the low temporal-spatial resolution, few spectrum band of RS images and field observations, most of natural grassland classification were applied in land use types (such as non-grassland, grassland, woodland, etc.) [33], different biophysics characteristics (for example, grassland with high, medium and low coverage) [34] and types with different climatic zones (e.g., groups and types of grassland) [35]. The most frequently used classification methods are visual interpretation, maximum likelihood classifiers, k-nearest neighbor and decision tree classification, and so on [65–67]. With the development of classification methods, the machine learning algorithm has obvious advantages in RS image classification [28,29]. However, the previous grassland classifications have been done at the vegetation type level, and few at the community level [8,9].

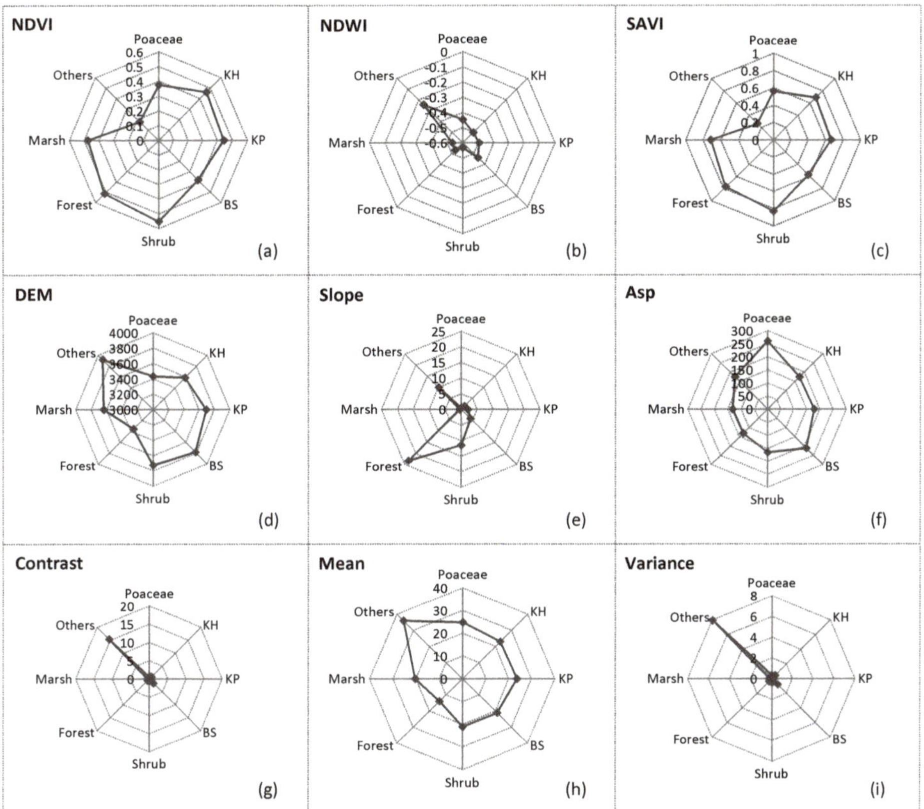

Figure 6. Characteristics of RS indices (**a–c**), topography (**d–f**) and texture metrics (**g–i**) in eight types of land covers: Poaceae, KH, KP and BS represent *Elymus nutans + Stipa silena + Festuca ovina*, *Kobresia humilis*, *Kobresia pygmaea* and black soil type, respectively.

Referenced with previously classification methods [20,58,68,69], the ANN, AVM and RF were used to distinguish the alpine meadow grass communities based on RS, texture and topographic indices in the QTP. Our results demonstrated that the RF algorithm had higher overall accuracy than other algorithms by using the same training samples (with 74.06% to 83.92%). Compared with other methods, RF is a data-driven algorithm. With the increase of input dataset, classification accuracy is improved correspondingly [66,70,71]. The RF algorithm can estimate complex nonlinear relationship and all the quantitative and qualitative information distributed within the models better; thus, these models are robust and fault-tolerant [69,70]. Moreover, the input classification indices can be acquired by different multi-spectral remote sensing images, and it helps to integrate multi-source remote sensing data [66,70,71]. However, it is difficult to train the RF model effectively with a small sample dataset. RF algorithm composes a large sample decision tree, and classification is performed based on the voting results of each decision tree, thus, has a strong tolerance for data error [40,41]. Constructing decision trees consumes more time while performing random forest classification [70].

5. Conclusions

Based on the band spectral, vegetation indices, texture feature of GaoFen 1/6 wide field view images, topographic indices and UAV field observation, this study examined four classification methods and evaluated their accuracy. Our results showed that the characteristics of RS indices in alpine meadow vegetation communities were very similar, and it was difficult to distinguish the alpine meadow grass communities based on single variable or simple combinations. The KP community could be distinguished through the RF method based on combination of RS, texture and topographic indices. The spatial distribution of KP community was fragmented with large spatial heterogeneity and small area in three counties. The area was 1434.07 km^2, which accounted for 7.20% of the whole study area. Our study demonstrated it was feasible to map at the community level using the satellite remote sensing, UAV surveying and machine learning methods. In future work, more detailed texture features derived from the high spatiotemporal resolution images are required to improve the grassland vegetation community classification.

Supplementary Materials: The following are available online at https://www.mdpi.com/article/10.3390/rs13132483/s1. Figure S1: Aerial photographs of alpine meadow vegetation communities. a, b, c and d were photographs taken at the height of 2 m. e, f, g and h were photographs taken at a height of 0.5 m; Poaceae, KH, KP and BS represent communities of Elymus nutans + Stipa silena + Festuca ovina, Kobresia humilis, Kobresia pygmaea and black soil type, respectively. Figure S2: Statistical analysis results of band1 to band8 of GF1/GF6 images; Poaceae, KH, KP and BS represent Elymus nutans + Stipa silena + Festuca ovina, Kobresia humilis, Kobresia pygmaea and black soil type, respectively. Figure S3: Statistical analysis results of texture indices of GF1/GF6 images; Poaceae, KH, KP and BS represent Elymus nutans + Stipa silena + Festuca ovina, Kobresia humilis, Kobresia pygmaea and black soil type, respectively. Figure S4: Statistical analysis results of vegetation and topography indices of GF1/GF6 images; Poaceae, KH, KP and BS represent Elymus nutans + Stipa silena + Festuca ovina, Kobresia humilis, Kobresia pygmaea and black soil type, respectively.

Author Contributions: All authors contributed significantly to this manuscript. B.M. and S.Y. designed this study; Z.Y., B.M., H.Y., Y.Q., Y.S., J.Z., J.C., Z.W., W.Z., M.L. and Y.L. were responsible for the field observation, data processing, analysis, and writing of the paper. All authors have read and agreed to the published version of the manuscript.

Funding: This research was funded by the National Key R&D Program of China (2017YFA0604801), the National Nature Science Foundation of China (42071056, 31901393, 41861016). The APC was funded by National Key R&D Program of China (2017YFA0604801).

Institutional Review Board Statement: Not applicable.

Informed Consent Statement: Not applicable.

Data Availability Statement: GF1 and GF6 WFV imager images were downloaded from the China Centre for Resources Satellite Data and Application (http://www.cresda.com/EN/, accessed on 5 May 2020); The DEM data were 90 m shuttle radar topography mission (SRTM) images (version V004) (http://srtm.csi.cgiar.org/, accessed on 5 May 2020).

Acknowledgments: The authors thanks the Three-River-Source National Park Administration for their help in field sampling. We also appreciate the editor's and reviewers' constructive suggestions to greatly improve the paper.

Conflicts of Interest: the authors declare no conflict of interest.

References

1. The Editorial Committee of Vegetation Map of China, Chinese Academy of Sciences. *Vegetation of China and Its Geographic Pattern: Illustration of the Vegetation Map of the People's Republic of China (1:1,000,000)*; Geological Publishing House: Beijing, China, 2007. (In Chinese)
2. Miehe, G.; Bach, K.; Miehe, S.; Kluge, J.; Yang, Y.; Duo, L.; Co, S.; Wesche, K. Alpine steppe plant communities of the Tibetan highlands. *Appl. Veg. Sci.* **2011**, *14*, 547–560. [CrossRef]
3. Wang, P.; Lassoie, J.; Morreale, S.; Dong, S. A critical review of socioeconomic and natural factors in ecological degradation on the Qinghai-Tibetan Plateau, China. *Rangeland J.* **2015**, *37*, 1–9. [CrossRef]

4.	Liu, S.; Zamanian, K.; Schleuss, P.; Zarebanadkouki, M.; Kuzyakov, Y. Degradation of Tibetan grasslands: Consequences for carbon and nutrient cycles. *Agric. Ecosyst. Environ.* **2018**, *252*, 93–104. [CrossRef]
5.	Cao, G.M.; Long, R.J. System Stability and its Self-maintaining Mechanism by Grazing in Alpine Kobresia Meadow. *Chin. J. Agrometeorol.* **2009**, *30*, 553–559.
6.	Lin, L. Response and Adaptation of Plant-Soil System of Alpine Meadows in Different Successional Stages to Grazing Intensity. Ph.D. Thesis, Gansu Agricultural University, Lanzhou, China, 2017.
7.	Li, X.; Gao, J.; Brierley, G.; Qiao, Y.; Zhang, J.; Yang, Y. Rangeland degradation on the Qinghai-Tibet Plateau: Implications for rehabilitation. *Land Degrad. Dev.* **2013**, *24*, 72–80. [CrossRef]
8.	Ma, Z.; Liu, H.; Mi, Z.; Zhang, Z.; Wang, Y.; Xu, W.; Jiang, L.; He, J.S. Climate warming reduces the temporal stability of plant community biomass production. *Nat. Commun.* **2017**, *8*, 15378. [CrossRef] [PubMed]
9.	Li, L.; Zhang, Y.; Liu, L.; Wu, J.; Li, S.; Zhang, H.; Zhang, B.; Ding, M.; Wang, Z.; Paudel, B. Current challenges in distinguishing climatic and anthropogenic contributions to alpine grassland variation on the Tibetan Plateau. *Ecol. Evol.* **2018**, *8*, 5949. [CrossRef] [PubMed]
10.	Miehe, G.; Schleuss, P.; Seeber, E.; Babel, W.; Biermann, T.; Braendle, M.; Chen, F.; Coners, H.; Foken, T.; Gerkenj, T.; et al. The Kobresia pygmaea ecosystem of the Tibetan highlands—origin, functioning and degradation of the world's largest pastoral alpine ecosystem Kobresia pastures of Tibet. *Sci. Total Environ.* **2019**, *648*, 754–771. [CrossRef]
11.	Beard, J.S. The vegetation survey of Western Australia. *Vegetatio* **1975**, *30*, 179–187. [CrossRef]
12.	Cawsey, E.; Austin, M.; Baker, B.L. Regional vegetation mapping in Australia: A case study in the practical use of statistical modelling. *Biodivers. Conserv.* **2002**, *11*, 2239–2274. [CrossRef]
13.	Naidoo, R.; Hill, K. Emergence of indigenous vegetation classifications through integration of traditional ecological knowledge and remote sensing analyses. *Environ. Manag.* **2006**, *38*, 377–387. [CrossRef]
14.	Bartha, S. A grid-based, satellite-image supported, multi-attributed vegetation mapping method (MTA). *Folia Geobot.* **2007**, *42*, 225–247.
15.	Su, Y.; Guo, Q.; Hu, T.; Guan, H.; Jin, S.; An, S.; Chen, X.; Guo, K.; Hao, Z.; Hu, Y.; et al. An Updated Vegetation Map of China (1:1000000). *Sci. Bull.* **2020**, *65*, 1125–1136. [CrossRef]
16.	Palm, C. Color texture classification by integrative co-occurrence matrices. *Pattern Recognit.* **2004**, *37*, 965–976. [CrossRef]
17.	Pietikäinen, M.; Mäenpää, T.; Viertola, J. Color texture classification with color histograms and local binary patterns. Presented at the Workshop on Texture Analysis in Machine Vision, Oulu, Finland, 14–15 June 1999; pp. 14–15.
18.	Arif, M.S.M.; Gülch, E.; Tuhtan, J.A.; Thumser, P.; Haas, C. An investigation of image processing techniques for substrate classification based on dominant grain size using rgb images from uav. *Int. J. Remote Sens.* **2017**, *38*, 2639–2661. [CrossRef]
19.	Wang, L.; Yang, R.; Tian, Q.; Yang, Y.; Zhou, Y.; Sun, Y.; Mi, X. Comparative Analysis of GF-1 WFV, ZY-3 MUX, and HJ-1 CCD Sensor Data for Grassland Monitoring Applications. *Remote Sens.* **2015**, *7*, 2089–2108. [CrossRef]
20.	Chen, Q.; Yu, R.; Hao, Y.; Wu, L.; Zhang, W.; Zhang, Q.; Bu, X. A new method for mapping aquatic vegetation especially underwater vegetation in lake ulansuhai using GF-1 satellite data. *Remote Sens.* **2018**, *10*, 1279. [CrossRef]
21.	Yang, Y.; Zhan, Y.; Tian, Q.; Gu, X.; Yu, T.; Wang, L. Crop classification based on GF-1/WFV NDVI time series. *Trans. Chin. Soc. Agric. Eng.* **2015**, *31*, 155–161.
22.	Yi, S.; Chen, J.; Qin, Y.; Xu, G.W. The burying and grazing effects of plateau pika on alpine grassland are small: A pilot study in a semiarid basin on the Qinghai-Tibet Plateau. *Biogeosciences* **2016**, *13*, 6273. [CrossRef]
23.	Ge, J.; Meng, B.P.; Liang, T.G.; Feng, Q.S.; Gao, J.L.; Yang, S.X.; Huang, X.D.; Xie, H.J. Modeling alpine grassland cover based on MODIS data and support vector machine regression in the headwater region of the Huanghe River, China. *Remote Sens. Environ.* **2018**, *218*, 162–173. [CrossRef]
24.	Meng, B.P.; Gao, J.; Liang, T.; Cui, X.; Ge, J.; Yin, J.; Feng, Q.; Xie, H. Modeling of alpine grassland cover based on unmanned aerial vehicle technology and multi-factor methods: A case study in the east of Tibetan Plateau, China. *Remote Sens.* **2018**, *10*, 320. [CrossRef]
25.	Sun, Y.; Yi, S.H.; Hou, F.J. Unmanned aerial vehicle methods makes species composition monitoring easier in grasslands. *Ecol. Indic.* **2018**, *95*, 825–830. [CrossRef]
26.	Yi, S. FragMAP: A tool for long-term and cooperative monitoring and analysis of small-scale habitat fragmentation using an unmanned aerial vehicle. *Int. J. Remote Sens.* **2017**, *38*, 8–10. [CrossRef]
27.	Guo, X.; Yi, S.; Qin, Y.; Chen, J. Habitat environment affects the distribution of plateau pikas:A study based on an unmanned aerial vehicle. *Pratacultural Sci.* **2017**, *34*, 1306–1313.
28.	Liang, T.; Yang, S.; Feng, Q.; Liu, B.; Zhang, R.; Huang, X.; Xie, H. Multi-factor modeling of above-ground biomass in alpine grassland: A case study in the Three-River Headwaters Region, China. *Remote Sens. Environ.* **2016**, *186*, 164–172. [CrossRef]
29.	Meng, B.P.; Yi, S.H.; Liang, T.G.; Yin, J.P.; Cui, X.; Ge, J.; Hou, M.J.; Lv, Y.Y. Modeling Alpine Grassland Above Ground Biomass Based on Remote Sensing Data and Machine Learning Algorithm: A Case Study in East of the Tibetan Plateau, China. *IEEE J. Sel. Top. Appl. Earth Obs. Remote Sens.* **2020**, *13*, 2986–2995. [CrossRef]
30.	Dong, S.W.; Sun, D.F.; Li, H. Crop Decision Tree Classification Extraction Based on MODIS NDVI in Beijing. *Adv. Mater. Res.* **2014**, *955–959*, 787–790. [CrossRef]
31.	Li, D.; Yang, F.; Wang, X. Study on Ensemble Crop Information Extraction of Remote Sensing Images Based on SVM and BPNN. *J. Indian Soc. Remote Sens.* **2016**, *45*, 1–9. [CrossRef]

32. Hou, M.J.; Ge, J.; Gao, J.L.; Meng, B.P.; Li, Y.C.; Yin, J.P.; Liu, J.; Feng, Q.S.; Liang, T.G. Ecological risk assessment and impact factor analysis of alpine wetland ecosystem based on LUCC and boosted regression tree on the Zoige Plateau, China. *Remote Sens.* **2020**, *12*, 368. [CrossRef]

33. Hao, C.; Wu, S. The effects of land-use types and conversions on desertification in Mu Us Sandy Land of China. *J. Geogr. Sci.* **2006**, *16*, 57–68. [CrossRef]

34. Wiesmair, M.; Feilhauer, H.; Magiera, A.; Otte, A.; Waldhardt, R. Estimating Vegetation Cover from High-Resolution Satellite Data to Assess Grassland Degradation in the Georgian Caucasus. *Mt. Res. Dev.* **2016**, *36*, 56–65. [CrossRef]

35. Feng, Q.; Liang, T.; Huang, X.; Lin, H.; Xie, H.; Ren, J. Characteristics of global potential natural vegetation distribution from 1911 to 2000 based on comprehensive sequential classification system approach. *Grassl. Sci.* **2013**, *59*, 87–99. [CrossRef]

36. Xu, K. Maximum Likelihood Estimate. *Encycl. Syst. Biol.* **2013**, *38*, 1–4.

37. He, Q. *Neural network and its application in IR*; Graduate School of Library and Information Science, Urbana-Champaign: Champaign, IL, USA, 2008.

38. Vapnik, V.N. *The Nature of Statistical Learning Theory*; Springer: New York, NY, USA, 1995.

39. Chang, C.C.; Lin, C.J. LIBSVM: A library for support vector machines. *ACM Trans. Intell. Syst. Technol.* **2011**, *2*, 27. [CrossRef]

40. Breiman, L. Bagging predictors. *Mach. Learn.* **1996**, *24*, 123–140. [CrossRef]

41. Breiman, L. Random Forests. *Mach. Learn.* **2001**, *45*, 5–32. [CrossRef]

42. Van der Linden, S.; Rabe, A.; Held, M.; Jakimow, B.; Leitão, P.J.; Okujeni, A.; Schwieder, M.; Suess, S.; Hostert, P. The EnMAP-Box—A Toolbox and Application Programming Interface for EnMAP Data Processing. *Remote Sens.* **2015**, *7*, 11249–11266. [CrossRef]

43. Miehe, G.; Miehe, S.; Kaiser, K.; Liu, J.; Zhao, X. Status and dynamics of the Kobresia pygmaea ecosystem on the Tibetan Plateau. *Ambio* **2008**, *37*, 272–279. [CrossRef]

44. Schleuss, P.-M.; Heitkamp, F.; Sun, Y.; Miehe, G.; Xu, X.; Kuzyakov, Y. Nitrogen uptake in an alpine Kobresia pasture on the Tibetan Plateau: Localization by 15N labeling and implications for a vulnerable ecosystem. *Ecosystems* **2015**, *18*, 946–957. [CrossRef]

45. Sun, Y.; Schleuss, P.M.; Pausch, J.; Xu, X.; Kuzyakov, Y. Nitrogen pools and cycles in Tibetan Kobresia pastures depending on grazing. *Biol. Fertil. Soils* **2018**, *54*, 569–581. [CrossRef]

46. Babel, W.; Biermann, T.; Coners, H.; Falge, E.; Seeber, E.; Ingrisch, J.; Schleuss, P.; Gerken, T.; Leonacher, J.; Leipold, T. Pasture degradation modifies the water and carbon cycles of the Tibetan highlands. *Biogeosciences* **2014**, *11*, 6633–6656. [CrossRef]

47. Wu, Y.; Tan, H.; Deng, Y.; Wu, J.; Xu, X.L.; Wang, Y.; Tang, Y.; Higashi, T.; Cui, X. Partitioning pattern of carbon flux in a Kobresia grassland on the Qinghai-Tibetan Plateau revealed by field 13C pulse-labeling. *Glob. Chang. Biol.* **2010**, *16*, 2322–2333. [CrossRef]

48. Hafner, S.; Unteregelsbacher, S.; Seeber, E.; Lena, B.; Xu, X.; Guggenberger, G.; Miehe, G.; Kuzyakov, Y. Effect of grazing on carbon stocks and assimilate partitioning in a Tibetan montane pasture revealed by 13CO2 pulse labeling. *Glob. Chang. Biol.* **2012**, *18*, 528–538. [CrossRef]

49. Ingrisch, J.; Biermann, T.; Seeber, E.; Leipold, T.; Li, M.; Ma, Y.; Xu, X.; Guggenerger, G.; Foken, T. Carbon pools and fluxes in a Tibetan alpine Kobresia pygmaea pasture partitioned by coupled eddy-covariance measurements and $^{13}CO_2$ pulse labeling. *Sci. Total Environ.* **2015**, *505*, 1213–1224. [CrossRef]

50. Li, J.M.; Ehlers, T.A.; Werner, M.; Mutz, S.G.; Steger, C.; Paeth, H. Late Quaternary climate, precipitation ^{18}O, and Indian monsoon variations over the Tibetan Plateau. *Earth Planet. Sci. Lett.* **2016**, *457*, 412–422. [CrossRef]

51. Li, R.; Luo, T.; Molg, T.; Zhao, J.; Li, X.; Cui, X.; Du, M.; Tang, Y. Leaf unfolding of Tibetan alpine meadows captures the arrival of monsoon rainfall. *Sci. Rep.* **2016**, *6*, 20985. [CrossRef] [PubMed]

52. Miehe, G.; Miehe, S. Environmental changes in the pastures of Xizang. *Marbg. Geogr. Schr.* **2000**, *135*, 282–311.

53. Lehnert, L.W.; Wesche, K.; Trachte, K.; Reudenbach, C.; Bendix, J. Climate variability rather than overstocking causes recent large scale cover changes of Tibetan pastures. *Sci. Rep.* **2016**, *6*, 24367. [CrossRef]

54. Wang, Y.; Heberling, G.; Görzen, E.; Miehe, G.; Seeber, E.; Wesche, K. Combined effects of livestock grazing and abiotic environment on vegetation and soils of grasslands across Tibet. *Appl. Veg. Sci.* **2017**, *20*, 327–339. [CrossRef]

55. Niu, Y.; Zhu, H.; Yang, S.; Ma, S.; Zhou, J.; Chu, B.; Hua, R.; Hua, L. Overgrazing leads to soil cracking that later triggers the severe degradation of alpine meadows on the Tibetan Plateau. *Land Degrad. Dev.* **2019**, *30*, 1243–1257. [CrossRef]

56. Zhou, H.; Zhao, X.; Tang, Y.; Gu, S.; Zhou, L. Alpine grassland degradation and its control in the source region of the Yangtze and Yellow Rivers, China. *Grassl. Sci.* **2005**, *51*, 191–203. [CrossRef]

57. Feng, X.; Park, D.; Walker, C.; Peterson, A.; Merow, C.; Papes, M. A checklist for maximizing reproducibility of ecologicla niche model. *Nat. Ecol. Evol.* **2019**, *3*, 1382–1395. [CrossRef]

58. Pedrotti, F. *Plant and Vegetation Mapping*; Springer: Berlin, Germany, 2013.

59. Wen, Q.; Zhang, Z.; Liu, S.; Wang, X.; Wang, C. Classification of Grassland Types by MODIS Time-Series Images in Tibet, China. *IEEE J. Sel. Top. Appl. Earth Obs. Remote Sens.* **2010**, *3*, 404–409. [CrossRef]

60. Yu, K.F.; Lehmkuhl, F.; Falk, D. Quantifying land degradation in the Zoige Basin, NE Tibetan Plateau using satellite remote sensing data. *J. Mt. Sci.* **2017**, *14*, 77–93. [CrossRef]

61. Guo, D.; Song, X.; Hu, R.; Cai, S.; Zhu, X.; Hao, Y. Grassland type-dependent spatiotemporal characteristics of productivity in Inner Mongolia and its response to climate factors. *Sci. Total. Environ.* **2021**, *775*, 145644. [CrossRef]

62. Zhang, J.; Liu, D.; Meng, B.; Chen, J.; Wang, X.; Jiang, H.; Yu, Y.; Yi, S. Using UAVs to assess the relationship between alpine meadow bare patches and disturbance by pikas in the source region of Yellow River on the Qinghai-Tibetan Plateau. *Glob. Ecol. Conserv.* **2021**, *26*, e01517. [CrossRef]
63. Ma, W.W. *Study on Methods for Grassland Classification and Quality Estimation by Remote Sensing: A Case Study in the Region around Qinghai Lake*; Chinese Academy of Science: Beijing, China, 2015.
64. Bolstad, P.V.; Lillesand, T.M. Improved Classification of Forest Vegetation in Northern Wisconsin Through a Rule-Based Combination of Soils, Terrain, and Landsat Thematic Mapper Data. *For. Sci.* **1992**, *1*, 5–20.
65. Settle, J.J.; Briggs, S.A. Fast maximum-likelihood class-ification of remotely sensed imagery. *Int. J. Remote Sens.* **1987**, *8*, 723–734. [CrossRef]
66. Zhu, H.W.; Basir, O. An adaptive fuzzy evidential nearest neighbor formulation for classifying remote sensing images. *IEEE Trans. Geosci. Remote Sens.* **2005**, *43*, 1874–1889.
67. Navin, M.S.; Agilandeeswari, L. Comprehensive review on land use/land cover change classification in remote sensing. *J. Spectr. Imaging.* **2020**, *9*, a8. [CrossRef]
68. Jiang, L.C.; Wang, Y. Mapping zoige alpine wetland using random forest classification and landsat 5 TM data. *Stud. Surv. Mapp. Sci. (SSMS)* **2016**, *4*, 24–27.
69. Qiao, Y.; Chen, T.; Qiang, W.; Feng, T. Method of grassland information extraction based on multi-level segmentation and cart model. *Int. Arch. Photogramm. Remote Sens. Spatial Inf. Sci.* **2018**, *XLII-3*, 1415–1420. [CrossRef]
70. He, T.; Wang, S. Multi-spectral remote sensing land-cover classification based on deep learning methods. *J. Supercomput.* **2020**, *77*, 2829–2843. [CrossRef]
71. Yuan, H.; Yang, G.; Li, C.; Wang, Y.; Liu, J.; Yu, H.; Feng, H.; Xu, B.; Zhao, X.; Yang, X. Retrieving Soybean Leaf Area Index from Unmanned Aerial Vehicle Hyperspectral Remote Sensing: Analysis of RF, ANN, and SVM Regression Models. *Remote Sens.* **2017**, *9*, 309. [CrossRef]

 *remote sensing*

Article

Assessment of Rangeland Degradation in New Mexico Using Time Series Segmentation and Residual Trend Analysis (TSS-RESTREND)

Melakeneh G. Gedefaw [1], Hatim M. E. Geli [2,3,*] and Temesgen Alemayehu Abera [4]

[1] Water Science and Management Program, New Mexico State University, Las Cruces, NM 88003, USA; melakene@nmsu.edu
[2] New Mexico Water Resources Research Institute, New Mexico State University, Las Cruces, NM 88003, USA
[3] Department of Animal and Range Sciences, New Mexico State University, Las Cruces, NM 88003, USA
[4] Department of Geosciences and Geography, University of Helsinki, P.O. Box 68, FI-00014 Helsinki, Finland; temesgen.abera@helsinki.fi
* Correspondence: hgeli@nmsu.edu; Tel.: +1-575-646-1640

Abstract: Rangelands provide significant socioeconomic and environmental benefits to humans. However, climate variability and anthropogenic drivers can negatively impact rangeland productivity. The main goal of this study was to investigate structural and productivity changes in rangeland ecosystems in New Mexico (NM), in the southwestern United States of America during the 1984–2015 period. This goal was achieved by applying the time series segmented residual trend analysis (TSS-RESTREND) method, using datasets of the normalized difference vegetation index (NDVI) from the Global Inventory Modeling and Mapping Studies and precipitation from Parameter elevation Regressions on Independent Slopes Model (PRISM), and developing an assessment framework. The results indicated that about 17.6% and 12.8% of NM experienced a decrease and an increase in productivity, respectively. More than half of the state (55.6%) had insignificant change productivity, 10.8% was classified as indeterminant, and 3.2% was considered as agriculture. A decrease in productivity was observed in 2.2%, 4.5%, and 1.7% of NM's grassland, shrubland, and ever green forest land cover classes, respectively. Significant decrease in productivity was observed in the northeastern and southeastern quadrants of NM while significant increase was observed in northwestern, southwestern, and a small portion of the southeastern quadrants. The timing of detected breakpoints coincided with some of NM's drought events as indicated by the self-calibrated Palmer Drought Severity Index as their number increased since 2000s following a similar increase in drought severity. Some breakpoints were concurrent with some fire events. The combination of these two types of disturbances can partly explain the emergence of breakpoints with degradation in productivity. Using the breakpoint assessment framework developed in this study, the observed degradation based on the TSS-RESTREND showed only 55% agreement with the Rangeland Productivity Monitoring Service (RPMS) data. There was an agreement between the TSS-RESTREND and RPMS on the occurrence of significant degradation in productivity over the grasslands and shrublands within the Arizona/NM Tablelands and in the Chihuahua Desert ecoregions, respectively. This assessment of NM's vegetation productivity is critical to support the decision-making process for rangeland management; address challenges related to the sustainability of forage supply and livestock production; conserve the biodiversity of rangelands ecosystems; and increase their resilience. Future analysis should consider the effects of rising temperatures and drought on rangeland degradation and productivity.

Keywords: NDVI; precipitation; drought; breakpoints and timeseries analysis; ecosystem structural change; BFAST

Citation: Gedefaw, M.G.; Geli, H.M.E.; Abera, T.A. Assessment of Rangeland Degradation in New Mexico Using Time Series Segmentation and Residual Trend Analysis (TSS-RESTREND). *Remote Sens.* **2021**, *13*, 1618. https://doi.org/10.3390/rs13091618

Academic Editor: Elias Symeonakis

Received: 11 March 2021
Accepted: 14 April 2021
Published: 21 April 2021

Publisher's Note: MDPI stays neutral with regard to jurisdictional claims in published maps and institutional affiliations.

1. Introduction

Land degradation affects ecosystem productivity and threatens its capacity to sustain human, livestock, and wildlife population specially in dryland environments. Drylands that are susceptible to desertification occupy 39.7% (~5.2 billion ha) of the global terrestrial ecosystems (~13 billion ha) [1,2]. Of this, sever land degradation is prevalent in over 10–20% of the dryland ecosystems [1–3]. For these reasons, land degradation in dryland ecosystems is recognized as one of the major environmental and socioeconomic challenges that can alter ecosystem services and human wellbeing [4–7]. Thus, understanding the rate, expansion, and severity of drylands degradation has received (and will continue to receive) considerable attention, due to their pivotal role in food production and water availability for more than 2 billion people in the world [3,8–11].

The main causes of drylands degradation are principally associated with population growth, overgrazing, inappropriate land and water use practices, and climate change impacts [1,11,12]. As a result, noticeable and persistent loss of vegetation cover and biomass productivity, reduction in forage and crop production, and decline in carrying capacity of rangelands are becoming common in these ecosystem [3,13,14]. Future predictions showed that climate change impacts (i.e., increase in surface air temperature and evapotranspiration, and decrease in precipitation) are expected to worsen poverty and inequality in developing countries [15,16].

Changes in above-ground biomass in drylands on which forage production and other life securing ecosystem services depend on can be measured by the net primary production (NPP) [13,17]. Several studies used temporal changes in NPP as an indicator of land degradation [2,18–20]. Nevertheless, the measurements and estimation of land degradation have been characterized by arbitrary assumptions, qualitative and inconsistent judgment, unreliable and spurious estimations [7,21,22]. Several studies have been criticized for underestimating the extent and severity of rangeland degradation [3,8,23]. The main reasons being the climate of dryland ecosystems (i.e., low annual precipitation and its increased interannual variability) [19,24], the successive occurrence of degradation for many decades at regional or continental scales [20], and the lack of objective measurements that quantify the extent and severity of all forms of degradation [4]. These reasons further complicate the identification of changes in dryland productivity either due to human activities (grazing and cropping) or natural variability [13,17]. Hence, the assessment of drylands degradation requires techniques that encompass spatial and temporal properties that strongly adhere to the measurement principles of repetitiveness, objectivity, and consistency [21,25]. In this regard, earth observations (OE) proved to be the only feasible means for long time and large-scale monitoring of vegetation productivity [8,9,18,24].

Vegetation indices (VIs) such as the normalized difference vegetation index (NDVI) have been used as proxy to detect and quantify long-term land degradation in drylands [6,7,11,17,18,21,25]. Two main categories of methodology have been employed to assess land degradation in drylands [7,22]—the first one observes changes in relationships between climate variables (i.e., precipitation and temperature) and VIs (i.e., measure of vegetation greenness and productivity) [8,22,23]. The second category analyzes vegetation phenology to detect structural change caused by processes such as deforestation and long-term trends [22,24]. Within each category, various methods have been applied. For example, the methods in the first category aim to control the influence of precipitation and/or temperature trends to detect permanent degradation [7,23]. The rain use efficiency (RUE) as proposed by [19]—the ratio of NPP to rainfall—has been used as an indicator of degradation [7,17,20,26,27] and ecosystem functioning [18,28]. The residual trend analysis (RESTREND) method as proposed by [11] was used frequently to estimate changes in productivity by relating annual maximum NDVI and precipitation [11,13,25,29]. In the second category, the breaks for additive and seasonal and trend (BFAST) developed by [30,31], have been employed to assess change in vegetation phenology with time (e.g., [32]).

Both RESTREND and BFAST, however, have limitations in accurately detecting land degradation when the relationship between vegetation and climate breaks and due to over sensitiveness in areas where natural climate variabilities are prevalent, respectively [5,8,22]. Over sensitiveness is represented by the detection of false breakpoints using BFAST when dryland vegetation skips phonological cycles in response to drought while the ecosystem is healthy. The time series segmented residual trend analysis (TSS-RESTREND) method was developed by [8,22,23] to address this limitation by combining BFAST and RESTREND to detect the breakpoints where the relationship between vegetation and precipitation or temperature changes [8,23]. The TSS-RESTREND leverages the ability of BFAST in detecting breakpoints in NDVI long timeseries [5,8]. The TSS-RESTREND was applied to detect land degradation in Australia and its results were qualitatively evaluated using wildfire data [8]. However, the TSS-RESTREND lacked in providing a quantitative assessment of detected breakpoints [8].

About 40% of the United States of America (USA) land is classified as drylands [33]. The largest portion of these drylands is rangelands—comprising 31% of the USA land [34]—which are mostly situated in the Western USA [35]. Rangelands support livestock production particularly beef cattle and major local economies in the Great Plains, the Intermountain West, and the Southwest rely heavily on this industry [36]. Since the pre-settlement era, about 34% of the USA rangelands have been permanently modified by human activities [37]. Woody plants encroachment, invasive species, prolonged drought, low resilience and management of rangelands could be the most critical factors that can affect current and future rangeland productivity [37,38]. Further, urban development, energy extraction (i.e., oil and gas), and expansion in agricultural lands are expected to contribute to rangelands fragmentation [39]. The degree to which climate variables (e.g., interannual variability of precipitation) can cause structural changes and affect forage productivity in New Mexico's (NM) rangeland ecosystems has not been adequately studied. Structural change in an ecosystems can be described as the significant change in the pattern of organization of the ecosystem necessary for functioning [35]. A structural change of rangeland was considered as when a pixel has a significant break in its vegetation- precipitation relationships (VPR) that leads to irreversible degradation or significant increase in productivity. Moreover, quantitative assessment of rangeland degradation and its impacts on NM's food-water-energy systems is critical for the sustainable use of its resources.

The goal of this study was to identify long-term structural and productivity changes in rangeland ecosystems due to interannual climate variability based on precipitation in New Mexico during the 1982–2015 period. Taking advantage of the consistent NDVI timeseries of Global Inventory Modeling and Mapping Studies (GIMMS), the study tested the hypothesis that there is no significant change in rangeland productivity in NM between 1982–2015 by evaluating the prevalence of significant difference in NM's rangelands mean productivity before and after the years where the VPR breaks were occurred. The specific objectives were to: (1) characterize changes in rangeland productivity in terms of (a) direction of, and (b) type of structural changes in NPP as represented by NDVI using TSS-RESTREND; (2) develop an assessment framework for the identified changes in productivity; (3) use the assessment framework along with independent productivity data to identify areas affected by significant structural changes in NM's rangeland ecosystems.

2. Data

2.1. Study Area

The study area was the state of New Mexico, USA, which covers a total land area of 314,918 km^2 (Figure 1). NM's climate is dominated by arid and semi-arid conditions with an average annual precipitation ranging from less than 254 mm in the southern desert to more than 500 mm in higher elevations in the northern part of the state [40]. The mean annual air temperature ranges from less than 4.4 °C in high mountains and valleys in the northern to 18 °C in the southern parts of the state.

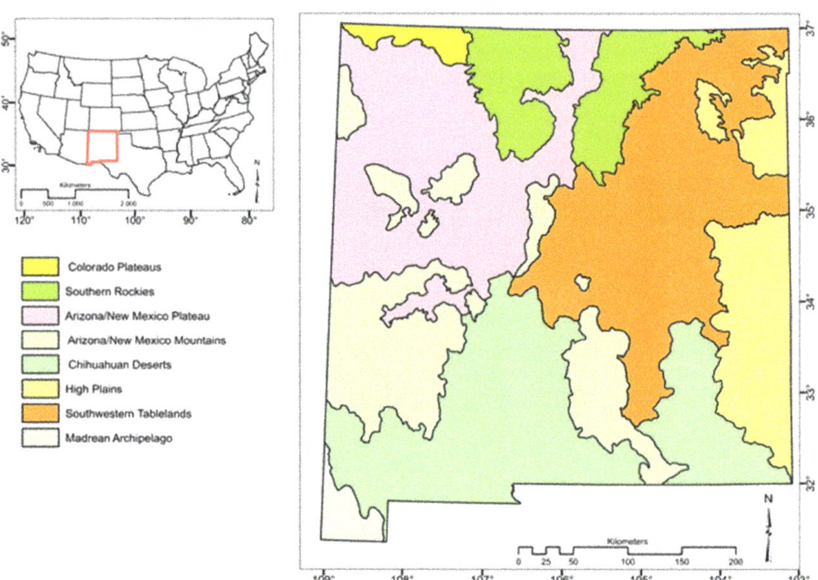

Figure 1. A map showing the location of the New Mexico, the contiguous USA (red polygon), and the nine ecoregions within New Mexico [41].

NM's land cover encompasses nine ecoregions that include the Chihuahua Deserts (22%) in south, southwestern part of NM, Southwestern Tablelands (22%) and Western High Plains (10%) in central, West and in the northwest parts of NM, Arizona/New Mexico Mountains (14%), Arizona/New Mexico Plateau (19%), Madrean Archipelago and southern Rockies [41] (Figure 1). Three major types of vegetation biomes (i.e., forest, shrubland and grassland) comprise the respective ecoregions [42].

2.2. Vegetation Cover

NM's grassland biomes, particularly in the Desert Grassland Association—desert plain grassland and mixed grassland or mixed prairie—are dominated by black grama grass (*Bouteloua eriopoda*) and tobosa grass (*Hilaria mutica*) in the desert plains. Bluestem (*Andropogon scoparius*), san blue stem (*A. halli*), and Indian Grass (*Sorghastrum nutans*) are parts of mixed grassland or mixed prairie. Woodland biome can be found within a range of 1371 m to 2286 m elevation (amsl) and consists of one-seed juniper (*Juniperus monosperma*) and pinon pine (*Pinus edulis*) sometimes with oak (Quercus spp.) with an understory of grassland, forbs, and shrubs. Coniferous forest biomes can be found within a range of 2590 to 3658 m amsl in Petran subalpine and petran montane dominated by Enngelman spruce (*Picea engel- mannii*) and subalpine fir (*Abies lasiocarpa*). Petran montane forest association (2591.8 to 2896 m a.m.s.l) covers an extensive area of the state dominated Douglas fir, and white fir species. The land use/land cover data used in this analysis included the National Land Cover Dataset of 2011 [43] as well as the state's ecoregions [41] and quadrants. These datasets were overlayed with the identified breakpoints to be related to changes in productivity over these different land cover types.

2.3. GIMMS NDVI

To study temporal changes in rangelands' structure (i.e., significant changes in NPP trend), long-term NDVI timeseries was used as a proxy to NPP. Specifically, the GIMMS NDVI product based on the third generation Global Inventory Modeling and Mapping Studies NDVI (GIMM NDVIv3.1g) data was used in this study. These data was based on the Advanced Very High Resolution Radiometer (AVHRR) sensors [5,8] and it is of the most accurate datasets currently available. This version of the data corrected for calibration errors in a previous one [22]. The GIMMS NDVI data is available for the 1981–2015 period only at spatial and temporal resolutions of 8 km and 16 days, respectively. The GIMMS NDVI dataset was obtained from ECOCAST [44].

2.4. Rangeland Productivity

Rangeland productivity for the USA was obtained from the Rangeland Production Monitoring Service (RPMS) dataset developed by the United State Forest Service [45]. The dataset was prepared using NDVI from the Thematic Mapper for the 1984–2020 period at 250-meter pixels. The dataset provides estimates of annual production of rangeland vegetation in pounds per acre, which is useful in understanding trends and variability of rangeland forage resources [45]. The RPMS dataset was coupled with the assessment framework (Section 3.2) to evaluate the significance of rangeland productivity changes compared to those identified by the TSS-RESTREND method.

2.5. Precipitation, Drought, and Fire

Gridded precipitation data developed by Parameter-elevation Regressions on Independent Slopes Model (PRISM) was obtained from PRISM Climate Group [46]. Monthly precipitation at ~4 km pixels was used. To evaluate the accuracy of breakpoints in terms of timing, extent, and direction of change relative to potential disturbances, historical drought and fire events were compared with the breakpoints. Mean monthly and annual self-calibrated Palmar Drought Severity Index (PDSI) was acquired from [47]. Fire data were obtained from New Mexico Resources Geographic Information System (RGIS) [48].

3. Methods

This analysis followed three main steps—data preparation, the application of TSS-RESTREND, and the development and application of an assessment framework to evaluate the detected breakpoints and changes in productivity (Figure 2). The first step involved data acquisition, projection, resampling, and extraction of pixel values of GIMMS NDVI, PRISM precipitation, and RPMS productivity. The second step (Section 3.1) involved the application of the TSS-RESTREND method to identify breakpoints in space and time, their significance, and their structural changes. The last step provided a framework (Section 3.2) to evaluate the detected breakpoints using the RPMS—an approach that was lacking in the work by [8,23].

3.1. Characterization of Change in Productivity Using TSS-RESTREND
3.1.1. NDVI and Precipitation Relationships

The Bimonthly GIMMS NDVI data were filtered using a quality control (QC) procedure to remove non-reliable values based on a quality flag [49,50]. Pixels with at least 75% reliable values were used. Non-vegetated pixels were excluded based on a threshold of a median NDVI of less than 0.1 [50,51]. Out of 4474 pixels, a total of 4454 pixel were used in the analysis. Complete mean monthly timeseries of NDVI (ctsNDVI) was assessed over each pixel. The PRISM precipitation data was resampled using a bilinear method to match the spatial resolution of the GIMMS NDVI and assessed at monthly time scale.

An ordinary Least Square regression (OLS) was used to develop the relationships between the ctsNDVI and the complete timeseries of optimal accumulated precipitation for each pixel for the 1982–2015 period (referred to as ctsVPR) [8]. The optimal accumulated precipitation was determined by identifying the pairs of ctsNDVI and accumulated precipitation that provide the highest correlation coefficient. The correlation coefficients of all possible OLS relationships between ctsNDVI and a matrix of precipitation accumulation period (1–12 months) with offset period (0–3 months) were evaluated following [8]. The relationships with the highest correlation coefficients were used to calculate the residual between the observed and predicted ctsNDVI (referred to as ctsVPR-residual).

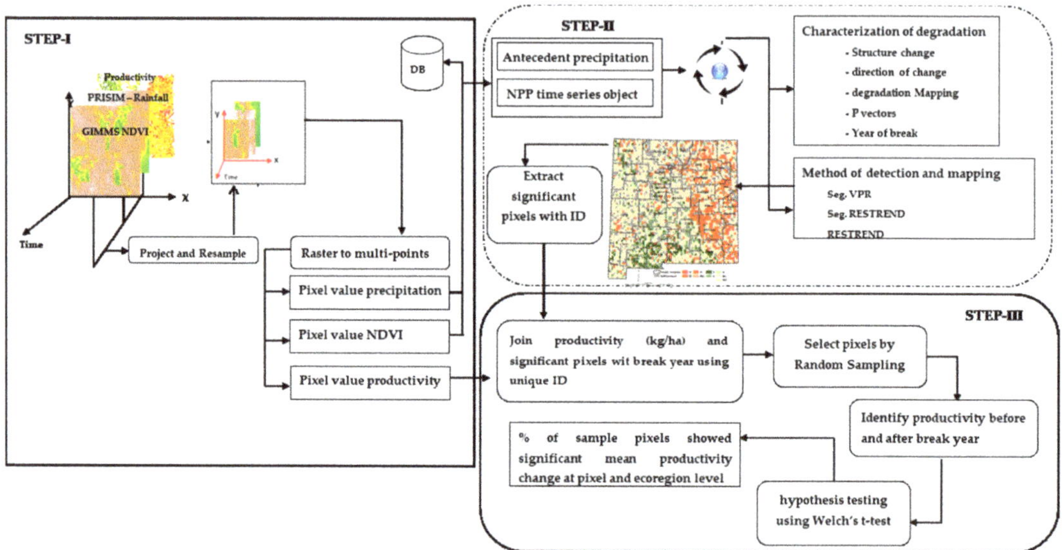

Figure 2. Depreciation of TSS-RESTRND analysis and validation of breakpoints. In step 1, Data preparation which includes (i.e., acquisition of NPP data, precipitation, and productivity data), projection, resampling, and extraction of pixel level values. In Step II (TSS-RESTREND): method of residual fit, *p*-vector, trend of productivity change (i.e., decreasing, increasing, non-significance change and indeterminant), and ecosystem structure change were detected. In Step III (Evaluation), include stratified random sampling for validation, Welch's *t* test on randomly selected pixels (i.e., to test the significant change of productivity before and after the break years).

3.1.2. Identification of Breakpoints

The BFAST method was applied on the ctsVPR-residual to list potential breakpoints. Briefly, the BFAST method decomposes the timeseries into season, trend, and reminder components—an approach that allows to detect changes in the season and trend components [52,53]. The list of potential breakpoints identified by the BFAST method [8] based on the ctsVPR-Residuals were further evaluated for their significance in the VPR—allowing to assess their impact on NPP as represented by the maximum NDVI timeseries. A Chow test was applied on the VPR-Residuals on all pixels with significant VPR (α = 0.05). Based on this test, all pixels that have no significant breakpoints in the VPR-Residual (α = 0.05) but have significant VPR (α = 0.05) were further assessed using the standard RESTREND by developing a regression between the VPR-Residuals and time (Equation (1)) [8].

$$y_i = \beta_0 + \beta_1\, x \ (\text{RESTREND}) \tag{1}$$

where β_0 is intercept, β_1 is slope, and x is year

Pixels that showed significant breakpoints in VPR-Residuals (based on the Chow test above) and had significant VPR were further evaluated for the significance of these breakpoints but in the VPR also using Chow test. Pixels with significant breakpoints in VPR-Residuals but not in the VPR were further assessed using the Segmented RESTREND by developing a multivariate regression between VPR-Residual, time, and a dummy variable (Equation (2))

$$y_i = \beta_0 + \beta_1 x_i + \beta_2 z_i + \beta_3 x_i z_i \ (\text{Segmented RESTREND}) \tag{2}$$

where β_0 is intercept, x is year, z is value of dummy variable (0 or 1), β_1 is slope, β_2 is offset at the breakpoint, and β_3 is the change in the slope at the breakpoint.

Pixels that showed significant breakpoints in VPR-Residuals and in VPR may indicate the presence of significant structural changes and thus the assumption of stationarity of the accumulation and offset periods used in the calculations of optimal accumulated precipitation before and after a breakpoint [5,54]. The set if the NDVImax and precipitation timeseries before and after a breakpoint was separated to recalculate new and independent VPR on either side of the breakpoints [8]. The precipitation data were standardized to account for the differences in the accumulation and offset periods among the breakpoints (Equation (3)).

$$z_i = \frac{(x_i - u)}{\delta} \tag{3}$$

where z is the standard score, x_i is observed values, μ is the mean, and δ is the standard deviation. The NDVImax and standard score timeseries were further evaluated by fitting a multivariant regression (Equation (4)) [5,8].

$$Yi = \beta_0 + \beta_1 x_i + \beta_2 z_i + \beta_3 x_i z_i \text{ (Segmented VPR)} \tag{4}$$

where x is the standardized precipitation for year i, z the value of the dummy variable (0 or 1), β_0 is intercept, β_1 is slope, β_2 is the offset at the breakpoint, and β_3 the change in the slope at the breakpoint.

Pixels that did not meet any of the above conditions were classified as indeterminant—met the following conditions: (1) had no significant VPR and no significant breakpoints in VPR; or (2) no significant VPR, significant breakpoints in VPR, and no significant breakpoints in segmented VPR.

3.1.3. Identification of Structural Changes

Structural changes of each pixel within NM ecosystems were identified based on three properties that include the significance of the breakpoints, direction of change (i.e., increasing and decreasing in productivity), and method of detection (as described in Section 3.1.2). With the regard to the significance level and direction of change, all pixels were classified into nine categories as shown in (Table 1) following [55] and similar to other dryland degradation studies in Australia [8] and China [5].

Table 1. Categories of threshold for vegetation change.

Category	Direction of Change	Significance	Description
I1		$p < 0.01$	Pixels with significant increasing trend of residual at four classes of p levels (0.01, 0.025, 0.05 and 0.1)
I2	Slope > 0	$0.01 \leq p < 0.025$	
I3		$0.025 \leq p < 0.05$	
INC		$0.05 \leq p < 0.1$	
DI		$p < 0.01$	Pixels with significant decreasing trend of residual at four classes of p levels (0.01, 0.025, 0.05 and 0.1)
D2	Slope < 0	$0.01 \leq p < 0.025$	
D3		$0.025 \leq p < 0.05$	
DNC		$0.05 \leq p < 0.1$	
NSC		$p \geq 0.1$	No significant change in productivity

INC = Increase No Change, DNC = Decrease No Change, I1 = increasing trend in productivity level 1, I2 = increasing trend in productivity level 2, I3 = increasing trend in productivity level 3, D1 = Decreasing trend in productivity level 1, Decreasing trend in productivity level 2, D3 = Decreasing trend in productivity level 3, NSC = Non-significant Change.

Consequently, a pixel can be considered having:

- Non-reversable degradation or initiation of degradation if it exhibited a breakpoint with $p < 0.05$ and a negative change (Table 1) in productivity as detected by Segmented VPR or Segmented RESTREND, respectively,

- Reversal or initiation of reversal in degradation, if it met the previous conditions except with a positive as detected by Segmented VPR or Segmented RESTREND, respectively,
- Stable increase in productivity, if it exhibited a breakpoint with $p < 0.05$ and a positive change as detected by RESTREND,
- Stable decrease in productivity if the opposite of the previous condition was detected (a condition that can be considered as an initiation of degradation) [11,13].
- Non-significant change (NSC) in productivity irrespective of the detection technique used (i.e., Segmented VPR, Segmented RESTREND, RESTREND, or Indeterminant), if it had a breakpoint with $p > 0.1$ and a constant direction change (i.e., 0),
- Indeterminate change if it had $p = 0$ and slope = 0.

A summary of the pixels based on these structural change categories was presented relative to NM's total area, quadrant, ecoregions, and major land use/land cover classes.

3.2. Breakpoints Assessment Framework

The TSS-RESTREND method by [8] provided only a qualitative assessment of breakpoints with other independent data in terms of significance and direction of changes. These two properties are important to properly characterize changes in productivity. A framework was developed in this study to address this gap by proposed means to quantitatively assess these properties. This framework consisted of four steps: (1) Develop random samples within the identified significant breakpoints; (2) Select and use independent productivity data; (3) Evaluate the random samples at the pixel level; and (4) Group and evaluate all random samples that fall within identified ecoregions—allowing to identify whether the changes in productivity at the individual pixels are reflective of consistent regional changes.

Random Samples: A set of random samples in terms of size and location can be developed following [56,57]. To estimate the size of random samples, a prior knowledge of image accuracy/variability is required [57]. The degree of variability in the RPMS data is unknown, thus it was assumed that the maximum variability would be about 50% with 95% confidence level and ±5% precision [56]. This criterion helps to determine representative sample pixels using a stratified random sampling approach [58]. The strata were developed based on ecoregions, land use/land cover, and direction of change. Based on the formula from [57], the sample size was calculated as (Equation (5)).

$$n_{0=}\frac{Z^2\,pq}{e^2} \tag{5}$$

where n_0 is the sample size, Z is the selected critical value of desired confidence level (1.96), p is the estimated proportion of an attribute that is present in the population (0.5), and $q = 1 - p$ (0.5) with p and e (0.5) represent the desired level of precision.

The allocation of the random sample was based on the majority of the identified breakpoints in each method (Section 3.1.2). The random samples can be broadly allocated based on the direction of change and land cover within the identified significant breakpoints in the two main categories—the Segmented VPR and Segment RESTREND. Ecoregions and land use/land cover (NLCD 2011) [43] maps were overlaid to identify the locations of the pixels within these regions for further assessment.

Selection of independent data: The productivity data from the RPMS [45] was used to evaluate the accuracy of the identified breakpoints based on the TSS-RESTREND in terms of direction and significance of change in productivity. The data was resampled to match that of GIMMS NDVI [44,58]. Using the coordinates of GIMMS NDVI pixels, the corresponding RPMS productivity was extracted. The rangeland productivity was converted from pound per acre per year to kilogram per hectare per year. this process allowed to prepare a dataset that include RPMS productivity and with the corresponding TSS-RESTREND estimates of timing and direction of change, significance of breakpoints, and method of detection for each random sample.

Pixel level assessment: Using the location of the randomly selected samples (breakpoint pixels) and their identified year of breaks, the mean annual productivity for each pixel based on the RPMS data can be calculated for before and after the break years. Over each pixel, the number of years before and after the breaks were different and thus these mean values had different sample sizes. These mean values before and after the breaks over each pixel were then statistically compared using the Welch's student-test for the significance in their difference and direction of change. The Welch's *t* test was conducted assuming that the variances were not equal before and after the breaks [58,59]. The obtained results based on RPMS data using the Welch's test were compared with those from the TSS-RESTREND method. A summary of the agreement of this comparison was provided over these pixels as well as over the representative land cover classes.

Ecoregion level assessment: a similar approach was followed in this step of the framework except in this case the randomly selected pixels were grouped to represent ecoregions—all sampled pixels within each ecoregion represent a single entity (analysis unit). The mean values of productivity based on RPMS of each group before their individual breaks were further averaged and compared to the corresponding average after the break. The averages were assessed for their significance and direction of change using the Welch's student-test. The results were then compared with those obtained based on the TSS-RESTREND over the equivalent ecoregions.

4. Results
4.1. Characteristics of Change
4.1.1. Breakpoints and Direction of Change

The number of significant breakpoints detected (increased and decreased productivity) and precipitation anomalies between 1982 and 2015 were shown in Figure 3. Out of all the analyzed pixels (i.e., 4454), 814 (18.3% or ~50,000 km^2) had significant breakpoints—with either increasing or decreasing trends—were detected over different land cover types. The remaining pixels that showed insignificant change (NSC), indeterminant, or identified as Agricultural represented about 55.6% (158,591 km^2), 10.8% (30,656 km^2) and 3.2% (9122 km^2) of NM's area (~315,900 km^2). The distribution of these pixels is shown in Figure 4. Out of the 814 significant breakpoints, 52.3% (26,176 km^2) and 46.5% (23,232 km^2) had negative (decreasing) and positive (increasing) change in productivity, respectively (Table 2). The areas that showed significant increasing trends in productivity as I1, I2 and I3 were about 10,752 km^2 (3.8%), 9408 km^2 (3.3%), and 3072 km^2 (1.1%) respectively. The areas that showed decreasing trend in productivity with D1, D2, and D3 level of significance were about 9088 km^2 (3.2%), 13,056 km^2 (4.6%), and 3776 km^2 (1.3%), respectively.

Table 2. A summary of the identified direction of change in productivity based on the TSS-RESTREND method in New Mexico during the 1982–2015 period.

Method	Direction of Change (% Relative to 4454 Pixels)					Total
	Deceasing	Increasing	NSC *	Agriculture	Indeterminant	
TSS-RESTREND	17.6	12.8	55.6	3.2	10.8	100

* NSC = Non-Significant Change.

From all significant breakpoints, 58%, 20%, and 15.7% were observed in shrubland, grassland, and evergreen forest ecosystems, respectively (Figure 5). The highest number of detected breakpoints was 250 (or 31% out of 814) in 2005. Among them, shrublands and grasslands combined accounted for 87.6% (219), out of which 72.8% and 14.8% were over shrublands and grasslands, respectively.

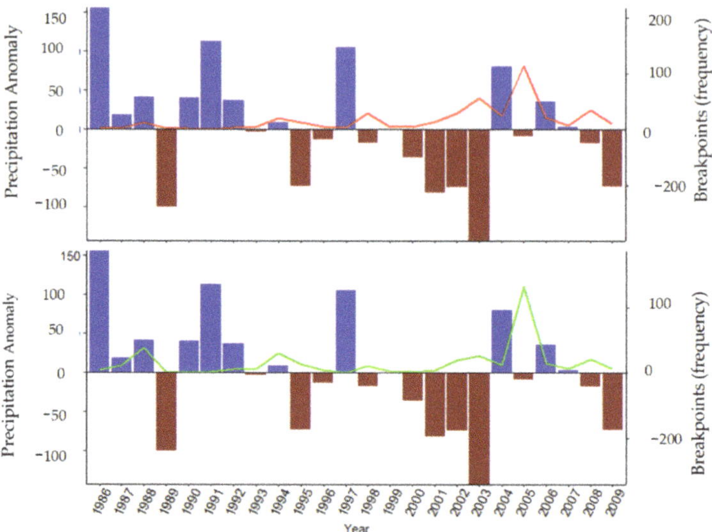

Figure 3. A summary of the number of detected significant breakpoints along with corresponding precipitation anomaly (mm) based on a 33 year mean from PRISM. Top and bottom panels show pixels with decreasing (red line) and increasing (green line) trends in productivity, respectively.

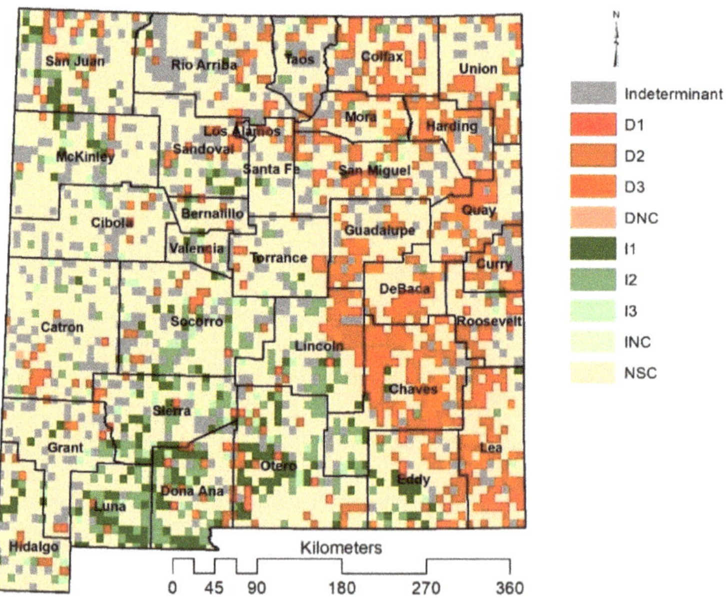

Figure 4. The spatial distribution, significance, and direction of change in productivity using TSS-RESTREND (1982–2015). Bands of red (D1, D2, D3, D4) and green (I1, I2, I3, and INC) pixels indicate those with decreasing and increasing trends in productivity, respectively. NSC (yellow) and indeterminant (gray) pixels were those with non-significant and unidentified change, respectively.

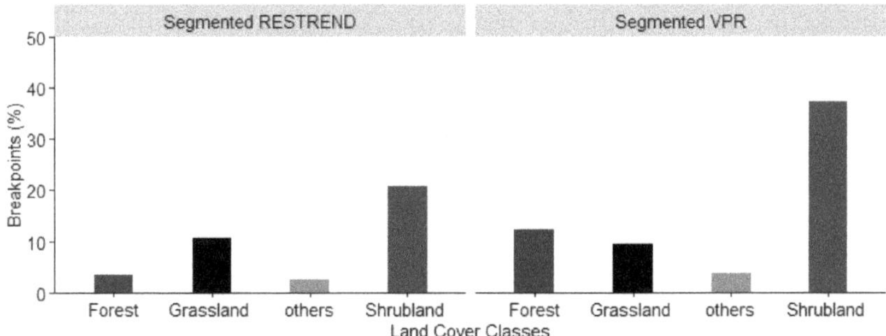

Figure 5. The percentages of breakpoints that were identified within shrubland, grassland, evergreen forest, and other land cover types between 1982 and 2015 in NM categorized by the Segmented RESTREND and Segmented VPR methods.

4.1.2. Observed Types of Structural Changes

In general, identified breakpoints by the Segmented VPR method indicate significant ecosystem structural change—either decrease (i.e., irreversible degradation) or increase in productivity. Those identified with the Segmented RESTREND method indicate changes— either decrease (initiation of degradation) or increase (reversal from degradation)—in productivity that are not significant enough to alter the ecosystem structure.

From all obtained 814 significant breakpoints, 62.7% and 37.3% were identified using Segmented VPR and Segmented RESTREND methods, respectively. From those identified by Segmented RESTREND method, 20.8%, 10.5%, 3.4%, and 2.6% were over shrubland, grassland, evergreen forest, and other land cover classes, respectively. From those identified by Segmented VPR method, 37.3%, 12.3%, 9.5%, and 3.6% were over shrubland, evergreen forest, grassland, and other land cover classes, respectively.

The total number of pixels identified by the different methods and their direction of change were presented in Table 3. Those showed decreased (significant or insignificant) productivity were 786 or 17.6% (46,976 km^2) out of 4454. About 4.9% (~14,016 km^2) were identified using the Segmented VPR method; 4.3% (~12,160 km^2) were identified using the Segmented RESTREND method; and the remaining 8.5% were identified using the RESTREND method (insignificant decrease or increase in productivity).

Table 3. A summary of the pixels (in % relative to 4454 pixels) and methods used to identify the direction of change in productivity in New Mexico during the 1982–2015 period.

Method of Change Detection	Decreasing	Increasing	NSC	Total
RESTREND	8.5	4.6	54.8	67.9
Segmented RESTREND	4.3	2.5	0.0	6.8
Segmented VPR	4.9	5.7	0.8	11.4
Total	17.6	12.8	55.6	86.1

NSC = Non-Significant Change.

From all pixels that showed increased productivity (12.8% or 570 pixels, 34,816 km^2), 5.7% (16,320 km^2) were identified using the Segmented VPR method (i.e., significant gradual increase), 4.6% (12,992 km^2) were identified using the RESTREND method (i.e., insignificant increase), and the remaining 2.5% (7168 km^2) were identified using the Segmented RESTREND method (i.e., reversal of degradation).

More than half of NM's area (55.6%~158,592 km^2) show insignificant change in productivity based on the total number of pixels that fell within the NSC category ($p \geq 0.1$). The pixels identified with the RESTREND method that accounted for 25.8% (~73,472 km^2), and the remaining 29% were considered indeterminate.

Regionally, non-reversible decrease in productivity (i.e., degradation) was mostly observed in the northeastern (1.8%~5184 km^2), northwestern (1.2%~3456 km^2), and southeastern (0.99%~2816 km^2) quadrants NM (Figure 6). Moreover, the initiation of degradation (i.e., decreasing trend based on the Segmented RESTREND method) was mostly detected in northeastern (1.8%~4992 km^2), and southeastern (2.1%~5888 km^2) quadrants, with a negligible initiation of degradation in the northwestern and southwestern NM (Figure 6).

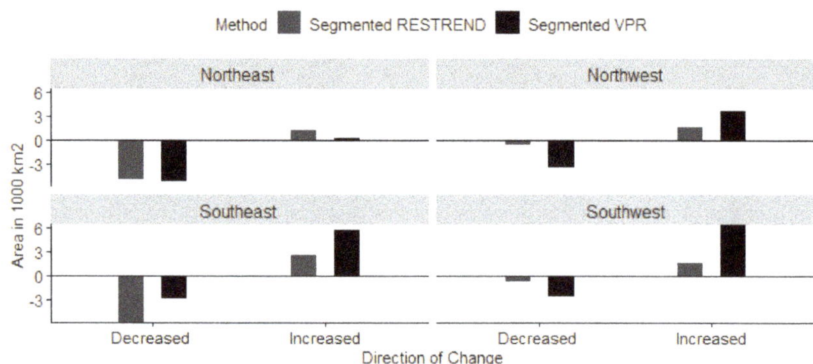

Figure 6. Area and direction of change in productivity in northeastern, southeastern, northwestern, and southwestern quadrants of New Mexico as identified using the Segmented VPR (degraded vegetation or gradual increase) and Segmented RESTREND (reversal and initiation of degradation) methods.

From all analyzed pixels (i.e., 4454 pixels), the ones that showed increasing trends in productivity based on the Segmented VPR method (i.e., 5.7% or ~16,320 km^2) were detected mostly in the southwestern (2.27%~6464 km^2), southeastern (2%~5824 km^2), and northwestern (1.3%~3712 km^2) quadrants (Figure 6). Furthermore, 2.51% of the pixels (7168 km^2) revealed reversal from degradation—increased productivity based on the Segmented RESTREND method (Figure 6). From which, 0.45% and 0.92% were in the northeastern and southeastern quadrants, respectively (Figure 6). The largest number of pixels that showed NSC was detected in northwestern (46,144 km^2), while southeastern quadrant exhibited the least (32,256 km^2~11.1%). Northeastern (40,896 km^2) and southwestern (39,296 km^2) NM revealed an equal number of NSC pixels during the study period.

4.2. Dominant Land Cover Class Changes

Significant trends (increasing or decreasing) in productivity (irreversible degradation, initiation in degradation, reversal in degradation, and initiation in reversal of degradation) were observed on NM's dominant land cover classes that include shrubland, grassland, and evergreen forest (Figure 7).

From all analyzed pixels (i.e., 4454), 2.2% of NM's grassland (6336 km^2), 4.5% of the shrubland (12,800 km^2), and 1.7% of the evergreen forest pixels (4800 km^2) showed significant decreasing trends in productivity–either with a complete or initiation of degradation. From all analyzed pixels, ~1.4% (3840 km^2) and 0.88% (2496 km^2) of NM's grasslands that showed significant decrease in productivity were attributed to initiation of degradation (i.e., the Segmented RESTREND method) and non-reversal degradation (i.e., the Segmented VPR method), respectively (Figure 7).

Similarly, from all analyzed pixels, 2.2% (6336 km^2) and 2.3% (6464 km2) of NM's shrublands that showed significant decrease in productivity were attributed to initiation of degradation and non-reversal degradation, respectively. The degradation of shrublands was dominant in the northwestern (2048 km^2) and southeastern (2112 km^2) quadrants, while the degradation in grasslands was notably observed in northwestern (512 km^2) and northeastern (1664 km^2) quadrants.

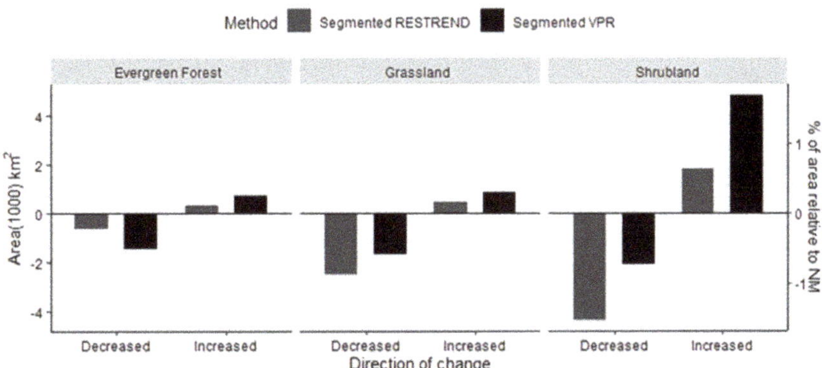

Figure 7. Area and direction of change in productivity in grasslands, shrublands, and evergreen forests in New Mexico as identified using the Segmented VPR (degraded vegetation or gradual increase) and Segmented RESTREND (reversal and initiation of degradation) methods.

On the other hand, 5.7% (out of the 4454 pixels) of NM's shrubland pixels (16,248 km^2), 1.3% of the grassland pixels (3648 km^2), and 0.92% of the evergreen forest pixels (2624 km^2) experienced either significant gradual increase in productivity or reversal of degradation (Figure 8). From all analyzed pixels, 1600 km^2 (0.56%) and 2048 km^2 (0.72%) of the grasslands that showed significant increase in productivity were attributed to reversal in degradation (using the Segmented RESTREND method) and significant increase in productivity (using the Segmented VPR method), respectively.

4.3. Assessment of Breakpoints

This section provides a summary of the results obtained based on the breakpoints assessment framework described in Section 3.2.

4.3.1. Identified Random Samples

The total number of breakpoints based on the Segmented RESTREND and Segmented VPR methods were about 304 and 510, respectively. From which, 384 samples were randomly selected with 165 and 219 were based on the Segmented RESTREND and Segmented VPR methods, respectively (Tables A1 and A2 in Appendix A). Since the Segmented VPR had more significant breakpoints with a noticeable change in productivity, only those pixels were subjected to the random selection. Only 155 samples out of the 219 (or 71%) showed significant difference in productivity before and after the break years (either decreasing or increasing) (Figure 9). The remaining 64 samples did not meet the criteria for significance and were not considered. From the 155 samples, 24% and 76% were obtained over grasslands and shrublands, respectively. The distribution of the grassland samples (i.e., 24%) represented the Southwestern (SW) Tablelands (46%), Arizona/New Mexico (AZ/NM) Plateau (35%), and Chihuahua Desert (16%) ecoregions. Similarly, 58%, 25%, and 11% of the shrubland samples (i.e., 76%) represented the Chihuahua Desert, AZ/NM Plateau, and SW Tablelands ecoregions, respectively.

4.3.2. Changes in Productivity at Pixel Level

A summary of the comparison of the significance of the differences in mean productivity before and after the breakpoints between the Segmented VPR and the RPMS data was shown in Table 4 along with the direction (increase or decrease) of change in productivity. The results were presented as the percent of pixels that fell within each category relative to the number of the random samples (i.e., 155).

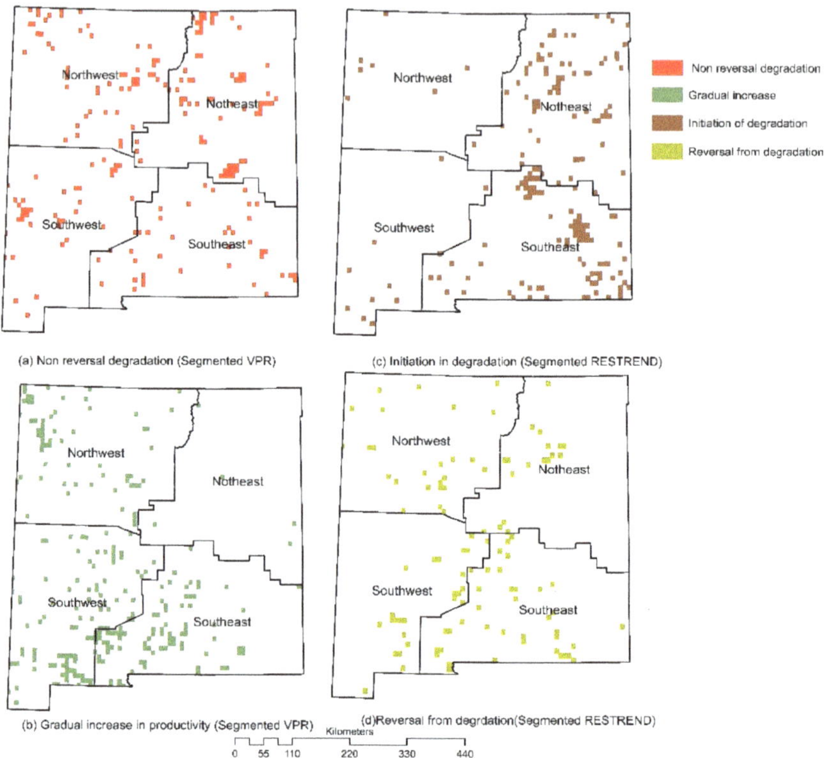

Figure 8. Direction of change in productivity in New Mexico's quadrants during the 1982–2015 period indicating (**a**) non reversal degradation, (**b**) gradual increase, (**c**) initiation of degradation, and (**d**) initiation of reversal from degradation along with the corresponding methods used.

Based on the Segmented VPR method, all the 155 randomly selected samples indicated significant difference in productivity before and after the break years. However, based on the RPMS only 55% of them showed significant differences in mean productivity before and after the break years until 2019, respectively (Table 4). From the 155 samples, 37% and 18% showed persistent decreasing and increasing in mean productivity after the break years until 2019 on RPMS data, respectively. About 36% and 9% showed increasing and decreasing trends before and after the break years on the RPMS data, respectively.

From the 37% of the sampled pixels that exhibited persistent decrease in mean productivity after the break years, 13% were in the Chihuahua Desert, and 18% in the SW Tablelands ecoregions. From the 18% of the sampled pixels that exhibited significant and consistent increase in mean productivity after break years, AZ/NM Mountains and AZ/NM Plateau ecoregions accounted for 3% and 15%, respectively. From the 36% of the sampled pixels with increased but insignificant difference in mean productivity before and after the break years, 35% were over the Chihuahua Desert ecoregion, and the remaining were equally obtained over AZ/NM Mountains and AZ/NM Plateau (Table 4). From the 9% of the sampled pixels with decreasing trend but insignificant difference in productivity before and after the break years, 5% were over AZ/NM Plateau.

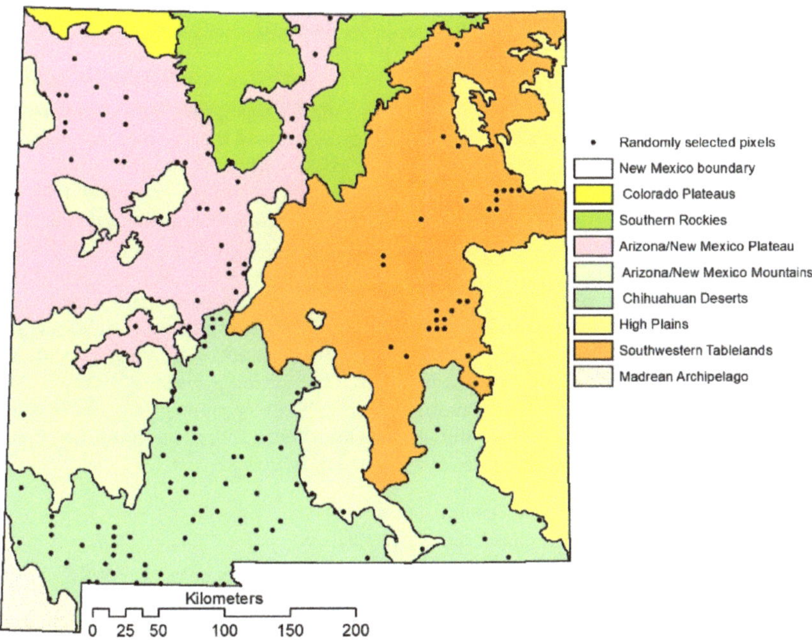

Figure 9. The distribution of the randomly selected samples from the identified breakpoints based on the Segmented VPR method along with the ecoregions in New Mexico.

Table 4. The percentages of pixels with decreasing and increasing trends in productivity as estimated from the RPMS data relative to randomly selected pixels (155) based on the Segmented VPR method categorized based on their significance of difference in mean productivity before and after the break years over major ecoregions in New Mexico.

Ecoregion	Insignificant Difference		Significant Difference		Total
	Decrease	Increase	Decrease	Increase	
Arizona/New Mexico Mountains	2	0.5	0	3	5.5
Arizona/New Mexico Plateau	5	0.5	6	15	26.5
Chihuahua Desert	1	35	13	0	49
Southwest Tablelands	1	0	18	0	19
Total	9	36	37	18	100

In the sampled grasslands and shrublands pixels (i.e., 24% and 76% of the 155 samples, respectively) with either decreased or increased productivity, the difference in mean productivity before and after the break years was insignificant in 5% and 40%, respectively on RPMS (Table A3 in Appendix B). The grassland (12% of the samples) and shrubland pixels (26% of the samples) with decreased productivity had significant lower mean productivity before the break years on RPMS data. Similarly, out of the pixels that showed increased trend on productivity, 7% from grasslands and 11% from shrublands had significantly higher mean production than that before the break years on RPMS (Table A3).

4.3.3. Changes in Productivity at Ecoregion Level

A summary of the comparison between the Segmented VPR and RPMS at the ecoregion level (Section 3.2) including grassland and shrubland cover classes based on the sampled pixels (i.e., 155) was provided in Table A5 (Appendix C)—allows to highlight whether the changes at the individual pixels are reflective of those at the regional level.

Based on the RPMS data, a continuous and significant decrease in shrublands' mean productivity after the break years in the Chihuahua (Welch's test $p \leq 0.0001$) and the AZ/NM Plateau ($p = 0.0397$) ecoregions was observed. Similarly, significant decline in mean productivity of grasslands of the SW Tablelands ($p \leq 0.0001$) and AZ/NM ($p = 0.019$) ecoregions were observed after break years. The shrublands within the AZ/NM Mountains and the SW Tableland ecoregions and the grassland in Chihuahua and the AZ/NM Plateau ecoregions exhibited insignificant differences in mean productivity between before and after the break years—stable ecosystem productivity during the study period.

In contrary, the sampled pixels over the shrublands in the Chihuahua Desert ($p = 0.0194$) and the AZ/NM ($p = 0.00765$). Mountain ecoregions showed significant increase in mean productivity after the break years (Table A5). There was a significant increase in mean productivity in the grasslands within AZ/NM Plateau ecoregions after the break years. However, sampled shrubland and grassland pixels in the AZ/NM Plateau and the Chihuahua Desert, respectively, showed insignificant difference in mean productivity before and after the break years, suggesting a negligible increase in mean productivity after the break years during the study period.

5. Discussion

5.1. Characteristics of Change

The significance of the breakpoints as identified by the TSS-RESTREND methods can be interpreted relative to observed ecosystem structural changes [8,13]. Out of 67.9% of the pixels that met the criteria of RESTREND [13], 8.5% and 4.6% showed decreased and increased productivity, respectively. These pixels exhibited gradual change as their VPR remined consistent over time with no major ecosystem structural changes [13,60]. The behavior of the pixels that were identified by the Segmented VPR method (11.6%) experiencing irreversible degradation (4.9%) or increased productivity (5.7%) can partially be attributed to abrupt land use changes (decreeing or increasing) induced by human activities and climate variability [13,61] such as overgrazing or easing of drought conditions [62]. More than half of NM that did not experience significant change in productivity (NSC = 55.6%—Table 2) was dominantly in the western part of the state. This can partially be explained by the fact that western NM is the driest region in the state. Thus, it experiences weaker interactions related to climate variability and human activities—thus minimal effects on productivity. The pixels that were identified as indeterminant (10.8%) were those that none of the methods was able to fit their observed behavior [8,55] and there was no clear explanation for such behavior.

There were relatively higher human activities in northeastern, northwestern and southeastern NM represented by crude oil and natural gas production and livestock grazing practices. The prevalence of irreversible degradation and initiation of degradation were apparent in these regions [63]. On the other hand, significant changes in productivity (i.e., increase) was observed mostly in southwestern and northwestern NM, where forest were minimally influenced by human activities. On landscapes where human activities were dominant, i.e., southeastern NM, significant increase in productivity can be associated with parallel restoration activities and land use management practices [55,64].

Overall, there was a consistent increase in the number of detected significant breakpoints during the 1980s, 1990s, and 2000s, with 9.5%, 19.2%, and 71.4%, respectively (Figure 10). Again, the combined effects of climate change and anthropogenic factors can explain this increase [65]. For example, NM's precipitation remained close to the long-term average (only recently showed increased variability) since 1920 while air temperature showed an increasing trend since 1970s [63,66,67] with a simultaneous increase in fossil fuel production activities since 1980s.

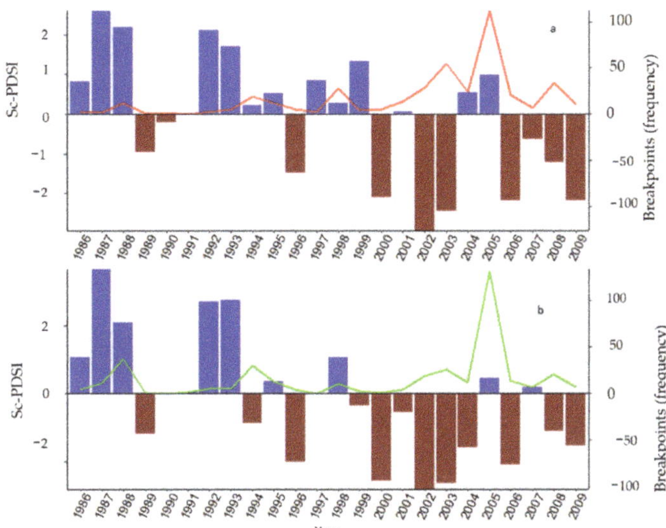

Figure 10. A summary of breakpoints with (**a**) decreasing and (**b**) increasing trends in productivity along with the weighted average of self-calibrated Palmer Severity Drought Index (sc-PSDI).

It is challenging to identify a single factor that can be considered as the direct cause of these breakpoints and ecosystem structural changes [8,68]. However, it was possible to compare these breakpoints with some observed ecosystem structural changes whether gradual (e.g., land use management and climate change) or abrupt (e.g., wildfire). Such a comparison can help in evaluating the accuracy of these breakpoints in terms of timing, distribution, and direction of change [8,53].

5.2. Land Cover Changes Relative to Drought and Wildfire

Some of the major drivers of change in NM's dryland ecosystems were identified as climate (e.g., drought) which is influenced by increased concentration of atmospheric greenhouse gases (GHG) (e.g., CO_2); wildfire; grazing practices (i.e., livestock density); and land use conversion [21,35,68] were used here to highlight their effects on the breakpoints.

5.2.1. Detected Changes Compared to Previous Studies and Current Restoration Activities

The significant decrease in grasslands and shrublands was mostly observed in northwestern, northeastern, and southeastern NM—consistent with [4] that indicated that degradation was mostly over grassland-savanna. Degradation of grasslands, in some cases, was directly linked to increased productivity of shrublands. In NM, increased productivity in shrublands was noticed southwestern deserts and plains. This increase was attributed to land cover conversion from black grama and other valuable grasses-dominated areas to bushes due to over grazing and drought [69]. The evident climate warming, expected increase interannual precipitation variability, and projected increase in aridity can further enhance the growth of shrubs (i.e., encroachments of woody plants) over grasses since the later are heavily depended on transient surface moisture [70,71]. It was also argued that increased CO_2 concentration in the atmosphere can improve plant CO_2 uptake and reduce water loss through plant's stomata thus increasing the photosynthesis process and this mechanism actually favor woody vegetation over non-woody ones [72,73].

The general notion that was suggested in some studies was that encroachments of shrublands to grasslands can be considered as one stage of degradation that can threaten the integrity of the rangelands [74]. However, it was argued that, shrublands are not necessarily degraded, nor do they necessarily represent "degradation" due to their ability to support valued ecosystem services [71,75], and they also have a long-term mean annual above ground NPP—equivalent to that of grasslands [76]. In NM, there have been a number of restoration and brush management efforts undertaken to control invasive species, and retain favored shrub species [64]. This suggested that the attribution of drivers to changes in productivity in environmental studies remains challenging owing to the diversity of driving forces and limited sources of ground truth data for validation [77].

The increased productivity in grasslands that was observed in western NM can partially be attributed to local scale successful restoration efforts by the Bureau of Land Management [64]. These efforts targeted the replacement of Creosote and mesquite by healthy grasslands, and reclamation of surfaces resulted from oil and gas extraction operations. According to Powell [78], gradual improvements in range conditions (increased productivity) in southwestern NM was not only related to the results of better moisture, but also cumulative efforts on rangeland management, such as proper stocking, vastly improved grazing distribution, and brush management.

5.2.2. Breakpoints and Drought

The sc-PSDI for NM and the detected significant breakpoints with decreasing and increasing trends were shown in Figure 10. A weighted average of the sc-PDSI was calculated over the breakpoints with decreasing and decreasing trends separately to evaluate their timing against drought events. Some previous findings suggested that extended periods of drought (or relatively dry conditions) can introduce lasting negative impacts (degradation) on rangelands and other ecosystems [8,79]. Based on Figure 10, it was noticed that from all the detected breakpoints during the study period (1982–2015), 67.9% were observed only in 2000s from which 38.2% exhibited decreased trend in productivity (Figure 10a). Coincidently, these breakpoints overlapped with frequent and extended periods of drought events in NM that were also observed regionally in the southwest USA during 2000s. The impact of this regional drought had resulted in a decrease in productivity in more than 30% of the coterminous USA, out of which ~15% (equivalent to more than 41 million ha) was rangelands [37].

The increased number of breakpoints after 2000 suggested a permanent degradation or damage to ecosystems which mechanistically can occur when these ecosystems have little to no time to recover from a previous consecutive drought event. A recovery period from a drought event as defined by [79] is the return of an ecosystem to pre-drought values of GPP and can vary from immediate to multiple years depending on vegetation, climate, disturbance, and drought. It was indicated that dryland ecosystems such as those in NM, had experienced increased recovery periods that were highly sensitive to precipitation and temperature [79]. With the expected future projections of rising temperature and drier conditions, the drought recovery of dry ecosystems would even be longer. When droughts have shorter return period (or more frequent) this makes ecosystems more susceptible to drought or ecosystem degradation continues and builds up until a threshold is reached and the degradation become irreversible (or permanent), a condition that was referred to in [79] as a tipping point. This study suggested that these thresholds may have been reached as represented by the detected breakpoints. This explains, to some extent, the appearance of breakpoints compared to drought events. The timing of these breakpoints suggested the accumulation of dry conditions that impact ecosystem functions until reaching tipping points. Further analysis is needed to understand the timing or emergence of the breakpoints and drought accumulation periods and the response of ecosystems relative to vegetation type and climate.

Moreover, these recent exceptionally drought conditions in NM (during 2000–2015) also varied in spatial extent, duration, and intensity [80]. These variations may have contributed to the occurrence of 38% breakpoints with decreasing trends (i.e., initiation and irreversible degradation) in southeastern (Eddy, Lea, Chaves, and Otero counties); northeastern (Colfax, De Bacca, and San Miguel counties); and northwestern NM (San Juan county). From all pixels that experienced significant decrease in productivity, 35% were observed in De bacca, San Miguel, San Juan, and Otero counties as drought might have a profound influence in reducing vegetation productivity (e.g., plant mortality) [81,82].

On the other hand, reversal from degradation and significant increase in productivity was dominantly observed in the northwestern, southeastern, and southwestern NM (Figure 8). In NM, wetter years were observed from 1986 to the end of 1990s, followed by drier years from 1999–2003 and 2005 (Figure 10). Significant breakpoints with increase in productivity were identified during 2000s dry years in Otero, Dona Anna, Luna, San Juan and Socorro counties. About 13.5% of these breakpoints were observed in the first four of counties. This can partially be attributed to the fact that precipitation remained near the long-term mean after the drier years over these counties that resulted in reversal from degradation or significant increase in productivity, respectively (Figure 10).

5.2.3. Breakpoints and Wildfire

Wildfire generally reduces plant cover, alters habitat structures, decrease rangeland conditions, and requires much longer recovery period [83]. The significant breakpoints with decreased productivity (i.e., irreversible degradation based on Segmented VPR) that coincided with fire events are shown in Figures 11 and 12. Major wildfire incidents occurred in 1989 and 1994 followed by a consistent increase in frequency each year since 2001. This increase was noticeably concurrent with drought events. The number and time of some breakpoints mostly followed these of the fire incidents. The alignment of fire incidents with breakpoints can indicate the ability of TSS-RESTREND to detect the timing of ecosystem changes as suggested by [8]. However, based on Figures 11 and 12, it appeared that only a small number of breakpoints coincided with fire incidents during the study period—suggested that fire may have a limited contribution to the development of breakpoints. While, the timing of these fire incidents can partly explain the occurrence of the breakpoints, it was not rationale to state that these fire incidents were the only direct cause of the breakpoints. Other factors need to be considered to provide a rational explanation of the remaining breakpoints such as vegetation types and their response to environmental and climate disturbance (drought). The accumulation of successive dry conditions can push an ecosystem to pass a threshold of permanent damage of vegetation— ideal fire-prone conditions. Further analysis is needed to identify the causes of breakpoints relative the fire and was out of the scope of this analysis.

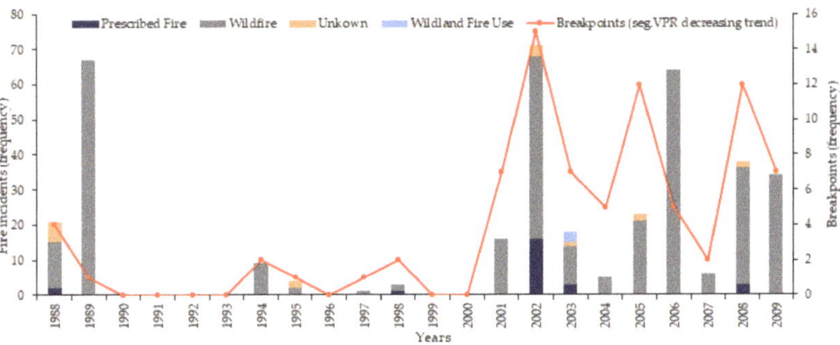

Figure 11. Timeseries of breakpoints concurrent different types of fire incidents in New Mexico during the 1984–2015 period.

5.3. Breakpoints and the RPMS

Based on the RPMS data, a decrease in the mean annual productivity was detected over the pixels with significant breakpoints (i.e., identified by Segmented VPR method) during the 1984–2019 period (Figure 13). The mean productivity of these pixels was lower than that of the long-term of the 36 years which was about 613.2 kg/ha. Moreover, the variability in the annual productivity of these pixels ranged from 588 kg/ha in 1995 to 641 kg/ha in 2006.

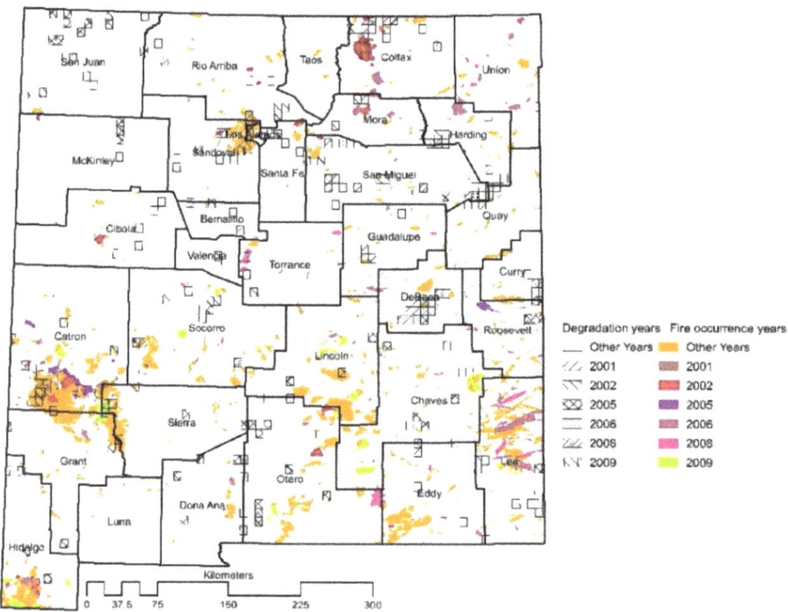

Figure 12. A summary of breakpoints and fire incidents indicated with the timing of some major incidents.

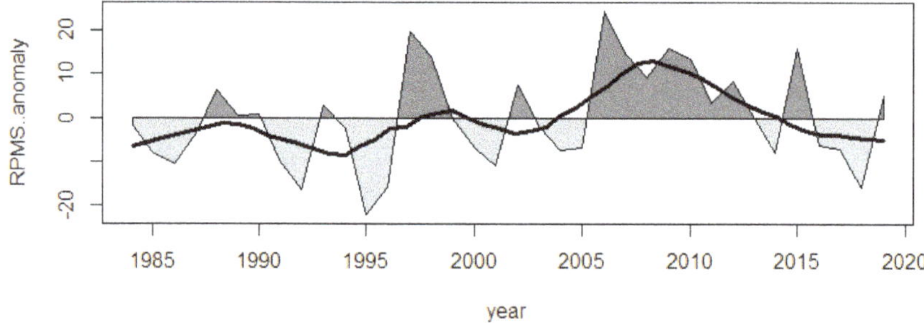

Figure 13. Productivity anomaly based on the Range land Production Monitoring Service (RPMS) [45] averaged over the pixels that were identified with the Segmented VPR method with decreasing trend during the 1984–2019 period.

Of the sampled pixels that were evaluated for changes in productivity before and after the break years, 38% with decreasing trend showed a significant difference in mean productivity after the break years with 12% in grasslands (Table A3) and 26% in shrublands (Table A4). Similar to this finding, [84,85] indicated that there was an apparent decline in NPP in the southwestern USA that was attributed to the response of these ecosystems to the combination of a warmer temperature and a decline in precipitation. The pixels that showed significant decrease in productivity in NM's rangeland (i.e., in Chihuahua Desert and AZ/NM plateau ecoregions) coincided with those from [45] that showed a similar behavior (Table A5). The difference in the productivity before and after the break years might be substantiated by the xeric nature of the southwestern USA and southeastern Great Plains in response to interannual variability of the precipitation [86]. On the other hand, 8% and 36% of sampled pixels with decreased and increased productivity (i.e., based on the Segmented VPR), respectively showed insignificant difference in mean productivity before and after the break years based on the RPMS data [45]. These findings were in agreement with those by [83] which suggested that the structure and composition of semi-arid and arid regions of the southwestern USA have undergone noticeable changes over the last two decades.

6. Limitation and Future Work

As the study aimed to detect degradation of vegetation, rangelands (i.e., shrubland and grasslands), the authors acknowledged some limitations that need to be considered when using these findings. The TSS-RESTREND was able to identify the breakpoints, their location, time, and type and direction of change in productivity at the state level. Due to the limited availability of ground biomass data, pixels that exhibited significant changes were validated using derived productivity from the RPMS [45] that showed only 55% agreement. Ground-based data can be more accurate means in assessing the identified breakpoints. Another important factor that can contribute to degradation is the temperature which can significantly affect NM's dryland ecosystems.

For future, the authors would address some of these limitations using ground observations (when available) for example from the long-term rangeland monitoring observatories at the Jornada Experimental Range. The obtained breakpoints seemed consistent with some of the disturbances (drought and fire). However, the authors would consider the combined use of temperature and precipitation to improve the detection of breakpoints such as the pixels that have been classified as indeterminant. Additional quantitative analysis would be conducted to determine how drought may affect the timing of breakpoints, rangeland condition, and productivity, which can be useful in developing rangeland management practices to adapt and mitigate drought impacts.

7. Conclusions

This study evaluated the degradation of New Mexico's rangelands during the 1984 2015 period with respect to climate using an NDVI timeseries—as a surrogate of NPP— and precipitation to represent climate variability. These datasets were evaluated using the TSS-RESTREN method to detect breakpoints in the NDVI timeseries, direction and significance of change in productivity at each pixel. The study developed a breakpoint assessment framework that allowed to quantitively evaluate the identified changes that used an independent productivity data (i.e., RPMS). A qualitative assessment of the changes against land management activities, drought, and wildfire was also conducted.

The study indicated that about 17.6% of New Mexico experienced a decrease in productivity while 12.8% of the state experienced an increase in productivity. More than half of the state (55.6%) had insignificant change productivity, 10.8% was classified as indeterminant, and the remaining 3.2% was considered as agriculture was not evaluated.

The degradation in productivity was observed in about 2.2%, 4.5%, and 1.7% of NM's grassland, shrubland, and evergreen forest land cover classes, respectively. Simultaneously, about 5.7%, 1.3%, and 0.92% of NM's shrublands, grasslands, and evergreen forests were characterized by an increase in productivity, respectively. Regionally, significant decrease in productivity was observed in the northeaster and southeastern quadrants of the state while significant increase was observed in northwestern, southwestern, and a small portion in southeastern quadrants. The timing and number of detected breakpoints coincided with NM's drought frequency and severity and some fire incidents.

The TSS-RESTREN showed 55% agreement with the RPMS data over areas with significant decrease in productivity based on randomly selected pixels. Regionally, there was an agreement between the TSS-RESTREND and RPMS on the occurrence of significant degradation in productivity over the grasslands and shrublands within the Arizona/New Mexico Tablelands and in the Chihuahua Desert ecoregions, respectively.

A long timeseries assessment of rangeland productivity in New Mexico is critical to support decisions related to ecosystems management and conservation. The findings of this study can be used to address some of the rangeland degradation challenges that directly impact rangeland conditions, forage supply in the region, and New Mexican's livelihood as these systems support livestock production. In areas where degradation was prevalent (i.e., northeastern, and southeastern NM), special intervention would be needed to conserve the biodiversity of rangeland and increase the resilience of these ecosystems. As this study considered only precipitation, it was noticed that the rising temperature can play a significant role in vegetation. Thus, future analysis should consider its effects on rangeland productivity. Future research should also pay more attention to the association of degradation and productivity with recurring droughts.

Author Contributions: Conceptualization, M.G.G. and H.M.E.G.; data curation, M.G.G. and T.A.A.; formal analysis, M.G.G.; funding acquisition, H.M.E.G.; investigation, M.G.G., H.M.E. and T.A.A.; methodology, M.G.G., H.M.E. and T.A.A.; project administration, H.M.E.G.; resources, H.M.E.G.; software, M.G.G. supervision, H.M.E.G.; validation, M.G.G., H.M.E. and T.A.A.; visualization, M.G.G., H.M.E.G. and T.A.A.; writing—original draft, M.G.G.; writing—review & editing, H.M.E.G. and T.A.A. All authors have read and agreed to the published version of the manuscript.

Funding: This research was partially funded by the National Science Foundation (NSF), awards #1739835 and#IIA-1301346, and New Mexico State University.

Data Availability Statement: The data used in this study are openly available in data [43–48].

Acknowledgments: The authors would like to thank two anonymous reviewers for their time and effort in providing valuable comments and suggestions that helped in improving the manuscript.

Conflicts of Interest: The authors declare no conflict of interest. The funders had no role in the design of the study; in the collection, analyses, or interpretation of data; in the writing of the manuscript, or in the decision to publish the results. Any opinions, findings, and conclusions or recommendations expressed in this material are those of the author(s) and do not necessarily reflect the views of the National Science Foundation.

Appendix A

Proportional allocation of the randomly sampled pixels as identified by Segmented VPR and Segmented RESTREND is shown in Table A1.

Table A1. Allocation of the randomly selected pixels as identified with Segmented VPR method over the different land cover classes based on the National Land Cover Dataset of 2011 (NLCD 2011) along the corresponding ecoregions.

Ecoregions	Decreasing			Increasing					NSC			Total
	52	71	82	31	42	52	71	82	42	52	71	
Arizona/New Mexico Mountains	1	1	0	0	1	10	1	0	0	0	1	16
Arizona/New Mexico Plateau	13	4	0	0	0	16	7	0	1	2	0	43
Chihuahua Desert	21	1	0	1	0	73	6	0	0	5	0	106
Colorado Plateaus	4	0	0	0	0	2	0	0	0	1	0	6
High Plains	4	1	1	0	0	1	1	0	0	0	2	11
Madrean Archipelago	0	0	0	0	0	3	1	1	0	1	0	5
Southwestern Tablelands	10	16	0	0	0	1	1	0	0	1	1	31
Total	53	24	1	1	1	106	18	1	1	9	4	219

31 = Bareland, 42 = Evergreen Forest, 52 = Shrub/Scrub, 71 = Grassland/Herbaceous, 82 = Cultivated Crops.

Table A2. Allocation of the randomly selected pixels as identified with Segmented RESTREND method over the different land cover classes based on the National Land Cover Dataset of 2011 (NLCD 2011) along with the corresponding ecoregions.

Ecoregions	Decreasing					Increasing				Total
	22	52	71	90	95	42	52	71	52	
Arizona/New Mexico Mountains	0	1	0	0	0	0	3	2	0	6
Arizona/New Mexico Plateau	0	1	2	1	0	0	6	4	0	13
Chihuahua Desert	1	37	1	0	1	0	25	4	0	68
Colorado Plateaus	0	1	0	0	0	0	1	0	0	1
High Plains	0	9	6	1	0	0	1	0	0	16
Madrean Archipelago	0	1	0	0	0	0	0	0	0	1

Table A2. Cont.

Ecoregions	Decreasing					Increasing				Total
	22	**52**	**71**	**90**	**95**	**42**	**52**	**71**	**52**	
Southern Rockies	0	0	1	0	0	0	0	0	0	1
Southwestern Tablelands	0	15	30	0	1	1	4	6	1	58
Total	1	65	39	1	1	1	40	16	1	165

22 = Low Intensity Developed, 52 = Shrub/Scrub, 71 = Grassland/Herbaceous, 90 = Woody Wetlands, 95 = Emergent Herbaceous Wetlands.

Appendix B

The percentages of the randomly selected pixels (i.e., 155 identified by Segmented-VPR method) over grasslands and shrublands that were evaluated for their significance of the difference in mean productivity between before and after the break years at the pixel level using the Rangeland Production Monitoring Service (RPMS) data.

Table A3. The percentages of the randomly selected grassland pixels with respect to their significance in the difference in mean productivity before and after the break years based on the RPMS data along with their identified direction of change over the major ecoregions in New Mexico.

Ecoregions	Insignificant Difference		Significant Difference		Total
	Decreasing	Increasing	Decreasing	Increasing	
Arizona/New Mexico Mountains	1	0	0	0	1
Arizona/New Mexico Plateau	0	0	1	7	8
Chihuahua Desert	0	3	1	0	4
Southwestern Tablelands	1	0	10	0	11
Total	2	3	12	7	24

Table A4. The percentages of the randomly selected shrublands pixels with respect to their significance in the difference in mean productivity before and after the break years based on the RPMS data along with their identified direction of change over the major ecoregions in New Mexico.

Ecoregions	Insignificant Difference		Significant Difference		Total
	Decreasing	Increasing	Decreasing	Increasing	
Arizona/New Mexico Mountains	1	1	0	3	5
Arizona/New Mexico Plateau	5	1	5	8	19
Chihuahua Desert	1	32	12	0	45
Southwestern Tablelands	0	0	8	0	8
Total	6	33	26	11	76

Appendix C

A summary of the statistical significance test result of the difference in mean in productivity before and after the break years using the randomly selected pixels (i.e., 155 identified by Segmented-VPR method) based on the Rangeland Production Monitoring Service (RPMS) data at the ecoregion level.

Table A5. Significance of the difference in mean productivity before and after the break years based on the on increasing and decreasing shrublands and grassland randomly selected pixels using RPMS data at the ecoregion level.

Trend	Ecoregion	Shrublands			Grasslands		
		Welch's t-Statistic	df	p	Welch's t-Statistic	df	p
Decreasing	Chihuahua Desert	−4.35	1379	≤0.0001 s	−1.23	39	0.227
	Arizona/New Mexico Mountains	0.33	119	0.742	−1.98	69.6	0.0516
	Southwestern Tablelands	−1.79	735	0.0745	−4.35	1053	≤0.0001 ***
	Arizona/New Mexico Plateau	2.06	1122	0.0397 *	2.38	126	0.019 *
Increasing	Chihuahua Desert	2.34	2178	0.0194 *	1.47	196	0.143
	Arizona/New Mexico Mountains	2.72	108	0.00765 **	-	-	-
	Southwestern Tablelands	-	-	-	-	-	-
	Arizona/New Mexico Plateau	1.42	857	0.157	2.4	736	0.065 *

*** p values less than or equal to 0.0001, ** p values less than 0.001, * p values less than 0.05.

References

1. Katyal, J.C.; Vlek, P.L.G. *Desertification: Concept, Causes and Amelioration*; ZEF Discussion Papers on Development Policy, N. 33; University of Bonn, Center for Development Research (ZEF): Bonn, Germany, 2000.
2. Hoover, D.L.; Bestelmeyer, B.; Grimm, N.B.; Huxman, T.E.; Reed, S.C.; Sala, O.; Seastedt, T.R.; Wilmer, H.; Ferrenberg, S. Traversing the Wasteland: A Framework for Assessing Ecological Threats to Drylands. *BioScience* **2020**, *70*, 35–47. [CrossRef]
3. World Resources Institute (Ed.) *Ecosystems and Human Well-Being: Desertification Synthesis: A Report of the Millennium Ecosystem Assessment*; World Resources Institute: Washington, DC, USA, 2005; ISBN 978-1-56973-590-9.
4. Noojipady, P.; Prince, S.D.; Rishmawi, K. Reductions in Productivity Due to Land Degradation in the Drylands of the Southwestern United States. *Ecosyst. Health Sustain.* **2015**, *1*, 1–15. [CrossRef]
5. Liu, C.; Melack, J.; Tian, Y.; Huang, H.; Jiang, J.; Fu, X.; Zhang, Z. Detecting Land Degradation in Eastern China Grasslands with Time Series Segmentation and Residual Trend Analysis (TSS-RESTREND) and GIMMS NDVI3g Data. *Remote Sens.* **2019**, *11*, 1014. [CrossRef]
6. Fensholt, R.; Langanke, T.; Rasmussen, K.; Reenberg, A.; Prince, S.D.; Tucker, C.; Scholes, R.J.; Le, Q.B.; Bondeau, A.; Eastman, R.; et al. Greenness in Semi-Arid Areas across the Globe 1981–2007—An Earth Observing Satellite Based Analysis of Trends and Drivers. *Remote Sens. Environ.* **2012**, *121*, 144–158. [CrossRef]
7. Higginbottom, T.; Symeonakis, E. Assessing Land Degradation and Desertification Using Vegetation Index Data: Current Frameworks and Future Directions. *Remote Sens.* **2014**, *6*, 9552–9575. [CrossRef]
8. Burrell, A.L.; Evans, J.P.; Liu, Y. Detecting Dryland Degradation Using Time Series Segmentation and Residual Trend Analysis (TSS-RESTREND). *Remote Sens. Environ.* **2017**, *197*, 43–57. [CrossRef]
9. Gibbs, H.K.; Salmon, J.M. Mapping the World's Degraded Lands. *Appl. Geogr.* **2015**, *57*, 12–21. [CrossRef]
10. UNEP. Global Drylands: A UN System-Wide Response. Report 1. 132. Available online: https://www.unep-wcmc.org/resources-and-data/global-drylands--a-un-system-wide-response (accessed on 17 April 2021).
11. Evans, J.; Geerken, R. Discrimination between Climate and Human-Induced Dryland Degradation. *J. Arid Environ.* **2004**, *57*, 535–554. [CrossRef]
12. Land Degradation by Main Type of Rural Land Use in Dryland Areas. Available online: http://www.fao.org/3/X5308E/x5308e04.htm#3.%20land%20degradation%20by%20main%20type%20of%20rural%20land%20use%20in%20dryland%20areas (accessed on 14 February 2020).
13. Wessels, K.J.; van den Bergh, F.; Scholes, R.J. Limits to Detectability of Land Degradation by Trend Analysis of Vegetation Index Data. *Remote Sens. Environ.* **2012**, *125*, 10–22. [CrossRef]
14. Nardone, A.; Ronchi, B.; Lacetera, N.; Ranieri, M.S.; Bernabucci, U. Effects of Climate Changes on Animal Production and Sustainability of Livestock Systems. *Livest. Sci.* **2010**, *130*, 57–69. [CrossRef]
15. IPBES. The Assessment Report on Land Degradation and Restoration. 748. Available online: https://www.ipbes.net/assessment-reports/ldr (accessed on 18 April 2021).
16. Mirzabaev, A.J.; Wu, J.; Evans, F.; García-Oliva, I.A.G.; Hussein, M.H.; Iqbal, J.; Kimutai, T.; Knowles, F.; Meza, D.; Nedjraoui, F.; et al. Weltz, 2019: Desertification. In *Climate Change and Land: An IPCC Special Report on Climate Change, Desertification, Land Degradation, Sustainable Land Management, Food Security, and Greenhouse Gas Fluxes in Terrestrial Ecosystems*; Shukla, P.R., Skea, J., Buendia, E.C., Masson-Delmotte, V., Pörtner, H.-O., Roberts, D.C., Zhai, P., Slade, R., Connors, S., van Diemen, R., et al., Eds.; in press; Available online: https://www.ipcc.ch/site/assets/uploads/sites/4/2019/11/06_Chapter-3.pdf (accessed on 18 April 2021).
17. Bai, Z.G.; Dent, D.L.; Olsson, L.; Schaepman, M.E. Proxy Global Assessment of Land Degradation. *Soil Use Manag.* **2008**, *24*, 223–234. [CrossRef]

18. Horion, S.; Prishchepov, A.V.; Verbesselt, J.; de Beurs, K.; Tagesson, T.; Fensholt, R. Revealing Turning Points in Ecosystem Functioning over the Northern Eurasian Agricultural Frontier. *Glob. Chang. Biol.* **2016**, *22*, 2801–2817. [CrossRef] [PubMed]
19. Houerou, H.N.L. Rain Use Efficiency: A Unifying Concept in Arid-Land Ecology. *J. Arid Environ.* **1984**, *7*, 213–247.
20. Fensholt, R.; Rasmussen, K.; Kaspersen, P.; Huber, S.; Horion, S.; Swinnen, E. Assessing Land Degradation/Recovery in the African Sahel from Long-Term Earth Observation Based Primary Productivity and Precipitation Relationships. *Remote Sens.* **2013**, *5*, 664–686. [CrossRef]
21. Andela, N.; Liu, Y.Y.; van Dijk, A.I.J.M.; de Jeu, R.A.M.; McVicar, T.R. Global Changes in Dryland Vegetation Dynamics (1988–2008) Assessed by Satellite Remote Sensing: Comparing a New Passive Microwave Vegetation Density Record with Reflective Greenness Data. *Biogeosciences* **2013**, *10*, 6657–6676. [CrossRef]
22. Burrell, A.L.; Evans, J.P.; Liu, Y. The Impact of Dataset Selection on Land Degradation Assessment. *ISPRS J. Photogramm. Remote Sens.* **2018**, *146*, 22–37. [CrossRef]
23. Burrell, A.L.; Evans, J.P.; Liu, Y. The Addition of Temperature to the TSS-RESTREND Methodology Significantly Improves the Detection of Dryland Degradation. *IEEE J. Sel. Top. Appl. Earth Obs. Remote Sens.* **2019**, *12*, 2342–2348. [CrossRef]
24. Heumann, B.W.; Seaquist, J.W.; Eklundh, L.; Jönsson, P. AVHRR Derived Phenological Change in the Sahel and Soudan, Africa, 1982–2005. *Remote Sens. Environ.* **2007**, *108*, 385–392. [CrossRef]
25. Herrmann, S.M.; Anyamba, A.; Tucker, C.J. Recent Trends in Vegetation Dynamics in the African Sahel and Their Relationship to Climate. *Glob. Environ. Chang.* **2005**, *15*, 394–404. [CrossRef]
26. Prince, S.D.; Wessels, K.J.; Tucker, C.J.; Nicholson, S.E. Desertification in the Sahel: A Reinterpretation of a Reinterpretation. *Glob. Chang. Biol.* **2007**, *13*, 1308–1313. [CrossRef]
27. Dardel, C.; Kergoat, L.; Hiernaux, P.; Grippa, M.; Mougin, E.; Ciais, P.; Nguyen, C.-C. Rain-Use-Efficiency: What It Tells Us about the Conflicting Sahel Greening and Sahelian Paradox. *Remote Sens.* **2014**, *6*, 3446–3474. [CrossRef]
28. Bernardino, P.N.; De Keersmaecker, W.; Fensholt, R.; Verbesselt, J.; Somers, B.; Horion, S. Global-scale Characterization of Turning Points in Arid and Semi-arid Ecosystem Functioning. *Glob. Ecol. Biogeogr.* **2020**, *29*, 1230–1245. [CrossRef]
29. Archer, E.R.M. Beyond the "Climate versus Grazing" Impasse: Using Remote Sensing to Investigate the Effects of Grazing System Choice on Vegetation Cover in the Eastern Karoo. *J. Arid Environ.* **2004**, *57*, 381–408. [CrossRef]
30. Verbesselt, J.; Zeileis, A.; Herold, M. Near Real-Time Disturbance Detection Using Satellite Image Time Series. *Remote Sens. Environ.* **2012**, *123*, 98–108. [CrossRef]
31. Verbesselt, J.; Hyndman, R.; Zeileis, A.; Culvenor, D. Phenological Change Detection While Accounting for Abrupt and Gradual Trends in Satellite Image Time Series. *Remote Sens. Environ.* **2010**, *114*, 2970–2980. [CrossRef]
32. Robin, P.W.; Nackoney, J. *Drylands, People, and Ecosystem Goods and Services: A Web-Based Geospatial Analysis*; PDF Version; World Resources Institute; Available online: http://pdf.wri.org/drylands (accessed on 18 April 2021).
33. Thomas, P.H.; Symeonakis, E. Identifying Ecosystem Function Shifts in Africa Using Breakpoint Analysis of Long-Term NDVI and RUE Data. *Remote. Sens.* **2020**, *12*, 1894. [CrossRef]
34. Schuman, G.E.; Janzen, H.H.; Herrick, J.E. Soil Carbon Dynamics and Potential Carbon Sequestration by Rangelands. *Environ. Pollut.* **2002**, *116*, 391–396. [CrossRef]
35. Havstad, K.; Peters, D.; Allen-Diaz, B.; Bartolome, J.; Bestelmeyer, B.; Briske, D.; Brown, J.; Brunson, M.; Herrick, J.; Huntsinger, L.; et al. The Western United States Rangelands: A Major Resource. In *ASA, CSSA, and SSSA Books*; Wedin, W.F., Fales, S.L., Eds.; American Society of Agronomy, Crop Science Society of America, Soil Science Society of America: Madison, WI, USA, 2015; pp. 75–93. ISBN 978-0-89118-194-1.
36. Holechek, J.L.; Geli, H.M.E.; Cibils, A.F.; Sawalhah, M.N. Climate Change, Rangelands, and Sustainability of Ranching in the Western United States. *Sustainability* **2020**, *12*, 4942. [CrossRef]
37. Reeves, M.C.; Mitchell, J.E. Extent of Coterminous US Rangelands: Quantifying Implications of Differing Agency Perspectives. *Rangel. Ecol. Manag.* **2011**, *64*, 585–597. [CrossRef]
38. Skaggs, R.K. Ecosystem Services and Western U.S. Rangelands. *Choices* **2008**, *23*, 37–41.
39. Mitchell, J.E. *Rangeland Resource Trends in the United States: A Technical Document Supporting the 2000 USDA Forest Service RPA Assessment*; U.S. Department of Agriculture, Forest Service, Rocky Mountain Research Station: Ft. Collins, CO, USA, 2000; p. RMRS-GTR-68.
40. New Mexico Topo Maps and Outdoor Places to Visit. Available online: https://www.anyplaceamerica.com/directory/nm/ (accessed on 24 November 2019).
41. Level III Ecoregions of New Mexico. Available online: https://hort.purdue.edu/newcrop/cropmap/newmexico/maps/NMeco3.html (accessed on 2 July 2020).
42. Castetter, E.F. The Vegetation of New Mexico. *New Mex. Q.* **1956**, *26*, 16.
43. Data | Multi-Resolution Land Characteristics (MRLC) Consortium. Available online: https://www.mrlc.gov/data (accessed on 21 September 2019).
44. Ecocast Data Drop Directory. Available online: https://ecocast.arc.nasa.gov/data/pub/gimms/3g.v1/ (accessed on 5 June 2020).
45. USDA. Forest Service FSGeodata Clearinghouse—Rangelands. Available online: https://data.fs.usda.gov/geodata/rastergateway/rangelands/index.php (accessed on 26 August 2020).
46. PRISM Climate Group, Oregon State U. Available online: http://www.prism.oregonstate.edu/ (accessed on 29 January 2020).
47. WWDT. Available online: https://wrcc.dri.edu/wwdt/time/ (accessed on 19 October 2019).

48. NM RGIS | New Mexico Resource Geographic Information System Program. Available online: http://rgis.unm.edu/ (accessed on 21 September 2019).
49. Quality Control Introducing the R "Gimms" Package. Available online: https://envin-marburg.gitbooks.io/introducing-the-r-gimms-package/content/chapter04/quality_control.html (accessed on 2 July 2020).
50. De Jong, R.; Verbesselt, J.; Zeileis, A.; Schaepman, M. Shifts in Global Vegetation Activity Trends. *Remote Sens.* **2013**, *5*, 1117–1133. [CrossRef]
51. Holben, B.N. Characteristics of Maximum-Value Composite Images from Temporal AVHRR Data. *Int. J. Remote Sens.* **1986**, *7*, 1417–1434. [CrossRef]
52. Verbesselt, J.; Hyndman, R.; Newnham, G.; Culvenor, D. Detecting Trend and Seasonal Changes in Satellite Image Time Series. *Remote Sens. Environ.* **2010**, *114*, 106–115. [CrossRef]
53. De Jong, R.; Verbesselt, J.; Schaepman, M.E.; de Bruin, S. Detection of breakpoints in global NDVI time series. In Proceedings of the 34th International Symposium on Remote Sensing of Environment (ISRSE), Sydney, Australia, 10–15 April 2011; Available online: https://www.zora.uzh.ch/id/eprint/77356/ (accessed on 18 April 2021).
54. Fensholt, R.; Horion, S.; Tagesson, T.; Ehammer, A.; Grogan, K.; Tian, F.; Huber, S.; Verbesselt, J.; Prince, S.D.; Tucker, C.J.; et al. Assessing Drivers of Vegetation Changes in Drylands from Time Series of Earth Observation Data. In *Remote Sensing Time Series*; Kuenzer, C., Dech, S., Wagner, W., Eds.; Springer International Publishing: Cham, Switzerland, 2015; Volume 22, pp. 183–202. ISBN 978-3-319-15966-9.
55. Li, X.B.; Li, R.H.; Li, G.Q.; Wang, H.; Li, Z.F.; Li, X.; Hou, X.Y. Human-Induced Vegetation Degradation and Response of Soil Nitrogen Storage in Typical Steppes in Inner Mongolia, China. *J. Arid Environ.* **2016**, *124*, 80–90. [CrossRef]
56. PEOD5/PD005: Sampling the Evidence of Extension Program Impact. Available online: https://edis.ifas.ufl.edu/pd005 (accessed on 16 April 2021).
57. Cochran, W.G. *Sampling Techniques*, 3rd ed.; Wiley Series in Probability and Mathematical Statistics; Wiley: New York, NY, USA, 1977; ISBN 978-0-471-16240-7.
58. R: The R Project for Statistical Computing. Available online: https://www.r-project.org/ (accessed on 5 February 2020).
59. Welch T-Test: Excellent Reference You Will Love—Datanovia. Available online: https://www.datanovia.com/en/lessons/types-of-t-test/unpaired-t-test/welch-t-test/ (accessed on 31 August 2020).
60. Jamali, S.; Seaquist, J.; Eklundh, L.; Ardö, J. Automated Mapping of Vegetation Trends with Polynomials Using NDVI Imagery over the Sahel. *Remote Sens. Environ.* **2014**, *141*, 79–89. [CrossRef]
61. Van Auken, O.W. Shrub Invasions of North American Semiarid Grasslands. *Annu. Rev. Ecol. Syst.* **2000**, *31*, 197–215. [CrossRef]
62. Browning, D.M.; Maynard, J.J.; Karl, J.W.; Peters, D.C. Breaks in MODIS Time Series Portend Vegetation Change: Verification Using Long-Term Data in an Arid Grassland Ecosystem. *Ecol. Appl.* **2017**, *27*, 1677–1693. [CrossRef]
63. Gedefaw, M.G.; Geli, H.M.E.; Yadav, K.; Zaied, A.J.; Finegold, Y.; Boykin, K.G. A Cloud-Based Evaluation of the National Land Cover Database to Support New Mexico's Food–Energy–Water Systems. *Remote Sens.* **2020**, *12*, 1830. [CrossRef]
64. Restore New Mexico—ScienceBase-Catalog. Available online: https://www.sciencebase.gov/catalog/item/573cda13e4b0dae0d5e4b15a (accessed on 21 January 2021).
65. Loftin, S.R.; Agllilar, R.; Chung, M.C.; Alice, L.; Robbie, W.A. Desert grassland and shrubland ecosystems. In *United States Department of Agriculture Forest Service General Technical Report Rm*; 1995; pp. 80–94. Available online: https://www.srs.fs.usda.gov/pubs/38859 (accessed on 16 April 2021).
66. Sawalhah, M.N.; Holechek, J.L.; Cibils, A.F.; Geli, H.M.E.; Zaied, A. Rangeland Livestock Production in Relation to Climate and Vegetation Trends in New Mexico. *Rangel. Ecol. Manag.* **2019**, *72*, 832–845. [CrossRef]
67. Johnson, L.E.; Geli, H.M.E.; Hayes, M.J.; Smith, K.H. Building an Improved Drought Climatology Using Updated Drought Tools: A New Mexico Food-Energy-Water (FEW) Systems Focus. *Front. Clim.* **2020**, *2*. [CrossRef]
68. Bond, W.J.; Midgley, G.F. Carbon Dioxide and the Uneasy Interactions of Trees and Savannah Grasses. *Philos. Trans. R. Soc. B Biol. Sci.* **2012**, *367*, 601–612. [CrossRef] [PubMed]
69. Allison, R.C.D.; Ashcroft, N. New Mexico Range Plants. 48. Available online: https://aces.nmsu.edu/pubs/_circulars/CR374/ (accessed on 18 April 2021).
70. Gremer, J.R.; Bradford, J.B.; Munson, S.M.; Duniway, M.C. Desert Grassland Responses to Climate and Soil Moisture Suggest Divergent Vulnerabilities across the Southwestern United States. *Glob. Chang. Biol.* **2015**, *21*, 4049–4062. [CrossRef]
71. Bestelmeyer, B.T.; Peters, D.P.C.; Archer, S.R.; Browning, D.M.; Okin, G.S.; Schooley, R.L.; Webb, N.P. The Grassland–Shrubland Regime Shift in the Southwestern United States: Misconceptions and Their Implications for Management. *BioScience* **2018**, *68*, 678–690. [CrossRef]
72. Ukkola, A.M.; Prentice, I.C.; Keenan, T.F.; van Dijk, A.I.J.M.; Viney, N.R.; Myneni, R.B.; Bi, J. Reduced Streamflow in Water-Stressed Climates Consistent with CO_2 Effects on Vegetation. *Nat. Clim. Chang.* **2016**, *6*, 75–78. [CrossRef]
73. Donohue, I.; Garcia Molinos, J. Impacts of Increased Sediment Loads on the Ecology of Lakes. *Biol. Rev.* **2009**, *84*, 517–531. [CrossRef] [PubMed]
74. Mangold, J.; Monaco, T.; Sheley, R.; Sosebee, R.; Svejcar, T. Invasive Rangeland Plants. In *Range and Animal Sciences and Resources Management, Vol. II*; ©Encyclopedia of Life Support Systems (EOLSS), 2010; Available online: https://www.eolss.net/sample-chapters/C10/E5-35-21.pdf (accessed on 18 April 2021).

75. Eldridge, D.J.; Bowker, M.A.; Maestre, F.T.; Roger, E.; Reynolds, J.F.; Whitford, W.G. Impacts of Shrub Encroachment on Ecosystem Structure and Functioning: Towards a Global Synthesis: Synthesizing Shrub Encroachment Effects. *Ecol. Lett.* **2011**, *14*, 709–722. [CrossRef]

76. Peters, D.P.C. Recruitment Potential of Two Perennial Grasses with Different Growth Forms at a Semiarid-Arid Transition Zone. *Am. J. Bot.* **2002**, *89*, 1616–1623. [CrossRef]

77. Cherlet, M.; Hutchinson, C.; Reynolds, J.; Hill, J.; Sommer, S.; von Maltitz, G. *World Atlas of Desertification Rethinking Land Degradation and Sustainable Land Management*; European Union: Luxembourg, 2018; ISBN 978-92-79-75350-3.

78. Powell, J.C. Rangelands of southwest New Mexico—An upside view. *Rangelands* **1992**, *14*, 265–268.

79. Schwalm, C.R.; Anderegg, W.R.; Michalak, A.M.; Fisher, J.B.; Biondi, F.; Koch, G.; Litvak, M.; Ogle, K.; Shaw, J.D.; Wolf, A. Global Patterns of Drought Recovery. *Nature* **2017**, *548*, 202–205. [CrossRef]

80. Preparedness Bureau Mitigation I NM Department of Homeland Security & Emergency. Available online: https://www.nmdhsem.org (accessed on 16 April 2021).

81. Lotsch, A. Response of Terrestrial Ecosystems to Recent Northern Hemispheric Drought. *Geophys. Res. Lett.* **2005**, *32*, L06705. [CrossRef]

82. Ponce-Campos, G.E.; Moran, M.S.; Huete, A.; Zhang, Y.; Bresloff, C.; Huxman, T.E.; Eamus, D.; Bosch, D.D.; Buda, A.R.; Gunter, S.A.; et al. Ecosystem Resilience despite Large-Scale Altered Hydroclimatic Conditions. *Nature* **2013**, *494*, 349–352. [CrossRef] [PubMed]

83. Snyman, H.A. Estimating the Short-Term Impact of Fire on Rangeland Productivity in a Semi-Arid Climate of South Africa. *J. Arid Environ.* **2004**, *59*, 685–697. [CrossRef]

84. Reeves, M.C.; Baggett, L.S. A Remote Sensing Protocol for Identifying Rangelands with Degraded Productive Capacity. *Ecol. Indic.* **2014**, *43*, 172–182. [CrossRef]

85. Polley, H.W.; Briske, D.D.; Morgan, J.A.; Wolter, K.; Bailey, D.W.; Brown, J.R. Climate Change and North American Rangelands: Trends, Projections, and Implications. *Rangel. Ecol. Manag.* **2013**, *66*, 493–511. [CrossRef]

86. Zhao, M.; Running, S.W. Drought-Induced Reduction in Global Terrestrial Net Primary Production from 2000 Through 2009. *Science* **2010**, *329*, 940–943. [CrossRef] [PubMed]

 *remote sensing*

Article

Persistent Vegetation Greening and Browning Trends Related to Natural and Human Activities in the Mount Elgon Ecosystem

Dan Wanyama [1,2,*], **Nathan J. Moore** [1] **and Kyla M. Dahlin** [1,3]

1   Department of Geography, Environment, and Spatial Sciences, Michigan State University, 673 Auditorium Road, Room 116, East Lansing, MI 48824, USA; moorena@msu.edu (N.J.M.); kdahlin@msu.edu (K.M.D.)
2   Remote Sensing and GIS Research and Outreach Services, Michigan State University, East Lansing, MI 48823, USA
3   Program in Ecology, Evolutionary Biology, and Behavior, Michigan State University, 103 Giltner Hall, 293 Farm Lane, Room 103, East Lansing, MI 48824, USA
*   Correspondence: wanyamad@msu.edu

Received: 5 June 2020; Accepted: 30 June 2020; Published: 1 July 2020

Abstract: Many developing nations are facing severe food insecurity partly because of their dependence on rainfed agriculture. Climate variability threatens agriculture-based community livelihoods. With booming population growth, agricultural land expands, and natural resource extraction increases, leading to changes in land use and land cover characterized by persistent vegetation greening and browning. This can modify local climate variability due to changing land–atmosphere interactions. Yet, for landscapes with significant interannual variability, such as the Mount Elgon ecosystem in Kenya and Uganda, characterizing these changes is a difficult task and more robust methods have been recommended. The current study combined trend (Mann–Kendall and Sen's slope) and breakpoint (bfast) analysis methods to comprehensively examine recent vegetation greening and browning in Mount Elgon at multiple time scales. The study used both Moderate Resolution Imaging Spectroradiometer (MODIS) Normalized Difference Vegetation Index (NDVI) and Climate Hazards group Infrared Precipitation with Stations (CHIRPS) data and attempted to disentangle nature-versus human-driven vegetation greening and browning. Inferences from a 2019 field study were valuable in explaining some of the observed patterns. The results indicate that Mount Elgon vegetation is highly variable with both greening and browning observable at all time scales. Mann–Kendall and Sen's slope revealed major changes (including deforestation and reforestation), while bfast detected most of the subtle vegetation changes (such as vegetation degradation), especially in the savanna and grasslands in the northeastern parts of Mount Elgon. Precipitation in the area had significantly changed (increased) in the post-2000 era than before, particularly in 2006–2010, thus influencing greening and browning during this period. The greenness–precipitation relationship was weak in other periods. The integration of Mann–Kendall and bfast proved useful in comprehensively characterizing vegetation greenness. Such a comprehensive description of Mount Elgon vegetation dynamics is an important first step to instigate policy changes for simultaneously conserving the environment and improving livelihoods that are dependent on it.

Keywords: bfast; Mann–Kendall; Sen's slope; East Africa; NDVI; breakpoint analysis; vegetation trends; greening; browning; Kenya; Uganda; trend analysis; land use; land cover

---

1. Introduction

Vegetation plays very important roles in ecosystem processes, including the mitigation of climate change effects [1] and the regulation of land surface temperatures, carbon and energy cycles [2–4].

At the same time, significant changes in terrestrial vegetation have been reported in the recent past [5,6]. Previous studies have concluded that these changes are driven by (1) slowly-changing natural processes, such as regional climate change (e.g., changes in temperature, precipitation, etc.), nitrogen disposition and increasing atmospheric CO_2 concentrations [7], and (2) more rapid anthropogenic activities, including land-use and land-cover (LULC) change (e.g., deforestation [8,9], overcultivation and overgrazing [8], afforestation, expanding green areas in cities [1] among others). These processes do not operate in isolation [7] but rather interact at multiple scales, with global-scale drivers interacting with processes at the regional and local scales [10], thus making vegetation change dynamics a complex phenomenon to examine. Understanding vegetation dynamics has attracted substantial attention in the past years, specifically in the wake of climate change and variability [11]. As such, knowledge of vegetation response to climate change is critically important in the effort to maintain the supported processes as well as the human livelihoods derived from them [12]. Such an analysis has been difficult in the past, due in part to limited access to consistent data both in space and time. However, with recent developments in Earth observation (EO) technologies, spatio-temporally contiguous remote sensing (RS) data have been collected, making it possible to investigate vegetation dynamics accurately and comprehensively in terms of other environmental processes, at multiple scales.

The Normalized Difference Vegetation Index (NDVI), which is the normalized sum of the difference in reflectance between near infrared and red bands, has extensively been used [5,10,13] as an indicator of photosynthetic activity and vegetation amount [14]. Previous studies used this vegetation index (VI) to characterize vegetation processes such as productivity decline [6], phenology [15,16] and greening and browning [1,5,10,12,17,18]. Variability in NDVI generally follows patterns of climatic conditions, mainly precipitation [11,19–22]. The NDVI–precipitation relationship is largely a strong positive one, particularly in locations where total annual precipitation is less than 1000 mm [19,22]. This relationship is further compounded in space and time by site-specific factors such as existing LULC, vegetation structure and composition [22], topography [10] and soil type [23]. As a result, vegetation greenness response to climate has shown significant spatial and temporal variability. For instance, contrasting natures and strengths of the NDVI–precipitation relationship have been reported: a strong linear relationship in the Sahel [19] and a log-linear one in East Africa [22]. In other studies, seasonal precipitation with different lag times was found to correlate strongly with NDVI [13,22]. This complexity implies that (i) results from a specific regional analysis are not transferable to another region; (ii) any slight spatial and/or temporal misspecification may lead to misleading results about vegetation greenness–precipitation relationships and patterns; and (iii) understanding vegetation change requires multiple images (time series, TS) as opposed to the extensively used traditional classification and change detection methods.

East Africa depends heavily on rainfed agriculture, which puts food security and rural livelihoods at risk [24]. The importance of land in this region cannot be overemphasized [24], yet land holdings are small and steadily declining [24,25]. At the same time, populations have been increasing thus necessitating expansion of food production while also navigating the effects of climate change in the area. Over recent decades, therefore, East Africa has witnessed significant landscape transformation due to both human and natural drivers [6,26]. Generally, natural vegetation has been converted to farmlands, grazing lands and human settlements [25]. This LULC transformation has been reported in the Mount Elgon ecosystem (MEE), a major water catchment tower supplying water to three major lakes in East Africa (Lake Turkana, Kenya, Lake Kyoga, Uganda and Lake Victoria, Kenya, Uganda and Tanzania [27]). The MEE is dominated by croplands in most locations, mixed vegetation (primarily savanna, grasslands, and shrubs) in the northern portion and the Afromontane forest (Figure 1). The high population growth and densities in the area have translated into need for more land [28]. Coupled with political interference and corruption among park and reserve staff, the need for more land has resulted in forest encroachment and deforestation as ecologically fragile land is cleared for agriculture and settlement [28–32]. The changes in LULC have, in part, altered the functioning of the ecosystem [29] and, as a result, the mountain area has experienced more frequent landslides, prolonged droughts, and flooding [30,31]. Evidence of a changing climate has been reported [33] and this may result in increased frequency of these events. As such, the livelihoods of at least 2 million people [27,31] are threatened. Despite this, the complex MEE landscape is currently understudied. A key study of LULC change was conducted by Petersson, Vedeld, and Sassen [27] and employed institutional theory in analyzing processes that led to deforestation within protected areas (PAs) in the transboundary MEE. It was found that, especially on the Kenyan side, it was challenging to correctly quantify LULC change due to the overlap between bamboo, plantation and Shamba system farms. Other studies have been conducted on relatively smaller spatial scales, mostly on the Ugandan side. Such studies have focused on the effect of LULC change on landslide occurrence [28], soil organic carbon, food security and climate change vulnerability [34], carbon stocks and climate variability [35] among others. However, some of the studies have reported contradictory results especially about the magnitude of LULC change within the agricultural land-use class. This may be due to the complex LULC orientation, which leads to persistent greening and browning in the MEE, thus making it difficult to correctly characterize vegetation dynamics especially using traditional classification and change detection methods. More robust methods are therefore needed, and TS analysis has been applied to comprehensively examine spatio-temporal landscape changes, particularly for constantly variable landscapes like the MEE.

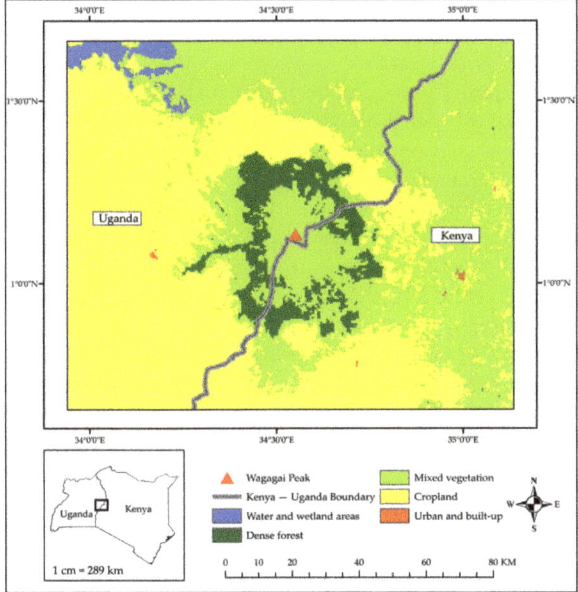

Figure 1. Mount Elgon ecosystem (MEE) land-use and land-cover (LULC) in 2018. This map was created by reclassifying Moderate Resolution Imaging Spectroradiometer (MODIS) land-cover data (MCD12Q1) created by the University of Maryland [36]. Mixed vegetation includes savanna, grasslands, and shrubs. Cropland includes cropland and the cropland-vegetation mosaic classes. The black-gray line represents the boundary between Kenya and Uganda and the red triangle is Wagagai Peak, the highest point on Mount Elgon (4321 m above sea level).

TS analysis of RS data has been used to characterize environmental phenomena by describing both trends and discrete change events [37]. In recent years, application of TS methods has increased and this has been driven by improved access to RS imagery (for instance, due to opening up of the Landsat archive in 2008 [38]), improvements in the integration of RS and GIS (Geographic Information Systems) and general advancements in computing power [39]. As such, analysis of LULC change, including vegetation greening and browning, has significantly evolved from the traditional bitemporal image analysis to using multiple and continuous observations. Common methods for TS analysis include Fourier analysis [40,41], principal components analysis, [42,43] and the Mann–Kendall statistic [44]. The Mann–Kendall statistic has been used to identify the presence and nature of monotonic trends in vegetation time series [5,10,45]. It is a non-parametric statistic; thus, data does not have to conform to any specific distribution [46]. Besides, Mann–Kendall compares relative magnitudes of sample data instead of raw data values and therefore missing values are allowed in the analysis. The Mann–Kendall analysis is often followed by the Sen's slope estimator [47], which quantifies the strength of the monotonic trends [12,45]. The Sen's slope estimator calculates the median of the set of slopes generated from Mann–Kendall [47]. Mann–Kendall and Sen's slope have been found to be more robust for TS with outlying values as compared to parametric statistics like ordinary least squares [12]. These two statistics identify and quantify any overall trends in a time series and are therefore well suited for examining vegetation greening and browning. Previous studies have successfully used these methods in assessing the consistency of greening and browning patterns across spatio-temporal scales in northern India [12] and assessing variability in greening and browning patterns caused by use of different RS imagery in the boreal forest of central Canada [48]. However, the assumption that the vegetation trend preserves its change rate throughout the period of study means that some greening and browning changes are masked [5]. For instance, later greening in a consistently browning vegetation may not be detected using Mann–Kendall and Sen's slope. To counter this issue, more TS decomposition methods have been proposed, including *Breaks for Additive Season and Trend*, bfast [16]. Using bfast, even subtle changes in vegetation can be monitored. This algorithm has successfully been used in delineating anthropogenic, fire and elephant damage within forest ecosystems in Kenya [49]. Complementing Mann–Kendall and Sen's slope with bfast can therefore be valuable in examining vegetation greening and browning trends, especially in a dynamic environment like the MEE. In such an analysis, the latter can be used to characterize changes detected by the former. Vegetation trends can then be characterized in more detail and this can be the basis for understanding effects of climate on terrestrial ecosystems [5] and the development of ultimate strategies for the sustainable management of ecosystems [10].

The goal of this study was to characterize comprehensively, over multiple time scales, recent patterns and trends in MEE vegetation greening and browning. Here, the main objectives were; (1) to assess and quantify the nature and magnitude of change in MEE greenness for the period 2001–2018; and (2) to characterize trends and variability in MEE precipitation as a way to disentangle nature- versus human-driven vegetation greening and browning. It is hypothesized that (1) changes in climate have forced local communities in the MEE to expand croplands at the expense of the natural vegetation thus leading to deforestation and degradation; and (2) the high variability exhibited in the MEE landscape requires integration of both general (such as Mann–Kendall and Sen's slope) and sequential (such as bfast) TS analysis methods to be fully characterized. To achieve these goals, the study analyzed spatio-temporal trends and patterns in Moderate Resolution Imaging Spectroradiometer (MODIS) NDVI (2001–2018) and Climate Hazards group Infrared Precipitation with Stations (CHIRPS) precipitation (1986–2018) at multiple temporal scales (dekadal (10-day), 16-day, seasonal), using an integration of Mann–Kendall, Sen's slope and bfast algorithms. The study also incorporates inferences from a field study conducted in the MEE in 2019 to explain some of the vegetation change dynamics in the area. This analysis thus produces a more comprehensive characterization of vegetation dynamics in the MEE, which would not be possible using traditional classification and change detection methods. Results would help fill in some of the existing gaps in literature about nature and magnitude of LULC change and the stability of LULC in the MEE. Such a comprehensive description of MEE vegetation dynamics is an important first step to initiate dialogue aimed to instigate policy changes for simultaneously conserving the environment and improving livelihoods dependent on it.

2. Study Area Description

The MEE is located in western Kenya and eastern Uganda (Figures 1 and 2). The studied area covers approximately 15,000 km^2 and extends from 1°37′42.82″ N, 33°55′45.07″ E and 0°42′15.76″ N, 35°14′18.84″ E. Mount Elgon is a solitary volcano and is among the oldest in East Africa [28,32]. The highest point, Wagagai Peak, is 4321 m [28,50] and is found on the Ugandan side. This area rises from a plateau that lies 1850–2000 m in the east and 1050–1350 m to the west with a caldera that extends 8 km wide [50]. Vegetation in this area is zoned by altitude and mountain forest, farmland and Afro-Alpine heath and moorland are the common land covers [27]. Declared a protected area in 1968 and 1992 in Kenya and Uganda, respectively [31,51], Mount Elgon Forest, a montane rainforest [52], is home to many important indigenous tree species [27].

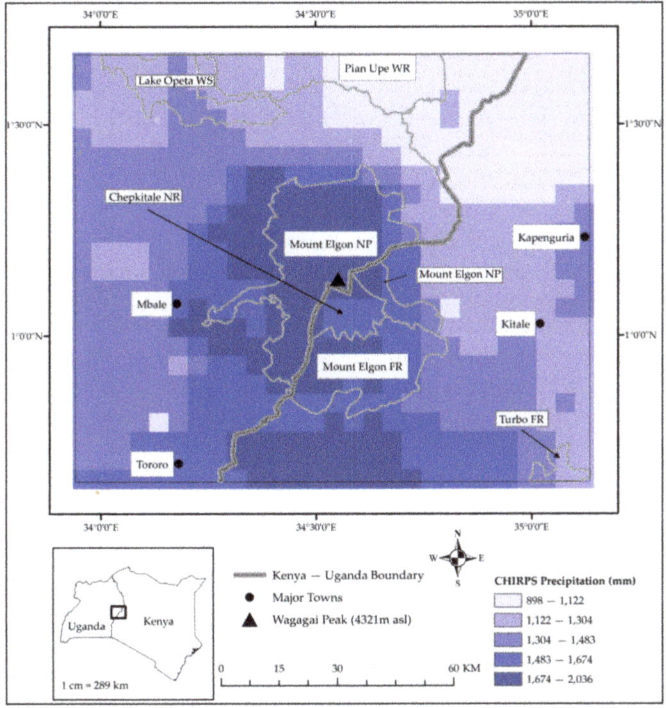

Figure 2. Map of the MEE in eastern Uganda and western Kenya showing long-term (1986–2018) mean annual total precipitation (CHIRPS [53]). Generally, the driest parts are the grasslands in northeastern MEE. The study area is wettest in the south and around the mountain region. Major protected areas and towns are shown for reference. FR is shorthand for forest reserve, NR is national reserve, NP is national park, WS is wetland system and WR is wildlife reserve.

The MEE receives rainfall in a bimodal pattern (two rainy seasons) and, according to Mugagga, Kakembo and Buyinza [28], most of the rainfall is received between April and October on the Ugandan side (with mean annual amounts ranging from 1500 to 2000 mm). The Kenyan side receives long rains between March and June and short rains between September and November—average annual rainfall ranging from 1400 to 1800 mm [54,55]. There is minimal temperature variation for the area—an average minimum of 15°C and a maximum of 23°C on the Ugandan side [28] and 14 and 24°C on the Kenyan side [55]. However, temperatures and precipitation have a strong variation with changes in altitude [27].

3. Materials and Methods

The present study integrated Mann–Kendall, Sen's slope and bfast in the analysis of NDVI and precipitation trends to characterize recent greening and browning patterns and trends within the MEE.

3.1. Data and Sources

This study utilized the following datasets (Table 1) in the analysis of greening and browning trends in the MEE.

Table 1. Description of original datasets used in the study.

Dataset	Spatial Resolution	Temporal Resolution	Duration	Source
MODIS MOD13Q1.V6	250 m	16-day	2001–2018	https://lpdaacsvc.cr.usgs.gov/appeears/
CHIRPS	5 km	5-day	1986–2018	https://earthengine.google.com/

3.1.1. MODIS NDVI and CHIRPS Precipitation

This study used MODIS NDVI data in the TS analysis of spatio-temporal changes in MEE vegetation greenness signal. The 16-day NDVI composite MOD13Q1.V6 [56] with a 250-meter spatial resolution was obtained through AppEEARS (https://lpdaacsvc.cr.usgs.gov/appeears/) [57]. Data for the period 2001–2018 were used. While Landsat data [58] have a better spatial resolution and historical coverage, and are therefore better suited for this study, they were limited by the many data gaps in the TS over the MEE due to persistent cloud cover and Landsat's long (16-day) revisit time. As such, MODIS NDVI, which has previously been used successfully in similar studies in East Africa [6,22,59], was used in this study. This study also used CHIRPS precipitation data [53]. These data have a spatial resolution of 5 km and provide global daily and pentad records from 1981 to present. For this study, these data for the period 1986–2018 were obtained and preprocessed within Google Earth Engine (GEE) [60]. It is worth noting here that there were no observation data to assess the accuracy of the CHIRPS dataset over the MEE. However, this dataset has been used extensively in similar studies [12,61,62]. Moreover, this dataset was recently validated [63] and the results showed that the CHIRPS data are reasonably accurate in estimating rainfall over east and South Africa.

NDVI preprocessing involved quality assessment using the associated VI quality files. Pixels with low quality, high aerosol content, cloud cover and possible shadows were excluded during this exercise. NDVI and precipitation composites for specific time scales were then generated. First computed were mean NDVI TS during the wet season for the period 2001–2018. Here, imagery in April, May and June was used. The TS of mean NDVI for each 16-day period during the season were also created. As a result, there were generally two TS for each month and were labeled h1 and h2 for TS created from NDVI composites recorded roughly in the first and second half of the month, respectively. For precipitation, the TS of total amount over the wet season were computed for different periods; 1986–2018 (33 years); 1986–1996 (first 11 years); 1997–2007 (median 11 years); 2008–2018 (last 11 years); 1986–2005 (first 20 years); 1999–2018 (last 20 years); and 2001–2018 (18 years, similar to the NDVI TS length). While the present study was aimed at characterizing changes in vegetation greenness from 2001 to 2018, analysis of longer precipitation TS was necessary to understand any longer-term precipitation patterns in the MEE that may influence the vegetation patterns. TS were also generated of total precipitation amounts for each dekad over the wet season. In this analysis, these dekads for the month of April were labelled April d1 (for dates 1–10), April d2 (for dates 11–20) and April d3 (for the rest of the month). The same nomenclature was used for dekad TS in May and June. These were used to assess nature and strength of relationship between vegetation greenness and precipitation variability over a shorter time scale, in which case results from dekad precipitation TS were compared to results from 16-day NDVI. The 16-day NDVI composites were used here due to unavailability of data to compute dekad NDVI composites. Finally, monthly precipitation composites were generated for the period 1986–2018, for use in analyzing breakpoints in MEE precipitation.

3.1.2. Field-Collected Data

Environmental data from local communities and government officers within the MEE were collected using semi-structured interviews and direct observation in July–September 2019 (IRB Number: STUDY00002404). Most of the interviewees were from significantly changing landscapes (areas showing significant changes in vegetation greenness). The participants in this study were interviewed about perceptions of climate change and land-use change, and historical patterns of agriculture and land-use change. Interviews were conducted and written responses recorded using Qualtrics software [64]. The fusion of such qualitative data with quantitative RS data is important because indigenous and historical accounts of LULC change in such a constantly variable landscape can be used to fill in gaps that may not be fully explained using RS and GIS alone.

3.2. Methods

This section describes the methods and analyses performed on NDVI and precipitation TS to identify areas and characterize patterns of vegetation greening and browning within the MEE. These two analyses were performed in R [65] and trends were assessed at the 95% significance level. After analysis of monotonic trends in greenness and precipitation, it was necessary to perform breakpoint analysis on these time series, to detect any significant breaks within the data. These two types of analyses are described in more detail below. Figure 3 highlights major methods used in this study.

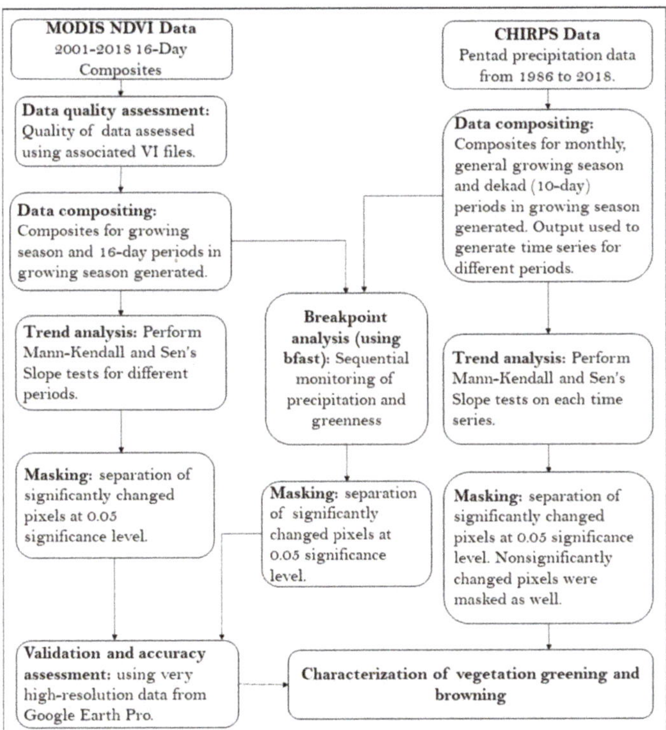

Figure 3. Flowchart of analysis methods used in the study.

3.2.1. TS Analysis: Mann–Kendall and Sen's Slope

To analyze initial spatiotemporal changes in vegetation greening and browning trends in the MEE, the current study borrowed from methodologies presented by Landmann and Dubovyk [6]. However, since NDVI TS often do not meet assumptions for parametric analysis [48], the current study used the nonparametric Mann–Kendall test rather than linear regression. To calculate the Mann–Kendall test statistic, data values are evaluated as an ordered TS [46] and each value is then compared to all subsequent values [46,66]. The Mann–Kendall S statistic is initially assumed to be 0 and a value of 1 is added to (subtracted from) the test statistic if the value of an observation is higher (lower) than that of the previous observation [46,66]. There is no change to the statistic if the values are equal. Equation (1) below shows the Mann–Kendall test equation. High positive and low negative values of S, respectively, indicate increasing and decreasing trends, but the strength of the trend is statistically quantified by computing probability associated with S and the size of the data sample [46].

$$S = \sum_{k=1}^{n-1} \sum_{j=k+1}^{n} sign\left(x_j - x_k\right) \tag{1}$$

where $\quad sign\left(x_j - x_k\right)$

$$= 1 \quad \text{if } x_j - x_k > 0$$
$$= 0 \quad \text{if } x_j - x_k = 0$$
$$= -1 \quad \text{if } x_j - x_k < 0 \quad \textbf{Source: } \text{Khambhammettu [46]}$$

The magnitude of the trends was quantified using Sen's slope estimator [47]. These algorithms were used in this study first to characterize and quantify any general patterns in MEE greenness during the growing season (April, May, and June). The analysis then investigated more subtle changes in vegetation greenness over shorter (16-day) periods during the growing season. From these, areas of increasing (decreasing) greenness were mapped as areas of vegetation greening (browning).

To disentangle changes due to human and natural forcings, monotonic trends were assessed in the precipitation TS and magnitude of the trends quantified. The analysis was performed on precipitation series at different temporal scales—based on the length of the time series (including 33-, 20-, 18- and 11-year periods) and TS resolution (including dekad totals, and general growing season totals). From these analyses, areas with significant positive (negative) precipitation trends were mapped as areas experiencing significant and consistent wetting (drying) over the analysis period. A cross examination of results from greenness and precipitation trend analysis was conducted to distinguish any human- from nature-driven vegetation greening and browning. Using available very-high-resolution imagery from Google Earth Pro [67], the nature of LULC conversion was ascertained. Besides, the statistics of LULC change (including greening, browning pixels) were calculated.

3.2.2. Breakpoint Analysis: bfast

To further understand any temporal patterns in MEE greenness and precipitation, the bfast algorithm was applied using the 'bfastSpatial' package (http://www.loicdutrieux.net/bfastSpatial/). There exists a predictable annual cycle of greening an browning in vegetation, and these coincide with the occurrence of rainy and dry seasons [49]. bfast, created by Verbesselt et al. [16] and Verbesselt and Zeileis [68], creates a best-fit seasonal regression model with a trend component for the time series [69]. This approach follows three main steps (i.) fitting a harmonic model based on a historical ('stable') period, (ii.) testing observations that follow the historical period for any structural breaks from the fitted model, and (iii.) calculating the magnitude of change which is the median residual between observed and expected values in the monitoring period [49,70]. While this approach has been applied on forested landscapes, the current study applied it to the whole of the MEE, whose LULC comprises forested, savanna, grassland, and cropland LULC (Figure 1). This was prompted by the unique LULC orientation in the MEE that makes it difficult to detect LULC changes via satellite image change detection analysis [27]. bfast breakpoint analysis can sequentially monitor vegetation change on a yearly basis, thus making it a suitable method for assessing changes in such LULC.

The models used in this part of the study were parameterized based primarily on the study by DeVries et al. [70]. First, the time series was divided into historical and monitoring periods. A minimum of two years for the historical period is recommended when using MODIS 16-day data [49,68]. In this study, therefore, the period 2001–2004 was used as the initial historical period and was assumed to be generally stable before the start of the monitoring period. First-, second- and third-order harmonic models were fit on the NDVI data, the results were assessed, and the first-order model was finally selected as the suitable instance to use. The single-order model has been used previously with the assumption that vegetation phenology generally follows a similar trend [70]:

$$y_t = \alpha + \gamma \sin(\frac{2\pi t}{f} + \delta) + \varepsilon_t \tag{2}$$

where
- yt dependent variable
- t independent variable
- f temporal frequency
- α model intercept
- γ model amplitude
- δ model phase
- ε_t error term

As in the study by DeVries et al. [70], the trend component was excluded from the time series to reduce chances of yielding false breakpoints.

The change magnitude was calculated as the median of the residuals in the monitoring period. The median is thought of as a conservative measure, unlike the sum, that minimizes the chances of getting inflated magnitudes and therefore false breakpoints [70]. However, increasing the number of observations before and after a change event, by including long monitoring periods, yielded very high magnitudes and unrealistically numerous breakpoints. As such, this study elected to use sequential non-overlapping one-year-a-time monitoring periods. Here, the TS was trimmed to include the historical period plus one-year monitoring period. For monitoring the period from January to December 2005, for example, the TS for January 2001 to December 2005 was used, and the monitoring period was set to start in January 2005. This sequential monitoring approach is illustrated in DeVries et al. [70]. The approach is advantageous as it enables the assessment of subtle changes in vegetation within the MEE, especially alternating degradation and restoration that would go undetected using other methods. The default values for h, the minimal segment size between potentially detected breaks in the trend model given as a fraction relative to the sample size [49], were used in this study.

Breakpoint analysis was also performed on the monthly total precipitation TS. The analysis was performed on both 1986–2018 and 2001–2018 time series, to understand any longer-term patterns as well as recent changes in the precipitation. Here, the third-order harmonic model with the trend term was the best fit. While there was no direct application on record of bfast for precipitation breakpoint analysis, it is noted that the algorithm can be used for this purpose [71,72].

3.3. Validation of Results

A hybrid validation of analysis results was performed in this study. The collection of reference data borrowed from the methodology used previously by Landmann and Dubovyk [6]. The seasonal greening and browning map from Mann–Kendall and Sen's slope analysis (Figure 4) was linked to very-high-resolution imagery in Google Earth Pro and both qualitative and quantitative accuracy assessment of the results was performed. Locations of greening, browning and no change were investigated by visually interpreting historical imagery in Google Earth Pro. The selection of these testing points was based on the minimum size of detectable change (at least 250 m, the size of a MODIS NDVI pixel) and the availability of sufficient imagery to interpret change. As such, only locations with very-high-resolution imagery in one of the three initial and final years (2001–2003, 2016–2018, respectively) were used. By visually assessing historical imagery spanning these periods, vegetation change, or lack thereof, could be observed. A total of 153 visually interpreted points (51 browned, 50 greened and 52 no change) were used. Greenness change in pixels at these locations was first recorded. The seasonal greening and browning map from Mann–Kendall and Sen's slope analysis (Figure 4) was also reclassified into greened, browned and no change classes. The testing points were then compared to the reclassified raster and accuracy measures including overall, producer's, and user's accuracies were calculated. The final points were locations with pixels that indicated highly discernible change (like deforestation, reforestation etc.).

It was challenging to evaluate sequentially monitored vegetation change since it was not possible to gather testing points due to many temporal gaps in images in Google Earth Pro. In most instances, one can rarely find images for successive years, thus making it difficult to validate year-to-year change results. Since the bfast algorithm detects even subtle changes in vegetation greenness, changes could not be discerned with high confidence and therefore calculating accuracy assessment statistics would be misleading due to inconsistencies in testing data [73]. As such, these results were only visually and qualitatively assessed, and a general trend of change in the pixels was interpreted rather than year-to-year change. Evaluating such time series results has been found to be challenging previously [37,74]. In this study, inferences drawn from the field interviews were also qualitatively incorporated in this exercise to explain some of the trends detectable in both trend and breakpoint analyses.

4. Results

This study highlights and characterizes, using the Mann–Kendall, Sen's slope and bfast algorithms, recent vegetation greening and browning trends and patterns in the MEE at multiple time scales. The results highlight portions of the MEE that experienced persistent and significant changes in vegetation greenness, as indicated by changing NDVI. The results from similar analyses of precipitation TS are also presented and, together, attempt to disentangle nature- from human-driven changes observed over the MEE landscape.

4.1. Trend Analysis Results

4.1.1. Persistent Vegetation Greening and Browning in the MEE

Greening (browning) was defined as any significant increase (decrease) in NDVI as shown by either Kendall τ (Mann–Kendall and Sen's slope) or the magnitude of change (bfast). The results indicate various greening and browning patterns during the growing season (Figure 4) and near-half month (Figure 5) periods. During the growing season, greenness significantly increased in approximately 27% (3400 km^2) of the study area. Here, NDVI increased at rates up to 0.025 per year. These locations were concentrated mostly within croplands, grasslands, and savanna (Figure 1). Significant browning was also evident, with NDVI in more than 1400 km^2 (about 11% of the total area) decreasing by up to 0.035 per year over the analysis period. These areas were mostly located in the southwestern part of the MEE in Uganda. This location includes the Namatala wetland, which has experienced intensive conversion to agriculture and settlement in the past years [75]. Browning was also evident in other locations around the Mount Elgon ecotone and elsewhere in the MEE. Based on our visual assessment, most of these corresponded to areas where deforestation occurred over the analysis period. No significant trends were found in the rest of the MEE (62%, 7800 km^2).

Analysis of greening and browning trends in NDVI for every 16-day period in the months of April, May and June showed similar patterns; most browning occurred in the southwestern MEE and greening elsewhere. Most of the changed areas experienced greening and browning at rates of up to 0.03 and -0.04, respectively. The highest proportions of land where greening occurred was found in the month of May, especially the May h1 period, in which about 20% (2400 km^2) of the MEE experienced significant greening (Figure 5). Greenness increased in more than 1600 km^2 (13%) of the study area during May h2. More greening was experienced in the June h2 and April h2 (9% of MEE greened during both periods). The greatest proportion of browned areas was observed in June h1, where about 600 km^2 (5%) of the MEE experienced vegetation greenness decline (Figure 5E). Most of the land within the MEE did not experience any significant change during these 16-day periods.

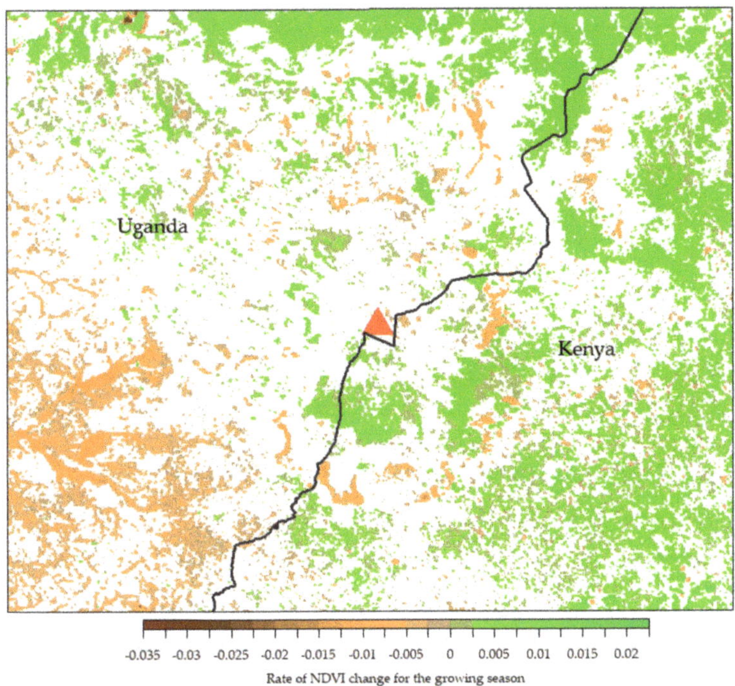

Figure 4. Map of significant changed (greened and browned) locations during the growing season. Slope values (Kendall τ), indicative of magnitude of change per time step, are shown here. White pixels indicate no significant change.

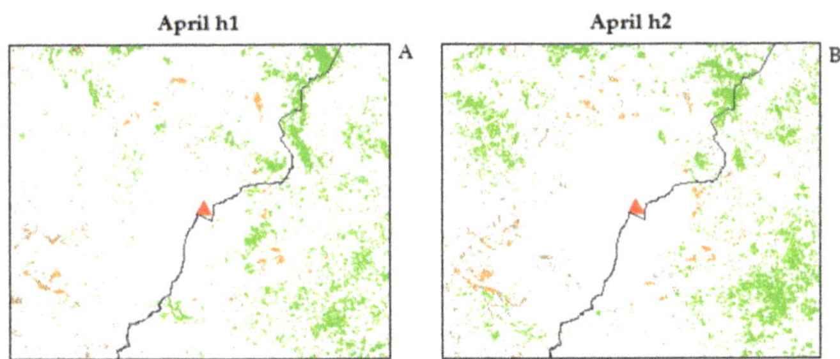

Figure 5. *Cont.*

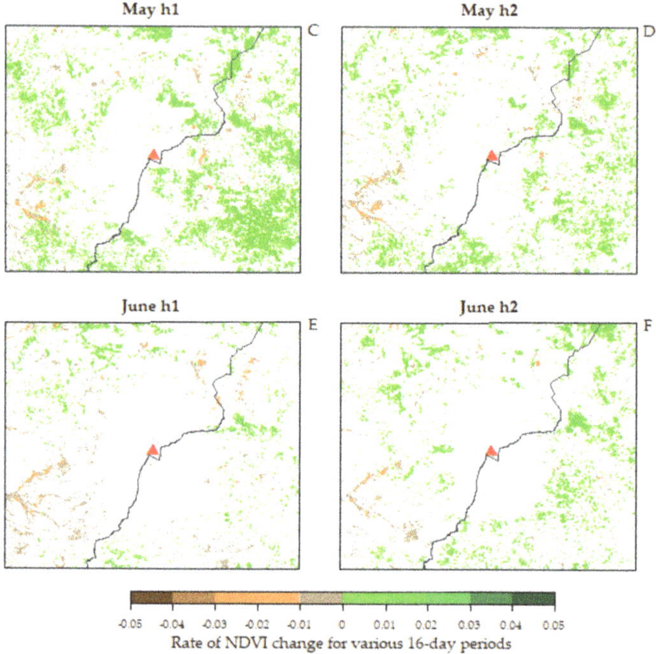

Figure 5. (**A–F**) Greening and browning trends in the Normalized Difference Vegetation Index (NDVI) for every 16-day period in the months of April, May and June for 18 years (2001–2018).

4.1.2. Precipitation Variability in the MEE

The test for monotonic trends in precipitation revealed various patterns of consistently increasing (wetting) and decreasing (drying) precipitation within the MEE (Figure 6). Most of the areas experienced increased precipitation over the years and only negligible proportions of the study area became drier. The greatest proportions of land during which consistently wetter conditions prevailed include 1999–2018, 1986–2018 and 2001–2018 (82%, 58% and 38%, respectively). The areas experiencing change covered approximately 10,300, 7200 and 4800 km^2, respectively. These periods also showed the greatest magnitude of change in precipitation amount. Here, precipitation increased by at least 13, 4, and 5 mm per year for 1999–2018, 1986–2018 and 2001–2018 periods. In addition, about 27% of the MEE also experienced wetter conditions during 2008–2018.

There were very small portions of the MEE with wetter conditions during the earlier years in the analysis; 1.61% of land (200 km^2) recorded wetter conditions in 1986–1996 while there was no significant increase in precipitation during the 1986–2005 and 1997–2007 periods. Consistently drier conditions were observed in two time periods (1986–1996 and 1986–2005). However, only negligible proportions of land (up to about 2%) experienced this change, although at substantial magnitudes of up to 9 mm per year. Different spatial patterns existed from 2000; significantly increasing precipitation in 2008–2018 was observed mostly on the Ugandan side, and the mountain forest and wetland reserves seemed to be excluded (Figure 6B). In 2001–2018, the changed pixels were mostly found within the mountain area, in the north and some areas in the west (Figure 6E). On the other hand, areas in the northeastern portion of the study area did not experience changes in precipitation during the 1986–2018 period (Figure 6F). Lastly, the 1999–2018 period had significant increases in total seasonal precipitation with an exception of a few eastern and southeastern portions of the MEE (Figure 6D). Overall, there was an increase in precipitation for most areas in the study area, with an increasing magnitude of change, especially in the post-2000 era.

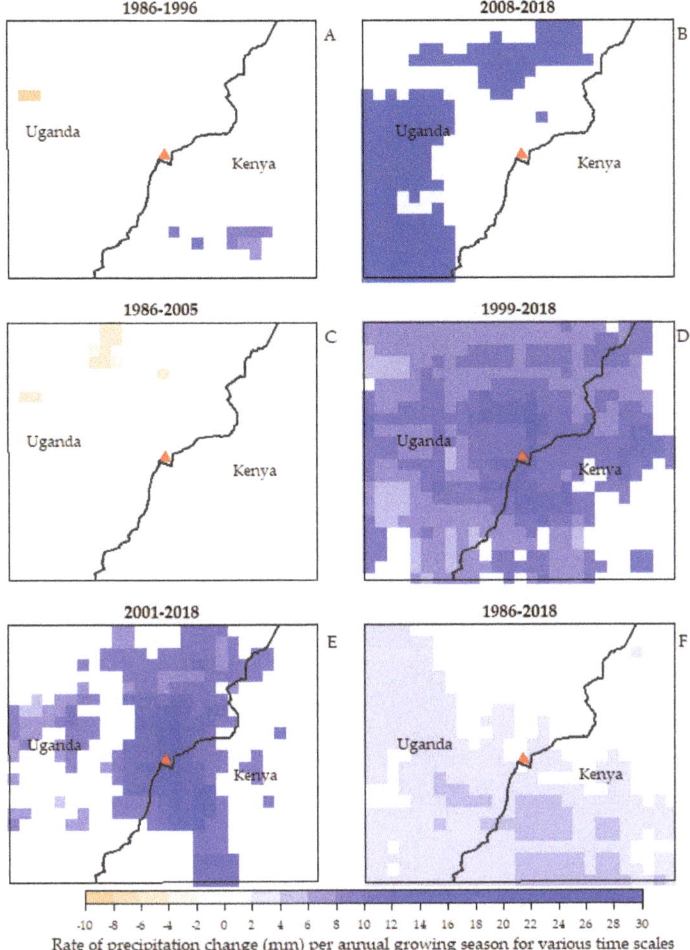

Figure 6. (**A–F**) Maps of precipitation change for each analyzed time period.

The trend analysis for each dekad in the growing season found that, based on rates of change, the 2008–2018 period recorded the highest change in precipitation in April d1 and May d1, at rates of up to 13 mm per yearly dekad, see Figure 7). Areas where this change was experienced were located mostly in western MEE. An increase in precipitation was also recorded in April d2 in 2001–2018, June d3 in 1986–1996 and June d3 in 2008–2018. Negative trends in MEE precipitation were also observed in some dekads, mostly in 1986–2005. During this period, precipitation decreased in April d1, May d2 and June d1, at the rate of up to 3 mm per annual dekad.

Based on the proportions of land with significant change in precipitation, the study found that most of MEE precipitation significantly changed during May d1 in 1986–2018 (90%, 11,300 km^2) and May d1 in 1999–2018 (43%, 5400 km^2, Figure 8). From these results, some patterns are clear. For the months of May and June, annual dekad precipitation increased significantly for all time periods, although at varying rates and proportions. Precipitation in most of the MEE was more stable in the earlier years in the study (as in 1986–1996) or generally depicted significant and consistently drier conditions (as in 1986–2005). Based on these results and those in Figure 6 above, it was necessary to conduct a breakpoint analysis to provide a better characterization of any specific spatio-temporal breaks in precipitation and greenness trends.

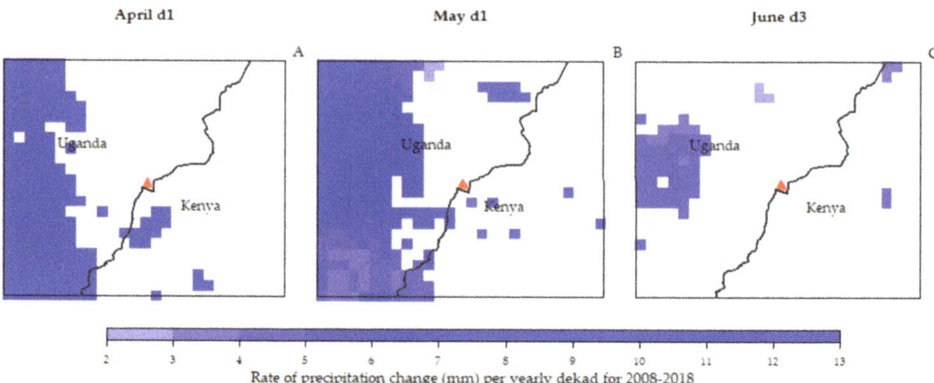

Figure 7. (**A–C**) Significantly changed precipitation for dekads in the period 2008 to 2018.

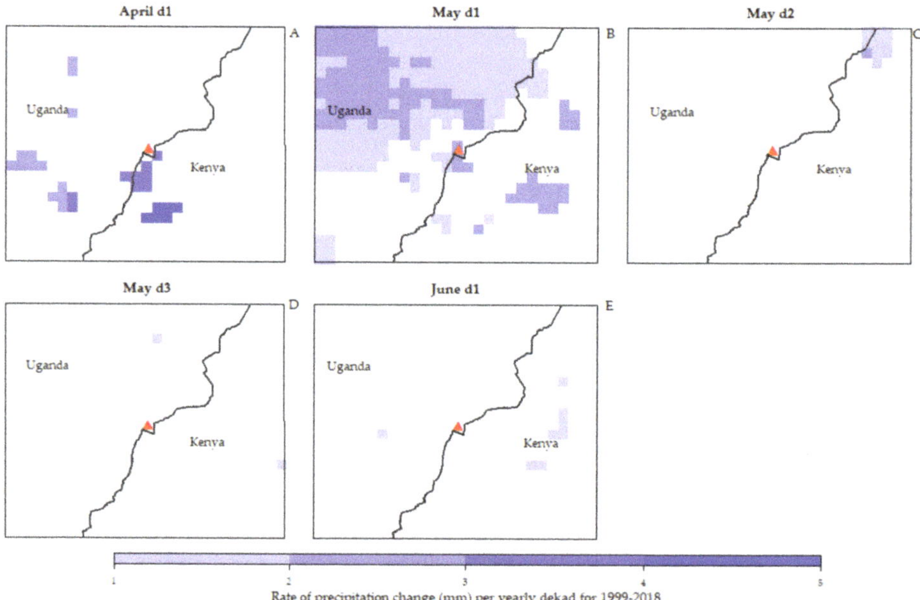

Figure 8. (**A–E**) Significantly changed precipitation for dekads in the period 1999 to 2018.

4.2. Breakpoint Analysis Results: bfast

While breakpoint analysis was performed for both 1986–2018 and 2001–2018 precipitation TS, no significant breaks were found in the former. Therefore, results from 2001–2018 are presented in this study.

4.2.1. MEE Precipitation

The breakpoint analysis in bfast revealed no significant breaks in MEE average mean total monthly precipitation. However, breakpoints monitored for each pixel from January 2005 revealed very interesting patterns. In most configurations, the analysis revealed significant breakpoints in 2006 and 2007 for most of the MEE. To reduce the influence of post-change detection observations on change magnitude, the same analysis was performed for the period 2001–2008, with 2001–2004 set as the historical period. Therefore, change maps and statistics provided are based on this adjusted analysis. Precipitation changed significantly in the two years, with about 9800 km^2 (66% of the area) and 4800 km^2 (32%) of the MEE experiencing wetter conditions in 2006 and 2007, respectively (Figure 9). Most of the breakpoints in 2006 were detected in the months of July–October, while a great proportion of changes in 2007 were detected in March and April. Similar wetter conditions were detected in some locations in the months of May and June for both years. No drier conditions were detected. The magnitude of precipitation change ranged from approximately 10 to 53 mm during the monitored period (Figure 10A) and the bfast models used could explain up to 70% of the variance in the precipitation breakpoints (Figure 10B). Precipitation was also monitored sequentially for the period 2005–2018 and no breakpoints were detected except for a few pixels in 2005–2007.

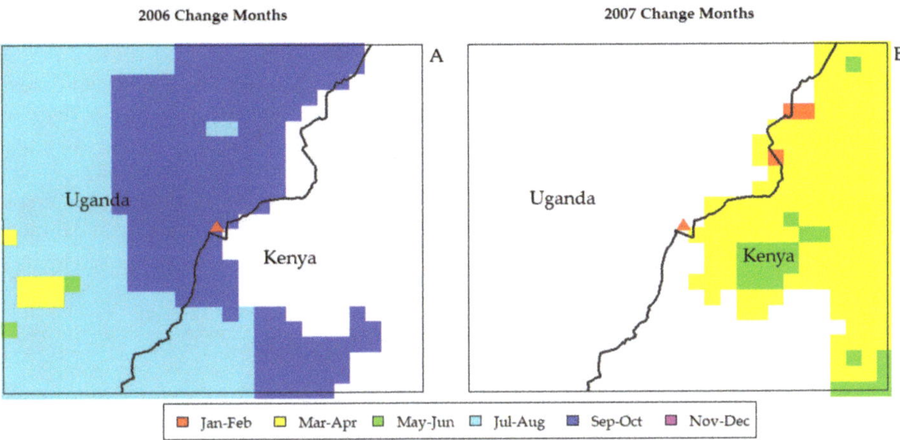

Figure 9. (**A**,**B**) Months when major breaks were detected in the precipitation time series (2001–2008) when the period 2005–2008 was monitored.

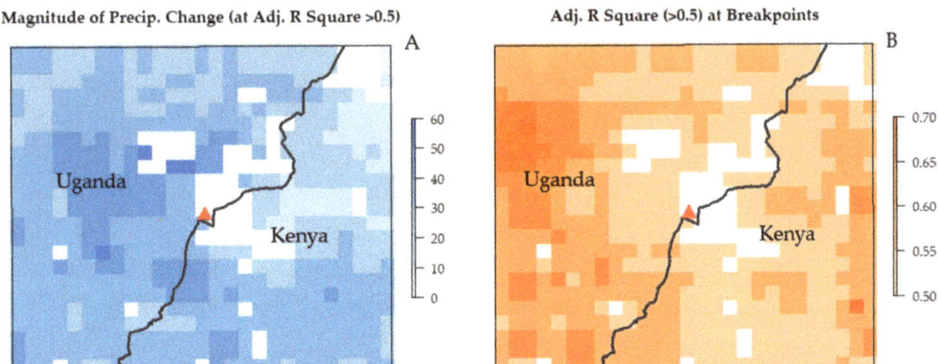

Figure 10. (**A**,**B**) Adjusted R2 (greater than 0.5) and magnitude of change in the precipitation time series (2001–2008) using 2005–2008 as the monitoring period.

4.2.2. MEE Greenness

The analysis of breakpoints revealed that some significant breaks existed in the NDVI time series for each year (2005–2018). Having been sequentially monitored, the results include, for each monitored year, months when the breakpoints were observed, the magnitude of change at the breakpoints, the length of historical data used and adjusted R^2. In this analysis, only highly statistically significant changes ($p < 0.05$) are reported. This decision was based on two reasons: (i.) There were no ground data for validating results from sequentially monitored bfast. To ensure that only accurate results are reported, breakpoints from models with less than 50% adjusted R^2 were excluded. Thus, only breakpoints from models with average to high explanatory power were reported. This 50% threshold has been used elsewhere by Landmann and Dubovyk [6]. (ii.) Vegetation greenness in the MEE showed significant interannual variability. As such, to ensure a long-enough historical period used by 'bfastSpatial', the study excluded any breakpoints from models using a historical period of less than two years.

The 'bfastSpatial' model was able to detect changes in vegetation greenness, especially in the grasslands of the northeastern MEE (Figures 1 and 11). The observed changes could be due to both natural factors (such as variability in precipitation) and/or human activities (for example, clearing of land for agriculture and settlement, deforestation for charcoal burning and construction etc.). The maximum magnitude of changes ranged from −0.24 and 0.21, thus indicating only subtle changes in the MEE vegetation. The breakpoints were detected in each of the years, but most of them were observed in 2013, 2007 and 2010 (greening) and 2009 and 2017 (browning), as shown in Figure 11. Overall, there were more areas where significant greening trends were detected compared to browning, with about 17% of the MEE (about 2500 km^2) and 10% (about 1500 km^2) showing greening and browning, respectively, over the monitoring period. These represent annual greening and browning rates of about 1.2% and 0.7%, respectively.

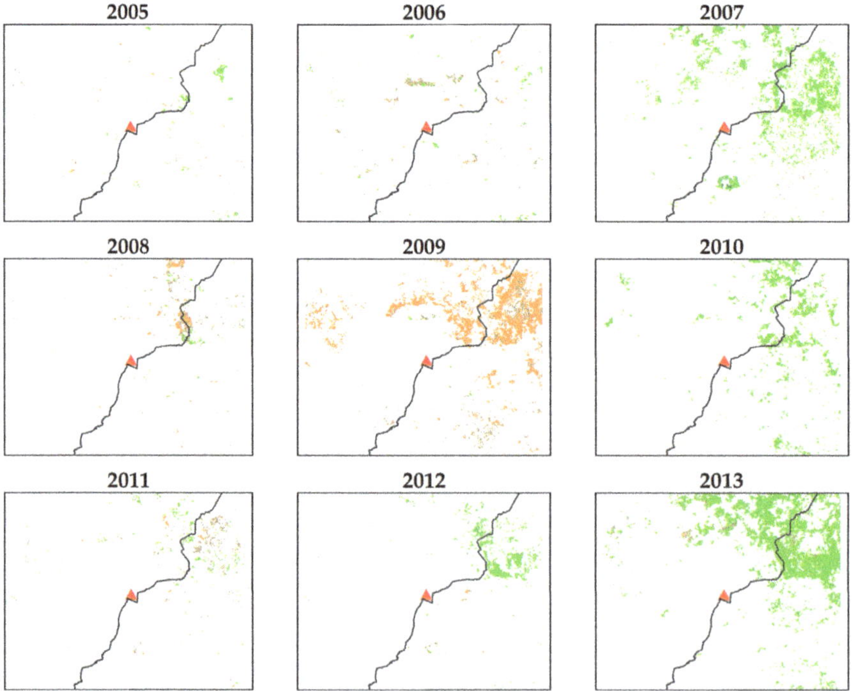

Figure 11. *Cont.*

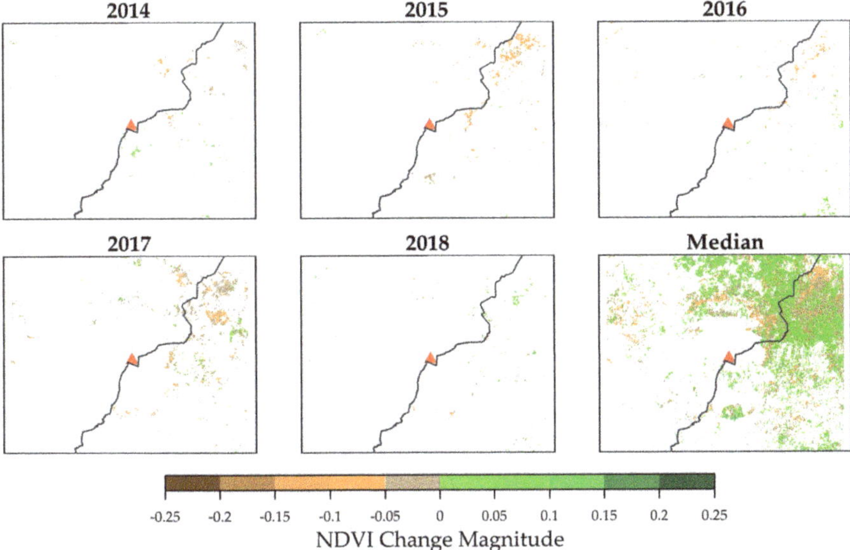

Figure 11. Greening and browning in the MEE. The results from bfastSpatial using NDVI time series (2001–2018) while sequentially monitoring each year in 2005–2018. These maps show magnitudes of change for pixels with significant breakpoints.

Based on the yearly changes, differentiated greening and browning patterns were observed during the monitoring period. Cumulatively, the greatest proportion of land with changed greenness was in 2013 in which significant breakpoints were observed in approximately 13% of the MEE (approximately 1850 km^2). This was followed by 2009 (1350 km^2), 2007 (1200 km^2) and 2010 (650 km^2) (Figure 12). For 2013, 2010 and 2007, the majority (over 95%) of the changed locations experienced greening. The year 2009 recorded browning in over 99% of the changed locations.

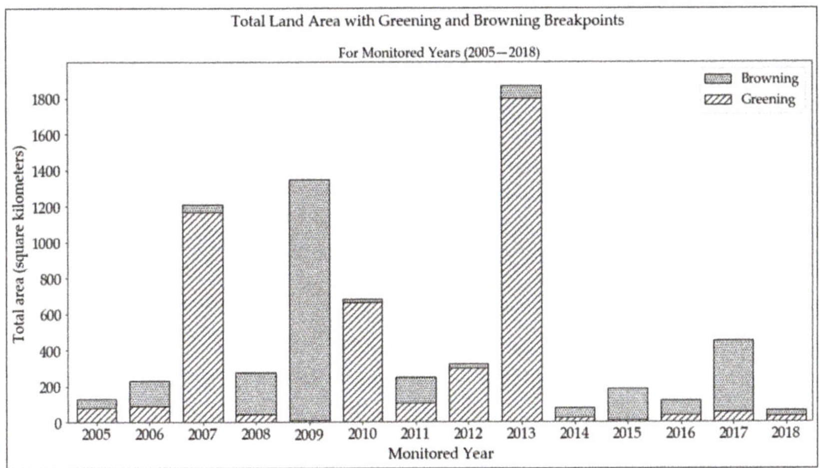

Figure 12. Greening and browning in the MEE. The results from bfastSpatial using NDVI time series (2001–2018) while sequentially monitoring each year in 2005–2018. The graphs show the total area of land that experienced greening and browning for each monitored year.

4.2.3. MEE Greenness vs Precipitation

There was an increase in precipitation in 2006 compared to previous years (Figure 13). There followed a steady decrease in 2007–2009 and finally an increase in 2010. Similar changes were observed in greenness during the five years. First, there was greening at most breakpoints in 2007 following the significant wetting in 2006. Significant browning followed, most of which occurred in 2009. This was also observable in an aspatial bfast breakpoint analysis performed on MEE mean NDVI, in which a sudden increase in NDVI was observed towards the end of 2006, followed by a consistently reducing trend until 2009/2010, in which another sudden increase was found (Figure 14). No significant breakpoints were found from a similar analysis using mean MEE precipitation. However, results from 'bfastSpatial' indicate that most of the MEE had very significant increases in precipitation in the second half of 2006 and the first half of 2007. Therefore, the corresponding changes in vegetation greenness can be attributed to this change in precipitation.

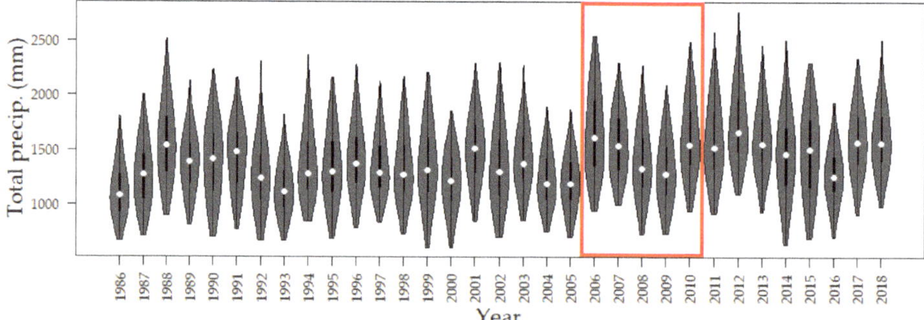

Figure 13. Violin plot of average total annual precipitation in the MEE. The red box highlights the period 2006–2010.

The significant and extensive browning observable mostly in 2009 (Figure 11) may be explained in terms of vegetation regaining its 'normal' greenness levels following sudden greening due to an above-normal precipitation. However, no significant breaks were found in precipitation around 2013, the year with the highest proportion of land with detected change in vegetation greenness. Thus, such browning and greening trends can be explained with regards to other factors, including temperature and anthropogenic activities that may have altered vegetation greenness during the monitoring period. The 'bfastSpatial' model did not find any significant breakpoints, in precipitation TS, in 2012. This may be due to the unstable historical data prior to the monitoring period. However, the violin plot (Figure 13) indicates that this year had some of the highest precipitation recordings. As such, there is a high chance that the greening following in 2013 was influenced by this increase in precipitation.

A visual inspection of these results in Google Earth Pro indicates that the subtle changes are indicative of various LULC changes. There was evidence of human settlement being introduced into the grasslands in the northeastern MEE. Information gathered from the field corroborates this finding and further explains the implications for landscape greenness. Inhabited by nomadic pastoralist communities, this part of the MEE is susceptible to degradation, especially when these communities move, driven by rainfall patterns, to settle within the grasslands. Based on fieldwork and data from Google Earth Pro, reduced natural vegetation and tree density were evident in these locations. In other instances, the new inhabitants converted natural vegetation to agriculture and, although this would result in environmental degradation, some planted crops were greener than the natural vegetation, and, therefore, these areas were shown to exhibit greening trends. In other locations outside of the grasslands, visible greening trends were indicative of some afforestation practices. Data from the field revealed that this kind of greening was driven by the cultivation of evergreen early maturing blue gum (*Eucalyptus globulus*) tree species, sometimes together with and other times in the place of the maize crop. This was especially common in the eastern and southeastern parts of the domain and parts of Uganda.

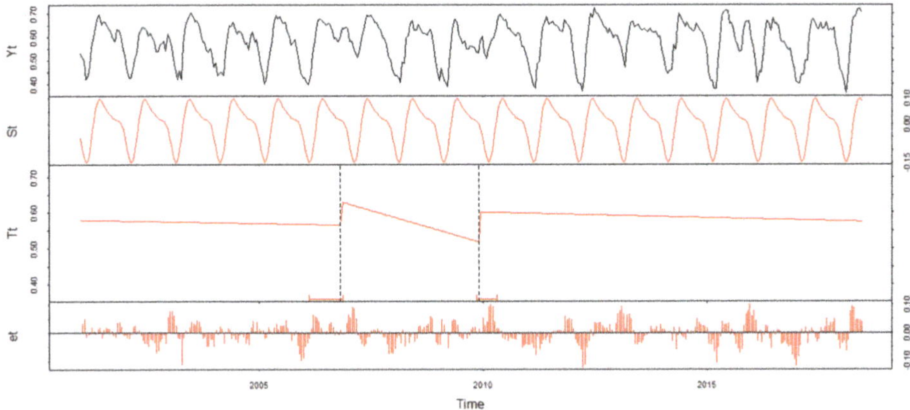

Figure 14. bfast mean 16-day NDVI time series decomposition (2001–2018). Breakpoints in trend were found between 2006 and 2010.

4.3. Accuracy Assessment

Validation of Man-Kendall and Sen's slope results revealed an overall accuracy of 98.04% and user's accuracies of 100%, 98% and 96.2% for browned, greened, and unchanged locations, respectively (Table 2). Producer's accuracies of 98% (for both browned and greened locations) and 98.1% (no change) were also obtained. The visual inspection of bfast results using Google Earth Pro also revealed that most of the detected subtle changes occurred. However, no more detailed information like year of change could be discerned due to the lack of available high-resolution imagery.

Table 2. Mann–Kendall and Sen's slope accuracy assessment statistics.

	Browned	Greened	No Change	User's Accuracy
Browned	50	0	0	100
Greened	0	49	1	98
No change	1	1	51	96.2
Producer's Accuracy	98	98	98.1	
Overall Accuracy				**98.04**

5. Discussion

5.1. Precipitation and Vegetation Change in the MEE

Climate change continues to affect economies in most developing nations, especially those relying heavily on natural processes for their livelihoods. Agriculture remains the backbone of many of these nations [76–82], yet agriculture's vulnerability to climate change effects cannot be contested. Climate-related natural hazards, including extensive flooding, extended droughts [83], landslides [28] among others, have significantly impacted agricultural production and endangered lives. The frequency of these events shows an increasing trend [84,85], thus trapping many agriculture-dependent communities in an unending struggle for survival. Therefore, existing food insecurity in these developing nations can be attributed to their overreliance on rainfed agriculture [83]. Moreover, the high population growth and densities in these nations and elsewhere have translated into need for more agricultural land [28]. Coupled with political interference and corruption among forest park and reserve staff, this need for more land has resulted in forest encroachment and deforestation as ecologically fragile land is cleared for agriculture and settlement [28–32]. Therefore, understanding the major drivers of landscape change is an important first step to inform better decisions to simultaneously conserve the natural environment and improve the livelihoods dependent on it. First, being able to separate nature- from human-induced landscape changes would be valuable in this endeavor.

Changes in the landscape occur at varying rates and magnitudes across space and time, from very subtle tree damages to forest clearings [49]. In this study, two major forms of landscape change— browning (areas of declining vegetation) and greening (areas of increasing vegetation) [12]—were studied. The results show that MEE greenness exhibited substantial variability, and some form of the greening and browning change was recorded each year. As expected, these changes varied by scale, with the highest proportions of greened and browned locations observed over the growing season rather than any of the individual 16-day periods. Importantly, there was a lot of activity in areas bordering the mountain forest, as expected. Clearly, both greening and browning trends were observable, particularly on the Kenya side. Significant deforestation occurred as a result of encroaching fertile land on the high slopes of the mountain, for agriculture and settlement (examples in Figure 15). This finding has been reported in previous studies [27,28]. The Shamba system, thought of as a win-win arrangement, enabled local communities to farm in protected areas while tending to the growing trees in their early stages of growth [27]. This was a government effort originally meant to convert native to plantation forests and later to replant trees on harvested forest land. However, the Shamba system farms overlap with plantation forest and bamboo vegetation thus making it more difficult to conclusively identify, characterize and quantify the nature and magnitude of LULC change, especially by the use of traditional classification and change detection methods [27]. The significant and consistent browning in the southwestern MEE are attributed to the conversion of the Namatala wetland to agriculture (mostly paddy rice farming) and settlement (due to the growth of Mbale town) [75]. Since the early 2000s, 80–90% of this wetland has been converted. The wetland area is an Important Bird Area (IBA) [86] and therefore the reported LULC conversion caused many nature–human and human–human conflicts, including, respectively, bird poisoning and competition among people to own and utilize the wetland.

Disentangling nature- from human-induced vegetation change is an important yet complex task. In this study, patterns of change in precipitation varied with the TS duration (33-, 11-, 20- and 18-years) and resolution (dekad, seasonal). Previous studies indicated that precipitation in the area exhibited significant temporal variability, with both positive [63] and negative [80] trends observed over time. It can be inferred that precipitation had changed more (increased) over the period after 2000 than earlier. This implies that some of the vegetation greening and browning should be linked to this MEE wetting. This is true for the period 2006–2010, where bfast reveals that greening and browning events in the MEE follow significant wetting and drying events. However, a visual interpretation of Mann–Kendall analysis maps for both MODIS NDVI and CHIRPS precipitation did not show much similarity. Besides, bfast did not reveal any breakpoints in precipitation around 2013, the year when most breakpoints in greenness were detected. While this may be a fault in the bfast model used (because of, for instance, instability in historical data used), precipitation alone may not be a reliable predictor of vegetation change [49]. Davenport and Nichols [22] concluded that the NDVI–precipitation relationship varies both spatially and temporally, is not linear, generally exhibits a three-month lag, and depends on underlying LULC types. Therefore, it is likely that these complex relationships were not well revealed in both analyses in this study. Besides, the observed greening and browning patterns may be driven by other climate factors (such as temperature), which were not examined in this study, due in part to data unavailability for the MEE. Moreover, the spatial scale mismatch between the CHIRPS and MODIS data make more quantitative comparisons challenging.

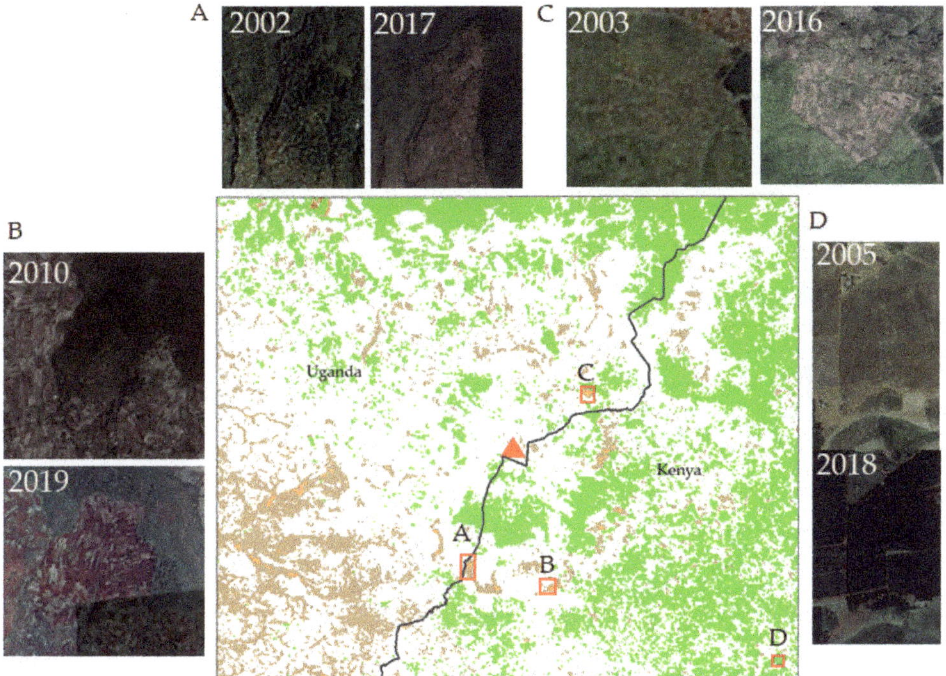

Figure 15. (**A–D**) Some of the greened and browned areas identified by Mann–Kendall and Sen's slope. Images (**A–C**) indicate the conversion of natural vegetation (mostly forest) to cropland and settlement. Image (**D**) indicates afforestation. Source: Google Earth Pro [67].

Visual inspections of the results from both analyses revealed the applicability of Mann–Kendall and bfast methods in detecting changes in vegetation greenness. Most people in the MEE are small-scale farmers, owning sometimes less than an acre (0.4 hectares) of land. This represents less than 10 percent of the pixel size used (250 by 250-meter MODIS), meaning that some subtle land conversions were missed. The choice to use these data was necessitated by the frequent availability of MODIS NDVI data. The ability of Mann–Kendall and bfast in delineating areas of change in the MEE using this coarse imagery is important, as free finer RS imagery like Landsat is limited by both cloud cover across tropical regions and their coarse temporal resolutions. Looking forward, combining Sentinel-2 and Landsat imagery may help address this issue, but this will not resolve the problem for historical studies like this one.

The use of Mann–Kendall and bfast algorithms proved to be a valuable integration. Mann–Kendall performed well in mapping locations that had experienced significant change in the entire TS (2001–2018). Consistently greened and browned locations were mapped with high confidence (95% significance level) and similar patterns existed for both 16-day and average growing season periods: significant and consistent browning in southwestern MEE and greening elsewhere. The best and most accurate results were obtained using average growing season NDVI TS. Here, browned locations (especially around mountain boundaries and southwestern MEE) and greening elsewhere were clearly demarcated. This indicates that the growing season period, rather than the shorter 16-day period, is the best temporal scale for monitoring vegetation change in the MEE. bfast on the other hand performed well in mapping changed greenness locations for sequentially monitored periods from 2005 to 2018. This part of the analysis revealed only subtle changes in vegetation greenness in most locations, indicative of vegetation degradation rather than deforestation [70]. Importantly, areas with breakpoints across multiple years were effectively identified, a finding that would likely be omitted if using traditional post-classification change detection methods. These findings demonstrate that the process of vegetation greening and browning can be studied more thoroughly by fusing these two methods. Using Mann–Kendall and Sen's Slope, one can assess and quantify monotonic trends in their TS to get the general picture in the series. This can be complimented with an assessment of the temporal occurrence of significant breaks in the series using bfast. Using this integrated approach, vegetation greening and browning can be fully characterized and understood to provide more information for better decision making.

5.2. Sources of Uncertainty

Due to a lack of ground precipitation measurements, the present study did not validate the CHIRPS data over the MEE. The GHCN data [87] would potentially be used, but only one data point existed in the study area and was therefore deemed insufficient for this task. While this is a potential source of uncertainty, these data have been used extensively in similar studies. Moreover, CHIRPS data are based on ground station data since they are created through "smart interpolation" procedures that incorporate both satellite information and gauge data [53] thus making daily CHIRPS imprecise, and reliable only at dekad or higher aggregations. Validation of bfast results was also not possible due to a similar lack of data. Frequent, high-resolution imagery was not freely available to download and process, and the study relied on imagery from Google Earth Pro which had a lot of spatial and temporal gaps. While no accuracy statistics were calculated for these results, the results were interpreted based in part on data collected from the MEE field study.

6. Conclusions

The MEE of eastern Uganda and western Kenya was found to exhibit significant variability in vegetation dynamics and precipitation regimes. This variability was attributed to the existing LULC orientation especially in eastern MEE and climate change and variability. As such, it is highly probable that analysis of only a few images to ascertain MEE landscape change would yield inconsistent results. In this study, greening and browning in the MEE was examined using both TS trend and breakpoint analysis methods. The MEE had experienced significant and persistent greening and browning at different time scales and this change was attributed to both natural factors (including changing precipitation) and anthropogenic factors (especially the vegetation-to-cropland conversion). The southwestern MEE had consistently browned due to the conversion of the Namatala swamp to paddy rice farming and settlement. A lot of activity was also observed around the mountain forest boundary as people encroached and converted the forest LULC to agriculture and settlement. There were breakpoints in the vegetation greenness TS, particularly in the savanna and grassland land covers in northeastern MEE. The breakpoints were detected in each of the monitored years (2005–2018), but most of them were observed in 2013, 2007 and 2010 (greening) and 2009 and 2017 (browning). The study also concluded that MEE precipitation had significantly changed (increased) in the post-2000 era. More specifically, total precipitation significantly increased in 2006 and 2009–2010 with a consistently decreasing trend in between. We therefore concluded that these precipitation changes influenced significant greening and browning patterns observed in the same period. The greenness–precipitation relationship was weak in other periods as greening and browning changes were not strongly influenced by changing precipitation. This may be attributed to the complex nature of the MEE landscape and/or the spatial and temporal scale mismatch between MODIS NDVI and CHIRPS precipitation data. The integration of Mann–Kendall, Sen's slope and bfast proved useful in comprehensively characterizing recent changes in vegetation greenness within the MEE. Having a comprehensive description of vegetation change is an important first step, especially for such a variable landscape, to effect policy changes aimed at simultaneously conserving the environment and improving livelihoods that are dependent on it.

Author Contributions: Conceptualization, D.W. and N.J.M.; methodology, D.W., N.J.M. and K.M.D.; software, D.W.; validation, D.W. and K.M.D.; formal analysis, D.W.; investigation, D.W. and N.J.M.; resources, D.W. and N.J.M.; data curation, D.W.; writing—original draft preparation, D.W.; writing—review and editing, D.W., N.J.M. and K.M.D.; visualization, D.W. and N.J.M.; funding acquisition, D.W. and N.J.M.; supervision, N.J.M. All authors have read and agreed to the published version of the manuscript.

Funding: This research was supported by funds from the 2019 Graduate Office Fellowship (GOF) awarded by College of Social Science and Department of Geography, Environment, and Spatial Sciences at Michigan State University.

Acknowledgments: Multiple people made this publication a success and authors would like to express their gratitude especially to Eliud Akanga for your valuable support in the fieldwork portion of the study.

Conflicts of Interest: The authors declare no conflict of interest.

References

1. Gemitzi, A.; Banti, M.; Lakshmi, V. Vegetation greening trends in different land use types: Natural variability versus human-induced impacts in Greece. *Environ. Earth Sci.* **2019**, *78*, 1–10. [CrossRef]

2. Alavipanah, S.; Wegmann, M.; Qureshi, S.; Weng, Q.; Koellner, T. The role of vegetation in mitigating urban land surface temperatures: A case study of Munich, Germany during the warm season. *Sustainability* **2015**, *7*, 4689–4706. [CrossRef]

3. Forzieri, G.; Alkama, R.; Miralles, D.G.; Cescatti, A. Response to Comment on "Satellites reveal contrasting responses of regional climate to the widespread greening of Earth". *Science* **2018**, *360*, 1180–1184. [CrossRef] [PubMed]

4. Ballantyne, A.; Smith, W.; Anderegg, W.; Kauppi, P.; Sarmiento, J.; Tans, P.; Shevliakova, E.; Pan, Y.; Poulter, B.; Anav, A.; et al. Accelerating net terrestrial carbon uptake during the warming hiatus due to reduced respiration. *Nat. Clim. Chang.* **2017**, *7*, 148–152. [CrossRef]

5. Pan, N.; Feng, X.; Fu, B.; Wang, S.; Ji, F.; Pan, S. Increasing global vegetation browning hidden in overall vegetation greening: Insights from time-varying trends. *Remote Sens. Environ.* **2018**, *214*, 59–72. [CrossRef]

6. Landmann, T.; Dubovyk, O. Spatial analysis of human-induced vegetation productivity decline over eastern Africa using a decade (2001-2011) of medium resolution MODIS time-series data. *Int. J. Appl. Earth Obs. Geoinf.* **2014**, *33*, 76–82. [CrossRef]

7. Zhu, Z.; Piao, S.; Myneni, R.B.; Huang, M.; Zeng, Z.; Canadell, J.G.; Ciais, P.; Sitch, S.; Friedlingstein, P.; Arneth, A.; et al. Greening of the Earth and its drivers. *Nat. Clim. Chang.* **2016**, *6*, 791–795. [CrossRef]

8. Olsson, L.; Eklundh, L.; Ardö, J. A recent greening of the Sahel - Trends, patterns and potential causes. *J. Arid Environ.* **2005**, *63*, 556–566. [CrossRef]

9. Le Quéré, C.; Raupach, M.R.; Canadell, J.G.; Marland, G.; Bopp, L.; Ciais, P.; Conway, T.J.; Doney, S.C.; Feely, R.A.; Foster, P.; et al. Trends in the sources and sinks of carbon dioxide. *Nat. Geosci.* **2009**, *2*, 831–836. [CrossRef]

10. Mishra, N.B.; Mainali, K.P. Greening and browning of the Himalaya: Spatial patterns and the role of climatic change and human drivers. *Sci. Total Environ.* **2017**, *587–588*, 326–339. [CrossRef]

11. Fensholt, R.; Langanke, T.; Rasmussen, K.; Reenberg, A.; Prince, S.D.; Tucker, C.; Scholes, R.J.; Le, Q.B.; Bondeau, A.; Eastman, R.; et al. Greenness in semi-arid areas across the globe 1981-2007—An Earth Observing Satellite based analysis of trends and drivers. *Remote Sens. Environ.* **2012**, *121*, 144–158. [CrossRef]

12. Murthy, K.; Bagchi, S. Spatial patterns of long-term vegetation greening and browning are consistent across multiple scales: Implications for monitoring land degradation. *L. Degrad. Dev.* **2018**, *29*, 2485–2495. [CrossRef]

13. Chamaille-Jammes, S.; Fritz, H.; Murindagomo, F. Spatial patterns of the NDVI-rainfall relationship at the seasonal and interannual time scales in an African savanna. *Int. J. Remote Sens.* **2006**, *27*, 5185–5200. [CrossRef]

14. Tucker, C.J. Red and photographic infrared linear combinations for monitoring vegetation. *Remote Sens. Environ.* **1979**, *8*, 127–150. [CrossRef]

15. Vrieling, A.; de Beurs, K.M.; Brown, M.E. Variability of African farming systems from phenological analysis of NDVI time series. *Clim. Change* **2011**, *109*, 455–477. [CrossRef]

16. Verbesselt, J.; Hyndman, R.; Zeileis, A.; Culvenor, D. Phenological change detection while accounting for abrupt and gradual trends in satellite image time series. *Remote Sens. Environ.* **2010**, *114*, 2970–2980. [CrossRef]

17. Guan, Q.; Yang, L.; Pan, N.; Lin, J.; Xu, C.; Wang, F.; Liu, Z. Greening and browning of the Hexi Corridor in northwest China: Spatial patterns and responses to climatic variability and anthropogenic drivers. *Remote Sens.* **2018**, *10*, 1270. [CrossRef]

18. Emmett, K.D.; Renwick, K.M.; Poulter, B. Disentangling Climate and Disturbance Effects on Regional Vegetation Greening Trends. *Ecosystems* **2019**, *22*, 873–891. [CrossRef]

19. Malo, A.R.; Nicholson, S.E. A study of rainfall and vegetation dynamics in the African Sahel using normalized difference vegetation index. *J. Arid Environ.* **1990**, *19*, 1–24. [CrossRef]

20. Schmidt, H.; Gitelson, A. Temporal and spatial vegetation cover changes in Israeli transition zone: AVHRR-based assessment of rainfall impact. *Int. J. Remote Sens.* **2000**, *21*, 997–1010. [CrossRef]

21. Fabricante, I.; Oesterheld, M.; Paruelo, J.M. Annual and seasonal variation of NDVI explained by current and previous precipitation across Northern Patagonia. *J. Arid Environ.* **2009**, *73*, 745–753. [CrossRef]

22. Davenport, M.L.; Nicholson, S.E. On the relation between rainfall and the Normalized Difference Vegetation Index for diverse vegetation types in East Africa. *Int. J. Remote Sens.* **1993**, *14*, 2369–2389. [CrossRef]

23. Farrar, T.J.; Nicholson, S.E.; Lare, A.R. The influence of soil type on the relationships between NDVI, rainfall, and soil moisture in semiarid Botswana. I. NDVI response to rainfall. *Remote Sens. Environ.* **1994**, *50*, 107–120. [CrossRef]

24. Guzha, A.C.; Rufino, M.C.; Okoth, S.; Jacobs, S.; Nóbrega, R.L.B. Impacts of land use and land cover change on surface runoff, discharge and low flows: Evidence from East Africa. *J. Hydrol. Reg. Stud.* **2018**, *15*, 49–67. [CrossRef]

25. Maitima, J.M.; Mugatha, S.M.; Reid, R.S.; Gachimbi, L.N.; Majule, A.; Lyaruu, H.; Pomery, D.; Mathai, S.; Mugisha, S. The linkages between land use change, land degradation and biodiversity across East Africa. *African J. Agric. Res.* **2009**, *3*, 310–325.

26. Vrieling, A.; De Leeuw, J.; Said, M.Y. Length of growing period over africa: Variability and trends from 30 years of NDVI time series. *Remote Sens.* **2013**, *5*, 982–1000. [CrossRef]

27. Petursson, J.G.; Vedeld, P.; Sassen, M. An institutional analysis of deforestation processes in protected areas: The case of the transboundary Mt. Elgon, Uganda and Kenya. *For. Policy Econ.* **2013**, *26*, 22–33. [CrossRef]

28. Mugagga, F.; Kakembo, V.; Buyinza, M. Land use changes on the slopes of Mount Elgon and the implications for the occurrence of landslides. *Catena* **2012**, *90*, 39–46. [CrossRef]

29. Muhweezi, A.B.; Sikoyo, G.M.; Chemonges, M. Introducing a Transboundary Ecosystem Management Approach in the Mount Elgon Region. *Mt. Res. Dev.* **2007**, *27*, 215–219. [CrossRef]

30. EAC; UNEP; GRID-Arendal. *Sustainable Mountain Development in East Africa in a Changing Climate*; EAC: Arusha, Tanzania, 2016; ISBN 9788277011547.

31. Nakakaawa, C.; Moll, R.; Vedeld, P.; Sjaastad, E.; Cavanagh, J. Collaborative resource management and rural livelihoods around protected areas: A case study of Mount Elgon National Park, Uganda. *For. Policy Econ.* **2015**, *57*, 1–11. [CrossRef]

32. Bamutaze, Y.; Tenywa, M.M.; Majaliwa, M.; Vanacker, V.; Bagoora, F.; Magunda, M.; Obando, J.; Wasige, J.E. Infiltration characteristics of volcanic sloping soils on Mount Elgon, Eastern Uganda. *Catena* **2010**, *80*. [CrossRef]

33. Myhren, S.M. *Rural Livelihood and Forest Management in Mount Elgon, Kenya*; Norwegian University of Life Sciences: Ås, Norway, 2007.

34. Barasa, B.; Kakembo, V. *The Impact of Land Use/Cover Change on Soil Organic Carbon and Its Implication on Food Security and Climate Change Vulnerability on the Slopes of Mt. Elgon, Eastern Uganda*; Nelson Mandela Metropolitan University: Port Elizabeth, South Africa, 2013.

35. Mugagga, F.; Nagasha, B.; Barasa, B.; Buyinza, M. The Effect of Land Use on Carbon Stocks and Implications for Climate Variability on the Slopes of Mount Elgon, Eastern Uganda. *Int. J. Reg. Dev.* **2015**, *2*, 58. [CrossRef]

36. Sulla-Menashe, D.; Friedl, M.A. *User Guide to Collection 6 MODIS Land Cover (MCD12Q1 and MCD12C1) Product*; USGS: Reston, VA, USA, 2018.

37. Gómez, C.; White, J.C.; Wulder, M.A. Optical remotely sensed time series data for land cover classification: A review. *ISPRS J. Photogramm. Remote Sens.* **2016**, *116*, 55–72. [CrossRef]

38. Woodcock, C.E.; Allen, R.; Anderson, M.; Belward, A.; Bindschadler, R.; Cohen, W.; Gao, F.; Goward, S.N.; Helder, D.; Helmer, E.; et al. Free Access to Landsat Imagery. *Science* **2008**, *320*, 1011–1012. [CrossRef] [PubMed]

39. Badjana, H.M.; Helmschrot, J.; Selsam, P.; Wala, K.; Flugel, W.-A.; Afouda, A.; Akpagana, K. Land Cover Changes assessment using object-based image analysis in the Binah River watereshed (Togo and Benin). *Earth Sp. Sci.* **2015**, *2*, 403–416. [CrossRef]

40. Bradley, B.A.; Jacob, R.W.; Hermance, J.F.; Mustard, J.F. A curve fitting procedure to derive inter-annual phenologies from time series of noisy satellite NDVI data. *Remote Sens. Environ.* **2007**, *106*, 137–145. [CrossRef]

41. Hermance, J.F.; Jacob, R.W.; Bradley, B.A.; Mustard, J.F. Extracting phenological signals from multiyear AVHRR NDVI time series: Framework for applying high-order annual splines with roughness damping. *IEEE Trans. Geosci. Remote Sens.* **2007**, *45*, 3264–3276. [CrossRef]

42. Crist, E.P.; Cicone, R.C. A Physically-Based Transformation of Thematic Mapper Data-The TM Tasseled Cap. *IEEE Trans. Geosci. Remote Sens.* **1984**, *22*, 256–263. [CrossRef]

43. Anyamba, A.; Eastman, J.R. Interannual variability of ndvi over africa and its relation to el niño/southern oscillation. *Int. J. Remote Sens.* **1996**, *17*, 2533–2548. [CrossRef]

44. Mann, H.B. Nonparametric Tests Against Trend. *Econometrica* **1945**, *13*, 245–259. [CrossRef]

45. Lamchin, M.; Lee, W.K.; Jeon, S.W.; Wang, S.W.; Lim, C.H.; Song, C.; Sung, M. Long-term trend of and correlation between vegetation greenness and climate variables in Asia based on satellite data. *MethodsX* **2018**, *5*, 803–807. [CrossRef]

46. Khambhammettu, P. *Annual Groundwater Monitoring Report, Appendix D - Mann-Kendall Analysis for the Fort Ord Site*; HydroGeoLogic, Inc.: Monterey County, CA, USA, 2005.

47. Sen, P.K. Estimates of the Regression Coefficient Based on Kendall's Tau. *J. Am. Stat. Assoc.* **1968**, *63*, 1379–1389. [CrossRef]

48. Alcaraz-Segura, D.; Chuvieco, E.; Epstein, H.E.; Kasischke, E.S.; Trishchenko, A. Debating the greening vs. browning of the North American boreal forest: Differences between satellite datasets. *Glob. Chang. Biol.* **2010**, *16*, 760–770. [CrossRef]

49. Morrison, J.; Higginbottom, T.P.; Symeonakis, E.; Jones, M.J.; Omengo, F.; Walker, S.L.; Cain, B. Detecting vegetation change in response to confining elephants in forests using MODIS time-series and BFAST. *Remote Sens.* **2018**, *10*, 1075. [CrossRef]

50. Hamilton, A.C.; Perrott, R.A. A study of altitudinal zonation in the montane forest belt of Mt. Elgon, Kenya/Uganda. *Plant Ecol.* **1981**, *45*, 107–125. [CrossRef]

51. *The Struggle over Land in Africa: Conflicts, Politics and Change*; Anseeuw, W.; Alden, C. (Eds.) HSRC Press: Cape Town, South Africa, 2010; ISBN 9780796923226.

52. Doumenge, C.; Gilmour, D.; Pérez, M.R.; Blockhus, J. Tropical Montane Cloud Forests: Conservation Status and Management Issues. In *Tropical Montane Cloud Forests*; Springer: New York, NY, USA, 1995; pp. 24–37.

53. Funk, C.; Peterson, P.; Landsfeld, M.; Pedreros, D.; Verdin, J.; Shukla, S.; Husak, G.; Rowland, J.; Harrison, L.; Hoell, A.; et al. The climate hazards infrared precipitation with stations—A new environmental record for monitoring extremes. *Sci. Data* **2015**, *2*, 1–21. [CrossRef] [PubMed]

54. Okello, S.V.; Nyunja, R.O.; Netondo, G.W.; Onyango, J.C. Ethnobotanical study of medicinal plants used by the Sabaots of Mt. Elgon, Kenya. *African J. Tradit. Contemp. Altern. Med.* **2010**, *7*, 1–10. [CrossRef] [PubMed]

55. Musau, J.; Sang, J.; Gathenya, J.; Luedeling, E. Hydrological responses to climate change in Mt. Elgon watersheds. *J. Hydrol. Reg. Stud.* **2015**, *3*, 233–246. [CrossRef]

56. Didan, K. *MOD13Q1 MODIS/Terra Vegetation Indices 16-Day L3 Global 250m SIN Grid V006*; NASA EOSDIS Land Processes DAAC: Sioux Falls, SD, USA, 2015.

57. AppEEARS. *AppEEARS Team Application for Extracting and Exploring Analysis Ready Samples (AppEEARS)*; LP DAAC: Sioux Falls, SD, USA, 2019.

58. USGS Landsat Satellite Missions. Available online: https://www.usgs.gov/land-resources/nli/landsat/landsat-satellite-missions?qt-science_support_page_related_con=2#qt-science_support_page_related_con (accessed on 12 May 2020).

59. Hawinkel, P.; Thiery, W.; Lhermitte, S.; Swinnen, E.; Verbist, B.; Van Orshoven, J.; Muys, B. Vegetation response to precipitation variability in East Africa controlled by biogeographical factors. *J. Geophys. Res. Biogeosci.* **2016**, *121*, 2422–2444. [CrossRef]

60. Gorelick, N.; Hancher, M.; Dixon, M.; Ilyushchenko, S.; Thau, D.; Moore, R. Google Earth Engine: Planetary-scale geospatial analysis for everyone. *Remote Sens. Environ.* **2017**, *202*, 18–27. [CrossRef]

61. Georganos, S.; Abdi, A.M.; Tenenbaum, D.E.; Kalogirou, S. Examining the NDVI-rainfall relationship in the semi-arid Sahel using geographically weighted regression. *J. Arid Environ.* **2017**, *146*, 64–74. [CrossRef]

62. Chen, C.; Li, T.; Li, J.; Fu, W.; Wang, G. Vegetation Change Analyses Considering Climate Variables and Anthropogenic Variables in the Three-River Headwaters Region. *EPiC Ser. Eng.* **2018**, *3*, 419–427.

63. Muthoni, F.K.; Odongo, V.O.; Ochieng, J.; Mugalavai, E.M.; Mourice, S.K.; Hoesche-Zeledon, I.; Mwila, M.; Bekunda, M. Long-term spatial-temporal trends and variability of rainfall over Eastern and Southern Africa. *Theor. Appl. Climatol.* **2019**, *137*, 1869–1882. [CrossRef]

64. Qualtrics LLC. *Qualtrics*; Qualtrics: Provo, UT, USA, 2019.

65. R Core Team. *R: A Language and Environment for Statistical Computing*; R Core Team: Vienna, Austria, 2018.

66. De Beurs, K.M.; Henebry, G.M. A land surface phenology assessment of the northern polar regions using MODIS reflectance time series. *Can. J. Remote Sens.* **2010**, *36*, S87–S110. [CrossRef]

67. Google LLC. *Google Earth Pro*; Google LLC: Menlo Park, CA, USA, 2020.

68. Verbesselt, J.; Zeileis, A.; Herold, M. Near real-time disturbance detection using satellite image time series. *Remote Sens. Environ.* **2012**, *123*, 98–108. [CrossRef]

69. Smith, V.; Portillo-Quintero, C.; Sanchez-Azofeifa, A.; Hernandez-Stefanoni, J.L. Assessing the accuracy of detected breaks in Landsat time series as predictors of small scale deforestation in tropical dry forests of Mexico and Costa Rica. *Remote Sens. Environ.* **2019**, *221*, 707–721. [CrossRef]

70. DeVries, B.; Verbesselt, J.; Kooistra, L.; Herold, M. Robust monitoring of small-scale forest disturbances in a tropical montane forest using Landsat time series. *Remote Sens. Environ.* **2015**, *161*, 107–121. [CrossRef]

71. Verbesselt, J.; Herold, M.; Hyndman, R.; Zeileis, A.; Culvenor, D. A robust approach for phenological change detection within satellite image time series. In Proceedings of the 2011 6th International Workshop on the Analysis of Multi-Temporal Remote Sensing Images, Multi-Temp 2011 - Proceedings, Trento, Italy, 12–14 July 2011; pp. 41–44.

72. Verbesselt, J.; Hyndman, R.; Newnham, G.; Culvenor, D. Detecting trend and seasonal changes in satellite image time series. *Remote Sens. Environ.* **2010**, *114*, 106–115. [CrossRef]

73. Foody, G.M. Status of land cover classification accuracy assessment. *Remote Sens. Environ.* **2002**, *80*, 185–201. [CrossRef]

74. Kennedy, R.E.; Yang, Z.; Cohen, W.B. Detecting trends in forest disturbance and recovery using yearly Landsat time series: 1. LandTrendr - Temporal segmentation algorithms. *Remote Sens. Environ.* **2010**, *114*, 2897–2910. [CrossRef]

75. Ministry of Water and Environment Uganda. *Uganda Wetlands Atlas*; Ministry of Water and Environment Uganda: Kampala, Uganda, 2016; Volume II.

76. FEWS Net. *Kenya Food Security Brief*; USAID: Washington, DC, USA, 2013.

77. Getachew Tesfaye Ayehu, S.A. Land Suitability Analysis for Rice Production: A GIS Based Multi-Criteria Decision Approach. *Am. J. Geogr. Inf. Syst.* **2015**, *4*, 95–104.

78. UNEP. *Green Economy Sector Study on Agriculture in Kenya*; UNEP: Nairobi, Kenya, 2015.

79. Salami, A.O.; Kamara, A.B.; Brixiova, Z. Smallholder Agriculture in East Africa: Trends, Constraints and Opportunities. In *Working Paper Series*; African Development Bank Group: Abidjan, Côte d'Ivoire, 2010.

80. Wanyama, D.; Koti, F. *A Spatial Analysis of Climate Change Effects on Maize Productivity in Kenya*; University of North Alabama: Florence, AL, USA, 2017.

81. Wanyama, D.; Mighty, M.; Sim, S.; Koti, F. A spatial assessment of land suitability for maize farming in Kenya. *Geocarto Int.* **2019**. [CrossRef]

82. NAAIAP; KARI. *Soil Suitability Evaluation for Maize Production in Kenya*; Ministry of Agriculture Kenya: Nairobi, Kenya, 2014.

83. Kotikot, S.M.; Flores, A.; Griffin, R.E.; Sedah, A.; Nyaga, J.; Mugo, R.; Limaye, A.; Irwin, D.E. Mapping threats to agriculture in East Africa: Performance of MODIS derived LST for frost identification in Kenya's tea plantations. *Int. J. Appl. Earth Obs. Geoinf.* **2018**, *72*, 131–139. [CrossRef]

84. Li, C.; Chai, Y.; Yang, L.; Li, H. Spatio-temporal distribution of flood disasters and analysis of influencing factors in Africa. *Nat. Hazards* **2016**, *82*, 721–731. [CrossRef]

85. Ayugi, B.; Tan, G.; Rouyun, N.; Zeyao, D.; Ojara, M.; Mumo, L.; Babaousmail, H.; Ongoma, V. Evaluation of meteorological drought and flood scenarios over Kenya, East Africa. *Atmosphere* **2020**, *11*, 307. [CrossRef]

86. BirdLife International Important Bird and Biodiversity Areas (IBAs). Available online: https://www.birdlife.org/worldwide/programme-additional-info/important-bird-and-biodiversity-areas-ibas (accessed on 1 May 2020).

87. National Centers for Environmental Information. *Global Historical Climatology Network—Daily (GHCN-Daily)*; National Centers for Environmental Information: Asheville, NC, USA, 2012.

Article

Earth Observation-Based Detectability of the Effects of Land Management Programmes to Counter Land Degradation: A Case Study from the Highlands of the Ethiopian Plateau

Esther Barvels * and Rasmus Fensholt

Department of Geosciences and Natural Resource Management, University of Copenhagen, DK-1350 Copenhagen, Denmark; rf@ign.ku.dk
* Correspondence: lbm692@alumni.ku.dk

Abstract: In Ethiopia land degradation through soil erosion is of major concern. Land degradation mainly results from heavy rainfall events and droughts and is associated with a loss of vegetation and a reduction in soil fertility. To counteract land degradation in Ethiopia, initiatives such as the Sustainable Land Management Programme (SLMP) have been implemented. As vegetation condition is a key indicator of land degradation, this study used satellite remote sensing spatiotemporal trend analysis to examine patterns of vegetation between 2002 and 2018 in degraded land areas and studied the associated climate-related and human-induced factors, potentially through interventions of the SLMP. Due to the heterogeneity of the landscapes of the highlands of the Ethiopian Plateau and the small spatial scale at which human-induced changes take place, this study explored the value of using 30 m resolution Landsat data as the basis for time series analysis. The analysis combined Landsat derived Normalised Difference Vegetation Index (NDVI) data with Climate Hazards group Infrared Precipitation with Stations (CHIRPS) derived rainfall estimates and used Theil-Sen regression, Mann-Kendall trend test and LandTrendr to detect changes in NDVI, rainfall and rain-use efficiency. Ordinary Least Squares (OLS) regression analysis was used to relate changes in vegetation directly to SLMP infrastructure. The key findings of the study are a general trend shift from browning between 2002 and 2010 to greening between 2011 and 2018 along with an overall greening trend between 2002 and 2018. Significant improvements in vegetation condition due to human interventions were found only at a small scale, mainly on degraded hillside locations, along streams or in areas affected by gully erosion. Visual inspections (based on Google Earth) and OLS regression results provide evidence that these can partly be attributed to SLMP interventions. Even from the use of detailed Landsat time series analysis, this study underlines the challenge and limitations to remotely sensed detection of changes in vegetation condition caused by land management interventions aiming at countering land degradation.

Keywords: developing countries; Google Earth Engine; land degradation; Landsat time series analysis; semi-arid areas; sustainable land management programmes

Citation: Barvels, E.; Fensholt, R. Earth Observation-Based Detectability of the Effects of Land Management Programmes to Counter Land Degradation: A Case Study from the Highlands of the Ethiopian Plateau. *Remote Sens.* **2021**, *13*, 1297. https://doi.org/10.3390/rs13071297

Academic Editor: Elias Symeonakis

Received: 30 January 2021
Accepted: 9 March 2021
Published: 29 March 2021

Publisher's Note: MDPI stays neutral with regard to jurisdictional claims in published maps and institutional affiliations.

1. Introduction

Degradation of land and soil affects approximately one third of the global land area that is used for agriculture [1], involving livelihoods of more than 1.5 million people [2]. Land degradation is of particular concern in developing countries, as this issue poses a threat to food security for a large number of poor people and to local economic activities [3]. In the United Nation's Convention to Combat Desertification (UNCCD) (art. 1f) land degradation is defined as "reduction or loss, in arid, semi-arid and dry sub-humid areas, of the biological or economic productivity and complexity of rainfed cropland, irrigated cropland, or range, pasture, forest and woodlands resulting from land uses or from a process or combination of processes, including processes arising from human activities and habitation patterns". Among those processes are soil erosion or long-term loss of natural vegetation [4]. In Ethiopia, land degradation results mainly from soil erosion by water [5] and occurs particularly in the Ethiopian highlands which are inhabited by 88% of the national population, cover 60% of the national livestock resources and encompass 90% of the area suitable for agriculture [6]. The country's natural physical conditions are one of the underlying drivers of land degradation. Ethiopia has always been prone to soil erosion and droughts due to high rainfall variability, which causes reduced vegetation cover in dry years and soil loss in subsequent wet years [7]. In Ethiopia's highlands, soil erosion is facilitated by steep terrain with slopes in excess of 30% [5]. Moreover, population pressure, increasing livestock (and with it deforestation), overgrazing (due to uncontrolled free grazing) and the expansion of agricultural fields into marginal land are underlying drivers of soil erosion [5,8]. The shortage of fertile cropland led to a shifting of cattle and livestock grazing activities to areas that are specifically vulnerable to soil erosion such as deforested, ecologically fragile hillsides with steep slopes. Gully formation, the removal of soil along drainage lines (channels) by surface water runoff, is one of the apparent consequences of soil erosion in Ethiopia [8].

The restoration of degraded land and soil, the implementation of sustainable land management (SLM) and resilient agricultural practices are targeted in the United Nation's Sustainable Development Goals (SDG). To address in particular target 15.3 which aims to combat desertification and restore degraded and soil, UNCCD adopted the Land Degradation Neutrality (LDN) Target Setting Programme. Achieving LDN will also contribute to reaching other SDGs including those on poverty reduction, food and water [9]. To counter desertification and land degradation UNCCD and affected developing countries have set voluntary targets and implemented National Action Programmes that are supported by international cooperation, including financial and technical resources [4]. In this context, developing countries have implemented land management and land restoration projects on both national and local levels in collaboration with multilateral and bilateral development partners. To rehabilitate degraded landscapes and scale up SLM in Ethiopia, the Ethiopian government launched the Sustainable Land Management Programme (SLMP) in 2009 in collaboration with a range of international donors including the World Bank and the German Development Bank (KfW) [10]. The impact of Sustainable Land Management (SLM) in Ethiopia has been studied through runoff and soil loss measurements [11] and analyses within the economics field, for instance by examining household data to assess the effect on crop yields [12]. A study by Ali et al. used earth observation derived vegetation indices to estimate the impact of SLM in a single watershed in Ethiopia [13].

Monitoring of land surface dynamics, such as land degradation ('browning') or land recovery ('greening') is widely done by implementing earth observation time series analysis. A plethora of studies have examined long term trends by applying ordinary least square (OLS) linear regression models, e.g., regressing vegetation indices with time, based on high temporal resolution data such as from Advanced Very High Resolution Radiometer onboard National Oceanic and Atmospheric Administration (AVHRR-NOAA) or from Moderate Resolution Imaging Spectroradiometer (MODIS). For the Sahel, linear trends of yearly NDVI anomalies derived from AVHRR [14] or linear trends of the seasonal NDVI amplitude and integral [15] have been examined to monitor vegetation. MODIS NDVI data have been used to detect land degradation and regeneration processes in the Sahel [16], Mongolia [17] and Ethiopia [18].

Studies of semi-arid areas [19–22] and of Ethiopia [23] have demonstrated a strong relationship between rainfall and NDVI. Therefore, methods have been developed to disentangle rainfall-related effects from human-induced effects on land degradation. The rain-use efficiency (RUE), defined as the ratio of above-ground net primary production (ANPP) to annual precipitation [24], has been used to detect non-precipitation related land degradation. The basic assumption involved in the use of RUE is the existence of a constant linear relationship between vegetation productivity (or ANPP) and precipitation in areas where land is not affected by human-induced degradation [25]. By normalising for the effect of interannual rainfall variability on ANPP, human-induced changes can then be singled out [26]. RUE has been used as a measure in several studies, e.g., of the Sahel [25,27], South Africa [19], global drylands [2] and Northern Eurasia [26]. Furthermore, residual trend analysis, i.e., analysing the residuals from a NDVI-rainfall regression model, was found to be effective in disentangling the climate effects from human-induced land degradation [28,29]. In areas with steep terrain land degradation can be driven by climate effects such as high rainfall variability. Hermans-Neumann et al. combined NPP trends, precipitation variability and census data to identify areas in Ethiopia where high in-migration is coupled with land degradation, proposing the latter is likely occurred due to human activities [18].

Since the opening of the Landsat archive by the United States Geological Survey (USGS) in 2008, an increasing amount of studies that exploit medium/high resolution data for time series analysis has been published [30,31]. In parallel, new change detection methods that not only account for linear (in this case gradual trends), but also for abrupt occurrences by separating time series into individual segments, have been developed. Amongst those are Landsat-based detection of Trends in Disturbance and Recovery (LandTrendr) and Breaks For Additive Seasonal and Trend (BFAST) which have been widely used for vegetation monitoring. BFAST has for example been used to detect gradual and abrupt changes in NDVI and rain-use efficiency [26,32–34] and water-use efficiency [35]. LandTrendr has been widely used for monitoring forest disturbances (fire or stand clearing) and forest regrowth [36–38], forest biomass [39] and for agricultural and land abandonment mapping [40].

The aim of this study was to temporally and spatially analyse vegetation dynamics in degraded land areas in Ethiopia between 2002 and 2018 in relation to management programmes implemented to counter land degradation. Due the heterogeneity of the landscapes of the areas examined in this study and the fact that human-induced changes are expected to take place at a small spatial scale, this study aimed at exploring the value of using medium resolution Landsat data derived from different sensors for trend analysis at a spatial and temporal scale compatible with the scale of SLMP interventions. The investigated areas had gone through interventions aiming at avoiding further land degradation and increasing vegetation cover. In this context, associated human-induced and climate-related factors of land degradation and land recovery were examined. The specific objectives of this study were:

1. The examination of spatiotemporal vegetation trends using Landsat time series and to analyse their forcing mechanisms (climate-related vs. human-induced).
2. The assessment of the detectability of the impact from typical SLMP interventions on vegetation conditions from the use of relevant remote sensing data sources available at no costs.

2. Materials and Methods

2.1. Study Area

The study area consists of 21 major watersheds which are distributed in three different zones of Ethiopia, in Amhara, Oromia and Tigray, and have mean altitudes between 1200 and 3100 m (Figure 1), mean slopes up to 14.4 degrees and mean annual rainfall between approx. 600 and 1900 mm per watershed (calculations based on CHIRPS data). The watersheds are located in semi-arid and sub-humid agro-ecological zones where temperate to cool climate prevails and are surrounded by low-lying tropical warm to hot savannas and semidesert regions [18,41]. They are characterised by heterogeneous landscapes, with croplands and grasslands as the most dominant land cover and hillsides, which have been degraded and closed for farming and grazing. Agriculture is dominated by small-scale subsistence mixed farming systems, i.e., crop production mixed with livestock rearing activities [41]. Crops are mainly grown during the wet period from March through September with harvest taking place mostly in October to December [18].

The watersheds constitute intervention areas of the Sustainable Land Management Programme, a multi-donor supported project, first implemented by the Government of Ethiopia in 2009 and phased out in 2019. SLMP's overall goal was to "reduce land degradation and improve land productivity in selected watersheds in targeted regions in Ethiopia" [10]. It's first component, watershed and landscape management, aimed at reforesting and afforesting degraded communal land, increasing agricultural and livestock productivity, reducing carbon emission, building climate resilience and increasing water availability. To achieve these goals, activities such as hillside communal land treatment, including the prohibition of free grazing, gully rehabilitation and cropland treatment using biophysical measures, promoting agro-forestry and fodder production, and the construction of water harvesting structures were supported in the watersheds [10].

The 21 major watersheds each comprise between 6 to 20 micro-watersheds (Figure 1) with a total of 314 micro-watersheds and an average area of 7 km^2. 220 micro-watersheds (1541 km^2) received SLMP support from 2011 to 2019 by KfW with the technical assistance of the German Agency for International Coorporation (GIZ). In the following, the supported micro-watersheds are referred to as treatment areas while the remaining 94 micro-watersheds (543 km^2) that did not receive any support are referred to as control areas.

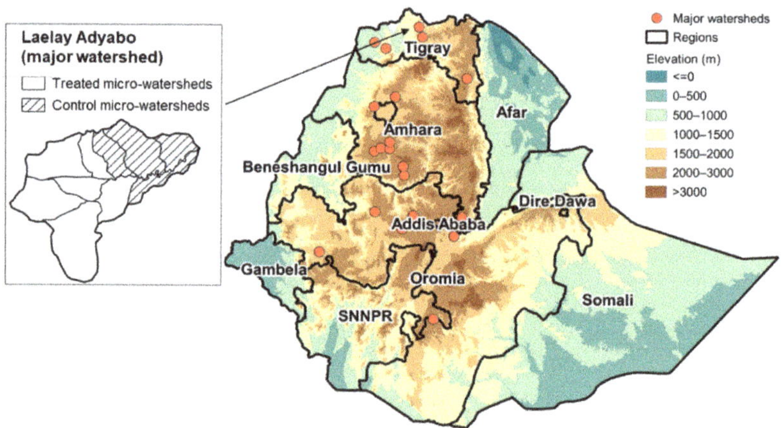

Figure 1. Overview map of the location of the study areas (i.e., major watersheds) and elevation.

2.2. Data

Landsat Collection 1 atmospherically corrected Surface Reflectance (SR) Tier 1 products were used for the period 2001–2019 including three different sensor systems: Landsat 5 Thematic Mapper (TM) for the epoch 2001–2012, Landsat 7 Enhanced Thematic Mapper Plus (ETM+) for the epoch 2001–2019 and Landsat 8 Operational Land Imager (OLI) for the epoch 2013–2019. Due to the failure of the Landsat-7 ETM+ Scan Line Corrector (SLC) in 2003, the ETM+ data are reduced by about 22% in each scene [42]. Rainfall estimates were derived from Climate Hazards group Infrared Precipitation with Stations (CHIRPS) data. The product is resampled to a spatial resolution of 0.05 degrees [43]. Following Funk et al. CHIRPS data have been largely used to examine rainfall trends and drought patterns in Ethiopia. The dataset is affected by uncertainties due to the inverse distance weighting function that is used for the blending procedure [43].

Polygon shapefiles for the micro-watersheds were provided by GIZ. Furthermore, between 2012 and 2018, the GFA Consulting Group collected georeferenced location data of soil and water conservation (SWC) measures that were implemented through SLMP to monitor the progress in the treated micro-watersheds. These data were used to relate vegetation development directly to SWC measure locations. The types of measures included in the dataset represent combinations of physical SWC constructions and biological activities (e.g., planting). Personal communication with SWC experts and project leaders during field visits in 2019 revealed that two types of measures included in the dataset, hillside terraces and check dams, should have a direct impact on the surrounding vegetation cover, within a radius of approx. 500–1500 m. See Appendix A, Table A1 for details regarding the purpose and the number of geolocations available in the dataset.

2.3. Methods

All remote sensing data were acquired in Google Earth Engine (GEE), a cloud-based platform for geospatial data processing that stores a large repository of publicly available data [44]. Data processing and analysis were conducted using Python packages such as NumPy and Rasterio and the GEE client library (Figure 2).

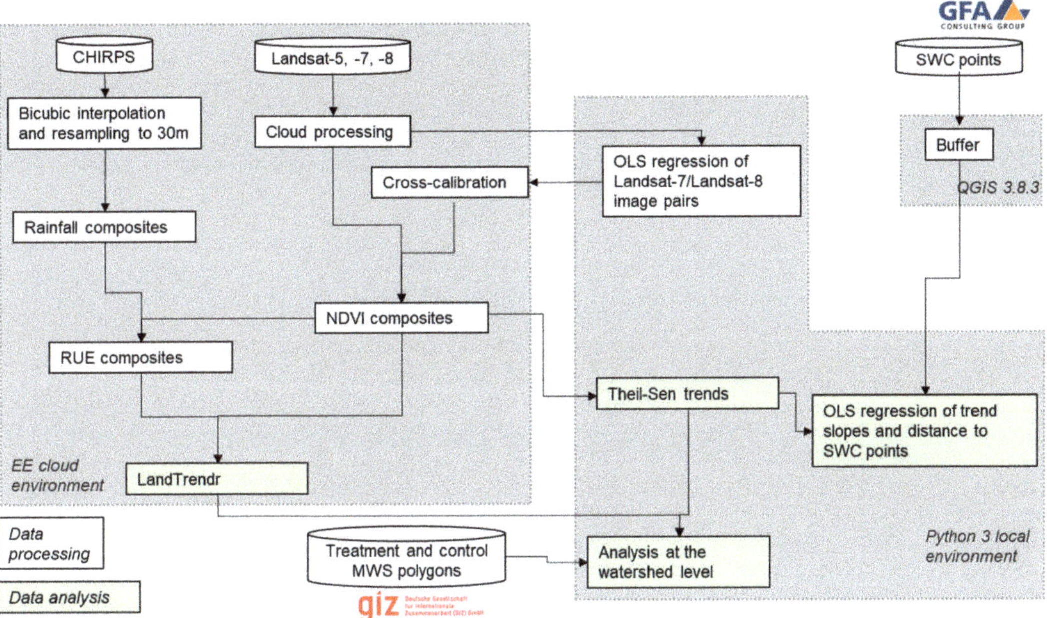

Figure 2. Workflow of the overall steps of the methodology.

2.3.1. Pre-Processing

For both cloud and cloud shadow masking, the C Function of Mask (CFMask), provided by USGS as pixel quality band as part of the Landsat products [45] was utilised. Visual inspection showed that using only CFMask proved to be insufficient. Therefore, clouds were additionally processed by the Google cloud score algorithm and cloud shadows by the Temporal Dark Outlier Mask (TDOM) [46]. Remote sensing analyses that integrate different sensor data require cross-calibration of the different datasets to ensure consistency [47–49]. As OLI spectral bands widths are narrower compared to ETM/ETM+, OLI NDVI values are on average higher [47] and it was therefore necessary to adjust OLI to ETM/ETM+ NDVI values before temporally aggregating the data. Transformation functions were developed using ordinary least squares (OLS) regression:

$$NDVIETM+ = a \times NDVIOLI + b \tag{1}$$

For this purpose, OLI and ETM+ NDVI images were paired based on the closest available dates. As OLI and ETM+ share the same orbit offset by 8 days, the western and eastern side of a sensor acquisition are overlapped by the eastern and western sides, respectively of the other sensor [47]. In these areas where the sensor acquisitions spatially overlap, image pairs are available with a temporal separation of only one day. To optimise the inter-comparison, pixels used to produce the OLS models (Figure A1) were therefore extracted for the study area where overlap existed. To minimise problems related to changing surface states and conditions such as different crop cycle stages [47], pairing was done during the dry season (November to February), i.e., for two seasons, 2013/14 and 2017/18, for an area including a large range of vegetation densities (including evergreen vegetation).

NDVI has been used extensively for analysis of dryland vegetation [50] as NDVI saturation that can occur in densely vegetated areas is rarely a concern in drylands [51]. NDVI was here used as a proxy for the vegetation condition during the period of analysis. Annual NDVI maximum value composites (MVC) were produced based on all cloud-free available pixels during the period August through October, which represents in the end of the growing season. As image coverage was insufficient, a three-year (+1/−1 year) moving maximum window was used to fill data gaps (Figure A2). Annual rainfall composites were computed using the seasonal rainfall sum (March to September). As agriculture is primarily rainfed and crop productivity highly dependent on rainfall (MOFED, 2002) [52], the rain-use efficiency (RUE) was used as a proxy for assessing non-climate related changes in vegetation conditions [25,26] inherent to the implementation of the SLMP activities to counter land degradation. As vegetation productivity in the study area is predominantly determined by seasonal rainfall, RUE was calculated as the ratio of the maximum NDVI, as an approximation of ANPP, and the seasonal rainfall sum. For this purpose, the rainfall composites were resampled from 0.05 degrees to Landsat resolution of 30 m using bicubic interpolation.

2.3.2. Theil-Sen Regression and Mann-Kendall (MK) Trend Test

The temporal development of NDVI was used as an indicator of land degradation ('browning') and land recovery ('greening'). Spatiotemporal patterns of NDVI and rainfall were examined using the non-parametric Theil-Sen median slope [53] to analyse changes in vegetation condition (climate-related vs. human-induced). The Mann–Kendall (MK) test [54] was applied to evaluate NDVI trends at the 99% ($p < 0.01$) and rainfall trends at the 95% ($p < 0.05$) significance level. A stricter criterion ($p < 0.01$) for NDVI trends was applied due to the temporal smoothing of the Landsat-based NDVI time series. Pixel-wise slope differences were calculated between the two periods of 2002–2010 and 2011–2018. The split of the two periods is determined by the timing of the SLMP implementation and length of time series.

Aggregated NDVI trends were calculated for treatment and control areas using the median. This was done for the total study area ("regional" scale) as well as for each major watershed that comprises both treatment and control micro-watersheds ("local" scale). To assess the effects of treatments, i.e., SLMP interventions, the significance of the difference in the distribution of per-pixel-NDVI trends between the two sample groups was evaluated using the Mann-Whitney U test, which is a non-parametric test for independent samples, non-normally distributed data and different sample sizes [55]. Spatial patterns were assessed by inspecting trend maps with the additional use of multi-temporal Google Earth VHR images.

In order to investigate the spatiotemporal relationship between vegetation and rainfall, the agreement of NDVI and rainfall trend directions was computed on a pixel-basis for each period. This method is based on a model which, following Horion et al. [26], interprets the nature of changes in ecosystem functioning based on the combination of growing season vegetation and rainfall trends. A decrease in growing season vegetation despite an increase in precipitation or vice versa is likely to be caused by human activities. In contrast, trend combinations with the same direction of change are likely to be caused by climate (see Figure A3 for further explanation). The stronger the magnitudes of change in a given direction, the more likely the cause attribution [26].

2.3.3. LandTrendr

In addition to the trend analysis based on fixed time periods (Section 2.3.2), we conducted an analysis based on the LandTrendr approach in GEE [56]. This was done to further explore the importance of analysing land degradation and SLMP interventions at the level of Landsat pixel resolution rather than at coarser spatial resolution (e.g., MODIS predominantly being used for time series analysis) where subtle vegetation changes at local scale is likely to go unnoticed. The implementation of SLMP interventions was conducted at different times during the eight-year epoch 2011–2018 in the different micro-watersheds. Therefore, human-induced changes through SLMP occurred presumably within a time period shorter than eight years. Horion et al. argue that abrupt changes in RUE can indicate significant changes in ecosystem response to precipitation through human activities [26]. To identify trends with a shorter duration than eight years and to obtain more information about the types and timing of the changes, LandTrendr was applied to NDVI and RUE composites. The algorithm was used to fit a model for the period 2002–2018 with a maximum number of three segments (for the sake of simplicity) and a confidence interval of 95% ($p \leq 0.05$) determining the significance of the fitted segments. For each significant segment, the algorithm returns the magnitude, duration and rate of change as well as the start year in which the segment was detected, the end year and the corresponding NDVI value defined by the identified vertices (Figure A4). The Pearson's correlation coefficient (r) was used to mask pixels where RUE correlated with rainfall over the overall period 2002–2018 using a confidence interval of 95% ($p < 0.05$). This was done, as the use of satellite-based RUE time series to identify non-precipitation related land degradation/recovery is problematic for pixels where RUE remains correlated with NDVI, as this suggests NDVI changes still to be controlled by changes in precipitation [25,57,58].

2.3.4. Effect of Soil and Water Conservation (SWC) Measures on Vegetation Trends

OLS regression was used to estimate the effect of SWC measures on vegetation trends where the trend represents the dependent variable and the distance to the SWC points the explanatory variable. To investigate the influence of SWC distance on the trends, buffers were created based on the geolocations of the different SWC types and implemented based on three different sizes: buffers with a 250 m and 500 m radius, both with a zone width of 50 m (Figure A5), and buffers with a 1000 m radius and a zone width of 100 m. Within each zone the trend results were aggregated in two ways; first, by using the median of all Theil-Sen trends and second, by using the proportion of the significant Theil-Sen as well as LandTrendr increases. The aggregated results were regressed against the corresponding distance value of the zone.

3. Results

MK trend test revealed that rainfall trends were not significant in any of the sub-periods apart from 0.3% of the study area with increasing trends in 2002–2010 ($p < 0.05$). Over the entire study period 2002–2018, monotonic increasing trends were found for 17% and decreasing trends for 1.3% of the total area. When considering all rainfall trends, increasing trends were observed in 91% of the entire study area for the sub-period 2002–2010, 54% for the sub-period 2011–2018 and 78% for the entire study period 2002–2018.

The increasing rainfall trends during 2002–2010 had a median annual change of 15.2 mm and occurred with mainly decreasing NDVI trends (Figure 3a). During 2011–2018 rainfall experienced almost no or little changes (median annual change of 1.5 mm) while NDVI was predominately increasing (Figure 3b). Over the entire study period 2002–2018, minimal increases in rainfall with a median annual change of 4.2 mm occurred with increases in NDVI (Figure 3c). When considering only significant rainfall trends for this period (Figure 3d), the agreement shows a more pronounced pattern of positive significant monotonic NDVI and rainfall increases with an annual change rate of 15.8 mm per year.

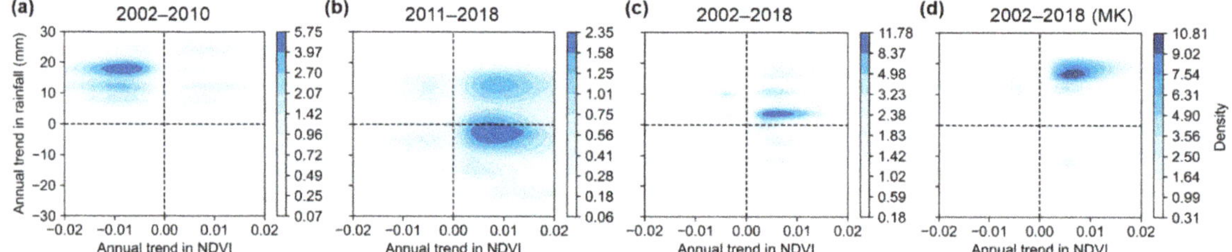

Figure 3. Spatial agreement of all annual rainfall trends and significant NDVI trends ($p < 0.01$) as density plots for the total study area for (**a**) the period 2002–2010, (**b**) 2011–2018 and (**c**) 2002–2018, and for (**d**) 2002–2018 with significant rainfall trends ($p < 0.05$).

3.1. Treatment and Control Areas

3.1.1. Theil-Sen Trends

In the period 2002–2010 the median trend in NDVI was negative with an annual decrease of -0.0065, while in 2011–2018 NDVI annually increased by 0.009 (Figure 4a). The proportion of pixels where significant NDVI trends occurred in both periods accounted for 2.6% (treatment) and 2.8% (control). The distributions of the trend differences of these pixels are left-skewed for both treatment and control areas and have a median of 0.019 (treatment) and 0.018 (control) (Figure 4b).

The most common trend was a shift from negative to positive (Figure 4c). The proportion of significant negative-positive trends was slightly higher for treatment areas (1.84%) than for control areas (1.79%) (Figure 4c). The overall trends (all trend types included), the negative-positive trends, and trends with the same sign in both sub-periods all had a positive median trend (Figure 4d). The median trends were similar for both sample groups.

The results of the individual watersheds reveal that cases where treatment areas have larger trends than control areas dominated (Table A2). When treatment and control microwatersheds from all major watersheds were treated as two large sample groups, the overall NDVI trend was larger for treatment (0.019) than for control (0.0183).

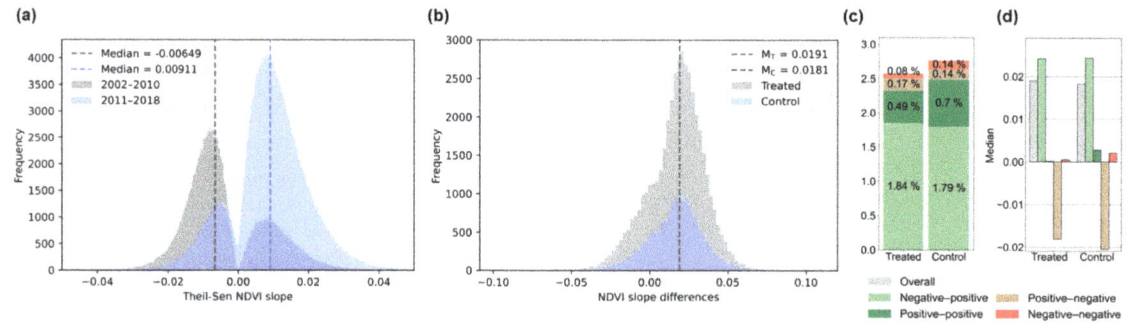

Figure 4. Theil-Sen NDVI trends. (**a**) Distribution of statistically significant ($p < 0.01$) trends for both sub-periods for the total study area; (**b**) Distribution of NDVI trend differences (slope of 2011–2018 minus slope of 2002–2010) for treatment and control areas; (**c**) Proportion of each type of change for all significant pixels and (**d**) Median of NDVI trend differences (slope of 2011–2018 minus slope of 2002–2010) for the overall trend and for each type of change constellation.

3.1.2. LandTrendr

A LandTrendr-based approach was used to refine the Theil-Sen trend analysis that was based on two fixed time periods defined by the major scheme of SLMP support. LandTrendr detected for 46% of both treatment and control areas statistically significant

($p \leq 0.05$) changes in NDVI for the entire study period 2002–2018, that were subsequently analysed as a function of change types. For 92% of the study area pixels showed negative correlation of RUE and rainfall and were therefore not included in the change type analysis (Section 2.3.3). The remaining significant RUE changes (that were not correlated with rainfall changes) accounted for 2% of treatment and 3% of control areas.

The most frequent NDVI change type (amongst the 46% of statistically significant pixels) consisted of one decreasing followed by an increasing trend for 29 and 28% of all significant pixels in treatment and control areas, respectively (Figure 5a). The second most frequent change type were two consecutive increasing trends (24 and 26%, respectively). For RUE, the most common trends were of the same type as for NDVI, however with a larger share of two consecutive increasing trends (44 and 42%, respectively).

To examine the timing of detected trend segments for treatment and control areas, the onset of the decreasing and increasing trend segments with the greatest rate were extracted from each pixel (i.e., from each trend sequence) and aggregated in two groups, respectively. Generally, treatment and control areas showed the same pattern in the timing of trends without any pronounced differences. For both sample groups, the timing of the onset of approx. 90% of both NDVI and RUE decreasing trends was in 2002 (Figure 5b). In contrast, only approx. 40% of the NDVI increasing trends for both groups occurred in this year. For RUE increasing trends the proportions amounted to 53% (treatment) and 51% (control). The remaining NDVI and RUE recoveries occurred proportionally more equally distributed over the years after 2002 (2–10% in the remaining years of the time series).

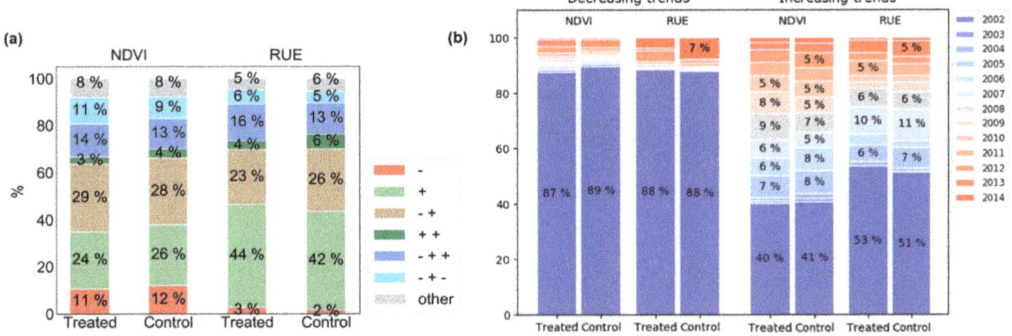

Figure 5. LandTrendr results for treatment and control areas, respectively; (**a**) Change types detected by LandTrendr with "+" indicating increasing and "-"decreasing trends. As the change types resulted from fitting either one, two or three segments into the time series, different combinations of trends existed. For example, while "-" indicates a decreasing trend over the entire study period, "- + +" indicates one decreasing trend followed by two increasing trends; (**b**) Timing of NDVI and RUE trends.

3.2. Visual Inspections of Trends Using Google Earth

Google Earth imagery was used to inspect all 21 major watersheds and showed that the limited amount of significant RUE trends to a wide extent occurred in areas where tree plantations have been established extensively. The watershed of Banja was selected as an exemplary case of illustration (Figure 6) to link interventions on the ground with the trends captured by the remote sensing time series data. Here, the positive trend differences represent pixels where trees have been planted between 2011 and 2018 and negative trend differences were mostly found in croplands, while grazing land showed no change in NDVI (Figure 6a). NDVI recoveries (an increasing trend segment) detected by LandTrendr were observed to be more dispersed in time (Figure 6c) and space (Figure 6d) as compared the RUE (Figure 6e). The higher uniformity of the RUE results at the pixel level can be explained by the coarser resolution of the CHIRPS rainfall data used to compute RUE time series at the scale of Landsat NDVI data. The strongest changes in RUE were therefore more likely to be found in the same year for different pixels as compared to changes in NDVI. RUE recoveries were mostly detected between 2013 and 2014 (Figure 6c) and represent pixels where trees have been grown on former cropland or grazing land (Figure 6e). This is exemplified in the time series (Figure 6b) at the point of interest (POI) where LandTrendr detected the strongest NDVI recovery in 2012 that continued until 2017 with a rate of 0.07 per year. The rather high NDVI and low rainfall in 2015 led to high RUE in this year; hence the detection of a positive RUE trend in 2013 with a rate of 0.09 per year. The VHR images show that while agricultural fields and grazing lands were largely unterraced in 2005 (Figure 6f), benches of terraces were fully established in 2013 (Figure 6g). From this year on, tree cover expanded until 2020 (Figure 6h).

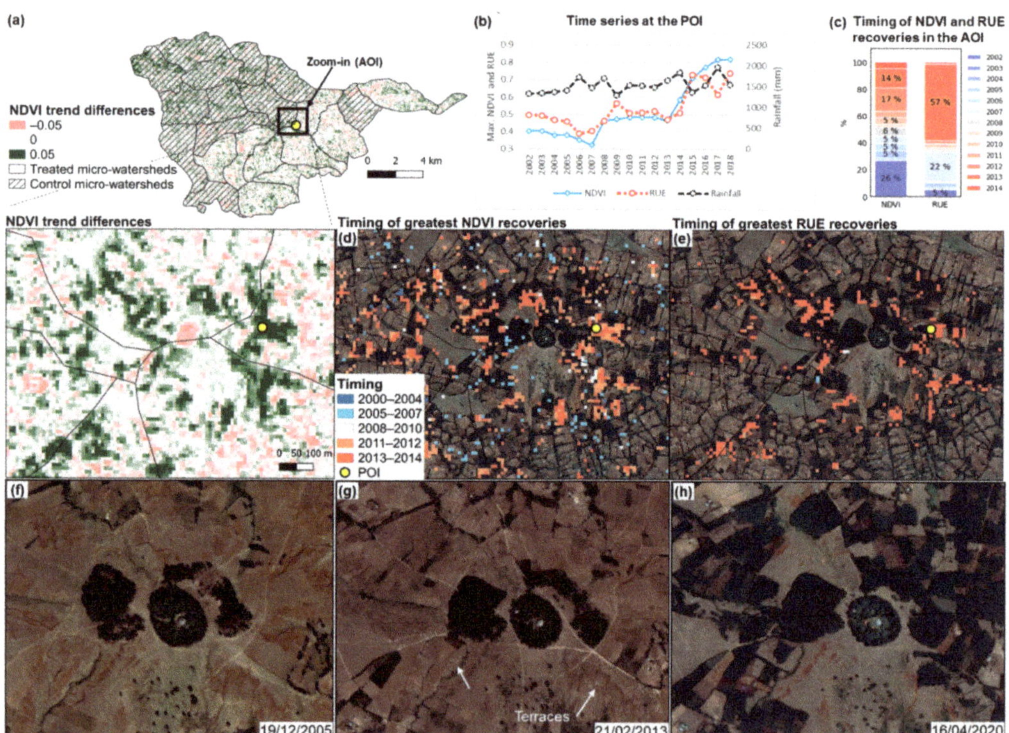

Figure 6. Watershed of Banja. (**a**) NDVI trend differences (slope of 2011–2018 minus slope of 2002–2010); (**b**) Time series at the POI; (**c**) Timing of the largest NDVI and RUE recoveries in the AOI; (**d**) Spatial distribution of the timing of the largest NDVI recoveries and of (**e**) RUE recoveries in the AOI; VHR images showing (**f**) mainly unterraced hillside in 2005, (**g**) the establishment of terraces in 2013 and (**h**) expanded tree cover area in 2020.

Furthermore, an inspection of Google Earth imagery from the watershed Yilmana Densa (Figure 7a) revealed that positive vegetation trends occurred mainly outside agricultural fields. Generally, the timing of NDVI recoveries in this watershed was equally distributed over the entire study period while RUE recoveries occurred mainly in 2009, 2011 and 2013 (Figure 7b). Recovery trends detected after 2010 were mainly detected along riverbanks, on steep hillsides, and on land affected by gully erosion. This is shown in area I (Figure 7c) and area II (Figure 7d) which accordingly showed positive trend differences that were associated with a trend shift of significant monotonic negative-positive trends. The two VHR images of the river in area I show that this development is related to a decline in eroded soil from 2013 to 2016. For area II, the VHR images show that while the hillside was largely covered by degraded soil in 2014, it was revegetated in 2019 which explains the positive NDVI trends. Furthermore, area III (Figure 7e), characterised by degraded land and affected by gully formation, increased in vegetation cover between 2005 and 2019 which is reflected by the positive NDVI trend differences. This result also coincides with the evaluation through the performance assessment by GFA Consulting Group in which the area of the corresponding micro-watershed was described as "reclaimed land from gully erosion transformed into forage production and other economic activities" (personal communication with GFA's SLMP team leader).

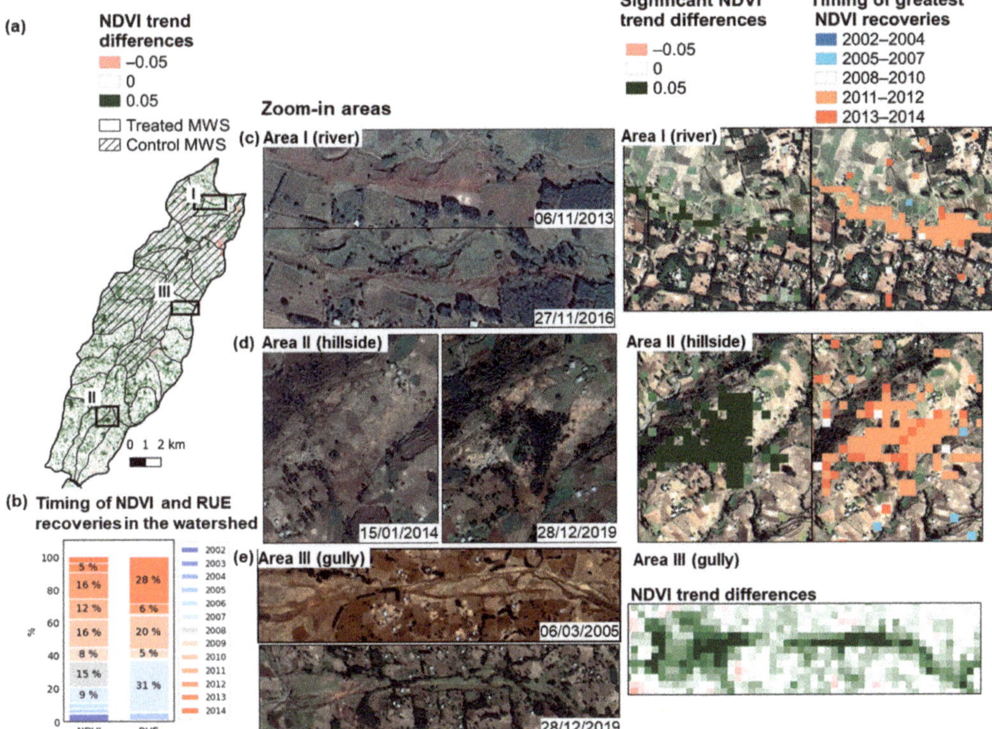

Figure 7. Watershed of Yilmana Densa. (**a**) NDVI trend differences (slope of 2011–2018 minus slope of 2002–2010); (**b**) Timing of the largest NDVI and RUE recoveries; Zoom-in areas showing (**c**) a decline of eroded soil at a riverbank (area I), (**d**) an increase in vegetation along a hillside (area II) and (**e**) an increase in vegetation in and nearby a gully (area III) with corresponding trend patterns.

3.3. Effect of SWC Measures on Vegetation Trends

The regional regression results (results based on the available SWC data in the entire study area) showed strong negative relationships of trends and distance particularly for check dams. The strongest relationships were found between the distance and the median trend differences (slope of 2011–2018 minus slope of 2002–2010) within 250 m ($r = -0.97$ **, $r^2 = 0.94$) and 500 m ($r = -0.98$ ***, $r^2 = 0.8$) (Figure A6a, Table 1) as well as between the distance and the proportion of significant increases 2011–2018 within 250 m ($r = -0.97$ **, $r^2 = 0.93$) and 500 m ($r = -0.95$ ***, $r^2 = 0.91$) (Figure A6b, Table 2). For the latter type of trends, terraces showed strong negative relationships within 500 m ($r = 0.87$ **, $r^2 = 0.76$) and 1500 m ($r = 0.85$ ***, $r^2 = 0.73$) radius (Table 2). Otherwise, linear relationships within 1500 m buffers were for both SWC types predominately weak.

Table 1. Regional OLS regression results for median trend differences (all trends).

SWC Type	Buffer Option 1 (250 m)			Buffer Option 2 (500 m)			Buffer Option 3 (1500 m)		
	Slope	r	r^2	Slope	r	r^2	Slope	r	r^2
Check dams	-1.7×10^{-5}	-0.97 **	0.94	-8.60×10^{-6}	-0.98 ***	0.8	-4.00×10^{-7}	0.23	0.05
Terraces	-2.0×10^{-7}	-0.02	0	-2.00×10^{-7}	-0.06	0	-8.00×10^{-7}	-0.58 *	0.34

Differences exist with different significance levels (* $p < 0.05$, ** $p < 0.01$, *** $p < 0.001$).

Table 2. Regional OLS regression results for the proportion of significant positive trends in the period 2011–2018.

SWC Type	Buffer Option 1 (250 m)			Buffer Option 2 (500 m)			Buffer Option 3 (1500 m)		
	Slope	r	r^2	Slope	r	r^2	Slope	r	r^2
Check dams	-0.046	-0.97 **	0.93	-0.03	-0.95 ***	0.91	-0.007	-0.79 ***	0.63
Terraces	-0.035	-0.64	0.42	-0.029	-0.87 **	0.76	-0.010	-0.85 ***	0.73

Differences exist with different significance levels (** $p < 0.01$, *** $p < 0.001$).

The results showed stronger correlations for a few local regressions models (results of the individual watersheds). An example is the watershed of Tahtay Koraro that experienced increasing trends particularly along the micro-watershed borders where steep slopes exist. Strong negative relationships of trends and the distance to both check dams and terraces were observed. Significant increases between 2011 and 2018 clustered at the location of two check dams and one terrace (Figure 8a). The strongest relationship was found within a 250 m radius ($r^2 = 0.98$) with a decreasing proportion of significant trends from 74% at 50 m, to 36% at 250 m and to 13% at 1500 m (Figure 8b). The VHR images reveal that the hillsides were, in accordance with the SWC data, terraced in 2016 and increased in vegetation cover up to 2019 (Figure 8c).

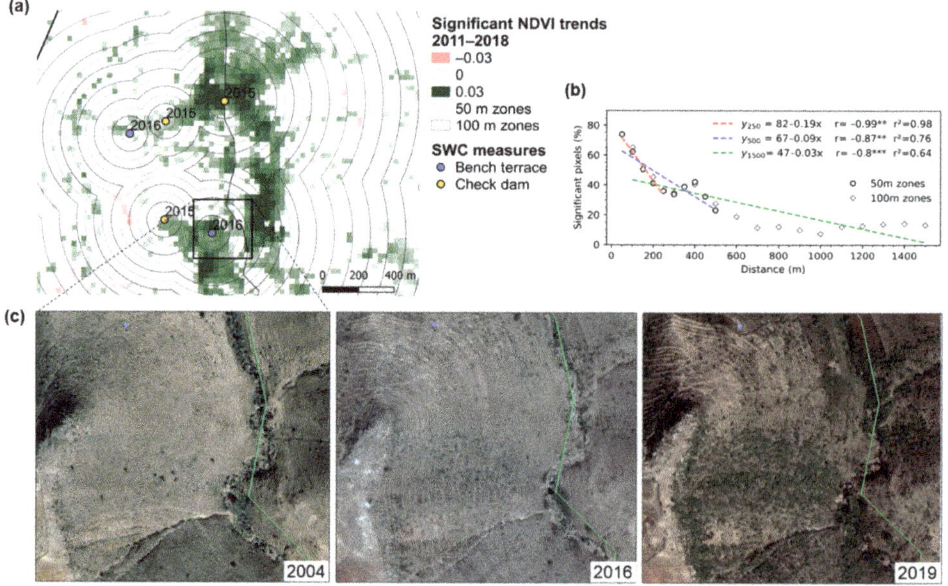

Figure 8. OLS regression of trends and the locations of SWC measures in the watershed of Tahtay Koraro. (**a**) Significant NDVI trends 2011–2018 with SWC points showing the year of completion; (**b**) Combined regression model for all check dams and terraces in the watershed for the proportion of significant increasing NDVI trends 2011–2018; (**c**) Multi-temporal Google Earth VHR images showing the zoom-in area with an increase in vegetation cover from 2004 to 2019 and the construction of terraces in 2016. ** $p < 0.01$, *** $p < 0.001$.

Another example of spatial correlations of increasing NDVI trends and SWC measure locations was observed for the watershed of Gudeyabila where the trend differences (Figure 9a) and recovery trends detected after 2010 (Figure 9b) clustered in close proximity to a check dam. Here, a decrease of the median NDVI from 0.015 at 50 m to 0.005 at 250 m ($r^2 = 0.85$) was observed (Figure 9c). The VHR images show the existence of a gully as well as mainly unterraced hillside in 2013 (Figure 9d). The construction of the check dam, which was completed at the gully in 2015, was accompanied by treatment through terraces of the surrounding hillside area including grazing land and cropland.

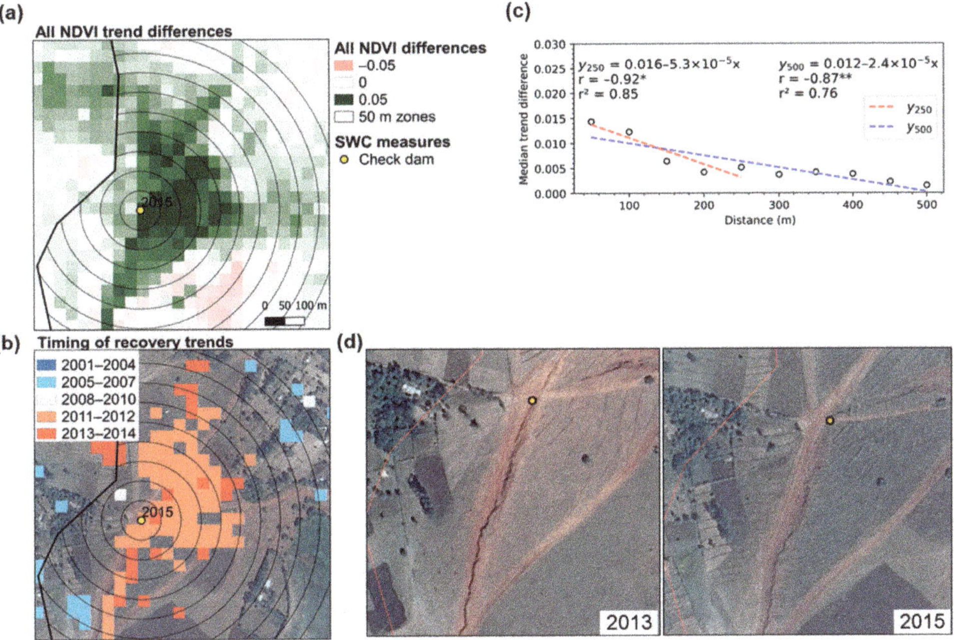

Figure 9. OLS regression of trends at a check dam location in the watershed of Gudeyabila; (**a**) Trend differences (slope of 2011–2018 minus slope of 2002–2010); (**b**) Timing of recovery trends; (**c**) Regression model of the median trend difference for the check dam location; (**d**) Google Earth VHR images showing mainly unterraced hillside in 2013 and the construction of terracing in 2015. * $p < 0.05$, ** $p < 0.01$.

4. Discussion

During the entire period 2002–2018 and during the second sub-period 2011–2018, the study area showed more pixels of NDVI increase than decrease and vice-versa during the sub-period 2002–2010 (Figures 3 and 4a). The browning trends in the latter period coincide with previous findings by Hermans-Neumann et al. who identified declining net primary production between 2000 and 2009 in the highlands of Amhara and Oromia [18]. Besides the greening trends between 2002 and 2018, several results of this study indicate a shift from browning to greening within the same period. On the one hand, this was indicated by the results from the Theil-Sen trend analysis with a shift from decreases to increases in 2011 (Figure 4c), on the other hand, browning to greening was also expressed by the LandTrendr results showing a relatively large number of the negative-positive change types (Figure 5a). For RUE, this change type was the second most common one. The general shift from negative to positive trends was also in agreement with the timing of increases and decreases, as decreases occurred mainly in 2002 and increases mainly after 2010 (Figure 5b).

14% of the study area experienced a significant increase in both NDVI and rainfall in the overall study period 2002–2018. This spatiotemporal pattern indicates climate-related greening [26]. In the first epoch 2002–2010, the study area presented mainly negative rates of change in NDVI despite increasing (but not significant) rainfall trends (Figure 3a). Even though this pattern indicates human-induced browning, evidence of this is not provided due to the insignificance of the rainfall trends and the results rather suggest that declining NDVI was not related to a long-term change in rainfall. Rainfall variability in North, West and central Ethiopia increased during the period 1983 to 2012 [59]. As extreme weather conditions pose major challenges to agricultural activities through reduced crop yields and intensified soil erosion [59], changing intra-annual patterns are overlooked by using seasonally summed rainfall and the results could possibly be improved by examining trends in rainfall intra-annual variability [60].

4.1. Treatment and Control Areas

The Mann–Whitney U test results showed larger median NDVI trends between 2011 and 2018 for treatment areas than for control areas at the regional level with most local test results (i.e., within each individual major watershed) indicating this as well (Table A2). However, the medians differed only marginally and the significant differences in the distributions of the per-pixel trends were not immediately apparent in the trend maps. The significance of test results may be attributed to large sample sizes facilitating distributions to be significant.

The LandTrendr results did not reveal any substantial differences in the types and timing of changes between treatment and control areas (Figure 5); hence did not provide evidence for an improved development of treatment than control areas. However, it should be considered that this result may be tied to the applied methodology, particularly the use of the maximum NDVI which may not fully capture effects of interventions. This may be the case if the impact of interventions show an increase in crop yield that might be reflected better as an increase in the integral of the phenological crop cycle curve (integrated NDVI) or if the impact is the reduction of eroded soil in the end of the rainy season which could lead to higher NDVI values in the end of the phenological cycle, rather than causing an increase in maximum NDVI. Future research should, provided that the data availability will be sufficient (e.g., Sentinel-2), therefore include seasonal metrics when examining changes in vegetation condition related to SLM interventions.

4.2. Visual Inspections of Trends Using Google Earth

Whereas the previously discussed results based on the analysis conducted and reported at the level of watersheds did not indicate strong differences in vegetation development between treatment and control areas, closer visual examination of the trend maps showed that human-induced land improvements can be detected from the Landsat-based approach developed, though at localised scales rather than consistently spread throughout entire watersheds. For a few major watersheds, particularly for Banja and Sekela in Amahara region, the VHR images supported that RUE increases between 2012 and 2014 were due to the establishment of tree plantations (Figure 6). Following Mekonnen et al., communal and private lands in rural areas in Amhara region have been extensively used for the expansion of Eucalyptus and Acia plantations due to the demand for wood resource [61]. Furthermore, the government has been promoting tree planting widely through the introduction of campaigns. Consequently, woodlots, home gardens, trees on cropland and farm boundary plantations have become common agroforestry practices [61] which may explain the general emergence of small tree patches in various watersheds.

Furthermore, from visual inspection it was observed that notable changes occurred outside cultivated fields as patches of increasing trends in vegetation cover, particularly at hillside locations and along streams and gullies (Figure 7). These clusters of pixels were characterised by a negative-positive trend shift (MK) and were in agreement with LandTrendr recoveries of which the largest were detected after 2010. Since in these instances

VHR images could verify the regrowth of vegetation, these negative-positive trend shifts can be linked to anthropogenic land improvement where degradation previously occurred. Moreover, findings of positive development could be confirmed by results from the GFA Consulting Group performance assessment (Section 3.2).

4.3. Effect of SWC Measures on Vegetation Trends

The OLS regression results demonstrated that positive changes in NDVI could be attributed to the impact of SLMP infrastructure. Visual inspection showed that this was visible in the spatial patterns for several locations at check dams and terraces, as shown in the case examples (Figures 8 and 9). This coincides with results from a study by Ali et al. who used NDVI and other satellite derived indices to evaluate the impact of SWC measures on different land use (i.e., on cultivated land and non-cultivated land such as degraded hillsides), and found that biophysical measures had a particular high impact on non-cultivated land [13].

In general, the most significant positive changes in NDVI were observed within the smallest buffer zone; i.e., the decrease in the medium change rate or density of change occurrence was mainly observed within 250 m distance from the location, in few cases within 500 m (Tables 1 and 2). At distances larger than this and up to 1500 m, trends levelled off or even increased again. Field visits showed that gullies typically pass through cultivated fields and in these cases revegetation efforts are conducted mainly directly at the gully only. Hence, for land restoration of gullies dense vegetation regrowth does not often take place across larger distances from the restoration activities. Land improvement in proximity to check dams could therefore also occur as a line type pattern in the trend map. In this case, circular buffer zones will not help explaining vegetation trends. Apart from this, it should be considered that rehabilitation activities can occur at a smaller scale than Landsat's spatial resolution of 30×30 m, in which cases changes would not easily be detected. The spatial scale of the impact of SWC interventions as implemented in Ethiopia therefore underlines the challenge of detecting changes related to improved land management in developing countries based on the use of traditional remote sensing methods for change detection.

5. Conclusions

The aim of this study was to examine vegetation dynamics between 2002 to 2018 in degraded areas in the Ethiopian highlands and assess the impact of SLMP interventions. To examine vegetation dynamics in complex landscapes in Ethiopia on a detailed spatial scale, we investigated the potential of combining remote sensing data from different Landsat sensors using cloud-based geospatial processing supporting a high-resolution time series analysis. The vegetation dynamics in the study areas showed a shift from browning (2002–2010) to greening (2011–2018) along with an overall greening trend over the full period (2002–2018). From the spatiotemporal patterns of NDVI and rainfall it could be concluded that the browning trend was not explained by long-term changes in rainfall. In contrast, the greening trend over the full period could—for 14% of the study area—be explained by increases in rainfall. Overall, no clear patterns of anthropogenic induced changes in vegetation were found when aggregating results at the catchment scale, as NDVI median trends did not clearly indicate better development in SLMP intervention areas than in control areas. Visual inspection based on multi-temporal Google Earth imagery showed that the changes in NDVI and rain-use efficiency did spatially overlap areas of small-scale land improvements related to human management, however, on a smaller scale than a micro-watershed (the smallest aggregation level). The OLS regression results provided evidence of land recovery that could be attributed particularly to SLMP infrastructure (check dams and terraces). Positive impacts on vegetation were found to be contributing to improving the rehabilitation of degraded hillside areas and gullies. These findings underline that the little differences found between treatment and control areas when aggregated to the level of (micro-)watersheds are rooted in a scale issue, and highlight the need for per-pixel trend analysis using sensor systems like Landsat, or higher spatial resolution, to be able to remotely capture the effect of SLMP interventions.

The ecological improvements through SLMP, identified here at the per-pixel level from the use of Landsat time series, are an important contribution to restore terrestrial ecosystems as targeted in the Sustainable Development Goals. Continuous efforts in developing means for improved monitoring of human-induced vegetation restoration of degraded lands will be essential to maintain rehabilitated land, prevent further land degradation and support environmental sustainability.

Author Contributions: Conceptualization, E.B. and R.F.; methodology, E.B. and R.F.; formal analysis, E.B.; writing—original draft preparation, E.B.; writing—review and editing, E.B. and R.F.; visualization, E.B. All authors have read and agreed to the published version of the manuscript.

Funding: R.F. acknowledges support by the Villum Foundation through the project 'Deep Learning and Remote Sensing for Unlocking Global Ecosystem Resource Dynamics' (DeReEco).

Institutional Review Board Statement: Not applicable.

Informed Consent Statement: Not applicable.

Data Availability Statement: The data presented in this study are available on request from the corresponding author.

Acknowledgments: The authors would like to thank GFA Consulting Group GmbH who provided the geolocation data on soil and water conservation measures and the German Agency for International Cooporation (GIZ) GmbH who provided the data on treatment and control areas.

Conflicts of Interest: The authors declare no conflict of interest. The funders had no role in the design of the study; in the collection, analyses, or interpretation of data; in the writing of the manuscript, or in the decision to publish the results.

Appendix A

Table A1. Description of soil and water conservation (SWC) measure data.

Type of SWC Measure	Purpose	Number of Geolocation Points
Hillside terraces	Terraces are built to stabilise cultivated land, or to stabilise area.	11
Check dams	Check dams are obstruction walls constructed at the bottom of a gully, small streams or trenches in order to reduce run-off volume and prevent further widening of the gully channel [62]. These treatment measures are typically combined with revegetation activities to gain higher run-off infiltration into the sediments.	43

(a)

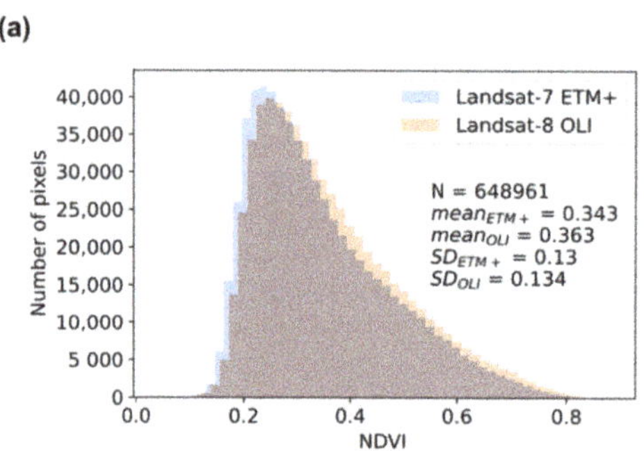

(b)

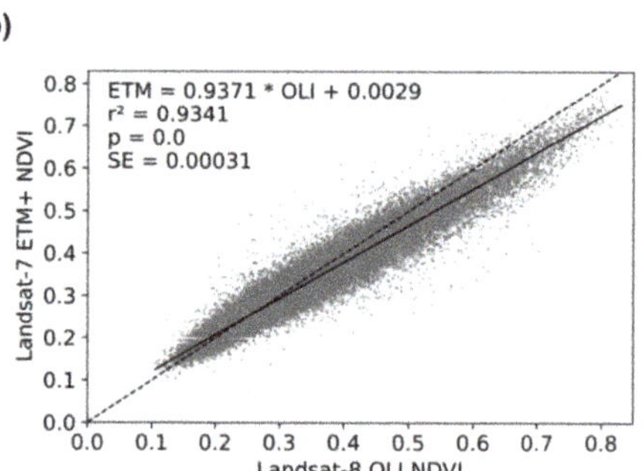

Figure A1. Landsat cross-calibration. (**a**) Distributions of ETM+ and OLI NDVI values; (**b**) OLS regression of ETM+ NDVI against OLI NDVI values. For a given OLI NDVI value, the corresponding ETM+ value is usually lower.

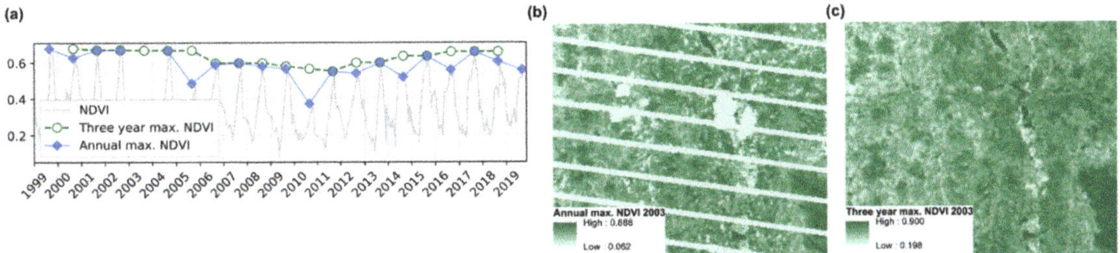

Figure A2. Interpolation. (**a**) Interpolation of a pixel's annual maximum NDVI time series (blue) through a three-year maximum moving window (green). The data gap in 2003 is interpolated; (**b**) Annual maximum value composite (MVC) with gaps produced by Landsat-7 ETM+ SLC error and clouds; (**c**) Interpolated MVC through applying a three-year maximum moving window.

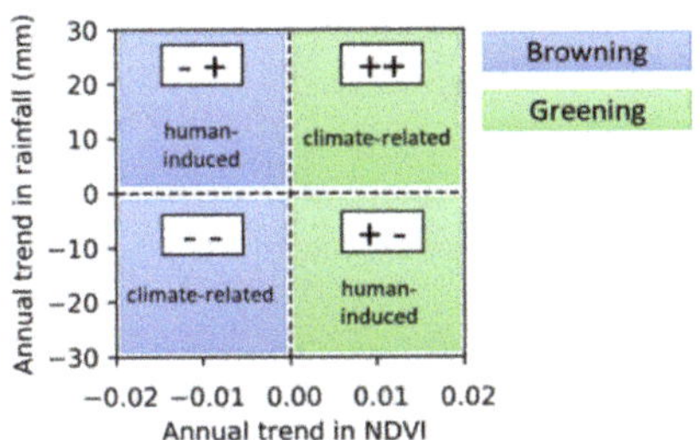

Figure A3. Interpretation of the nature of changes in ecosystem functioning based on the spatial agreement of vegetation (NDVI) and rainfall trends. A decrease in NDVI despite an increase in precipitation or vice versa is likely to be caused by human management. Trend combinations with the same slope direction are likely to be caused by climate change [26].

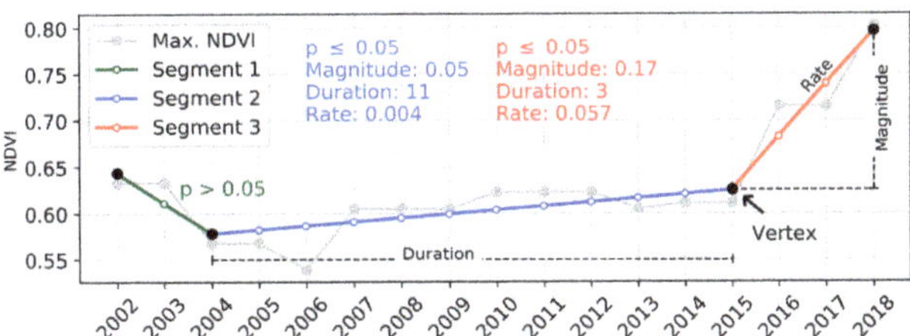

Figure A4. LandTrendr. Fitted segments into an annual maximum NDVI time series.

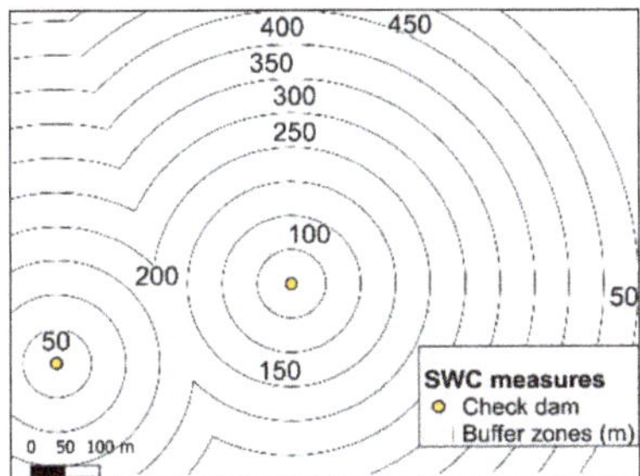

Figure A5. Example of buffer zones used for the OLS regression analysis to examine trends at soil and water conservation measure locations.

Table A2. Mann-Whitney U results with the sample size N (number of significant pixels), the median NDVI trend, and U indicating whether treatment areas have significantly [1] larger or smaller trends than control areas. Watersheds that did not include any control micro-watersheds are not included.

Major Watershed	Treatment		Control		U
	N	Median	N	Median	
Laelay Adyabbo	2091	0.0128	750	0.0116	-
Tahtay Koraro	6627	0.0203	604	0.0165	Larger ***
Emba Alaje	1080	0.0089	34	−0.0091	Larger ***
Gondar Zuriya	693	0.0137	491	0.0172	Smaller **
Takusa	2182	0.0208	2526	0.0130	Larger ***
West Estie	5248	0.0252	1203	0.0237	Larger ***
Hagere Mariam	1060	0.0065	60	0.0164	Smaller ***
Sinan	2135	0.0213	1324	0.0218	-
Aneded	3517	0.0292	2387	0.0270	Larger ***
Yilmana Densa	2539	0.0242	2204	0.0193	Larger ***
Sekela	2384	0.0212	972	0.0196	Larger **
Quarit	2000	0.0222	1220	0.0183	Larger ***
Banja	2370	−0.0003	2974	0.0052	Smaller ***
Ale	861	0.0112	131	0.0117	-
All watersheds	43,984	0.0190	16,880	0.0183	Larger *

[1] Differences exist with different significance levels (* $p < 0.05$, ** $p < 0.01$, *** $p < 0.001$).

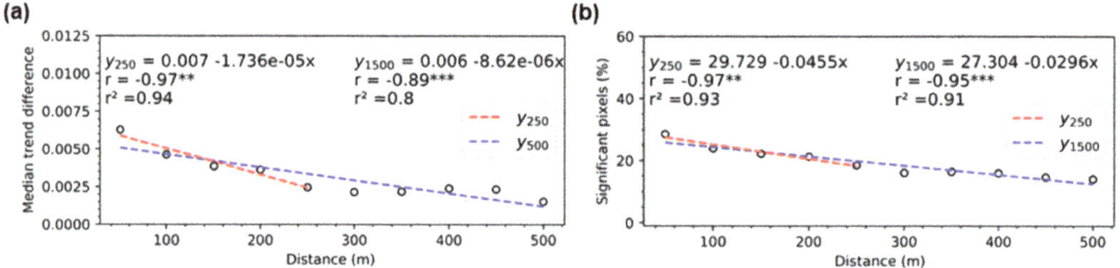

Figure A6. Regional OLS regression models for check dams with (**a**) the median of all trend differences and (**b**) the proportion of significant positive trends during 2011–2018 within the zones as dependent variable. The red line shows the linear slope when including zones up to 250 m distance, the blue line when including zones up to 500 m. ** $p < 0.01$, *** $p < 0.001$.

References

1. Hannam, I.; Boer, B. *Drafting Legislation for Sustainable Soils: A Guide*; World Conservation Union: Gland, Switzerland; Cambridge, UK, 2004; ISBN 2-8317-0813-3.
2. Bai, Z.G.; Dent, D.L.; Olsson, L.; Schaepman, M.E. Proxy global assessment of land degradation. *Soil Use Manag.* **2008**, *24*, 223–234. [CrossRef]
3. Scherr, S.J.; Yadav, S. *Land Degradation in the Developing World: Implications for Food, Agriculture, and the Environment to 2020*; Paper 14; International Food Policy Research Institute: Washington, DC, USA, 1996. [CrossRef]
4. United Nations Convention to Combat Desertification (UNCCD). *Text of the United Nations Convention to Combat Desertification*; 1994. Available online: https://www.unccd.int/sites/default/files/relevant-links/2017-01/UNCCD_Convention_ENG_0.pdf (accessed on 24 March 2021).
5. Bekele, W. *Economics of Soil and Water Conservation. Theory and Empirical Application to Subsistence Farming in the Eastern Ethiopian Highlands*; Wagayehu Bekele: Uppsala, Sweden, 2003; ISBN 9157664331.
6. Hurni, H. *Land Degradation, Famine, and Land Resource Scenarios in Ethiopia*; Cambridge University Press: Cambridge, UK, 1993; ISBN 9780511735394.
7. Berry, L.; Olson, J.; Campbell, D. *Assessing the Extent, Cost and Impact of Land Degradation At the National Level: Findings and Lessons Learned From Seven Pilot Case Studies*; 2003. Available online: https://rmportal.net/library/content/frame/land-degradation-case-studies-01-overview/at_download/file (accessed on 24 March 2021).
8. Lakew, D.; Menale, K.; Pender, J. *Land Degradation and Strategies for Sustainable Development in the Ethiopian Highlands: Amhara Region*; International Livestock Research Institute: Nairobi, Kenya, 2000; ISBN 9291460907.

9.   Khiari, H.; Jauffret, S.; Walter, S.; Almuedo, P.L.; Lhumeau, A. *Land Degradation Neutrality. Transformative Projects and Programmes: Operational Guidance for Country Support*; United Nations Convention to Combat Desertification (UNCCD): Bonn, Germany, 2019; ISBN 9789295117655.

10. World Bank Group. *Ethiopia—Second Phase of the Sustainable Land Management Project (English)*; Washington, DC, USA, 2013. Available online: https://documents.worldbank.org/en/publication/documents-reports/documentdetail/621921468257058351/ethiopia-second-phase-of-the-sustainable-land-management-project (accessed on 24 March 2021).

11. Ebabu, K.; Tsunekawa, A.; Haregeweyn, N.; Adgo, E.; Meshesha, D.T.; Aklog, D.; Masunaga, T.; Tsubo, M.; Sultan, D.; Fenta, A.A.; et al. Effects of land use and sustainable land management practices on runoff and soil loss in the Upper Blue Nile basin, Ethiopia. *Sci. Total Environ.* **2019**, *648*, 1462–1475. [CrossRef]

12. Schmidt, E.; Tadesse, F. The impact of sustainable land management on household crop production in the Blue Nile Basin, Ethiopia. *L. Degrad. Dev.* **2019**, *30*, 777–787. [CrossRef]

13. Ali, D.A.; Deininger, K.; Monchuk, D. Using satellite imagery to assess impacts of soil and water conservation measures: Evidence from Ethiopia's Tana-Beles watershed. *Ecol. Econ.* **2020**, *169*. [CrossRef]

14. Anyamba, A.; Tucker, C.J. Analysis of Sahelian vegetation dynamics using NOAA-AVHRR NDVI data from 1981–2003. *J. Arid Environ.* **2005**, *63*, 596–614. [CrossRef]

15. Eklundh, L.; Olsson, L. Vegetation index trends for the African Sahel 1982-1999. *Geophys. Res. Lett.* **2003**, *30*, 1–4. [CrossRef]

16. Rasmussen, K.; Fensholt, R.; Fog, B.; Vang Rasmussen, L.; Yanogo, I. Explaining NDVI trends in northern Burkina Faso. *Geogr. Tidsskr.* **2014**, *114*, 17–24. [CrossRef]

17. Eckert, S.; Hüsler, F.; Liniger, H.; Hodel, E. Trend analysis of MODIS NDVI time series for detecting land degradation and regeneration in Mongolia. *J. Arid Environ.* **2015**, *113*, 16–28. [CrossRef]

18. Hermans-Neumann, K.; Priess, J.; Herold, M. Human migration, climate variability, and land degradation: Hotspots of socio-ecological pressure in Ethiopia. *Reg. Environ. Chang.* **2017**, *17*, 1479–1492. [CrossRef]

19. Wessels, K.J.; Prince, S.D.; Malherbe, J.; Small, J.; Frost, P.E.; VanZyl, D. Can human-induced land degradation be distinguished from the effects of rainfall variability? A case study in South Africa. *J. Arid Environ.* **2007**, *68*, 271–297. [CrossRef]

20. Helldén, U.; Tottrup, C. Regional desertification: A global synthesis. *Glob. Planet. Change* **2008**, *64*, 169–176. [CrossRef]

21. Symeonakis, E.; Drake, N. Monitoring desertification and land degradation over sub-saharan africa. *Int. J. Remote Sens.* **2004**, *25*, 573–592. [CrossRef]

22. Nicholson, S.; Davenport, M.L.; Malo, A.R. A comparison of the vegetation response to rainfall in the Sahel and East Africa, using normalized difference vegetation index from NOAA AVHRR. *Clim. Chang.* **1990**, *17*, 209–241. [CrossRef]

23. Liou, Y.A.; Mulualem, G.M. Spatio-temporal assessment of drought in Ethiopia and the impact of recent intense droughts. *Remote Sens.* **2019**, *11*, 1828. [CrossRef]

24. Prince, S.D.; Brown De Colstoun, E.; Kravitz, L.L. Evidence from rain-use efficiencies does not indicate extensive Sahelian desertification. *Glob. Chang. Biol.* **1998**, *4*, 359–374. [CrossRef]

25. Fensholt, R.; Rasmussen, K.; Kaspersen, P.; Huber, S.; Horion, S.; Swinnen, E. Assessing land degradation/recovery in the african sahel from long-term earth observation based primary productivity and precipitation relationships. *Remote Sens.* **2013**, *5*, 664–686. [CrossRef]

26. Horion, S.; Prishchepov, A.V.; Verbesselt, J.; de Beurs, K.; Tagesson, T.; Fensholt, R. Revealing turning points in ecosystem functioning over the Northern Eurasian agricultural frontier. *Glob. Chang. Biol.* **2016**, *22*, 2801–2817. [CrossRef]

27. Hein, L.; De Ridder, N.; Hiernaux, P.; Leemans, R.; De Wit, A.; Schaepman, M. Desertification in the Sahel: Towards better accounting for ecosystem dynamics in the interpretation of remote sensing images. *J. Arid Environ.* **2011**, *75*, 1164–1172. [CrossRef]

28. Herrmann, S.M.; Anyamba, A.; Tucker, C.J. Recent trends in vegetation dynamics in the African Sahel and their relationship to climate. *Glob. Environ. Chang.* **2005**, *15*, 394–404. [CrossRef]

29. Ibrahim, Y.Z.; Balzter, H.; Kaduk, J.; Tucker, C.J. Land degradation assessment using residual trend analysis of GIMMS NDVI3g, soil moisture and rainfall in Sub-Saharan West Africa from 1982 to 2012. *Remote Sens.* **2015**, *7*, 5471–5494. [CrossRef]

30. Wulder, M.A.; Masek, J.G.; Cohen, W.B.; Loveland, T.R.; Woodcock, C.E. Opening the archive: How free data has enabled the science and monitoring promise of Landsat. *Remote Sens. Environ.* **2012**, *122*, 2–10. [CrossRef]

31. Zhu, Z. Change detection using landsat time series: A review of frequencies, preprocessing, algorithms, and applications. *ISPRS J. Photogramm. Remote Sens.* **2017**, *130*, 370–384. [CrossRef]

32. Verbesselt, J.; Hyndman, R.; Newnham, G.; Culvenor, D. Detecting trend and seasonal changes in satellite image time series. *Remote Sens. Environ.* **2010**, *114*, 106–115. [CrossRef]

33. De Jong, R.; Verbesselt, J.; Schaepman, M.E.; de Bruin, S. Trend changes in global greening and browning: Contribution of short-term trends to longer-term change. *Glob. Chang. Biol.* **2012**, *18*, 642–655. [CrossRef]

34. Higginbottom, T.P.; Symeonakis, E. Identifying ecosystem function shifts in Africa using breakpoint analysis of long-term NDVI and RUE data. *Remote Sens.* **2020**, *12*, 1894. [CrossRef]

35. Horion, S.; Ivits, E.; De Keersmaecker, W.; Tagesson, T.; Vogt, J.; Fensholt, R. Mapping European ecosystem change types in response to land-use change, extreme climate events, and land degradation. *L. Degrad. Dev.* **2019**, *30*, 951–963. [CrossRef]

36. Griffiths, P.; Kuemmerle, T.; Kennedy, R.E.; Abrudan, I.V.; Knorn, J.; Hostert, P. Using annual time-series of Landsat images to assess the effects of forest restitution in post-socialist Romania. *Remote Sens. Environ.* **2012**, *118*, 199–214. [CrossRef]

37. Grogan, K.; Pflugmacher, D.; Hostert, P.; Kennedy, R.; Fensholt, R. Cross-border forest disturbance and the role of natural rubber in mainland Southeast Asia using annual Landsat time series. *Remote Sens. Environ.* **2015**, *169*, 438–453. [CrossRef]

38. Meigs, G.W.; Kennedy, R.E.; Cohen, W.B. A Landsat time series approach to characterize bark beetle and defoliator impacts on tree mortality and surface fuels in conifer forests. *Remote Sens. Environ.* **2011**, *115*, 3707–3718. [CrossRef]
39. Main-Knorn, M.; Cohen, W.B.; Kennedy, R.E.; Grodzki, W.; Pflugmacher, D.; Griffiths, P.; Hostert, P. Monitoring coniferous forest biomass change using a Landsat trajectory-based approach. *Remote Sens. Environ.* **2013**, *139*, 277–290. [CrossRef]
40. Yin, H.; Prishchepov, A.V.; Kuemmerle, T.; Bleyhl, B.; Buchner, J.; Radeloff, V.C. Mapping agricultural land abandonment from spatial and temporal segmentation of Landsat time series. *Remote Sens. Environ.* **2018**, *210*, 12–24. [CrossRef]
41. Amede, T.; Auricht, C.; Boffa, J.; Dixon, J.; Mallawaarachchi, T.; Rukuni, M.; Teklewold-deneke, T. *A Farming System Framework for Investment Planning and Priority Setting in Ethiopia*; Canberra, 2017. Available online: https://aciar.gov.au/publication/technical-publications/farming-system-framework-investment-planning-and-priority-setting-ethiopia (accessed on 24 March 2021).
42. Zhang, C.; Li, W.; Travis, D. Gaps-fill of SLC-off Landsat ETM+ satellite image using a geostatistical approach. *Int. J. Remote Sens.* **2007**, *28*, 5103–5122. [CrossRef]
43. Funk, C.; Peterson, P.; Landsfeld, M.; Pedreros, D.; Verdin, J.; Shukla, S.; Husak, G.; Rowland, J.; Harrison, L.; Hoell, A.; et al. The climate hazards infrared precipitation with stations—A new environmental record for monitoring extremes. *Sci. Data* **2015**, *2*, 1–21. [CrossRef] [PubMed]
44. Gorelick, N.; Hancher, M.; Dixon, M.; Ilyushchenko, S.; Thau, D.; Moore, R. Google Earth Engine: Planetary-scale geospatial analysis for everyone. *Remote Sens. Environ.* **2017**, *202*, 18–27. [CrossRef]
45. Foga, S.; Scaramuzza, P.L.; Guo, S.; Zhu, Z.; Dilley, R.D.; Beckmann, T.; Schmidt, G.L.; Dwyer, J.L.; Hughes, M.J.; Laue, B. Remote Sensing of Environment Cloud detection algorithm comparison and validation for operational Landsat data products. *Remote Sens. Environ.* **2017**, *194*, 379–390. [CrossRef]
46. Housman, I.W.; Chastain, R.A.; Finco, M. V An Evaluation of Forest Health Insect and Disease Survey Data and Satellite-Based Remote Sensing Forest Change Detection Methods: Case Studies in the United States. *Remote Sens.* **2018**, *10*, 1184. [CrossRef]
47. Roy, D.P.; Kovalskyy, V.; Zhang, H.K.; Vermote, E.F.; Yan, L.; Kumar, S.S.; Egorov, A. Characterization of Landsat-7 to Landsat-8 reflective wavelength and normalized difference vegetation index continuity. *Remote Sens. Environ.* **2016**, *185*, 57–70. [CrossRef]
48. Tian, F.; Fensholt, R.; Verbesselt, J.; Grogan, K.; Horion, S.; Wang, Y. Evaluating temporal consistency of long-term global NDVI datasets for trend analysis. *Remote Sens. Environ.* **2015**, *163*, 326–340. [CrossRef]
49. Chastain, R.; Housman, I.; Goldstein, J.; Finco, M. Empirical cross sensor comparison of Sentinel-2A and 2B MSI, Landsat-8 OLI, and Landsat-7 ETM+ top of atmosphere spectral characteristics over the conterminous United States. *Remote Sens. Environ.* **2019**, *221*, 274–285. [CrossRef]
50. Fensholt, R.; Langanke, T.; Rasmussen, K.; Reenberg, A.; Prince, S.D.; Tucker, C.; Scholes, R.J.; Le, Q.B.; Bondeau, A.; Eastman, R.; et al. Greenness in semi-arid areas across the globe 1981–2007—An Earth Observing Satellite based analysis of trends and drivers. *Remote Sens. Environ.* **2012**, *121*, 144–158. [CrossRef]
51. Fensholt, R.; Horion, S.; Tagesson, T.; Ehammer, A.; Grogan, K.; Tian, F.; Huber, S.; Verbesselt, J.; Prince, S.D.; Tucker, C.J.; et al. Assessment of vegetation trends in drylands from time series of earth observation data. In *Remote Sensing Time Series. Remote Sensing and Digital Image Processing*; Springer: Cham, Switzerland, 2015; pp. 159–182.
52. The Federal Democratic Republic of Ethiopia. Ministry of Finance and Economic Development (MOFED). *Ethiopia: Sustainable Poverty Reduction Program*; 2002. Available online: https://www.imf.org/external/np/prsp/2002/eth/01/073102.pdf (accessed on 24 March 2021).
53. Sen, P.K. Estimates of the Regression Coefficient Based on Kendall's Tau. *J. Am. Stat. Assoc.* **1968**, *63*, 1379. [CrossRef]
54. Kendall, M.G. A New Measure of Rank Correlation. *Biometrika* **1938**, *30*, 81. [CrossRef]
55. McKnight, P.E.; Najab, J. Mann-Whitney U Test. In *The Corsini Encyclopedia of Psychology*; John Wiley & Sons, Inc.: Hoboken, NJ, USA, 2010; ISBN 9780470479216.
56. Kennedy, R.E.; Yang, Z.; Gorelick, N.; Braaten, J.; Cavalcante, L.; Cohen, W.B.; Healey, S. Implementation of the LandTrendr algorithm on Google Earth Engine. *Remote Sens.* **2018**, *10*, 691. [CrossRef]
57. Fensholt, R.; Horion, S.; Tagesson, T.; Ehammer, A.; Grogan, K.; TIan, F.; Huber, S.; Verbesselt, J.; Prince, S.D.; Tucker, C.J.; et al. Assessing Drivers of Vegetation Changes in Drylands from Time Series of Earth Observation Data. In *Remote Sensing Time Series. Remote Sensing and Digital Image Processing*; Springer: Cham, Switzerland, 2015; Volume 22, pp. 183–202, ISBN 978-3-319-15966-9.
58. Verón, S.R.; Oesterheld, M.; Paruelo, J.M. Production as a function of resource availability: Slopes and efficiencies are different. *J. Veg. Sci.* **2005**, *16*, 351–354. [CrossRef]
59. Suryabhagavan, K.V. GIS-based climate variability and drought characterization in Ethiopia over three decades. *Weather Clim. Extrem.* **2017**, *15*, 11–23. [CrossRef]
60. Zhang, W.; Brandt, M.; Tong, X.; Tian, Q.; Fensholt, R. Impacts of the seasonal distribution of rainfall on vegetation productivity across the Sahel. *Biogeosciences* **2018**, *15*, 319–330. [CrossRef]
61. Mekonnen, M.; Sewunet, T.; Gebeyehu, M.; Azene, B.; Melesse, A.M. GIS and remote sensing-based forest resource assessment, quantification, and mapping in Amhara region, Ethiopia. In *Landscape Dynamics, Soils and Hydrological Processes in Varied Climates. Springer Geography*; Springer: Cham, Switzerland, 2016; pp. 9–29. ISBN 9783319187877.
62. Engdayehu, G.; Fisseha, G.; Mekonnen, M.; Melesse, A.M. Evaluation of technical standards of physical soil and water conservation practices and their role in soil loss reduction: The case of debre mewi watershed, north-west Ethiopia. In *Landscape Dynamics, Soil and Hydrological Processes in Varied Climates. Springer Geography*; Springer: Cham, Switzerland, 2016; pp. 789–818. ISBN 9783319187877.

Article

Social-Ecological Archetypes of Land Degradation in the Nigerian Guinea Savannah: Insights for Sustainable Land Management

Ademola A. Adenle [1,2,*] **and Chinwe Ifejika Speranza** [1]

[1]   Institute of Geography, University of Bern, Hallerstrasse 12, 3012 Bern, Switzerland; chinwe.ifejika.speranza@giub.unibe.ch
[2]   Department of Geography, Federal University of Technology, Minna P.M.B.65, Niger State, Nigeria
*   Correspondence: ademola.adenle@giub.unibe.ch or ade.adenle@gmail.com or ade.ademola@futminna.edu.ng; Tel.: +41-31-631-59-09 or +234-81-432-769-18

Abstract: The Nigerian Guinea Savannah is the most extensive ecoregion in Nigeria, a major food production area, and contains many biodiversity protection areas. However, there is limited understanding of the social-ecological features of its degraded lands and potential insights for sustainable land management and governance. To fill this gap, the self-organizing map method was applied to identify the archetypes of both proximal and underlying drivers of land degradation in this region. Using 12 freely available spatial datasets of drivers of land degradation—4 environmental; 3 socio-economic; and 5 land-use management practices, the identified archetypes were intersected with the Moderate-Resolution Imaging Spectroradiometer (MODIS)-derived land-degradation status of the region, and the state administrative boundaries. Nine archetypes were identified. Archetypes are dominated by: (1) protected areas; (2) very high-density population; (3) moderately high information/knowledge access; (4) low literacy levels and moderate–high poverty levels; (5) rural remoteness; (6) remoteness from a major road; (7) very high livestock density; (8) moderate poverty level and nearly level terrain; and (9) very rugged terrain and remote from a major road. Four archetypes characterized by very high-density population, moderate–high information/knowledge access, and moderate–high poverty level, as well as remoteness from a major town, were associated with 61.3% large-area degradation; and the other five archetypes, covering 38.7% of the area, were responsible for small-area degradation. While different combinations of archetypes exist in all the states, the five states of Niger (40.5%), Oyo (29.6%), Kwara (24.4%), Nassarawa (18.6%), and Ekiti (17.6%), have the largest shares of the archetypes. To deal with these archetypical features, policies and practices that address increasing population in combination with poverty reduction; and that create awareness about land degradation and promote sustainable practices and various forms of land restoration, such as tree planting, are necessary for progressing towards land-degradation neutrality in the Nigerian Guinea Savannah.

Keywords: archetypes; self-organizing maps; land degradation; drivers; savannah; Nigeria

Citation: Adenle, A.A.; Ifejika Speranza, C. Social-Ecological Archetypes of Land Degradation in the Nigerian Guinea Savannah: Insights for Sustainable Land Management. *Remote Sens.* **2021**, *13*, 32. https://dx.doi.org/10.3390/rs13010032

Received: 17 November 2020
Accepted: 23 December 2020
Published: 23 December 2020

1. Introduction

With increasing global population, environmental change, and competing claims on land, the need to maintain land productivity and reduce land degradation has become even more critical. Various global initiatives reflecting this urgency include the United Nations Convention to Combat Desertification (UNCCD) goal of achieving a Land-Degradation Neutral (LDN) World [1]; the African Forest Landscape Restoration Initiative (AFR100) aiming to restore 100 million hectares of land in Africa by 2030 [2]; and the Great Green Wall Initiative across the Sahel [3]. Yet, land degradation (LD), i.e., the persistent reduction (negative trend) or loss of the biological productivity or ecological integrity of land or its value to humans, remains a diverse and complex issue [4,5].

The African Savannah is among the globally threatened landscapes [6], where climatic and edaphic conditions, as well as human activities, constrain vegetation regeneration [7,8].

In West Africa, the savannah ecozones are prone to LD and experience both anthropogenic and non-anthropogenic pressures [7,9]. These expose its 353 million inhabitants and their livelihoods to various impacts such as decline in ecosystem services, food insecurity, migration, and civil conflict [8]. In Nigeria, LD cases across its agroecological zones are mostly triggered by a singular factor or a combination of factors, which include: desertification, deforestation leading to biomass loss, land pollution caused by oil spillage and illegal mining, as well as extensive soil erosion [10]. Thus, agroecological zones like the Nigerian Guinean Savannah (NGS) are experiencing pressures from urbanization, agricultural expansion and an increasing population that is largely dependent on land resources for livelihoods [11,12]. The NGS, covering 49% of Nigeria, is widely degraded, its ecosystem services continue to decline, and livelihoods remain precarious [11,12]. Unsustainable land use and climate stress have been implicated in this widespread degradation [11,13–15]. While the indicators of LD drivers, their interplay, and implications at scale are generally acknowledged [14,16], there is a lack of knowledge of the constellation of factors characterizing specific degraded landscapes, such as the NGS, and their interplay [11]. The explanations of several studies in the NGS on LD drivers are often without an integrative perspective capable of exposing the interactions between potential LD drivers [16,17].

Recent studies thus stress the need for an integrative approach to enable a better understanding of the constellation and interplay of LD drivers in land systems [14,17]. One such integrative approach is archetypical clustering for identifying recurrent patterns in land conditions [18,19]. Archetypes, i.e., patterns or processes that occur repeatedly across space and time [19,20], have been found to provide a more holistic understanding of land system processes [20]. This understanding enables comparability across cases and helps identify strategic policy options to address land management across scales [18–20]. Archetype studies have been conducted on food security in the Peruvian Altiplano [21]; institutional analysis and climate change in the Peruvian Andes [22]; national analysis of ecosystem services in Germany [23]; water governance of river basins [24], and global land resources management [25]. An archetype approach thus helps to illuminate the associative patterns and influence of the complex drivers of global changes (such as LD) that have often been treated in isolation [19,20,25].

This paper thus aims to identify the characteristic patterns of social-ecological factors associated with LD in the NGS and to analyze the implications of the identified LD archetypes for land governance and sustainable land management (SLM) in the region. In the subsections, the description of the social-ecological conditions in the NGS, the study methods and hypothesis, the archetypical patterns of drivers, their thematic, and spatial characterization, including the links to different levels of LD, and the implications of the archetypes for land governance and SLM in the NGS are presented.

2. Materials and Methods

2.1. Study Area

The Nigerian Guinea Savannah (NGS) (Figures 1 and 2) is an ecological region found between 6.50°N and 9.62°N, 2.77°E and 13.18°E. Occupying central Nigeria, it is the country's most extensive ecological zone, referred to as the Middle Belt of Nigeria. The zone consists of parkland savannah, gallery forests, and derived savannah, including distinctive montane vegetation [26], with tropical dry and wet seasons. Rainfall in the wet season (April to October) is about 1240–1440 mm. The dry season lasts from November to March [27]. Average maximum annual temperature varies from 31 °C to 35 °C while the average minimum ranges from 20 °C to 23 °C [27]. The region is broadly divided into two sub-regions based on distinctive vegetation types, namely the Southern and Northern Guinea Savannahs. The Southern Guinea Savannah has taller trees, such as locust bean tree (*Parkia biglobosa*), and grasses such as Gamba grass (*Andropogon gayanus*); while the Northern Guinea Savannah is characterized by bush with some trees (e.g., *Isoberlinia* spp.) and relatively shorter grasses (e.g., *Hyparrhenia* spp.) [27]. The NGS has a high level of fauna and flora and hosts major perennial rivers such as River Niger and River Benue. Subsequent years of uncontrolled deforestation and poor land management has transformed the zone largely into a degraded landscape (Figure 2) [12].

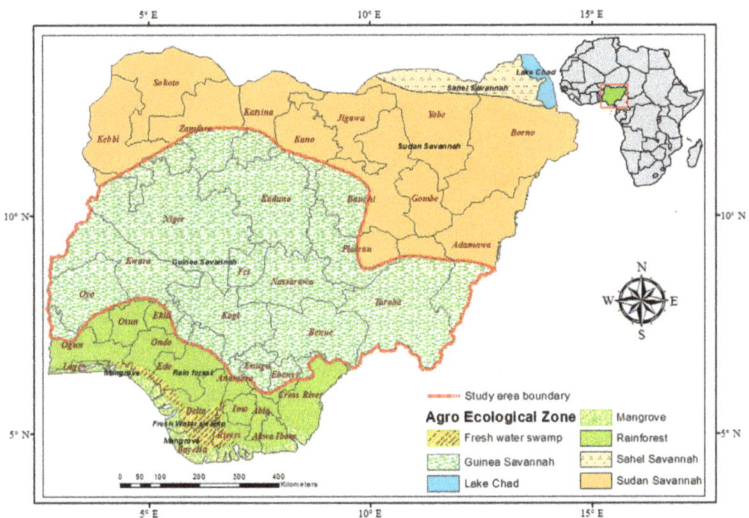

Figure 1. Map of the Nigerian Guinea Savannah including the administrative boundaries of states (Adapted from [26]).

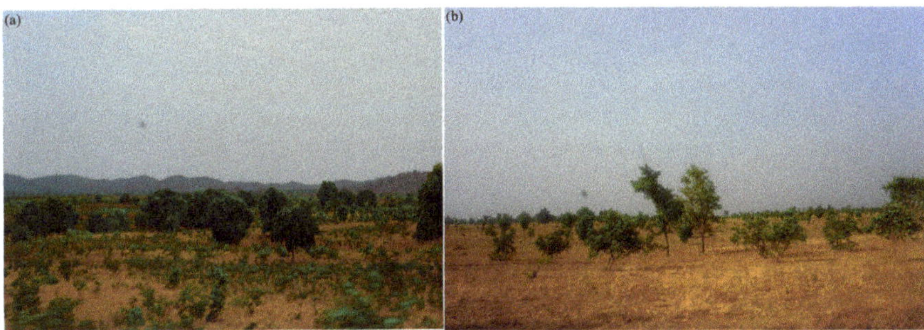

Figure 2. (**a**) Landscape showing a degraded patch of the Nigerian savannah vegetation in Shiroro local government area, Niger state, Nigeria; (**b**) a mix of degraded land with few bushes and trees interspersed in Borgu local government area, Niger state, Nigeria. (Source: own fieldwork, 2019).

2.2. Framing Land-Degradation Drivers

In this study, LD drivers were understood as determinants, i.e., reasons, factors, or actions, shaping the rate of human activities, leading to a decision to remove or reduce vegetation cover thereby causing a decline or loss of land resources' productivity [14,28]. Factors prompting LD are broadly categorized either as proximal or underlying drivers [16,17,29]. Proximate drivers are human activities or immediate actions, including the decision to directly use or alter the land cover [14]. Underlying drivers include indirect or underpinning factors, such as socio-economic factors (e.g., poverty) and biophysical factors (e.g., topographic variables) that trigger proximate causes of LD [14]. Based on literature [17,28,29] and a report on the LDN target for Nigeria in 2018 [30], three main categories of LD drivers of the Nigerian Guinea Savannah Archetypes (NGSA) namely: environmental, socio-economic, and land-use management practices were identified. Based on the available spatial and remote sensing data, 12 drivers comprising three environmental, four socio-economic, and five land-use management practice categories were selected (Table 1a–c).

The processes through which the variables (whether as proximate or underlying drivers), drive the LD are explained in Table 1a–c.

2.3. Datasets Selection

For this study, factors were considered based on their influence on land-use decisions and LD [16,29,31] (see Table 1 for details). Climatic variables such as rainfall were not included because human-induced activities are the prevailing cause of LD, especially in the NGS [11,12]. However, human-induced drivers are better managed than climatic drivers of LD [11,32].

2.3.1. Land-Use Management Practices

Land-use management practices have been linked to degraded land in Nigeria through land clearing and conversion including intensification [13]. Other unsustainable land-use management practices include rapid agricultural expansion, uncontrolled bush burning, deforestation, excessive wood extraction, unplanned infrastructure extension, and urban development, including overgrazing [12,13,33]. For this study and based on spatial data availability, data on land-use management practices were acquired and processed, respectively. The fire-occurrence density was derived from the National Aeronautics and Space Administration (NASA) active fire hotspot data of 2018 (firms.modaps.eosdis.nasa.gov) after running a spatial point-density analysis of the active fire spots at 250 m resolution. The livestock grazing intensity data for 2005, developed by Harvest Choice in 2018 (www.ifpri.org/project/harvestchoice) was used. The generated distance to the major road in 2016 for Nigeria, at a resolution of 3 arc-second (approximately 100 m at the equator) was acquired from (www.worldpop.org/project/categories?id=14) [34]. The Euclidean distance analysis of the extracted major town polygons from Google Earth (www.google.com/earth/) was used to determine the distance to major towns at 250 m resolution. Using the International Union for Conservation of Nature (IUCN) recognized protected area polygon for Nigeria (www.protectedplanet.net/country/NGA), the density of the protected area at 250 m resolution across the NGS was generated through point-density analysis.

2.3.2. Socio-Economic Drivers

Demographic and socio-economic factors such as population density, income, poverty, and illiteracy are drivers of LD in certain contexts [16,29]. In other contexts, they may even be drivers of improved land conditions [35]. However, in the case of Nigeria, its over 200 million inhabitants and population density of 226 km^2 place a huge demand on land resources [33,36]. Thus, gridded human population density constructed for 2018 in 2020, from random forest-based dasymetric redistribution at 3 arc-second (approximately 100 m at the equator) spatial resolution, was downloaded from www.worldpop.org [34]. Male and female literacy layers of high resolution at 1 km × 1 km gridded cells developed for 2003 in 2017, based on a geostatistics approach [37], were acquired [34]. These are necessary as they are proxies for access to agricultural extension information, as previous studies show that limited information on SLM drives LD [16,38]. The poverty headcount in percentages for Nigeria at 1 km for 2013 mapped through Bayesian model-based geostatistics analysis was downloaded from www.worldpop.org [34], because poverty can foster practices that cause LD, while LD can foster a poverty trap [38]. These drivers were selected in that they are known factors that influence decisions on land and SLM LD [28,29,38].

2.3.3. Environmental Drivers

Land can be sensitive to degradation due to its environmental and physiographic characteristics, influencing human decisions to use land [31]. Such influential characteristics include soil bulk density (BD), elevation, and slope [13,39]. For this study, the already-processed NASA Shuttle Radar Topography Mission (SRTM) based slope and elevation for Nigeria by [40] at a resolution of 3 arc-second (approximately 100 m at the equator) produced for 2020 was acquired from www.worldpop.org [34]. The spatial mapping of the bulk density (BD; soil compaction) in 2018 at 250 m resolution from 0 cm to 30 cm depth were downloaded from www.soilgrids.org and then averaged to give the overview of the BD for the study area [41–43].

2.4. Methods

2.4.1. Conceptual Framework

The workflow in Figure 3 was applied to the 12 drivers (Table 1a–c), which served as inputs for identifying archetypes. The intermediate outputs from the framework, such as correlation between drivers including cluster features, can be found in supplementary material Figures S1–S5.

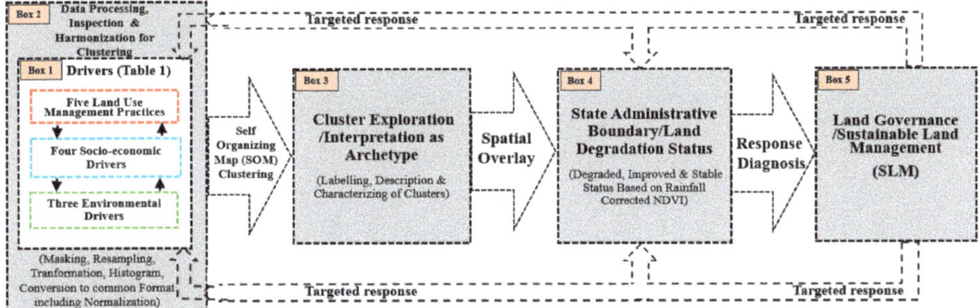

Figure 3. The conceptual framework.

2.4.2. Identifying Archetypes of Land-Degradation Drivers Using Self-Organizing Maps

To develop the Nigerian Guinea Savannah Archetypes (NGSA) of LD drivers, the 12 driver datasets were clipped to the boundary outline and resampled to a 250 m pixel size, using the nearest neighbor technique (Figure 3, Box 2), and then projected to a uniform coordinate system (i.e., Minna/UTM zone 31N). To enhance the clustering of the datasets, data were normalized (Z-score) to reformate into a common scale (Figure 3, Box 2). Correlation was calculated to identify relationships between the drivers (Figures S5 and S6) and the extent of interdependencies among the dataset that might limit the analysis. Then the Kohonen Self-Organizing Map (SOM) technique was applied to generate a single-layer map of the archetypes of LD drivers [54]. The SOM approach involves organizing data (in this case, the 12 spatial datasets) into patterns based on their inherent similarities and dissimilarities into different groups [54,55]. By testing the different combinations of clusters and using a performance analysis involving both the Davies–Bouldin index and the mean distance of the classified grid cells [23,25], nine clusters of LD drivers were identified as archetypes (see supplementary material, Figures S2–S5). The Z-score, i.e., standardized score from the SOM, was used to examine values of each driver in terms of distance from the mean and to explain the clusters. A Z-score that is equal to the mean is zero, while positive and negative Z-score, i.e., greater than or less than zero, depict distance way above or below the mean. Through an ordinal scaling, the relative standing of the Z-score from the mean was ranked. If the Z-scores = 0, this implies mean/low influence of a driver; Z-score $\leq \pm 1$ = moderate influence; $\pm 1 <$ Z-score $< \pm 2$ = high influence; and Z-score $\geq \pm 2$ = very high influence, respectively (Figure 3, Box 3). From the absolute values of the Z-scores of each cluster, the percentage dominance categories in the clusters were determined [23] (Figure 3, Box 3).

Table 1. List of driver's datasets for the development of archetypes.

(1a) Land-Use Management Practices				
Variables	**Type of Driver**	**Explanation**	**Hypothesized Effect**	**References**
Fire occurrence density derived from active fire and hotspot data	Proximate	Uncontrolled fire occurrences and bush burning destroy soil and biomass, leading to LD—while controlled fire could be a management strategy.	Frequent man-made fire activities lead to more LD. The fewer the man-made fire activities, the less the LD.	[44–46]
Livestock grazing intensity	Proximate	Overgrazing due to higher livestock intensity leads to LD, especially when the threshold limits of the support systems are exceeded, while lower livestock intensity does not lead to LD.	The higher the livestock grazing density, the higher the LD. The lower the grazing density, the less the LD.	[47,48]
Distance to major road	Proximate/Anthropogenic	Roads are a measure of infrastructural development that enhances access to markets and extension services. For instance, a good access road encourages land conversion, including the spread and the adoption of land management practices, while inadequate access discourages land conversion.	Proximity drives LD because nearer forest patches are easier to clear, while areas far away are not affected. The farther an area to a major road, the less the LD; the closer to a major road, the more the LD.	[16,29,38,49]
Distance to major towns	Proximate/Anthropogenic	Land that is close to major towns is prone to LD pushed by urban development; while land that is far from major towns is less prone to conversion and LD.	The farther an area to a major town, the less the LD; the closer to a major town, the more the LD. Areas near settlements or towns are more likely to degrade due to expanding settlements and resource extraction (e.g., wood).	[16,17,29]
Protected area status	Proximate	Human pressure that causes LD is more severe on unprotected land than on protected land.	Protected areas are less prone to degradation due to their protected status.	[50,51]
(1b) Socio-Economic Drivers				
Variables	**Type of Driver**	**Explanation**	**Hypothesized Effect**	**References**
Population density	Underlying	Population density can trigger better land management, but can also lead to LD due to overuse or poor management.	The higher the population density, the less the LD; the lower the population density, the more the degradation. Alternatively, the higher the population density, the higher the LD. The lesser the population density, the lesser the LD.	[16,17,35]
Female/Male Literacy	Underlying	High literacy indicates better access to information or knowledge for making informed decisions, while low literacy may limit people's ability to understand available information—hence they are likely to make poorer land management and investment decisions.	The higher the literacy of women/men, the better the access to knowledge on combating LD; the lower the literacy of women/men, the lesser the access to such information.	[38]

Table 1. *Cont.*

Variables	Type of Driver	Explanation	Hypothesized Effect	References
Poverty head count	Underlying	Diverging evidence. Poverty and LD are intertwined: while poverty could trigger LD, LD could exacerbate poverty. Poor land users are more likely to give up their land tenure security to other more powerful land users, for example losing their land to large-scale land investments for monocultures.	The higher the poverty, the more the prevalence of LD due to various intervening factors such as lack of capital and labor to invest in land management and insecure land tenure.	[16,29,38]

(1c) Environmental Drivers				
Variables	**Type of Driver**	**Explanation**	**Hypothesized Effect**	**References**
Slope	Proximate	Slope influences land-use decisions. Steep lands are often avoided for land-use activities such as cultivation. However, cultivation on non-terraced steep slopes and overgrazing can cause LD such as water and wind erosion.	The steeper the slope, the more susceptible to LD, but the less will be the land's attractiveness for use. The less steep the slope, the less susceptible to LD but the more the attractiveness for land-use activities.	[16,39,52]
Bulk Density (BD)	Underlying	Soils with low bulk density have favorable conditions in terms of soil pore space, texture and organic matter content that influence the choice of land for crop cultivation and biomass clearance.	Low soil bulk density encourages tillage and crop cultivation, while high soil bulk density discourages crop cultivation. The higher the bulk density, the higher the LD; The lower the bulk density, the less the LD.	[41–43]
Elevation/Topography	Proximate/Anthropogenic	Land uses such as farming, which promote degradation are often practiced on flat terrain while rough or high hilly terrains are avoided. Hence, flat terrains are more likely to be exposed to land-use pressure from crop farming.	The higher the elevation, the more susceptible is the land to LD such as erosion, but the less the land's attractiveness for use. The lower the elevation, the less susceptible is the land to LD, but the more attractive for land use because of soil attributes such as deeper soils.	[16,53]

2.4.3. Linking Archetypes of Land-Degradation Drivers to State Administrations and Land Status

In a previous study by [11], LD status was captured using rainfall-corrected vegetation greenness as a proxy (Figure 3, Box 4). By overlaying the archetypes cluster (Figure 3, Box 3) with the LD status of the area (i.e., degraded, stable, and improvement) (Figure 3, Box 4), the percentage share of each archetype per LD status was determined. This enabled the grouping of the nine archetypes as undergoing large-area or small-area degradation, respectively. Thus, archetypes with area coverage <10% of the total degraded area have a small-area degradation, while archetypes with area coverage ≥10% of the total degraded area are classified as large-area degradation clusters [56]. Through spatial overlay [23], the linking of the archetypes with the states' administrative boundaries to determine the share of each archetype per state was possible (Figure 3, Box 4). The emerging results were used to explain and discuss the implications for land governance and SLM in the NGS (Figure 3, Box 5).

3. Results

3.1. Land-Degradation Status

Using Normalized Difference Vegetation Index (NDVI) as a proxy for degradation status, Figure 4 shows the spatial distribution of LD in the NGS by [11]. About 38% (251,401 km^2) of the NGS is degraded, while 14% (91,258 km^2) and 48% (319,470 km^2) show improvement and remain stable, respectively. While improved and stable areas are mostly found in the north of the NGS and to a certain extent in the south of the NGS, large-area degradation is predominantly found in the centre of the NGS, ranging from its north-western to its eastern border.

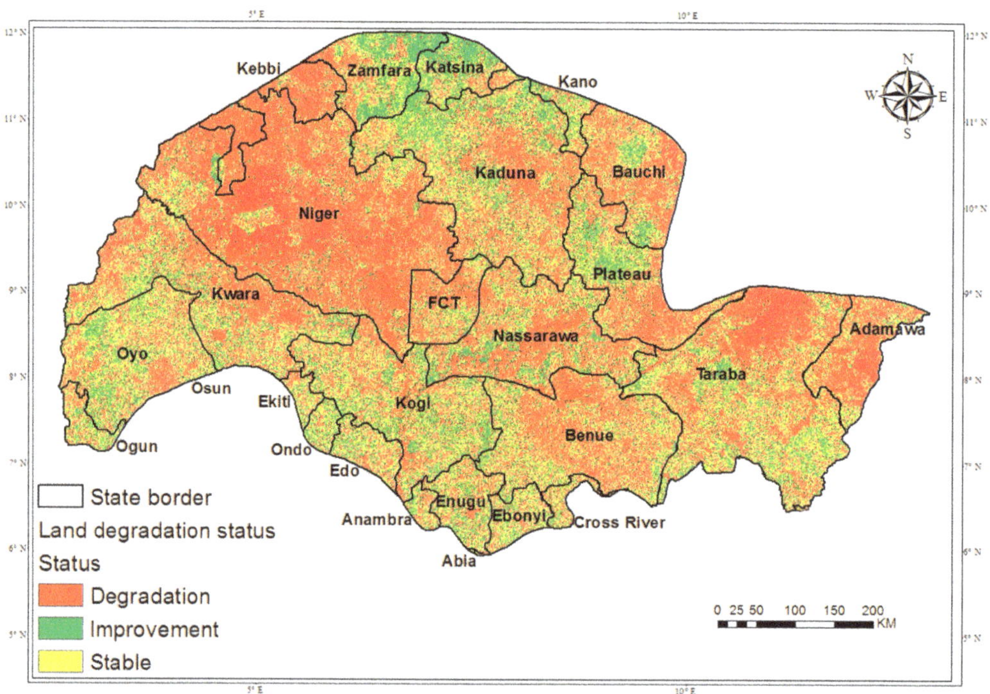

Figure 4. Land-degradation (LD) status for the Nigerian Guinea Savannah (NGS) (Source [11]) using Normalized Difference Vegetation Index (NDVI) as a proxy after correcting for rainfall.

3.2. Land-Degradation Archetypes

In this section, the archetypes and how they can improve the understanding of LD in the NGS are presented. Figure 5 displays each archetype according to the percentage contributions of driver categories, while Figure 6 shows the spatial distribution of the archetypes. Based on the 12 input drivers, nine archetypes of LD drivers were identified (Figure 5). Five archetypes were dominated by land-use management practices (NGSA 1, NGSA 5–8), and three dominated by socio-economic drivers (NGSA 2–4), while NGSA 9 was dominated by environmental drivers (Please refer to supplementary material, Table S1 for a description of the archetypes).

3.3. Spatial Distribution of Archetypes

The spatial distribution of the nine archetypes of LD drivers is shown in Figure 6. In Table 2, a brief description and share of each archetype in the NGS is provided.

From Table 2, six clusters (NGSA 2–5, NGSA 7 and NGSA 9) with individual total areas greater than 10% cover 78.5% of the total area, while the remaining three archetypes with individual total areas smaller than 10% of the area cover 21.5% of the NGS.

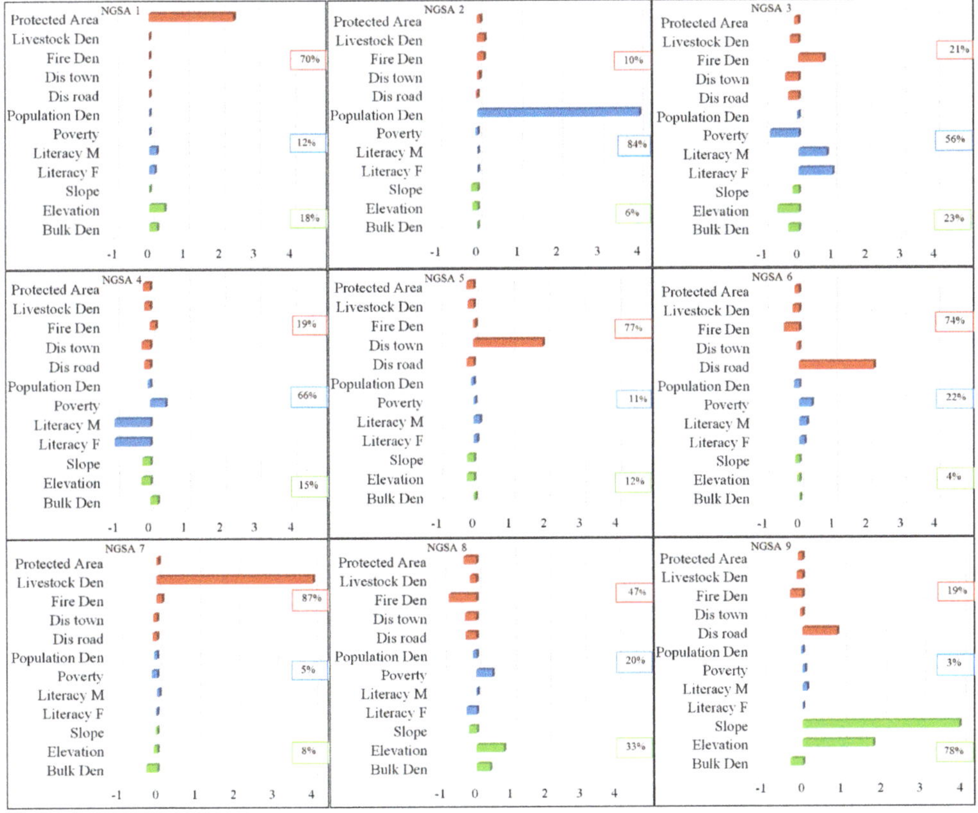

Figure 5. The Z-score normalized values of drivers characterizing the nine archetypes of LD drivers, (zero depicts the mean for the NGS; the percentage contribution of driver categories into archetype clusters is presented in the boxes: red: land-use management practice; blue: socio-economic; green: environmental drivers).

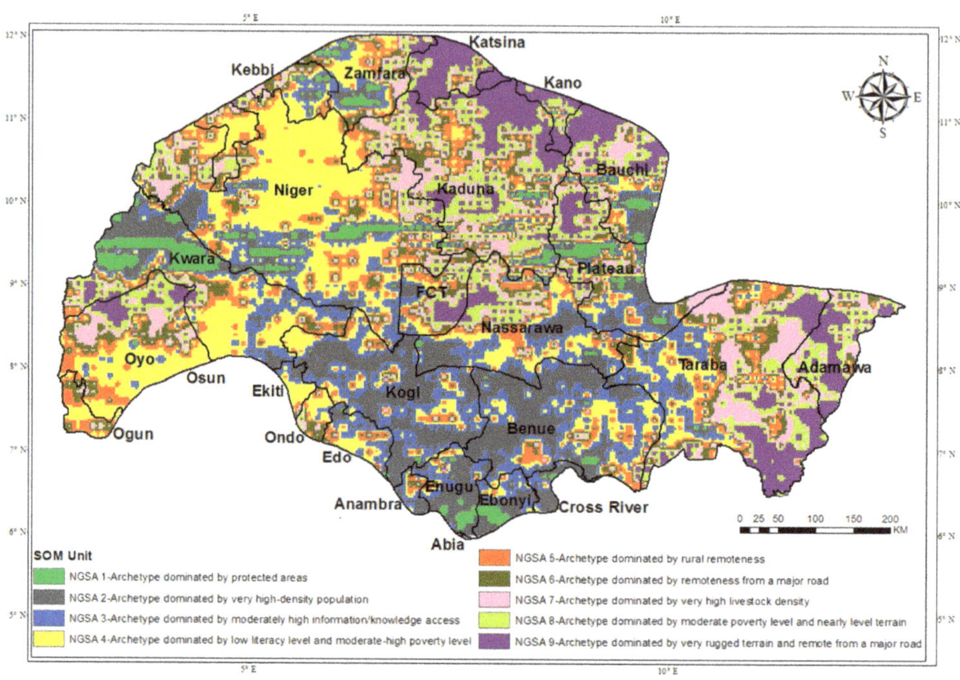

Figure 6. Spatial characterization of archetypes of LD drivers. See supplementary material, Section 6 for the full description and ranking of the archetypes in relation to administrative and land status.

Table 2. Description and share of the archetypes of LD drivers.

SOM	Brief Description	Area Share (%)	Area Share (km^2)
NGSA 1	**Archetype dominated by protected areas:** Areas with very high numbers of protected areas that are associated with the moderate–high influence of elevation, bulk density and high literacy.	3.3	14,412
NGSA 2	**Archetype dominated by very high-density population:** Areas with very high population density and with minimal influence of livestock and high fire activities.	15.3	67,169
NGSA 3	**Archetype dominated by moderately high information/knowledge access:** Mainly areas with a moderately high level of both male and female literacy, including fire-occurrence activities but with low poverty.	12.4	54,528
NGSA 4	**Archetype dominated by low literacy level and moderate–high poverty level:** Area characterized by moderate–high poverty and minimal fire activities, but with low levels of both male and female literacy.	20.1	88,036
NGSA 5	**Archetype dominated by rural remoteness:** Highly dominated by land-use management practices and remote from major towns but with a moderately low population density, protected area prevalence, and low livestock density.	10.1	44,408
NGSA 6	**Archetype dominated by remoteness from a major road:** Highly dominated by land-use management practices, which occur at a far distance away from major roads with moderately high poverty and literacy but with moderate fire and livestock activities.	8.7	38,273

Table 2. Cont.

SOM	Brief Description	Area Share (%)	Area Share (km^2)
NGSA 7	**Archetype dominated by very high livestock density:** Areas with a very high livestock density and moderate levels of other drivers.	10.1	44,044
NGSA 8	**Archetype dominated by moderate poverty level and nearly level terrain:** Collectively driven by all drivers' categories but fairly dominated by land-use management practices, but with a moderate elevation and moderate influence of bulk density and poverty.	9.4	41,297
NGSA 9	**Archetype dominated by very rugged terrain and remote from a major road:** Areas with moderate elevation, high slope, and distant from the major road.	10.5	45,880
		100.00	438,046.88

3.4. Categories of Archetypes According to State Administrative Boundaries and LD Status

3.4.1. Degree of Land-Degradation Status per Archetype

Figure 6 links the archetypes with the LD status (degradation, stable, improvement; Figure 4), to highlight their proportions and potential interplay (Figure 7). Four archetypes with very high population density (NGSA 2), moderate–high information/knowledge access (NGSA 3), and moderate–high poverty level (NGSA 4), as well as NGSA 5—very remote from a major town—are associated with 61.3% of the large-area LD, while the other archetypes account for 38.7% of small-area degradation (Figure 7). Six archetypes—NGSA 2 to NGSA 5, as well as NGSA 7, with very high livestock density, and NGSA 8 with dominant land-use management practices and nearly level terrain—are responsible for 78.4% of the large-area stable status and the other archetypes for 21.6% of the small-area stable status (Figure 7). For the large-area improvement, six archetypes, NGSA 2 to NGSA 4 and NGSA 7 to NGSA 9 with rugged terrain, i.e., very high slope and moderate elevation, covered 78.7% while other archetypes account for 21.3% of small-area improvement (Figure 7). For the complementary table, see supplementary material, Table S3; and Section 6 for the full description and ranking of the archetypes in relation to LD status.

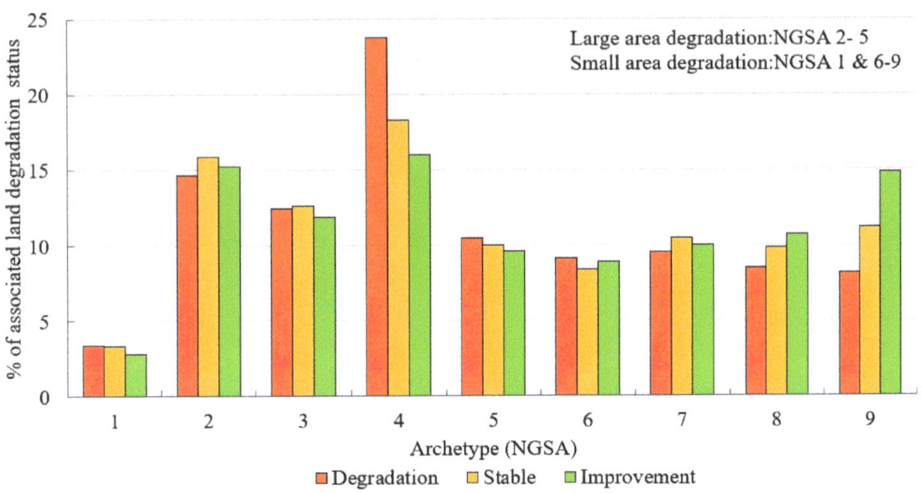

Figure 7. Archetypes in percentage of associated LD status (For full percentages, see supplementary material, Table S3).

3.4.2. Share of Land-Degradation Archetypes per State Administration Unit

State administrations manage land within their jurisdictions, hence they influence land-use decisions in Nigeria. Figure 8 shows the percentage share of the archetypes within a state's boundary (for the complementary table, see supplementary material, Table S2). Seven states, comprising Bauchi, Kaduna, Kwara, Nasarawa, Niger, Plateau, and Zamfara have all the nine archetypes. While four states, namely Benue, Kebbi, Taraba, and the Federal Capital Territory (FCT) are covered by eight of the nine archetypes. The remaining states have an uneven combination of all archetypes, with the portion of Abia state found within the NGS only embodying one archetype (i.e., NGSA 2). The five states comprising Niger (40.5%), Oyo (29.6%), Kwara (24.4%), Nassarawa (18.6%), and Ekiti (17.6%), have the largest shares of the archetype of NGSA 4, i.e., a moderate–high poverty level. The supplementary material Table S2 contains the grouping of the archetypes according to the state administrative units.

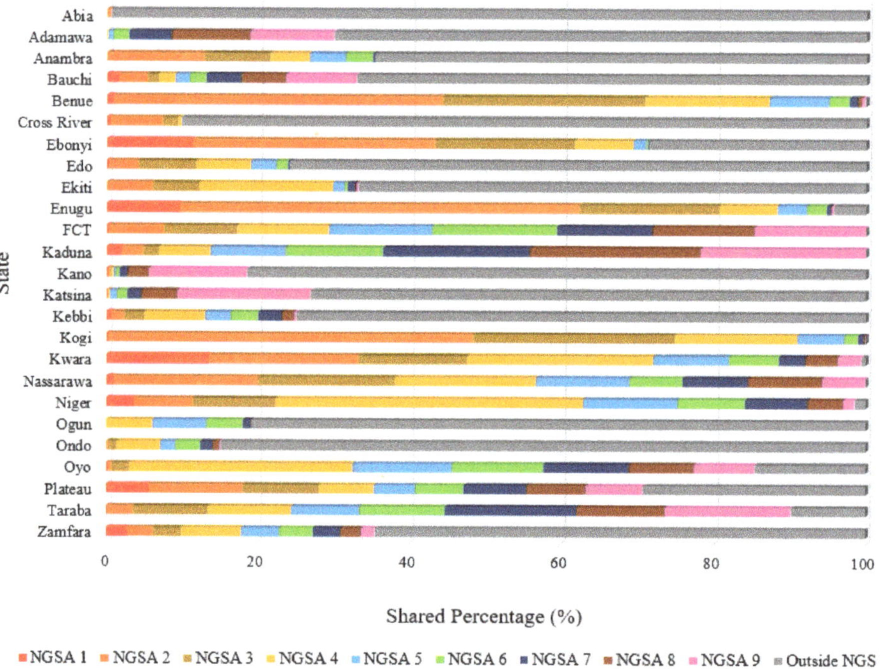

Figure 8. Share of LD archetypes by states in the NGS. Note: shares outside the NGS, that were not analyzed, are in grey. (For full percentages, see supplementary material, Table S2).

4. Discussion

4.1. Understanding the Archetypes of Large-Area Degradation

Areas identified to be under large-area degradation are archetypes with more than 10% of their areas experiencing biomass degradation (NDVI). Four major archetypes of large-area degradation were thus identified (Figures 5 and 7). Out of these, NGSA 3, with prolonged cases of fire occurrence, and NGSA 5, with rural remoteness from a major town (Figure 5), highlight land-use management practices as the drivers of large-area degradation in the NGS. NGSA 3 thus confirmed studies that have implicated fire-related activities such as charcoal making, farming, and hunting with bush burning as causes of LD [57,58]. The dominant characteristics of NGSA 5, on the other hand, contradicts the notion that land areas closer to major towns are more prone to LD than those farther away [16,17,29]. NGSA 5 rather reveals the rural areas and natural resource users in remote areas who adopt unsustainable land-use practices (of NGSA 3), such as continuous bush burning and deforestation, trigger LD [59].

Apart from archetype NGS 5 with high remoteness from a major town (Figure 5), the other three, i.e., archetypes with very high-density population (NGSA 2), moderately high information/knowledge access (NGSA 3), and moderate–high poverty level (NGSA 4), are dominated by high percentages of socio-economic drivers. Thus, socio-economic factors are major underlying causes that indirectly push other proximate drivers of large-area degradation in the NGS [29,60,61]. With the low population density of three (NGSA 3–NGSA 5) out of the four archetypes experiencing large-area degradation, high poverty, and low literacy, this factor can be inferred to be associated with large-area degradation in the NGS context [59,61,62]. This confirms studies such as [62,63], who reported that poverty intensifies a tendency to change vegetation cover, as many people deplete natural vegetation for fuel, food, and as a source of income because of fewer or no alternative livelihood options. This invariably points to the areas with high poverty and low population density, i.e., rural population, covering the northwest central and northeast of the NGS, encompassing the states of Kebbi, Niger, parts of northern Kwara, FCT (mainly around Abuja, see the area illustrated in Figure 5), and parts of Nasarawa, Plateau, Taraba, Kaduna, and Adamawa states, which have experienced extensive degradation [11]. Although, NGSA 2 with a very high-density population, hints that urban areas, with their high population density, are also associated with large-area degradation in the NGS, the degradation is not as extensive as in the low-population-density remote areas. Therefore, this result deviates from the general notion that a high-population density is the main cause of LD in Nigeria [13,33]. Hence, a high population density alone does not drive LD without certain complementary factors, such as poverty, illiteracy that restricts information or knowledge access, and poor national policies that are prevalent in Sub-Saharan African contexts [64]. Therefore, the three socio-economic sub-drivers—poverty, literacy and population density—are core interrelating drivers in large-area LD, that require attention in addressing LD in the NGS [29,62,65].

While the percentages of land-use management practices and socio-economic factors dominate as potential drivers of large-area LD, specific environmental drivers of the archetypes also underpin the large-area LD (Figure 5). The nearly level terrain condition, i.e., low-elevation–flat terrain of the four large-area archetypes, is known to encourage land cultivation in Nigeria [13]. In addition, the characteristics of low bulk density of NGSA 2, NGSA 3 and 5, and the high bulk density of NGSA 4 with moderate–high poverty level also highlight soil characteristics that encourage large-area degradation [42,43]. The low bulk density archetypes signify few areas in the southern part of the NGS with suitable soil for cultivation. The archetypes characterized by high bulk density on the other hand correspond to areas with the highest impact of agricultural management practices, such as machinery and high cropping impacts [42]. This represents 23% of the large-area degradation archetypes, and in turn reflects the widespread LD due to the high agricultural engagement by the rural dwellers in the zone [63].

4.2. Understanding the Archetypes of Small-Area Degradation

Five archetypes of small-area degradation were identified, that is, archetypes where degraded areas are less than 10% of the archetype area. LD and their drivers thus differ locally and are context-specific [5,60], and an archetype approach can help identify the socio-ecological contexts [66]. From the five small-area archetypes, three archetypes identified with very high presence of protected areas (NGSA 1), that are very remote from a major road (NGSA 6), have a very high livestock density (NGSA 7), and have high percentages of land-use practices (Figure 5). While the NGSA 1 reflects its conservation and restricted use status, the additional association of NGSA 1 with high and rugged elevations like NGSA 9, further explains its small extent of degradation. However, with its low proximity to major roads and major towns, degradation in protected areas as captured in NGSA 1 call for an investigation into the specific activities driving degradation in protected areas, such as encroachment by human activities [11,63], despite government regulations, particularly around communities that host protected areas [67,68].

In archetypes NGSA 6 and NGSA 9, the small-area degradation is highly driven by non-proximity to a major road, without much influence of other sub-drivers (Figure 5). Nearness to major roads is a measure of infrastructural development that influences accessibility and the spread of land-use management practices, including information [69]. Thus, NGSA 6 represents remote areas with restricted access, which can hamper the propagation of sustainable land-management initiatives [29,38,49]. As in NGSA 1 and NGSA 6, the very high livestock density configuration of NGSA 7 (Figure 5) is also associated with small-area degradation (Figure 7). The Guinea savannah in particular is currently under pressure from high grazing activities because the Sahel and Sudan savannahs have been extensively degraded by overgrazing [15,70], leading to competition for grazing resources and conflicts in the region [71]. Overgrazing has been associated with the disappearance of the typical savannah vegetation and the emergence of the Sudan–Sahelian Savannah in the NGS [15,72]. Thus, from the combination of drivers above, there is a critical need for an improved management of grazing resources, protected areas, and the governance of land resources in the NGS [11,36,68]. While all small-area archetypes are mostly dominated by the land-use management drivers, NGSAs 6 and 8 are the only small-area archetypes that are distinctly driven by the socio-economic drivers characterized by areas with low population density, i.e rural population with corresponding moderate information/knowledge access. Unlike other factors, poverty (high or low) is not a distinctive feature of these archetypes (Figure 5), hence this study cannot confirm the notion that 'the higher the poverty, the more the degradation' held by many studies of small-area degradation [38], as noted Table 1. Considering that poverty is widespread in the NGS, there is a need for integrating other social-demographic and social-relational data for a better understanding of the interactions between poverty and LD.

4.3. Archetypes and Policy Insights

As multiple factors are associated with LD, policy interventions aimed at achieving SLM need to be inter-sectoral. However, many policies in Nigeria, such as the Nigerian National Agricultural Policy, focus on single sectors and often do not have LD reduction as a primary objective [10]. Key policy topics related to these findings are sustainable use and management of natural resources, poverty reduction, environmental awareness and education, strategy to reduce dependence on land and natural resources for livelihoods, and inclusion of LD in land-use planning.

With the extensive LD in the NGS, policies for the sustainable management of natural resources including water, soil, and biodiversity, as well as their coherence, are essential [73]. While several response programmes such as the Nigeria National Policy on Environment are promulgated to address activities that cause LD [74], they remain reactive without an effective scaling up to tackle the drivers of LD. For example, the proposed national policy on the rediscovery of grazing routes and reserves remains unprepared to address LD because the advocates focus on the profit and the pressing need to tackle the farmers and herders clashes in Nigeria [67,71], without a recourse to the fact that spatial developments in Nigeria have overtaken several historical grazing spaces [67]. The polarized nature of Nigeria between sectional and ethnic divides further raised several counter notions with socio-political undertones to such policy moves by the government [75]. This subsequently affected the acceptance of related policies such as the 10-year National Livestock Transformation Plan (NLTP), ranching plan, and open grazing [67]. While it is obvious that degradation induces competition and tension among natural resources users, policy decisions on land-based issues require a special focus on LD and land restoration.

Although results did not explicitly show that only low/high poverty is associated with large-/small-area degradation, poverty contributes to the different archetypes identified. Nigeria has about 83 million (40%) of its population living below $1.90 per day, comprising 52% rural dwellers whose livelihoods are predominantly tied to agricultural activities [76,77]. About 30 million more Nigerians are expected to be added to the national population living in extreme poverty by 2030 [77]. Many of the poor depend on livelihood activities such as charcoal making and hunting with bush burning that promote LD [57,58]. Archetype NGS 4, which covers the largest proportion of the NGS (20%) and has moderate–high poverty and a low level of both male and female literacy, is characterized by large-area degradation. Two broad groups of the country's poverty alleviation programmes (PAPs) have been identified in Nigeria [78]: (a) the Core Poverty Alleviation Programmes (CPAPs) such as Better Life Programme for Rural Dwellers (BLP) and Family Economic Advancement Programme (FEAP), and (b) the Non-Core Poverty Alleviation Programmes (NCPAPs), which include the National Agricultural Land Development Authorities (NALDA). Such policies had no long-lasting effects [78,79] and did not focus on the intersections between poverty and LD. Other policies such as the Agricultural Development Projects (ADPs) and Vision 20-2020 are in a dying state [78,79], with most interventions aimed at increasing farmer revenue and reducing poverty, and mainly focus on improving input supply without giving attention to improving land management.

Nigeria has about 56.9% adult illiteracy [80], with variations across states and regions including urban and rural areas (i.e., urban 74.6% and rural 48.7%). NGSA 3 shows the linkage of low male and female literacy with large-area degradation. Most farmers and herders in Nigeria did not finish primary education and are less likely to access and understand the little knowledge and information disseminated through extension services or lack the resources to access this information themselves [79,81,82]. While the use of mobile phones is promoted to improve access to information, little or no information is provided on sustainable land-use and management. According to [83], technology is necessary to scale up the adoption of initiatives amongst resource-poor users. With the widespread poverty, weak industrial presence to absorb the increasing population, and the reliance on the primary sector, the quest to exploit environmental resources supersedes interest for environmental protection and management [84]. Hence, pathways to improving land management need to be sought both outside agriculture (e.g., creating employment opportunities outside agriculture) as well as within agriculture through improving farmers/herders' access to sustainable land management practices as well as motivating their adoption [85].

With the growing LD, effective policy on land-use planning is critical for degradation response in Nigeria as the current land-use policies and practices do not adequately consider sustainable land management [10,86]. The historic lapses in the National Land Use Act (LUA) of 1978 persist, whose focus only recognizes land ownership and promotes land access without a sustainable land-use plan or governance to cater for the pressure from the growing population. Calls to review the LUA to give room for a more sustainable policy for land-use planning and governance in Nigeria [10,13], remain unheeded. A challenging question is thus: what opportunities can be identified for promoting SLM [10]?

4.4. Archetypes and Sustainable Land Management (SLM)

Based on literature and the study results, sustainable fuelwood management/energy efficiency, reforestation and afforestation, sustainable pastoralism, and structural land management measures are potential interventions to address LD.

In Nigeria, over 70% of the population rely on wood fuel for cooking, which is an underlying driver of deforestation and associated LD [63]. In recognition of the reliance on fuel wood, sustainable fuelwood management (SFM) is being promoted under the United Nations Development Programme (UNDP)/Global Environment Facility (GEF) project on management of fuel wood in mitigating the effects of climate change such as in Kaduna State, Nigeria [87,88]. While such initiatives have made some progress in establishing woodlots, producing energy-efficient cooking stoves, and establishing local forest management committees, low community buy-in, land tenure, and governance remain key constraints [88]. A review of such initiatives can provide insights on how to improve states' SLM outcomes and out-scale them to other states in the NGS. Thus, SFM has some potential for promoting landscape stewardship in the region [87,89].

Tree-based programmes also hold potential to reduce LD and remain the principal focus of restoration programs [90]. Reforestation, afforestation, and agroforestry have been found advantageous and successful around the world and in Nigeria, particularly in LD response [91]. In Nigeria, successive governments at all levels have worked collaboratively to encourage and implement various afforestation projects [92]. For instance, among the frontline states of the Great Green Wall Afforestation Programme, tree planning campaigns with eucalyptus species and shelter belts for sand dune and degradation fixation are commonly practiced [74,92]. Therefore, land users in Nigeria can be incentivised to participate in tree-based initiatives, which have recorded successes elsewhere, such as in Kenya [64], to reawaken interest in combating LD. Such initiatives can focus on Niger, Nassarawa, Kwara, and Kogi, including Kaduna and Oyo states, due to the prevalence of large-area LD archetypes. Agroforestry, a multifunctional practice of cropping with trees and shrubs on arable land, is also a potential SLM practice that can improve land productivity, is a low-cost and adaptable tree-based initiative [91,92], that contributes to food security and land resource conservation [91].

With the evidence of high livestock grazing activities in archetype NGSA 7 (Figure 5), traditionally, pastoralism is predominantly practiced in northern Nigeria, with southward movement following the rain and in search of pasture and water during the dry season. Overgrazing causes LD, and indiscriminate overgrazing has caused negative stereotyping and fuelled tensions between pastoralists and non-pastoralist-actors, causing loss of lives and properties as well as communal crisis [67,71]. In some cases, overgrazing by livestock and excessive open grazing lead to the failure of afforestation programmes, including severe violation of protected areas across West Africa [67,93]. While pastoralism under a proper management system is ecologically, economically, and socially viable [94,95], climate change and poor land management in the face of growing national population and pressures from neighbouring herding countries make traditional pastoralism unsustainable in Nigeria [71,93,94]. Studies thus call for controlling open grazing to check indiscriminate overgrazing and secure livestock production in Nigeria [75,96].

In view of several environmental consequences of the archetype driven by terrain characteristics, avoiding degradation-prone rugged terrain is key to maintaining the remaining biomass of the zone. Investing in SLM structures such as land levelling, terracing, and contour farming are critical to tackling LD [97,98], particularly on agricultural landscapes like the NGS, where engagement in farming remains necessary for livelihood sustenance [12,13], and biodiversity and natural conditions are threatened largely by agricultural expansion [13]. Terracing and high-altitude afforestation for erosion control, for example, have been recognized to reduce loss of soil and LD on sloped terrain [99]. Similarly, contour farming and staggered contour trenching, which involves planting of crops across a slope based on elevation contour lines are also effective for managing degradation on rugged–steep terrain [99].

5. Conclusions

This study identified nine archetypes of LD drivers in the NGS, which are mostly dominated by social-economic, land-use management practices, and a slight influence from environment drivers. Specifically, four archetypes characterized by a very high-density population, moderately high information/knowledge access, and moderate–high poverty level, as well as remoteness from a major town, account for 61.3%, 78.4%, and 78.7% of total degraded, stable, and improvement areas, respectively. LD is mostly evident in states bordering the northwest to the central and northeast of the NGS, such as Niger state, which have predominantly large rural farming communities. Besides revealing the LD drivers, the archetypes characteristics provide a basis for determining and prioritizing relevant SLM policies and practices such as poverty reduction, creating environmental awareness and promoting sustainable pastoralism as well as robust land-use planning to strengthen land governance in Nigeria. Despite the limitations of spatial data on the driving factors, the outputs from this study provide a useful guide on how archetypes can serve as a tool for progressing Nigeria's LDN through SLM. Like most unsupervised classification techniques, field validation of the archetypes results is necessary because of the adopted self-organizing mapping techniques. However, this could not be conducted because mobility limitations and scarcity of spatially explicit data limited the number of variables that could be used for this study. As more spatially explicit data on Nigeria and Africa become available, they need to be integrated in future studies of archetypes as well as validating them with field observations.

Supplementary Materials: The following are available online at https://www.mdpi.com/2072-4292/13/1/32/s1, Figure S1: Correlation matrix between the selected drivers of land degradation archetypes, Figure S2: Training process for determining the nine cluster separations that are suitable for the land-degradation archetypes, Figure S3: The plot showing the aggregated profile of each neuron as synthesized by the selected drivers of land degradation, Figure S4: Quality of the nine clusters in the codebook, Figure S5: Neighbor distances between the neurons for the clusters, Table S1: Calculated Z-score normalized values of drivers characterizing the nine (9) distinctive archetypes of land degradation, Table S2: Archetypes and percentage share per state administration unit, Table S3: Archetypes and share of land status.

Author Contributions: A.A.A.: Conceptualization, Methodology, Software, Data curation, Writing—Original draft preparation. C.I.S.: Conceptualization of the research project, Supervision, Writing—Review & Editing. All authors have read and agreed to the published version of the manuscript.

Funding: Ademola. A. Adenle is funded by the UniBe International 2021, Initiative of the Vice-Rectorate Development, University of Bern, Switzerland and acknowledges the small grant funding from the Rufford foundation for the research project [27153-1] in Nigeria as well as the Small Equipment Grant from IDEA WILD toward the acquisition DJI Mavic Air Quadcopter with Remote Controller.

Acknowledgments: Ademola A. Adenle acknowledges the support from the UniBE International 2021, Initiative of the Vice-Rectorate Development, University of Bern, Switzerland. We are also grateful to the reviewers whose constructive feedback helped to improve this paper. This study contributes to the Programme on Ecosystem Change and Society (www.pecs-science.org) and the Global Land Programme (www.glp.earth).

Conflicts of Interest: The authors declare no conflict of interest.

References

1. Sims, N.C.; Barger, N.N.; Metternicht, G.I.; England, J.R. A land degradation interpretation matrix for reporting on UN SDG indicator 15.3.1 and land degradation neutrality. *Environ. Sci. Policy* **2020**, *114*, 1–6. [CrossRef]
2. Laestadius, L.; Maginnis, S.; Minnemeyer, S.; Potapov, P.; Saint-Laurent, C.; Sizer, N. Mapping opportunities for forest landscape restoration. *Unasylva* **2011**, *62*, 47–48.
3. Goffner, D.; Sinare, H.; Gordon, L.J. The Great Green Wall for the Sahara and the Sahel Initiative as an opportunity to enhance resilience in Sahelian landscapes and livelihoods. *Reg. Environ. Chang.* **2019**, *19*, 1417–1428. [CrossRef]
4. Olsson, L.; Barbosa, H.; Bhadwal, S.; Cowie, A.; Delusca, K.; Flores-Renteria, D.; Hermans, K.; Jobbagy, E.; Kurz, W.; Li, D.; et al. Land Degradation. In *Climate Change and Land: An IPCC Special Report on Climate Change, Desertification, Land Degradation, Sustainable Land Management, Food Security, and Greenhouse Gas Fluxes in Terrestrial Ecosystems*; IPCC Press Office: Geneva, Switzerland, 2019; pp. 345–436.
5. Akinyemi, F.O.; Tlhalerwa, L.T.; Eze, P.N. Land degradation assessment in an African dryland context based on the Composite Land Degradation Index and mapping method. *Geocarto Int.* **2019**, 1–17. [CrossRef]
6. Southworth, J.; Zhu, L.; Bunting, E.; Ryan, S.J.; Herrero, H.; Waylen, P.R.; Hill, M.J. Changes in vegetation persistence across global savanna landscapes, 1982–2010. *J. Land Use Sci.* **2016**, *11*, 7–32. [CrossRef]
7. Ellison, D.; Ifejika Speranza, C. From blue to green water and back again: Promoting tree, shrub and forest-based landscape resilience in the Sahel. *Sci. Total Environ.* **2020**, *739*, 140002. [CrossRef]
8. UNCCD. *The Global Land Outlook, West Africa Thematic Report*; United Nations Convention to Combat Desertification: Bonn, Germany, 2019.
9. Höhn, A.; Neumann, K. Shifting cultivation and the development of a cultural landscape during the Iron Age (0–1500 AD) in the northern Sahel of Burkina Faso, West Africa: Insights from archaeological charcoal. *Quat. Int.* **2012**, *249*, 72–83. [CrossRef]
10. Ifejika Speranza, C.; Adenle, A.; Boillat, S. Land Degradation Neutrality—Potentials for its operationalisation at multi-levels in Nigeria. *Environ. Sci. Policy* **2019**, *94*, 63–71. [CrossRef]

11. Adenle, A.A.; Eckert, S.; Speranza, C.I.; Adedeji, O.I.; Ellison, D. Human-induced land degradation dominance in the Nigerian Guinea savannah between 2003–2018. *Remote Sens. Appl. Soc. Environ.* **2020**, 100360. [CrossRef]

12. CILSS (Comité Permanent Inter-états de Lutte contre la Sécheresse dans le Sahel). *Landscapes of West Africa—A Window on a Changing World*; CILSS: Garretson, SD, USA, 2016.

13. Arowolo, A.O.; Deng, X. Land use/land cover change and statistical modelling of cultivated land change drivers in Nigeria. *Reg. Environ. Chang.* **2018**, *18*, 247–259. [CrossRef]

14. Lambin, E.F.; Geist, H.J.; Lepers, E. Dynamics of Land-Use and Land-Cover Change in Tropical Regions. *Annu. Rev. Environ. Resour.* **2003**, *28*, 205–241. [CrossRef]

15. Macaulay, B.M. Land degradation in Northern Nigeria: The impacts and implications of human-related and climatic factors. *Afr. J. Environ. Sci. Technol.* **2014**, *8*, 267–273. [CrossRef]

16. Mirzabaev, A.; Nkonya, E.; Goedecke, J.; Johnson, T.; Anderson, W. Global Drivers of Land Degradation and Improvement. In *Economics of Land Degradation and Improvement—A Global Assessment for Sustainable Development*; Nkonya, E., Mirzabaev, A., von Braun, J., Eds.; Springer: Cham, Switzerland, 2016; pp. 167–195. ISBN 978-3-319-19168-3.

17. Batunacun; Wieland, R.; Lakes, T.; Yunfeng, H.; Nendel, C. Identifying drivers of land degradation in Xilingol, China, between 1975 and 2015. *Land Use Policy* **2019**, *83*, 543–559. [CrossRef]

18. Sietz, D.; Fleskens, L.; Stringer, L.C. Learning from Non-Linear Ecosystem Dynamics Is Vital for Achieving Land Degradation Neutrality. *Land Degrad. Dev.* **2017**, *28*, 2308–2314. [CrossRef]

19. Oberlack, C.; Sietz, D.; Bürgi Bonanomi, E.; de Bremond, A.; Dell'Angelo, J.; Eisenack, K.; Ellis, E.; Epstein, G.; Giger, M.; Heinimann, A.; et al. Archetype analysis in sustainability research: Meanings, motivations, and evidence-based policy making. *Ecol. Soc.* **2019**, *24*. [CrossRef]

20. Levers, C.; Müller, D.; Erb, K.; Haberl, H.; Jepsen, M.R.; Metzger, M.J.; Meyfroidt, P.; Plieninger, T.; Plutzar, C.; Stürck, J.; et al. Archetypical patterns and trajectories of land systems in Europe. *Reg. Environ. Chang.* **2018**, *18*, 715–732. [CrossRef]

21. Sietz, D.; Mamani Choque, S.E.; Lüdeke, M.K.B. Typical patterns of smallholder vulnerability to weather extremes with regard to food security in the Peruvian Altiplano. *Reg. Environ. Chang.* **2012**, *12*, 489–505. [CrossRef]

22. Merino, M.V.; Sietz, D.; Jost, F.; Berger, U. Archetypes of Climate Vulnerability: A Mixed-method Approach Applied in the Peruvian Andes. *Clim. Dev.* **2019**, *11*, 418–434. [CrossRef]

23. Dittrich, A.; Seppelt, R.; Václavík, T.; Cord, A.F. Integrating ecosystem service bundles and socio-environmental conditions—A national scale analysis from Germany. *Ecosyst. Serv.* **2017**, *28*, 273–282. [CrossRef]

24. Oberlack, C.; Eisenack, K. Archetypical barriers to adapting water governance in river basins to climate change. *J. Inst. Econ.* **2018**, *14*, 527–555. [CrossRef]

25. Václavík, T.; Lautenbach, S.; Kuemmerle, T.; Seppelt, R. Mapping global land system archetypes. *Glob. Environ. Chang.* **2013**, *23*, 1637–1647. [CrossRef]

26. Iloeje, N.P. *A New Geography of Nigeria*; New Revised Edition; Longman Nig. Ltd.: Lagos, Nigeria, 2001; p. 200.

27. *Nigeria's First National Communication under the United Nations Framework Convention on Climate Change*; The Ministry of the Environment of the Federal 11 Republic of Nigeria: Abuja, Nigeria, 2003.

28. Mulinge, W.; Gicheru, P.; Murithi, F.; Maingi, P.; Kihiu, E.; Kirui, O.K.; Mirzabaev, A. Economics of Land Degradation and Improvement in Kenya. In *Economics of Land Degradation and Improvement—A Global Assessment for Sustainable Development*; Nkonya, E., Mirzabaev, A., von Braun, J., Eds.; Springer: Cham, Switzerland, 2016; pp. 471–498. ISBN 978-3-319-19167-6.

29. Vu, Q.M.; Le, Q.B.; Frossard, E.; Vlek, P.L.G. Socio-economic and biophysical determinants of land degradation in Vietnam: An integrated causal analysis at the national level. *Land Use Policy* **2014**, *36*, 605–617. [CrossRef]

30. FGN (Federal Government of Nigeria). *Final Report of the Land Degradation Neutrality Target Setting Programme*; FGN: Abuja, Nigeria, 2018.

31. Malek, Ž.; Douw, B.; Van Vliet, J.; Van Der Zanden, E.H.; Verburg, P.H. Local land-use decision-making in a global context. *Environ. Res. Lett.* **2019**, *14*, 083006. [CrossRef]

32. Zhu, Z.; Piao, S.; Myneni, R.B.; Huang, M.; Zeng, Z.; Canadell, J.G.; Ciais, P.; Sitch, S.; Friedlingstein, P.; Arneth, A.; et al. Greening of the Earth and its drivers. *Nat. Clim. Chang.* **2016**, *6*, 791–795. [CrossRef]

33. Olorunfemi, I.E.; Fasinmirin, J.T.; Olufayo, A.A.; Komolafe, A.A. GIS and remote sensing-based analysis of the impacts of land use/land cover change (LULCC) on the environmental sustainability of Ekiti State, southwestern Nigeria. *Environ. Dev. Sustain.* **2020**, *22*, 661–692. [CrossRef]

34. WorldPop. *Global 100m Covariates*; WorldPop: School of Geography and Environmental Science, University of Southampton; Department of Geography and Geosciences, University of Louisville; Departement de Geographie, Universite de Namur) and Center for International Earth Science Information Network (CIESIN), Columbia University 2018. Global High Resolution Population Denominators Project—Funded by the Bill and Melinda Gates Foundation (OPP1134076). Available online: https://dx.doi.org/10.5258/SOTON/WP00645 (accessed on 17 November 2020).

35. Tiffen, M.; Mortimore, M.; Gichuki, F. *More People, Less Erosion: Environmental Recovery in KENYA*; John and Wiley: Chichester, NY, USA, 1994; ISBN 978-0-471-94143-9.

36. Abdulaziz, H.; Johar, F.; Majid, M.R.; Medugu, N.I. Protected area management in Nigeria: A review. *J. Teknol.* **2015**, *77*. [CrossRef]

37. Bosco, C.; Alegana, V.; Bird, T.; Pezzulo, C.; Bengtsson, L.; Sorichetta, A.; Steele, J.; Hornby, G.; Ruktanonchai, C.; Ruktanonchai, N.; et al. Exploring the high-resolution mapping of gender-disaggregated development indicators. *J. R. Soc. Interface* **2017**, *14*, 20160825. [CrossRef] [PubMed]

38. Gnacadja, L.; Wiese, L. Land Degradation Neutrality: Will Africa Achieve It? Institutional Solutions to Land Degradation and Restoration in Africa. In *Climate Change and Multi-Dimensional Sustainability in African Agriculture*; Lal, R., Kraybill, D., Hansen, D.O., Singh, B.R., Mosogoya, T., Eik, L.O., Eds.; Springer: Cham, Switzerland, 2016; pp. 61–95. ISBN 978-3-319-41236-8.
39. Jendoubi, D.; Liniger, H.; Ifejika Speranza, C. Impacts of land use and topography on soil organic carbon in a Mediterranean landscape (north-western Tunisia). *Soil* **2019**, *5*, 239–251. [CrossRef]
40. Lloyd, C.T.; Chamberlain, H.; Kerr, D.; Yetman, G.; Pistolesi, L.; Stevens, F.R.; Gaughan, A.E.; Nieves, J.J.; Hornby, G.; MacManus, K.; et al. Global spatio-temporally harmonised datasets for producing high-resolution gridded population distribution datasets. *Big Earth Data* **2019**, *3*, 108–139. [CrossRef]
41. Salako, F.K.; Hauser, S.; Babalola, O.; Tian, G. Improvement of the physical fertility of a degraded Alfisol with planted and natural fallows under humid tropical conditions. *Soil Use Manag.* **2006**, *17*, 41–47. [CrossRef]
42. Enang, R.K.; Yerima, B.P.K.; Kome, G.K.; Van Ranst, E. Effects of Forest Clearance and Cultivation on Bulk Density Variations and Relationships with Texture and Organic Matter in Tephra Soils of Mount Kupe (Cameroon). *Commun. Soil Sci. Plant Anal.* **2017**, *48*, 2231–2245. [CrossRef]
43. Imbrenda, V.; D'emilio, M.; Lanfredi, M.; Macchiato, M.; Ragosta, M.; Simoniello, T. Indicators for the estimation of vulnerability to land degradation derived from soil compaction and vegetation cover: Indicators for the estimation of soil degradation. *Eur. J. Soil Sci.* **2014**, *65*, 907–923. [CrossRef]
44. Nadporozhskaya, M.A.; Chertov, O.G.; Bykhovets, S.S.; Shaw, C.H.; Maksimova, E.Y.; Abakumov, E.V. Recurring surface fires cause soil degradation of forest land: A simulation experiment with the EFIMOD model. *Land Degrad. Dev.* **2018**, *29*, 2222–2232. [CrossRef]
45. Miquelajauregui, Y.; Cumming, S.G.; Gauthier, S. Modelling Variable Fire Severity in Boreal Forests: Effects of Fire Intensity and Stand Structure. *PLoS ONE* **2016**, *11*, e0150073. [CrossRef] [PubMed]
46. Veldman, J.W. Clarifying the confusion: Old-growth savannahs and tropical ecosystem degradation. *Phil. Trans. R. Soc. B* **2016**, *371*, 20150306. [CrossRef] [PubMed]
47. Kosmas, C.; Detsis, V.; Karamesouti, M.; Kounalaki, K.; Vassiliou, P.; Salvati, L. Exploring Long-Term Impact of Grazing Management on Land Degradation in the Socio-Ecological System of Asteroussia Mountains, Greece. *Land* **2015**, *4*, 541–559. [CrossRef]
48. Bado, V.B.; Bationo, A. Integrated Management of Soil Fertility and Land Resources in Sub-Saharan Africa: Involving Local Communities. In *Advances in Agronomy*; Elsevier: Amsterdam, The Netherlands, 2018; Volume 150, pp. 1–33. ISBN 978-0-12-815175-4.
49. Deng, X.; Huang, J.; Huang, Q.; Rozelle, S.; Gibson, J. Do roads lead to grassland degradation or restoration? A case study in Inner Mongolia, China. *Environ. Dev. Econ.* **2011**, *16*, 751–773. [CrossRef]
50. Li, W.; Buitenwerf, R.; Munk, M.; Amoke, I.; Bøcher, P.K.; Svenning, J.-C. Accelerating savanna degradation threatens the Maasai Mara socio-ecological system. *Glob. Environ. Chang.* **2020**, *60*, 102030. [CrossRef]
51. De la Fuente, B.; Weynants, M.; Bertzky, B.; Delli, G.; Mandrici, A.; Bendito, E.G.; Dubois, G. Land Productivity Dynamics in and around Protected Areas Globally from 1999 to 2013. *PLoS ONE* **2020**, *15*, e0224958. [CrossRef]
52. Nabiollahi, K.; Golmohamadi, F.; Taghizadeh-Mehrjardi, R.; Kerry, R.; Davari, M. Assessing the effects of slope gradient and land use change on soil quality degradation through digital mapping of soil quality indices and soil loss rate. *Geoderma* **2018**, *318*, 16–28. [CrossRef]
53. Qasim, M.; Hubacek, K.; Termansen, M.; Fleskens, L. Modelling land use change across elevation gradients in district Swat, Pakistan. *Reg. Environ. Chang.* **2013**, *13*, 567–581. [CrossRef]
54. Kohonen, T. *Self-Organizing Maps*; Springer Series in Information Sciences; Springer: Berlin/Heidelberg, Germany, 2001; Volume 30, ISBN 978-3-540-67921-9.
55. Moosavi, V. Contextual mapping: Visualization of high-dimensional spatial patterns in a single geo-map. *Comput. Environ. Urban Syst.* **2017**, *61*, 1–12. [CrossRef]
56. Oldeman, L.R.; Hakkeling, R.T.A.; Sombroek, W.G.; International Soil Reference and Information Centre; United Nations. *Environment Programme World Map of the Status of Human-Induced Soil Degradation an Explanatory Note*; International Soil Reference and Information Centre: Wageningen, The Netherlands; Nairobi, Kenya, 1991.
57. Andela, N.; van der Werf, G.R. Recent trends in African fires driven by cropland expansion and El Niño to La Niña transition. *Nat. Clim. Chang.* **2014**, *4*, 791–795. [CrossRef]
58. Woollen, E.; Ryan, C.M.; Baumert, S.; Vollmer, F.; Grundy, I.; Fisher, J.; Fernando, J.; Luz, A.; Ribeiro, N.; Lisboa, S.N. Charcoal production in the Mopane woodlands of Mozambique: What are the trade-offs with other ecosystem services? *Phil. Trans. R. Soc. B* **2016**, *371*, 20150315. [CrossRef] [PubMed]
59. Barbier, E.B.; Hochard, J.P. Does Land Degradation Increase Poverty in Developing Countries? *PLoS ONE* **2016**, *11*, e0152973. [CrossRef] [PubMed]
60. Nkonya, E.; Mirzabaev, A.; von Braun, J. Economics of Land Degradation and Improvement: An Introduction and Overview. In *Economics of Land Degradation and Improvement—A Global Assessment for Sustainable Development*; Nkonya, E., Mirzabaev, A., von Braun, J., Eds.; Springer International Publishing: Cham, Switzerland, 2016; pp. 1–14. ISBN 978-3-319-19168-3.
61. Gadzama, N.M.; Ayuba, H.K. On major environmental problem of desertification in Northern Nigeria with sustainable efforts to managing it. *World J. Sci. Technol. Sustain. Dev.* **2016**, *13*, 18–30. [CrossRef]

62. Gadzama, N.M. Attenuation of the effects of desertification through sustainable development of Great Green Wall in the Sahel of Africa. *World J. Sci. Technol. Sustain. Dev.* **2017**, *14*, 279–289. [CrossRef]
63. Ayanlade, A.; Drake, N. Forest loss in different ecological zones of the Niger Delta, Nigeria: Evidence from remote sensing. *GeoJournal* **2016**, *81*, 717–735. [CrossRef]
64. Kiage, L.M. Perspectives on the assumed causes of land degradation in the rangelands of Sub-Saharan Africa. *Prog. Phys. Geogr. Earth Environ.* **2013**, *37*, 664–684. [CrossRef]
65. Pingali, P.; Schneider, K.; Zurek, M. Poverty, Agriculture and the Environment: The Case of Sub-Saharan Africa. In *Marginality: Addressing the Nexus of Poverty, Exclusion and Ecology*; von Braun, J., Gatzweiler, F.W., Eds.; Springer: Dordrecht, The Netherlands, 2014; pp. 151–168. ISBN 978-94-007-7061-4.
66. Eisenack, K.; Villamayor-Tomas, S.; Epstein, G.; Kimmich, C.; Magliocca, N.; Manuel-Navarrete, D.; Oberlack, C.; Roggero, M.; Sietz, D. Design and quality criteria for archetype analysis. *Ecol. Soc.* **2019**, *24*. [CrossRef]
67. Ducrotoy, M.J.; Majekodunmi, A.O.; Shaw, A.P.M.; Bagulo, H.; Bertu, W.J.; Gusi, A.M.; Ocholi, R.; Welburn, S.C. Patterns of passage into protected areas: Drivers and outcomes of Fulani immigration, settlement and integration into the Kachia Grazing Reserve, northwest Nigeria. *Pastoralism* **2018**, *8*, 1. [CrossRef] [PubMed]
68. Nchor, A.A.; Ogogo, A.U. Rapid assessment of protected area pressures and threats in Nigeria national parks. *Glob. J. Agric. Sci.* **2012**, *11*, 63–72. [CrossRef]
69. Mirzabaev, A. Land Degradation and Sustainable Land Management Innovations in Central Asia. In *Technological and Institutional Innovations for Marginalized Smallholders in Agricultural Development*; Gatzweiler, F.W., von Braun, J., Eds.; Springer: Cham, Switzerland, 2016; pp. 213–224. ISBN 978-3-319-25718-1.
70. Abbas, I.I.; Fasona, M.J. The Human Perception of Land Degradation in a Section of Niger Delta, Nigeria. *Mar. Sci.* **2012**, *2*, 94–100. [CrossRef]
71. Fasona, M.; Fabusoro, E.; Sodiya, C.; Adedayo, V.; Olorunfemi, F.; Elias, P.O.; Oyedepo, J.; Oloukoi, G. Some Dimensions of Farmers'-Pastoralists' Conflicts in the Nigerian Savanna. *J. Glob. Initiat. Policy Pedagog. Perspect.* **2016**, *10*, 7.
72. Naibbi, A.I.; Baily, B.; Healey, R.G.; Collier, P. Changing Vegetation Patterns in Yobe State Nigeria: An Analysis of the Rates of Change, Potential Causes and the Implications for Sustainable Resource Management. *IJG* **2014**, *5*, 50–62. [CrossRef]
73. Akhtar-Schuster, M.; Thomas, R.J.; Stringer, L.C.; Chasek, P.; Seely, M. Improving the enabling environment to combat land degradation: Institutional, financial, legal and science-policy challenges and solutions. *Land Degrad. Dev.* **2011**, *22*, 299–312. [CrossRef]
74. FGN Federal Government of Nigeria. *Great Green Wall for the Sahara and Sahel Initiative National Strategic Action Plan*; National Strategic Action Plan; Federal Ministry of Environment of Nigeria: Abuja, Nigeria, 2012.
75. Onele, J. The Grazing Bill and the Right to Property in Nigeria: Lest We Are Deceived! *SSRN J.* **2016**. [CrossRef]
76. World Bank. *Nigeria Agriculture and Rural Poverty: A Policy Note*; World Bank: Washington, DC, USA, 2014.
77. World Bank. *Poverty Reduction in Nigeria in the Last Decade*; World Bank: Washington, DC, USA, 2016.
78. Umo, J.O. *Escaping Poverty in Africa: A Perspective on Strastegic Agender for Nigeria*, 2nd ed.; Millennium Text Publishers Ltd.: Lagos, Nigeria, 2012.
79. Okoro, U.S.; Akpaeti, A.J.; Ekpo, U.N. Seasonality in income and poverty among rural food crop farmers in Akwaibom State, Nigeria. *Russ. Agric. Sci.* **2015**, *41*, 383–389. [CrossRef]
80. UNESCO. *Reaching the 2015 Literacy Target: Delivering on the Promise*; High Level International Round Table on Literacy: Paris, France, 2012.
81. Benjamin, A.M.N. Farmers' Perception of Effectiveness of Agricultural Extension Delivery in Cross-River State, Nigeria. *J. Agric. Vet. Sci. (IOSR-JAVS)* **2013**, *2*, 1–7.
82. Baloch, M.A.; Thapa, G.B. The effect of agricultural extension services: Date farmers' case in Balochistan, Pakistan. *J. Saudi Soc. Agric. Sci.* **2018**, *17*, 282–289. [CrossRef]
83. Lenhardt, A.; Glennie, J.; Intscher, N.; Ali, A.; Morin, G. *A Greener BurkinA Sustainable Farming Techniques, Land Reclamation and Improved Livelihoods*; Progress Case Study Report; Gallimard: Paris, France, 2014.
84. Atuo, F.A.; Fu, J.; O'Connell, T.J.; Agida, J.A.; Agaldo, J.A. Coupling law enforcement and community-based regulations in support of compliance with biodiversity conservation regulations. *Environ. Conserv.* **2020**, *47*, 104–112. [CrossRef]
85. Hughes, K.; Oduol, J.; Kegode, H.; Ouattara, I.; Vagen, T.; Winowiecki, L.A.; Bourne, M.; Neely, C.; Ademonla, A.D.; Carsan, S.; et al. *Regreening Africa Consolidated Baseline Survey Report*; World Agroforestry: Nairobi, Kenya, 2020.
86. Ifejika Speranza, C.; Ochege, F.U.; Nzeadibe, T.C.; Agwu, A.E. Agricultural Resilience to Climate Change in Anambra State, Southeastern Nigeria. In *Beyond Agricultural Impacts*; Elsevier: Amsterdam, The Netherlands, 2018; pp. 241–274. ISBN 978-0-12-812624-0.
87. Adewuyi, T.O.; Olofin, E.A. Sustainability of Fuel Wood Harvesting from Afaka Forest Reserve, Kaduna State, Nigeria. *JAS* **2014**, *7*, 129. [CrossRef]
88. Ndiboi, S.; Makinde, D. *Sustainable Fuelwood Management Project in Nigeria*; Final Draft Midterm Review (MTR) REPORT: Abuja, Nigeria, 2020.
89. Mononen, K.; Pitkänen, S. *Sustainable Fuelwood Management in West Africa*; University of Eastern Finland: Joensuu/Kuopio, Finland, 2016.

90. Chazdon, R.L.; Uriarte, M. Natural regeneration in the context of large-scale forest and landscape restoration in the tropics. *Biotropica* **2016**, *48*, 709–715. [CrossRef]
91. Idris Medugu, N.; Rafee Majid, M.; Johar, F.; Choji, I.D. The role of afforestation programme in combating desertification in Nigeria. *Int. J. Clim. Chang. Strat. Manag.* **2010**, *2*, 35–47. [CrossRef]
92. Ibrahim, K.M.; Muhammad, S.I. A Review of Afforestation Efforts in Nigeria. *Int. J. Adv. Res. Eng. Appl. Sci.* **2015**, *4*, 24–37.
93. Ibrahim, S.S.; Ozdeser, H.; Cavusoglu, B. Testing the impact of environmental hazards and violent conflicts on sustainable pastoral development: Micro-level evidence from Nigeria. *Environ. Dev. Sustain.* **2020**, *22*, 4169–4190. [CrossRef]
94. FAO Food and Agriculture Organization. *Improving Governance of Pastoral Lands. Governance of Tenure Technical Guides*; FAO: Rome, Italy, 2017; Available online: www.fao.org/3/a-i5771e.pdf (accessed on 17 November 2020).
95. Hannam, I. Governance of Pastoral Lands. In *International Yearbook of Soil Law and Policy 2017*; Ginzky, H., Dooley, E., Heuser, I.L., Kasimbazi, E., Markus, T., Qin, T., Eds.; Springer: Cham, Switzerland, 2018; Volume 2017, pp. 101–124. ISBN 978-3-319-68884-8.
96. Sule, P.E. Open grazing prohibitions and the politics of exclusivist identity in Nigeria. *Afr. Rev.* **2020**, 1–17. [CrossRef]
97. Liniger, H.; Studer, R.M.; Hauert, C.; Gurtner, M.; Gämperli, U.; Kummer, S.; Hergarten, C. *Sustainable Land Management in Practice—Guidelines and Best Practices for Sub-Saharan Africa*; TerrAfrica; World Overview of Conservation Approaches and Technologies (WOCAT) and Food and Agriculture Organization of the United Nations (FAO): Midrand, South Africa, 2011; Volume 13.
98. Liniger, H.; Harari, N.; van Lynden, G.; Fleiner, R.; de Leeuw, J.; Bai, Z.; Critchley, W. Achieving land degradation neutrality: The role of SLM knowledge in evidence-based decision-making. *Environ. Sci. Policy* **2019**, *94*, 123–134. [CrossRef]
99. Ferro-Vázquez, C.; Lang, C.; Kaal, J.; Stump, D. When is a terrace not a terrace? The importance of understanding landscape evolution in studies of terraced agriculture. *J. Environ. Manag.* **2017**, *202*, 500–513. [CrossRef]

Article

Using Sentinel-1 and Sentinel-2 Time Series for Slangbos Mapping in the Free State Province, South Africa

Marcel Urban [1,*], Konstantin Schellenberg [1], Theunis Morgenthal [2], Clémence Dubois [1], Andreas Hirner [3], Ursula Gessner [3], Buster Mogonong [4], Zhenyu Zhang [5], Jussi Baade [6], Anneliza Collett [2] and Christiane Schmullius [1]

[1] Department for Earth Observation, Friedrich-Schiller-University, 07743 Jena, Germany; konstantin.schellenberg@uni-jena.de (K.S.); clemence.dubois@uni-jena.de (C.D.); c.schmullius@uni-jena.de (C.S.)
[2] Department of Agriculture, Land Reform and Rural Development (DALRRD), Pretoria 0001, South Africa; TheunisM@Dalrrd.gov.za (T.M.); AnnelizaC@Dalrrd.gov.za (A.C.)
[3] German Aerospace Center, German Remote Sensing Data Center, 51147 Oberpfaffenhofen, Germany; andreas.hirner@dlr.de (A.H.); ursula.gessner@dlr.de (U.G.)
[4] South African Environmental Observation Network (SAEON), Arid Lands Node, Kimberley 0001, South Africa; buster@saeon.ac.za
[5] Karlsruhe Institute of Technology, University of Augsburg, 86159 Augsburg, Germany; zhenyu.zhang@kit.edu
[6] Department for Physical Geography, Friedrich-Schiller-University, 07743 Jena, Germany; jussi.baade@uni-jena.de
* Correspondence: marcel.urban@uni-jena.de

Citation: Urban, M.; Schellenberg, K.; Morgenthal, T.; Dubois, C.; Hirner, A.; Gessner, U.; Mogonong, B.; Zhang, Z.; Baade, J.; Collett, A.; et al. Using Sentinel-1 and Sentinel-2 Time Series for Slangbos Mapping in the Free State Province, South Africa. *Remote Sens.* **2021**, *13*, 3342. https://doi.org/10.3390/rs13173342

Academic Editor: Elias Symeonakis

Received: 18 June 2021
Accepted: 6 August 2021
Published: 24 August 2021

Publisher's Note: MDPI stays neutral with regard to jurisdictional claims in published maps and institutional affiliations.

Abstract: Increasing woody cover and overgrazing in semi-arid ecosystems are known to be the major factors driving land degradation. This study focuses on mapping the distribution of the slangbos shrub (*Seriphium plumosum*) in a test region in the Free State Province of South Africa. The goal of this study is to monitor the slangbos encroachment on cultivated land by synergistically combining Synthetic Aperture Radar (SAR) (Sentinel-1) and optical (Sentinel-2) Earth observation information. Both optical and radar satellite data are sensitive to different vegetation properties and surface scattering or reflection mechanisms caused by the specific sensor characteristics. We used a supervised random forest classification to predict slangbos encroachment for each individual crop year between 2015 and 2020. Training data were derived based on expert knowledge and in situ information from the Department of Agriculture, Land Reform and Rural Development (DALRRD). We found that the Sentinel-1 VH (cross-polarization) and Sentinel-2 SAVI (Soil Adjusted Vegetation Index) time series information have the highest importance for the random forest classifier among all input parameters. The modelling results confirm the in situ observations that pastures are most affected by slangbos encroachment. The estimation of the model accuracy was accomplished via spatial cross-validation (SpCV) and resulted in a classification precision of around 80% for the slangbos class within each time step.

Keywords: shrub encroachment; slangbos; land degradation; Earth observation; time series; Sentinel-1; Sentinel-2; Synthetic Aperture Radar (SAR); Soil Adjusted Vegetation Index (SAVI); machine learning

1. Introduction

Increasing woody cover and overgrazing in open semi-arid ecosystems are known to be one of the major factors driving land degradation [1]. In the context of this study, land degradation is defined as "the many human-caused processes that drive the decline or loss in biodiversity, ecosystem functions or ecosystem services in any terrestrial [...] ecosystems" [2] (p. 28).

During recent decades, woody cover encroachment in open ecosystems has significantly increased in southern Africa, which have led to crucial environmental, land cover and land-use changes [3–7]. Venter et al. [8] concluded that the intensification of woody cover is not only connected to rising CO_2 concentrations in the atmosphere, as a result of the human-induced global climate change, but it is also noticeable on regional aspects (e.g., extinction of large herbivore herds, and fires). However, CO_2 fertilization improves water-use efficiency, due to the reduced stomatal conductance of woody plants [9]. Consequently, even with constant water ingestion, rising CO_2 concentrations would lead to increased growth rates of woody biomass in relation to grass communities. Nevertheless, water availability and temperature are the key constraints on woody plant growth and are the crucial parameters in explaining encroachment patterns [5].

This study focuses on analyzing the spreading of the slangbos or bankrupt bush (*Seriphium plumosum*) in a selected test region in the Free State Province of South Africa. Though indigenous to South Africa, slangbos has been documented to be the main encroacher on the grassvelds (South African grassland biomes) in the provinces of the Free State, North West, Mpumalanga, Eastern Cape and Gauteng [10,11].

The rainfall optimum of *Seriphium plumosum* is between 620 and 750 mm per year [12], which lies between the semi-arid and mesic biome [13]. The shrub reaches a height and diameter of up to 0.6 m and has small light green leaves, which makes them perfectly adapted to long dry periods, compared to grass communities and the leaves are unpalatable to grazers due to their high oil content [14]. However, du Toit et al. [15] found that cattle graze young slangbos plants, which were regrown one year after active fire control measures. Slangbos prefers to grow on low fertile loamy soils, which makes hilltops and dry areas in the Free State vulnerable to this encroacher [16]. The encroachment of slangbos is responsible for a decrease in the grasslands productivity, pastures and livestock carrying capacity. The root system of *Seriphium plumosum*, which reaches a depth of up to 1.8 m and an extent of 1 m^2 around the plant [14], competes with the surrounding grass for water and the availability of nutrients, which leads consequently to a reduction in the grass layer [17,18]. As slangbos is unpalatable for grazers, the shrub puts bigger pressure on the existing open grassland, which becomes vulnerable to overgrazing, increasing the potential of land degradation [16]. Moreover, slangbos is found to have high allelopathic potential, which impedes other plant species from finding suitable growing conditions in the surrounding area of a slangbos [19]. This poses great challenges to farmers on the one hand and for the local biosphere on the other hand. Approximately 11 million hectares of rangeland could become unsuitable for grazing if measures to eradicate the plant do not succeed. This could result in an annual loss of about ZAR 760,000 (South African Rand) (about EUR 44,000) for a 1000 ha livestock farm with a slangbos infestation of 50% of the pasture area [20].

Field observations detecting shrub distribution are cost intensive, time consuming and often do not address the spatial heterogeneity of encroachment patterns [21]. Earth Observation (EO) data from different sources and across different wavelengths (e.g., from ESA's Copernicus Sentinel Programme) provide a suitable tool for mapping the extent and the velocity of woody encroachment [7,22–25]. The freely available optical and radar satellite data from the Copernicus Programme have short revisiting times (5 to 12 days) as well as high spatial resolution (10 m). Hence, they allow for monitoring woody cover encroachment in quasi near real-time and enable transferability and reproducibility in other regions. Bush encroachment mapping in southern Africa utilizing various sources of EO information (e.g., optical and radar) was investigated by recent studies [8,26–31]. These studies analyzed shrub cover increase, using high to low spatial resolution EO data investigating different approaches (e.g., land cover classification [27,31], random forest [8,26,29], trend analysis [28,30]). However, few studies have focused on one specific shrub species only. In general, an intensification of woody cover was present in all studies, especially in the open rangelands. Local studies revealed that protected areas with large herbivore populations (e.g., elephants) occasionally show a loss in woody or tree cover [27].

Remote sensing techniques are not likely to replace field measurements completely, as the validation of EO approaches deriving woody vegetation composition is still of high importance [21,32]. However, they allow for continuous wall-to-wall monitoring on a larger spatial extent of the parameter woody cover, which is considered an essential biodiversity variable [33].

The goal of this study is to monitor the slangbos encroachment on rangeland and pastures in the Free State Province, South Africa, between 2015 and 2020 by synergistically combining Synthetic Aperture Radar (SAR) (Sentinel-1) and optical (Sentinel-2) EO information. Both optical and radar satellite data are sensitive to the different vegetation properties and surface scattering or reflection mechanisms caused by the sensor specific characteristics. This study aims to (1) investigate the sensitivity of optical and radar remote sensing toward *Seriphium plumosum* and (2) to use combined SAR and optical dense time series to perform encroachment mapping on a regional scale in the South African grassland biomes. Utilizing time series, particular features in hyper-temporal radar and spectral data are examined to conclude the temporal variations observing *Seriphium plumosum*.

2. Materials and Methods

2.1. Study Area

The monitoring of slangbos encroachment was carried out on an approx. 90 km by 50 km (4500 km^2) area located in the Free State Province between the towns of Ladybrand in the east and Botshabelo in the west (Figure 1).

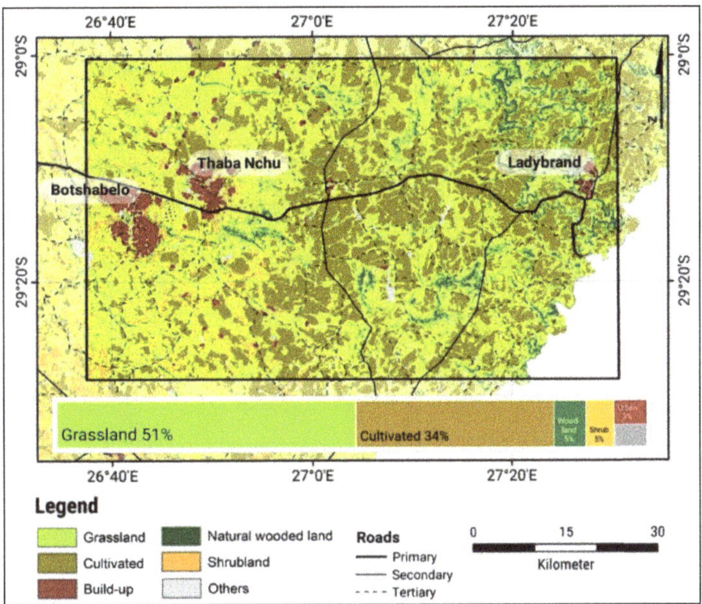

Figure 1. The study area is located in the Free State Province, South Africa, close to the border of the Kingdom of Lesotho (in white). (Source: National Land Cover Classification of 2018 [34], Roads of South Africa [35]).

The region is part of the Highveld grassland ecosystem (1300 m–1700 m a.s.l.), which is characterized by a continental climate with a mean annual temperature of around 14 °C. However, days with temperatures below the freezing point are common during the winter season. The mean annual precipitation ranges from 500 mm to 700 mm, and the majority of rain occurs in the summer season between November and March [11].

The study area is known for extensive plains, cultivated areas and plateaus, which form the relief. The region is sparsely populated and mostly used for cattle and arable farming as well as water reservoirs. The primary land cover types are rangelands for grazing, cultivated land, shrubland and open grassland. The Highveld grassland biome is influenced by three major land degradation phenomena, namely (1) changes in plant composition, (2) loss of vegetation cover with a subsequent increase in wind and water erosion occurrence and (3) bush encroachment [11]. Moreover, the study area is one of the regions with the highest impact of invasive alien plant species throughout all of South Africa [36].

2.2. Data

2.2.1. Sentinel-1 and Sentinel-2

The Sentinel-1 and Sentinel-2 missions of the European Space Agency (ESA) Copernicus Programme have led to the increased availability of open access EO information covering both the optical and the microwave spectra. This opens new possibilities for the analysis of data with high spatial as well as temporal resolution for various applications, e.g., agricultural monitoring, and vegetation change analysis. The synergetic use of these EO data is especially valuable, as both satellites acquire data in parallel but measure different properties of the Earth's surface. In comparison to Sentinel-2, Sentinel-1 acquires undisturbed images of the Earth's surface, regardless of atmospheric effects and sun illumination.

For this study, Sentinel-1A C-Band SAR (5.405 GHz—approx. 5 cm wavelength) dual-polarized (VV–vertical/vertical, and VH–vertical/horizontal) scenes were utilized, covering the time period between 2015 and 2020 with a spatial resolution of up to 10 m. Sentinel-1 has a revisit time of a few days (3 to 12 days), depending on the geographic location and acquisition of both Sentinel-1A and -1B. In South Africa, the image acquisition repetition is twelve days [37]. In the Free State Province, only Sentinel-1A data from the ascending orbit are available, collecting images at around 5:00 p.m., local time. The Sentinel-1 footprint is represented by relative orbit numbers 14 and 116, which were used jointly.

The Sentinel-2 constellation consists of two satellites, namely Sentinel-2A (start of acquisition: November 2015) and Sentinel-2B (start of acquisition: August 2017). Since Sentinel-2B became operational, the revisit time has been approximately five days. In this study, data acquired between 2016 and 2020 from the optical Sentinel-2 constellation were used. Each of the two satellites (Sentinel-2A and -2B) has 13 bands covering the spectrum from the visible to the Short-Wave-Infrared (SWIR) and a maximum spatial resolution of 10 m. The images are acquired at around 10:30 a.m. local time [38].

In this study, Sentinel-1 Single Look Complex (SLC) and Ground Range Detection (GRD) images were utilized. Whereas the SLC data comprise the phase information, which is used to derive the interferometric coherence, the GRD data contain only the amplitude and are already multilooked. In total, 313 Sentinel-1 GRD, 145 Sentinel-1 SLC (144 coherence pairs) as well as 503 Sentinel-2 images were utilized. For the coherence estimation, Sentinel-1 SLC scenes from the relative orbit 14 were used.

2.2.2. Agricultural Statistics 2014–2018

Spatial information for the different crop types planted in the study area was provided by the Department of Agriculture, Land Reform and Rural Development (DALRRD) [39]. The dataset contains information of crop types planted between 2014 and 2018, where each dataset represents an individual crop year. A crop year is defined from June until May of the next year, and includes both winter (June–September) and summer (October–May) planting seasons.

In general, the Free State Province has the highest number of farming units, which are summed up to almost 8000 entities, which represent approximately 20% of the national total. Approximately 460,000 km^2 are used for agriculture of which 80% were used for grazing and the remaining 20% were used mainly for dryland crop production [40].

Between 2014 and 2018, the cultivated land in the study area was dominated by pasture vegetation (Figure 2). Maize, sunflower and soybeans can be identified as major crops during the same period, whereas the amount of wheat and sorghum is negligible. However, fallow land areas, which are certainly prone to a slangbos invasion, have shown a significant increase between 2014 and 2018, with the largest expansion between 2016/17 and 2017/18.

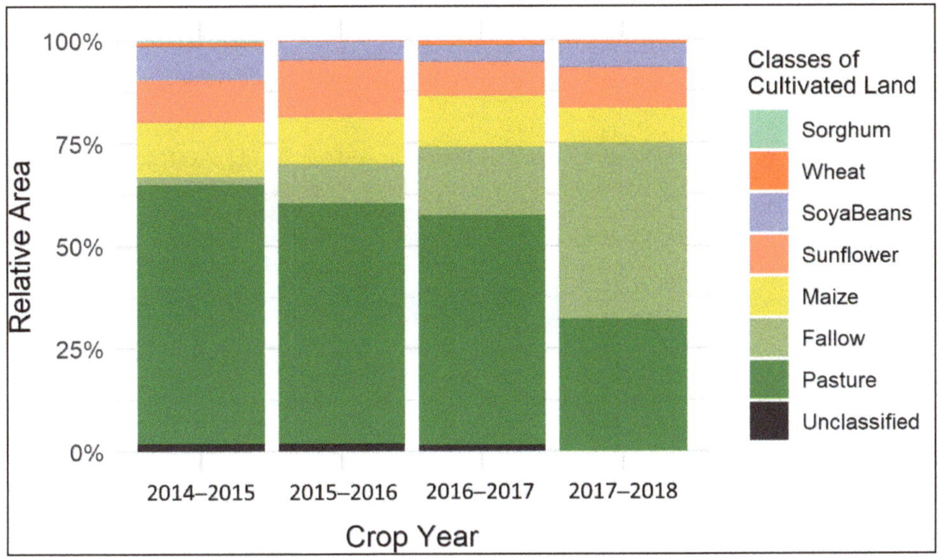

Figure 2. Area for different cover types of the cultivated land within the study area for the crop years between 2014 and 2018 (Source: [39]).

In 2014/15, croplands (approx. 380 km^2 (33%)) and pasture (approx. 730 km^2 (63%)) covered most of the cultivated land within the study area. At this time, the study region was characterized by 23 km^2 (2%) fallow land. In 2015/16, the area covered by fallow land increased by 7% to an area of 110 km^2, where both, pasture and cropland declined by equal percentages. Soybeans showed the largest decline during that time, which was followed by a slightly increasing trend toward 2018. In 2016/17, the statistics show that especially cropland turned into fallow land, which increased by approximately 7% to an area of 190 km^2. Sunflower showed the largest decline during that period (60 km^2 (5%)). In 2017/18, the class fallow showed the largest increase to 43%, which equals around 480 km^2. At this time, the area, which was covered with pasture beforehand, decreased by about 280 km^2 (23%), whereas the number of croplands had no significant loss. However, a large decline for maize was found in this year (50 km^2 (4%)), as these areas were changed to soybeans or sunflowers.

2.2.3. Reference Data

The essential ground reference sites were exploited via field exploration, aerial photo documentation and local expert knowledge. In order to scale the amount of ground validation, the Google Earth high-resolution time series imagery [41] and the National Geospatial Information (NGI) very high resolution aerial photos [42] were used. The previously identified slangbos sites were used as a blueprint for creating a set of labeled fields in the area, utilizing manual mapping in cooperation with local partners at DALRRD [39]. The ground references consist of binary spatial polygons indicating the occurrence or absence of slangbos. An oversampling of the critical slangbos and grassland land cover was performed to draw the focus of the classification algorithm to this discrimination task. Finally, the labeled dataset totals up to roughly 14 km^2, which makes up 1.2% of the agricultural area and 0.3% of the entire study area.

2.3. Methods

2.3.1. Sentinel-1 Pre-Processing

The Sentinel-1 GRD data were pre-processed using *PyroSAR* [43], which is designed for large-scale SAR satellite data processing within a Python framework. It offers a complete solution for organizing and processing SAR data for different historical and current satellite missions, with additional functionalities, which are available after the pre-processing of the SAR satellite images (e.g., mosaicking and resampling images to common pixel boundaries suited for time series analysis). *PyroSAR* offers the possibility to utilize the open-source ESA's Sentinel Application Platform (SNAP) as well as the GAMMA Remote Sensing software for licensed users.

The Sentinel-1 GRD data were pre-processed using GAMMA (Software Version: July 2018, GAMMA Remote Sensing AG, 3073 Gümligen, Switzerland) [44]. (1) To convert from digital numbers (DNs), which are recorded by the sensor, to physical units, a radiometric calibration was applied for each dataset. (2) The orthorectification of the data was carried out, using the precise orbit state vectors (precise orbit ephemerides, POE) as well as height information from the Shuttle Radar Topography Mission (SRTM) Digital Elevation Model (DEM) [45] at 30 m spatial resolution. (3) The conversion of beta naught (β^0) to gamma naught (γ^0) values was achieved by a terrain flattening [46], utilizing the SRTM DEM.

The interferometric coherence (γ) represents a measure of the correlation between two complex SAR images and can be described as follows:

$$\gamma = \frac{E\left[u_1\, u_2^{\,*}\right]}{\sqrt{E\left[|u_1|^2\right]}\,\sqrt{E\left[|u_2|^2\right]}} \tag{1}$$

where u_1 and u_2 are two complex (SLC) acquisitions, $|u_1|$ and $|u_2|$ are their amplitudes, and $E[x]$ is the expected value of the considered variable x. The coherence takes values between 0 and 1, where 0 represents no correlation and 1 represents a perfect correlation between the image phases [47,48]. Depending on the backscattering properties of the surfaces and their variation over time, the coherence is variable; therefore, it is useful for land cover classification [49,50], and biomass estimation [51,52], as well as for change detection (e.g., mapping deforestation) [53,54].

The Sentinel-1 SLC data were pre-processed, using SNAP. The coherences were only calculated for the co-polarized mode (VV) because of the stronger signal and thus higher signal-to-noise ratio. The coherence estimates were created for all adjacent date pairs with a temporal baseline of 12 days. However, in 2015 and 2016, eight interferometric coherence pairs (four in each year) were based on a 24-day temporal baseline, which was caused by the interruption in the acquisitions due to maintenance or other technical related issues at the early lifetime of the satellite.

(1) The first processing step was to apply the orbit state vector files to the SLC data, using the precise orbit state vectors (precise orbit ephemerides, POE). (2) As the SLC images are acquired in three sub-swaths (one image per swath and polarization) in TOPSAR mode (Terrain Observation with Progressive Scans SAR), the TOPSAR split function was applied to calculate the coherence based on each sub-swath. (3) In order to co-register the SLC swaths of an image pair, the back-geocoding function was accomplished utilizing the height information from the SRTM DEM [45] at 30 m spatial resolution. (4) The coherence between both SLC datasets was calculated using the interferogram module in SNAP. (5) TopSAR merge was utilized to merge the individual sub-swath before performing (6) the terrain correction, which also utilized the SRTM DEM dataset as the height information.

2.3.2. Sentinel-2 Pre-Processing

Sentinel-2 L1C products consist of TOA (top of atmosphere) reflectances in cartographic geometry. The Sentinel-2 data are provided in tiles with a fixed coverage of about 10,000 km^2 (100 by 100 km) projected in UTM/WGS84 along a single orbit [38].

In order to obtain the bottom of atmosphere reflectance, the Sen2Cor processor [55] was utilized with additional parameter settings, such as aerosol optical thickness and water vapor, but without the correction for cirrus and terrain. Each band of the pre-processed Sentinel-2 L2A products has the identical spatial resolution as the original L1C input band, namely 10 m for bands 2, 3, 4, 8, 20 m for bands 5, 6, 7, 8a, 11, 12 and 60 m for bands 1, 9, 10. In addition, the Python application [56] of the FMask algorithm [57–59] was used to produce cloud masks with a resolution of 20 m from the Sentinel-2 L1C data.

To assess the sensitivity of the satellite image data concerning vegetation processes, the NDVI (Normalized Difference Vegetation Index) [60] and SAVI (Soil Adjusted Vegetation Index) [61] were retrieved from each Sentinel-2 L2A scene with a spatial resolution of 10 m:

$$\text{NDVI} = \frac{(N - R)}{(N + R)} \tag{2}$$

$$\text{SAVI} = \frac{(N - R)}{(N + R + L)}(1 + L) \tag{3}$$

where N corresponds to the near-infrared (NIR) channel of Sentinel-2 (band 8), R corresponds to the red channel of Sentinel-2 (band 4) and L is a factor, which is used to minimize the influence of the soil brightness in the SAVI (here $L = 0.5$) [61].

Vegetation dynamics for a specific area are often analyzed using time series constructed from indices. Here, time series were constructed with the entire Sentinel-2 data archive from 2016 to 2020. To avoid spurious data due to atmospheric distortions, each index file was masked with the corresponding FMask product before stacking. In some regions, depending on the weather conditions in combination with the topography, each time series along a pixel stack can have substantial gaps lasting up to eight weeks. However, with the introduction of Sentinel-2B, gaps in areas with intensive cloud cover are considerably shorter, with at least one suitable acquisition each month.

2.3.3. Combined Time Series Analysis

For enhancing the interpretability of the dense time series, we employed the smoothing algorithm, Friedman's "super smoother" [62], an adaptive, variable-span linear smoother. It provides a smoothing prediction due to the adaptive spans and utilization of the parameter of least residuals. The default setting provided in the R-implementation was found to be the most suitable for the interpretation of the time series.

2.3.4. Predictive Modeling and Interpretation

In order to predict the spatial distribution of slangbos on cultivated land, a random forest model was fitted for each crop year and validated using spatial cross-validation (SpCV) [63]. Random forest models are widely used in remote sensing applications due to their robustness to overfitting and flexible approach to recognizing non-linear structures in feature spaces [64,65].

In this study, the predictor set consists of optical indices from Sentinel-2 (NDVI, SAVI), co-polarized (VV) and cross-polarized (VH) Sentinel-1 backscatter as well as coherences (VV) comprises 67, 154 and 119 features for each investigated crop year 2015/16, 2016/17 and 2017/18, respectively. The model was trained on a labeled set of 5000 randomly selected pixels, indicating slangbos and non-slangbos fields. Due to the anticipated spatial autocorrelation effects, the model was cross-validated with 100 iterations and 5 folds of k-nearest-neighbor spatially segmented test data sets in order to derive accuracte metrics of the model predictions presented.

The random forest models were assessed, using the overall accuracy (OA) throughout all model runs. The OA includes the model performance on both slangbos and non-slangbos sites and is, therefore, prone to an overoptimistic classification performance. Therefore, the averages of recall (how much slangbos were missed) and precision (how much slangbos were falsely assigned) were also estimated in order to assess the capacity of the model to distinguish slangbos. An independent validation approach, utilizing in-situ data, which are not part of the machine learning methodology, was not applicable, as no such additional ground data were available. Hence, this study relies on the results of the internal spatial cross-validation of the random forest approach, which is a feasible approach and was utilized by other studies to reduce the overoptimistic model performance when using only random distributed points instead of spatially separated folds [66].

Since the hyperparameter settings for random forests are relatively stable [67–69], no separate tuning was set up. The number of trees *ntrees* was set to 300 (large and stable averaging possible) with *mtry* = $\sqrt{P}$, where P is the number of features in the model [70]. These parameters were optimized prior to model fitting, as they were found to be sensitive with respect to model accuracy and computation time. The so-called model tuning was carried out to identify the best parameter set for the SpCV and training [71,72]. Other parameter settings were left as the default.

The random forest feature importance determines how much each feature decreases the impurity weight when kept out of the model. It was proven to be a reliable means of assessing feature importance on the model predictions while accounting for all other predictors in the model [73]. It must be mentioned that correlated predictors take away their ranks as features akin to compensate for the loss of information. The model setup, the retrieval of feature importance and the SpCV are implemented in the R software package mlr^3 [74] with bindings to *sperrorrest* [63].

3. Results

3.1. Combined Time Series Analysis of Sentinel-1 Backscatter and Coherence and Sentinel-2 NDVI and SAVI

The input variables for the random forest approach classifying the spreading of slangbos in the study area were (1) the Sentinel-1 backscatter (S1 VV and S1 VH), (2) Sentinel-1 coherence (S1 VV), (3) Sentinel-2 NDVI and SAVI time series. This section highlights the temporal profiles of these variables for different land cover classes (slangbos, grassland, woodland and cultivated areas) (Figure 3) in order to investigate which parameter could be best suited to distinguish between the slangbos class and the other classes. In addition, precipitation information from the CHIRPS (Climate Hazards Group Infrared Precipitation with Stations) product [75], which represents rainfall estimates from rain gauge and satellite observations, was utilized to account for the characteristics of the dry and wet season to the time series signal.

Areas with slangbos encroachment are characterized by a clear increasing trend of the cross-polarized VH backscatter intensities. This is especially true between 2016 and 2017. Afterwards it seems to be stabilized, as it is also noticeable for the other classes cultivated and grassland. At the beginning of the time series, the severe drought of 2015/2016, which had a vast impact on the vegetation development during these years [76–78], is clearly visible. The sparse precipitation indicated in the CHIRPS product in 2016 can be seen as an additional indicator of the drought. InSAR coherences similarly show low amplitudes in the first two years of the time series. Yet, this trend fades in years that were not drought-affected. Therefore, the drought may have affected a stronger slangbos growth. The coherence only shows increase and stabilization, compared to the other culture types where the coherence decreases between 2015 and 2016 and then increases afterwards. The SAVI time series is characterized by low amplitudes mainly in ranges between 0.1 and 0.2, while NDVI ranges between 0.2 and 0.6 at high amplitudes. During the drought, the amplitude of either optical index was lower. In comparison, the NDVI has higher values as well as higher amplitudes during the entire time series when compared to the SAVI.

Grasslands exhibit lower cross-polarized VH backscatter values, compared to the slangbos areas. These circumstances are due to the higher proportion of volumetric scattering effects in the slangbos shrubs compared to the vertically oriented grasses. However, the impact of the drought is also visible at the beginning of the time series. Thus, dry grasslands or grasslands fallen barren due to water scarcity result in reduced backscatter. The amplitude of the coherence is similar during the entire period when compared to slangbos but shows larger dynamics when related to the other classes. The SAVI indicates slightly higher temporal dynamics in comparison to areas infected by slangbos growth during the entire time series.

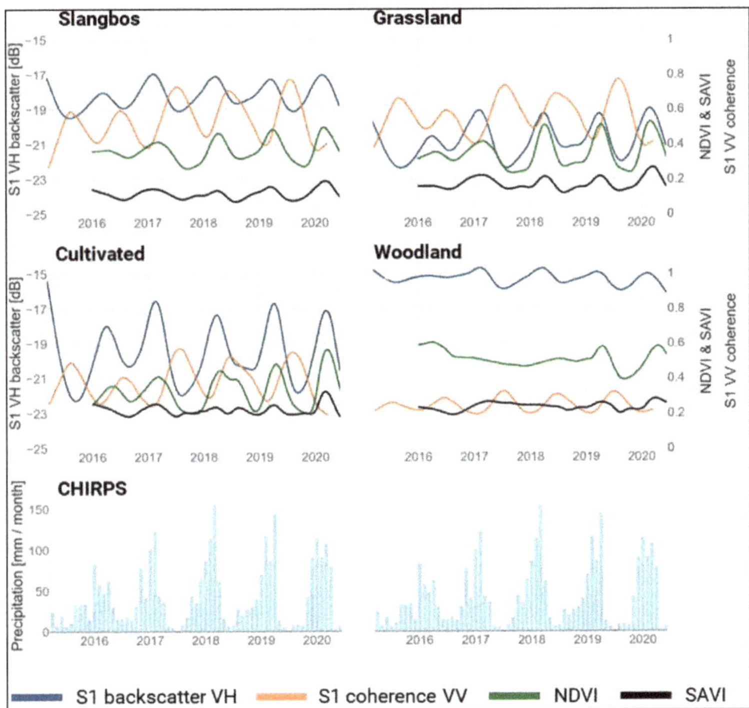

Figure 3. Temporal dynamics of the Sentinel-1 VH backscatter, Sentinel-1 VV coherence, Sentinel-2 NDVI and SAVI time series. In addition, precipitation information from the CHIRPS (Climate Hazards group Infrared Precipitation with Stations) product [75].

The woodland areas reveal higher backscatter values, compared to the other classes, which is a result of the volumetric scattering mechanism in the crones of upper tree canopies. The amplitude of the cross-polarized backscatter is low, which reflects almost no seasonality in the woodland areas. The temporal profile of the coherence is quite low (decorrelation) with also low seasonality.

The cultivated areas show the largest Sentinel-1 VH backscatter dynamics, which is caused by the harvest cycles. This is also true for the NDVI signal. The SAVI is somewhat comparable to the grassland class, with lower seasonal dynamics. The impact of the severe drought in 2015/2016 is visible between 2016 and 2017, which could be identified as repercussions of the drought to the vegetation growth.

The comparison of the temporal signature of the four different parameters for the classes slangbos, grassland, woodland and cultivated areas, have indicated that the Sentinel-2 derived SAVI index, as well as the Sentinel-1 VH backscatter, shows the highest potential in separating between the class slangbos from the others. Figure 4 shows the comparison between the NDVI and the SAVI as well as Sentinel-1 VH backscatter and VV coherence for the four classes. While the NDVI confirms no separability between slangbos, grasslands and cultivated areas, the SAVI provides a more potent means of discriminating these classes. In addition, the Sentinel-1 VH backscatter time series for slangbos delineate from the other classes when compared to the Sentinel-1 VV coherence.

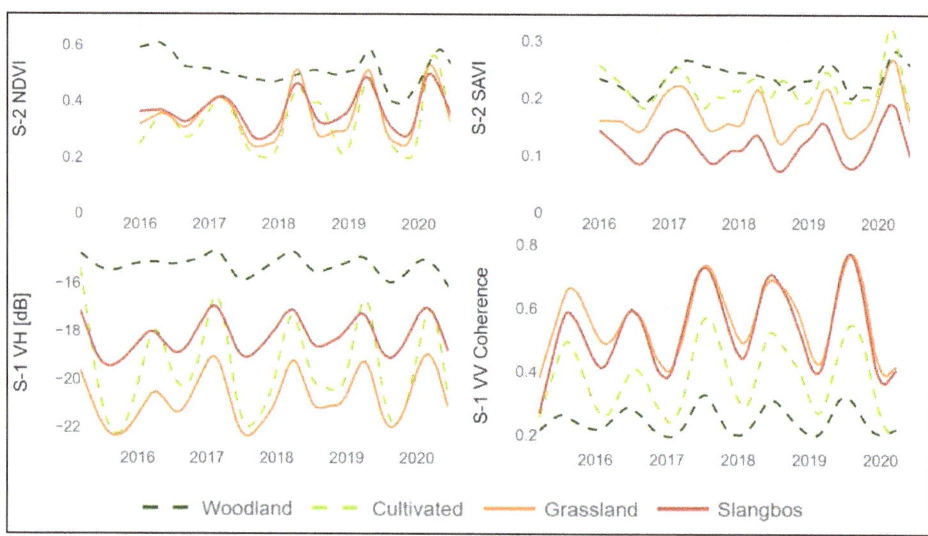

Figure 4. Separability of the class slangbos from other land cover classes for the variables Sentinel-1 VH backscatter, Sentinel-1 VV coherence, Sentinel-2 NDVI and SAVI time series.

3.2. Classifying Slangbos using Random Forest

3.2.1. Variable Importance

The variable importance was calculated for the crop years between 2015 and 2018 (Figure 5). The results show that the Sentinel-2 derived SAVI has the highest importance for the classification algorithm (refer to Section 3.2, Figure 4). This is especially true for the crop years 2015 to 2016. However, this year needs to be analyzed with caution, as the Sentinel-2 data are only available from early 2016, and thus are not available for the entire crop year. During the same crop year, the NDVI, as well as the cross-polarized Sentinel-1 VH data, have important contributions to the classification result. The Sentinel-1 VV coherence results in the lowest variable importance for all observed crop years. During 2016 and 2018, the variable Sentinel-1 VH had less significance for the classification algorithm when compared to the Sentinel-2 derived NDVI. This importance must be handled with care, as a higher abundance of a feature subset, e.g., the denser time series of Sentinel-1 compared to cloud-free Sentinel-2 acquisitions result in a relatively lower overall importance of that feature subset. However, the optical indices and cross-polarized SAR still dominate the bulk of the model performance. The inter-comparison between the years is further dominated by the number of predictors used for modeling, e.g., the crop year 2017/2018 shows the lowest importance, as the highest number of predictors was used.

3.2.2. Spatial Cross-Validation (SpCV)

The spatial cross-validation was carried out for each of the crop years between 2015 and 2018. Figure 6 shows the accuracies of the binary classification of the categories slangbos as well as non-slangbos. The overall accuracy exceeded 90% for the entire period. The precision measure indicates to what extent the slangbos was correctly classified (around 80%). The recall measure is even higher, which is an indicator of how much slangbos was missed during classification (13% to 16%). Slightly lower precisions were found for the years 2016/17 and 2017/18.

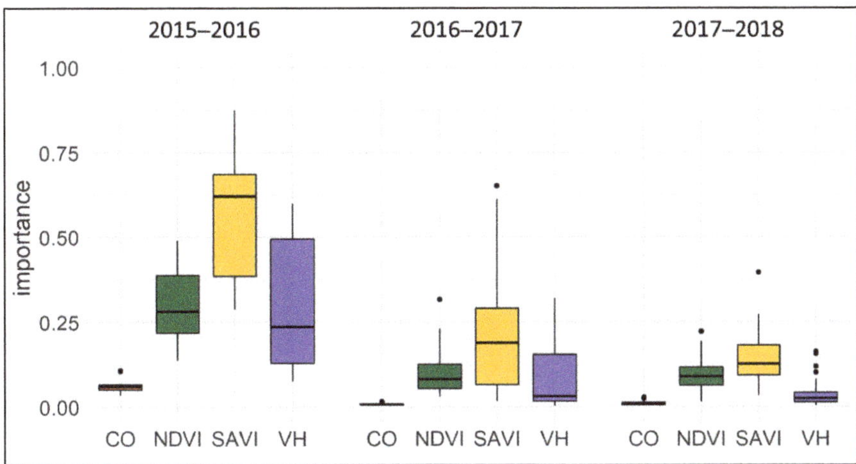

Figure 5. Variable importance for the input parameters Sentinel-1 VV coherence (CO), cross-polarized Sentinel-1 backscatter (VH) as well as Sentinel-2 derived NDVI and SAVI.

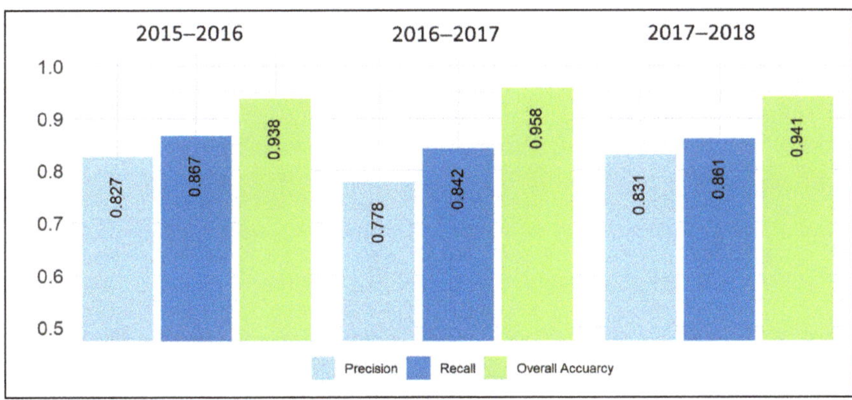

Figure 6. Classification accuracy metrics derived from spatial cross-validation for each of the crop years between 2015 and 2018.

3.2.3. Slangbos Probability Measures

The probability measure is an output of the random forest algorithm and describes the potential assignment of the individual pixel to the class slangbos between 2017 and 2018 [79] (Figure 7, left).

The figure gives a detailed insight into the different characteristics of agricultural fields being potentially infected by slangbos growth. The dark field in the center of the image is surrounded by fields with high probabilities for slangbos encroachment, which is also visible as brownish areas in the optical Sentinel-2 image (Figure 7, right). In combination with the Sentinel-2 images, it can be illustrated how the probability index reflects the heterogeneity within each of the fields. This information might be utilized to identify areas that need to be prioritized for slangbos-clearing investigations within a slangbos encroachment monitoring system.

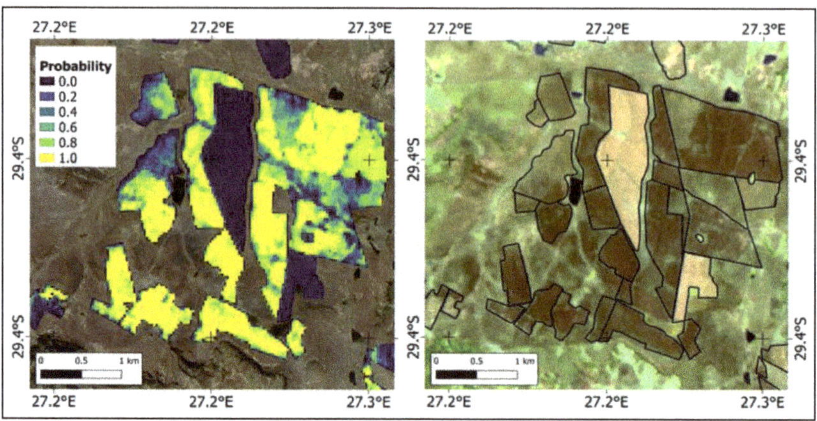

Figure 7. An example of the probability measure mapping slangbos encroachment in the project area. The comparison with an optical image illustrates how the probability index reflects the heterogeneity within fields. Left: probability measure for the assignment of individual pixels to the class slangbos between 2017 and 2018. Right: Sentinel-2 Short-Wave Infrared (SWIR) image from 30.08.2017 (RGB = bands 12-8A-4). (Field boundaries source: [39]; contains modified Copernicus Sentinel data [2017–2018]).

3.3. Mapping

3.3.1. Regional Scale Analysis

Figure 8 shows the distribution of the areas which were identified to be infected by slangbos encroachment for the entire study area between 2015 and 2020. The presented areas were not used for model training. Areas in red are dominant and relatively homogeneously distributed in the entire study area. These areas indicate that slangbos were prevalent during the beginning of the time series in 2015/2016. Blue areas are spread between the southern and the eastern part of the study area. The patches indicate slangbos and shrub cover during the second half of the time series, after 2017. These areas are likely to be those that were infected by the growth of slangbos or shrub encroachment during that time period. Few areas are classified as green, which indicate slangbos only found in 2017, which are often adjacent to orange areas, indicating slangbos or shrub cover also in 2015. These regions might be areas in which slangbos and shrub cover were cleared either by hand or due to fires.

Figure 9 highlights four areas, showing the classification result within the rangeland areas to describe the spatial pattern in more detail. Area 1 is a heavily encroached prone site. The local partners from LandCare confirmed this site to be cleared during our observation period. In particular, the red areas in the rangelands were cleared between 2015 and 2019. On the other hand, regrowth in recent (since 2017) years is found in other areas, which are shown in cyan. In the center, some white areas are visible, indicating slangbos infection during the entire period. In Area 2, a fire occurred in September 2017, which resulted in a massive loss of shrub cover (green areas) [80]. On the other side, shrub encroachment on native rangelands is found on the eastern part of the plateau (red areas). Area 3 is a large managed site, where also shrub clearing took place during the second half of the observed period. Area 4 is characterized by intensive shrub encroachment in the rangeland sites since 2017/2018. The classification algorithm was able to even detect these small areas.

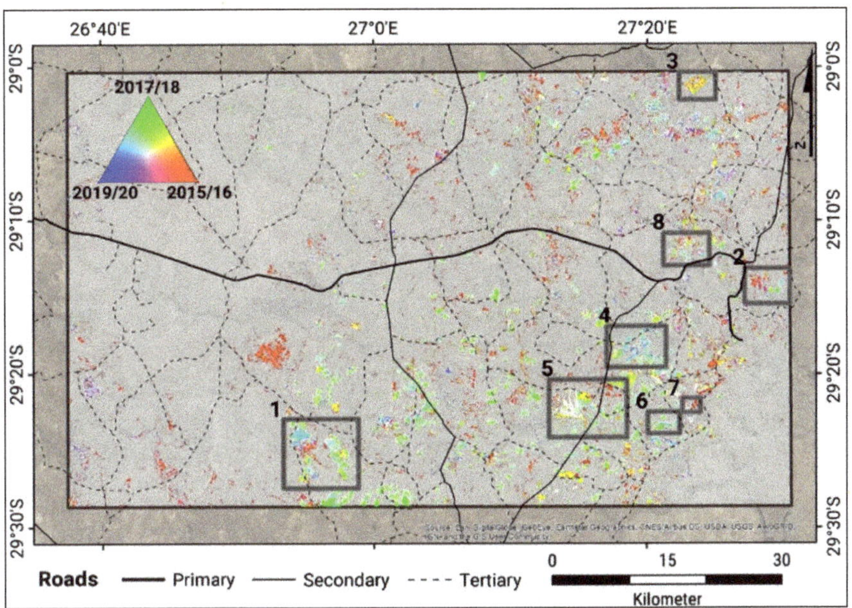

Figure 8. Spatial distribution of the regions affected by slangbos encroachment for the entire study area between 2015 and 2020. This map represents each crop year in a specific layer of the RGB composite (R: 2015/16, G: 2017/18, B: 2019/20). The numbered boxes show the zoom-in for a detailed analysis within the next sections. (Roads: [35], Map data source: Esri, DigitalGlobe, GeoEye, Earthstar Geographics, CNES/Airbus DS, USDA, USGS, AeroGRID, IGN, and the GIS User Community).

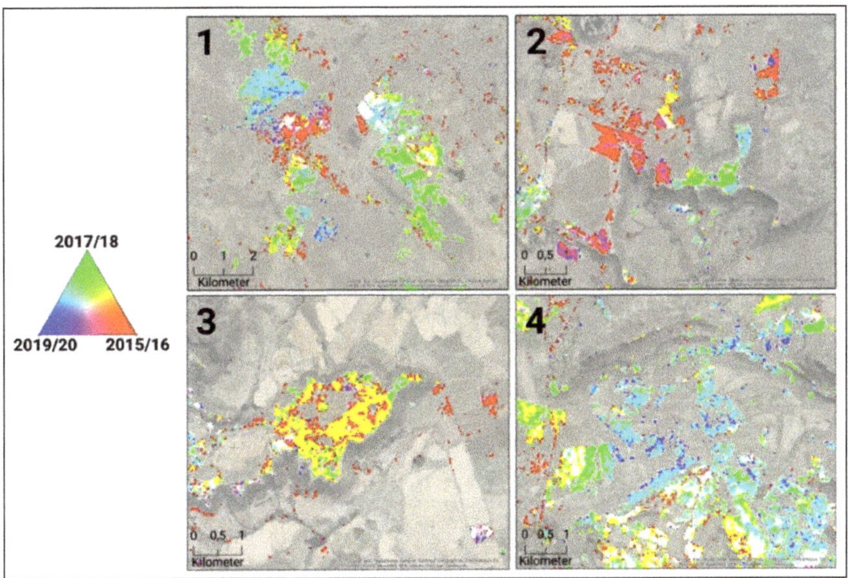

Figure 9. Subset of the boxes 1 to 4 (Figure 8) showing slangbos infected areas in the rangelands (R: 2015/16, G: 2017/18, B: 2019/20). (Map data source: Esri, DigitalGlobe, GeoEye, Earthstar Geographics, CNES/Airbus DS, USDA, USGS, AeroGRID, IGN, and the GIS User Community).

3.3.2. Field Boundary Scale Analysis

Figure 10 highlights four areas within the cropland boundaries, which were provided by the DALRRD [39]. The region inside the grey dashed line in area 5 is identical to the subset shown in Figure 7, which was utilized to show the classification results of the model. This region shows heavily encroached pastures, where some of them were found to be fallow. Large areas are shown in white, indicating that slangbos were growing on the rangelands during the entire time period. Area 6 illustrates areas that were rapidly encroached by slangbos in 2017/2018 (turquoise and blue). Green areas are likely caused due to misclassifications, as it would mean that slangbos only occurred in 2017/18 and suddenly disappeared. However, man-made clearing management or fires might also be causing this spatial pattern. Area 7 represents a vast shrub control side, where large areas were infected by slangbos during the entire time (white areas); some areas show the effect of clearing actions after 2015/16 (red areas). Area 8 shows a diverse spatial of different slangbos encroachment dynamics. Some areas indicate slangbos infections during the entire period. Large areas show slangbos clearing activities, which are shown in red (cleared in 2015/16) and yellow (cleared after 2017/18). Small areas indicate slangbos infections during the later stage of the time period (turquoise), which might be fields, which were cleared a few years before and when already affected by slangbos encroachment.

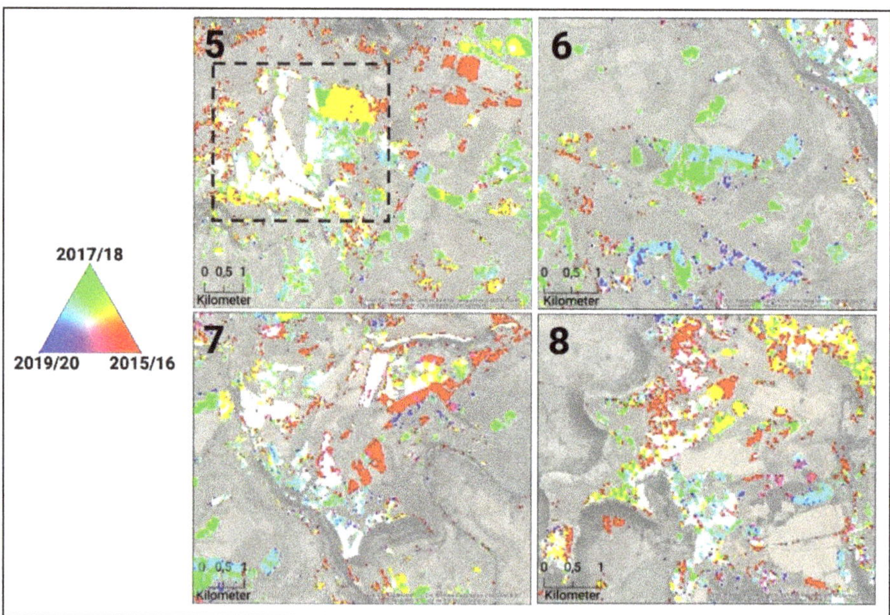

Figure 10. Subset of the boxes 5 to 8 (Figure 8) showing slangbos infected areas in the field boundaries by crop years (R: 2015/16, G: 2017/18, B: 2019/20). The grey dashed line in subset 5 represents the region shown in Figure 7. (Map data source: Esri, DigitalGlobe, GeoEye, Earthstar Geographics, CNES/Airbus DS, USDA, USGS, AeroGRID, IGN, and the GIS User Community).

The crop statistics were used to identify which crop type classes are affected the most by the encroachment of slangbos. Figure 11 indicates the relative area covered by slangbos on cropland, fallow land and pastures between 2015 and 2018. It needs to be mentioned that only time period for which crop statistics were available is covered. Thus, recent years were not accessible through the DALRRD, as these data were still classified for internal use only at the time of writing.

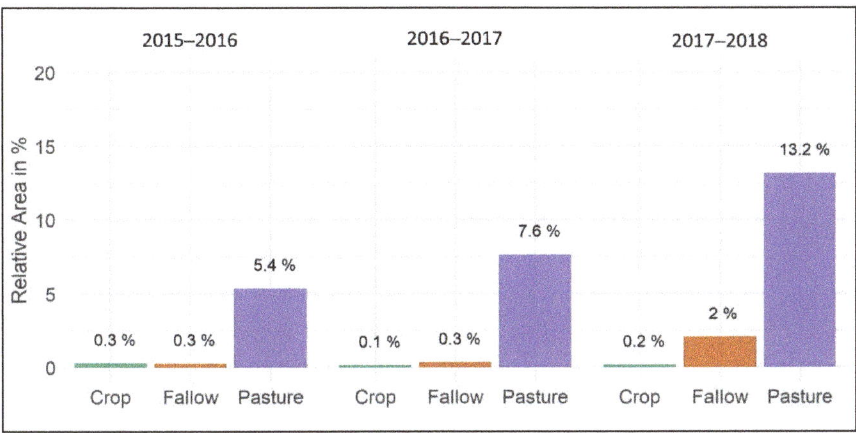

Figure 11. Relative area (in %) statistics on slangbos encroachment of different agricultural land types (cropland, fallow land and pastures).

The statistics on slangbos encroachment of different agricultural land types clearly demonstrate that pasture areas are highly affected by the encroachment of slangbos. Fallow areas also show some sensitivity to infection by the growth of slangbos. Croplands reveal almost no sign of being affected by the encroachment of slangbos, which is attributed to intensive management (i.e., plowing) throughout harvest cycles. These findings are an indication for the reliabilty of the classification approach, as we did not expect slangbos encroachment on croplands, due to the intensive management.

4. Discussion

This study focused on the classification of slangbos encroachment on agricultural land, using ESAs Copernicus Sentinel-1 and Sentinel-2 time series between 2015 and 2020 for a test area in the Free State Province, South Africa. The classification accuracies of over 80% indicate a solid approach for slangbos encroachment mapping, using optical and radar time series information from ESAs Sentinels data. Moreover, the random forest classifier was used as a framework for the multi-temporal and multi-sensor classification.

The paper investigated slangbos encroachment on a comparatively small study area of around 4500 km^2, where slangbos is known to be the main encroacher. The agricultural statistics revealed that fallow areas increased dramatically after 2015 and doubled until the end of 2018. In this year, fallow areas covered more than 50%, with the remaining areas being almost evenly distributed amongst croplands and pastures. This abandonment of agricultural land is a known phenomenon of the land-use change in South Africa, which is driven by climate as well as socio-economic factors [81]. The statistics on slangbos encroachment of different agricultural land types revealed that this process is more recent in abandoned crop areas as compared to regions near settlements [82].

Upcoming research activities might focus on larger areas, where other bush communities or shrub types represent the main invader. As an example, it is worth mentioning the black wattle (*Acacia mearnsii*), which also imposes great pressure on grassland communities in South Africa [83,84]. Hence, future studies could build upon this methodology to aim for the discrimination between different shrub communities in regions, where a mixture of different encroachers is present. If the goal is to classify different shrub communities, the presented binary approach has limitations in the separation of different encroachers. This might need additional data (e.g., hyperspectral [85]) for the training of more suitable machine-learning models.

However, classifying different shrub communities is likely to introduce other issues, such as spectral and scattering uncertainties, which can be attributed to the canopy spacing of the individual plants that need to be considered in future studies. Depending on the spatial resolution of the data, the influence of grass and soil between shrubs need to be analyzed in more detail, as it has major impacts on the reflectance and backscatter. Further research will increase the knowledge of those scattering mechanisms of slangbos and other bush communities in contrast to grassland. Regarding the classification of slangbos, it is likely that the Sentinel-2 derived SAVI as well as the Sentinel-1 VH backscatter are of higher potential than the NDVI and Sentinel-1 VV coherence. In the case of the SAVI, it might be attributed to the fact that slangbos patches grow more sparsely, and thus benefit from the bare soil consideration in the SAVI estimation, as the influence of the soil brightness is reduced in comparison to the NDVI [86]. As droughts might favor bush encroachment [87,88], it is of high importance to integrate information on precipitation dynamics in future studies. Here, we utilized coarse resolution CHIRPS data for analysing the yearly precipitation dynamics. Since rainfall events might occur locally, information from climate stations or data products with higher spatial resolution could be of great value for upcoming investigations.

To make the model performance more flexible and effective as well as applicable for investigations in other or larger regions, the training input should be limited to only important model features. In this study, more than 500 features were used as model training input, with around 100 for each individual crop year. These might be not applicable when transferring the methodology to other larger regions. To enhance the model performance and based on the known variable importance, Sentinel-1 VH backscatter, Sentinel-2 SAVI and NDVI time series might solely be utilized as input variables. Methods for interpretable machine learning and feature selection are desirable for future research, making the models comprehensible and transparent. In addition, the use of statistical metrics from hyper-temporal EO data might be a feasible solution to analyze time series in a cost-effective and descriptive way. Simple descriptive statistics, such as median, standard deviation or quantiles in conjunction with regression functions or temporal filters might improve the model performance as well as classification results.

The availability of agricultural statistics for the entire observation period is crucial for the identification of classes that were infected by slangbos encroachment. In this study, the statistics were available between 2014 and 2018 only. Hence, statistics for two crop years (2019 and 2020) were missing.

Future investigation might utilize different measures to assess the accuracy. In this study, we measured the accuracies based on a binary classification (slangbos and non-slangbos), which integrates issues related to the sample size of both classes. The overall accuracy might be not reliably interpretable in this case, whereas recall and precision are more valuable measures. The overall accuracy is prone to the imbalance of the sample size between the slangbos and non-slangbos classes [89]. The recall marks the rate of "missed" slangbos pixels, while the precision is the rate of falsely detected slangbos areas. Both numbers can be low, as they are only based on true positive, false negative and false positive values. The true negative sample size is extremely large, as is common in most remote sensing classifications since they include all classes, e.g., water, urban, and grassland, pixels.Upcoming investigations should emphasize the potential of mapping slangbos encroachment, using cloud-based solutions (e.g., Google Earth Engine) to minimize the data processing times for the users. This might result in a bush encroachment monitoring system, where products are directly accessible by the users without any data interaction. Such a monitoring system is likely to result in an early warning system using near-real-time classification approaches, advances in the potential of current methodologies identifying infected areas, as well as helping to plan optimal clearing strategies supporting sustainable land management strategies.

5. Conclusions

The objective of this paper was to monitor slangbos encroachment between 2015 and 2020 in a test region in the Free State Province, South Africa. The slangbos classification was carried out utilizing a synergetic combination of Sentinel-1 and Sentinel-2 time series within a machine-learning framework, applying a random forest classifier. Field inventory and high-resolution image analyses as well as spatial crop statistics and slangbos verified areas, which were provided by the DALRRD, were used for training and spatial cross-validation.

The time series analysis of the Sentinel-1 and Sentinel-2 data has shown that the Sentinel-1 VH (cross-polarization) and the Sentinel-2 SAVI (Soil Adjusted Vegetation Index) carry the highest separability between the shrub and other land cover classes. Moreover, random forest permutation-based feature importance showed that these parameters provide the largest contribution to the classifiers when accounting for the other variables in the model. The spatial interpretation revealed that the slangbos infected areas are well captured for the different crop years between 2015 and 2020. Pastures are particularly prone to slangbos encroachment, whereas cultivated areas are less affected. Even small patches of slangbos growth and the resulting heterogeneity on different fields could be identified with the used Sentinel-1 and Sentinel-2 data. This knowledge is an essential information source for a slangbos-encroachment-monitoring system to identify areas that need to be prioritized for slangbos-clearing investigations. The estimation of the classification accuracy was performed via spatial-cross validation and resulted in an overall accuracy of around 90% for each crop year, with a positive predictive value (precision) of around 80%. These accuracies indicate large potential for the transferability to other regions for monitoring shrub encroachment.

The study has shown the potential of using high-resolution optical and radar Earth observation time series to classify slangbos encroachment on pastures in the Free State Province of South Africa. Future studies might utilize these findings in other regions of shrub and bush encroachment.

Author Contributions: Conceptualization, M.U., K.S., T.M., C.D., and C.S.; methodology, M.U., K.S., and C.D.; data analysis and interpretation, all authors; validation, M.U., K.S., and T.M.; writing—original draft preparation, M.U.; writing—review and editing, all authors; visualization, M.U., and K.S.; project administration, J.B., U.G., and C.S.; funding acquisition, J.B., U.G., and C.S. All authors have read and agreed to the published version of the manuscript.

Funding: The authors received funding from the German Federal Ministry of Education and Research (BMBF) in the framework of the Science Partnerships for the Assessment of Complex Earth System Processes (SPACES2) under the grant 01LL1701 (South African Land Degradation Monitor (SALDi)).

Acknowledgments: The authors would like to thank the Department of Agriculture, Land Reform and Rural Development (DALRRD) for providing spatial information for the different crop types between 2014 and 2018 for this study. We acknowledge the support of Makzine Ranthimo, who unfortunately passed away in May 2021. She was instrumental as a very knowledgeable and warm-hearted contact person for our field studies in the Ladybrand region.

Conflicts of Interest: The authors declare no conflict of interest.

References

1. Dubovyk, O. The role of Remote Sensing in land degradation assessments: Opportunities and challenges. *Eur. J. Remote Sens.* **2017**, *50*, 601–613. [CrossRef]
2. Intergovernmental Science-Policy Platform on Biodiversity and Ecosystem Services (IPBES). *The IPBES Assessment Report on Land Degradation and Restoration*; IPBES: Bonn, Germany, 2018; Volume 744.
3. Eldridge, D.J.; Bowker, M.A.; Maestre, F.T.; Roger, E.; Reynolds, J.F.; Whitford, W.G. Impacts of shrub encroachment on ecosystem structure and functioning: Towards a global synthesis. *Ecol. Lett.* **2011**, *14*, 709–722. [CrossRef] [PubMed]
4. O'Connor, T.G.; Puttick, J.R.; Hoffman, M.T. Bush encroachment in southern Africa: Changes and causes. *Afr. J. Range Forage Sci.* **2014**, *31*, 67–88. [CrossRef]
5. Stevens, N.; Erasmus, B.F.N.; Archibald, S.; Bond, W.J. Woody encroachment over 70 years in South African savannahs: Overgrazing, global change or extinction aftershock? *Philos. Trans. R. Soc. B Biol. Sci.* **2016**, *371*, 20150437. [CrossRef]
6. Wigley, B.J.; Bond, W.J.; Hoffman, M.T. Bush encroachment under three contrasting land-use practices in a mesic South African savanna. *Afr. J. Ecol.* **2009**, *47*, 62–70. [CrossRef]
7. Cao, X.; Liu, Y.; Cui, X.; Chen, J.; Chen, X. Mechanisms, monitoring and modeling of shrub encroachment into grassland: A review. *Int. J. Digit. Earth* **2019**, *12*, 625–641. [CrossRef]

8. Venter, Z.S.; Cramer, M.D.; Hawkins, H.J. Drivers of woody plant encroachment over Africa. *Nat. Commun.* **2018**, *9*, 2272. [CrossRef] [PubMed]

9. Ainsworth, E.A.; Long, S.P. What have we learned from 15 years of free-air CO_2 enrichment (FACE)? A meta-analytic review of the responses of photosynthesis, canopy properties and plant production to rising CO_2. *New Phytol.* **2005**, *165*, 351–372. [CrossRef]

10. Morgenthal, T. Mapping *Seriphium plumosum* (Slangbos) using Sentinel data. In Proceedings of the Society of South African Geographers Conference, Bloemfontein, South Africa, 1–4 October 2018.

11. Rutherford, M.C.; Mucina, L. *The Vegetation of South Africa, Lesotho and Swaziland. Strelitzia 19*; South African Biodiversity Institute: Pretoria, South Africa, 2006; 816p.

12. Wepener, J. *The Control of Stoebe Vulgans Encroachment in the Hartbeesfontein area of the North West Province*; North-West University: Potchefstroom, South Africa, 2007.

13. Snyman, H.A. Control measures for the encroacher shrub Seriphium plumosum. *S. Afr. J. Plant Soil* **2012**, *29*, 157–163. [CrossRef]

14. Jordaan, D. Bankruptbush (Slangbos)—A silent threat to grasslands? *Grassroots Newsl. Grassl. Soc. South. Afr.* **2009**, *9*, 40–42.

15. du Toit, J.C.O.; Cronje, W.B.; Trollope, W.S.W. Towards low-input control of slangbos (*Seriphium plumosum*)—Quality and grazing interaction hypotheses. *Grootfontein Agric.* **2013**, *13*, 21–25.

16. Snyman, H.A. Habitat preferences of the encroacher shrub, *Seriphium plumosum*. *S. Afr. J. Bot.* **2012**, *81*, 34–39. [CrossRef]

17. Hare, M.L.; Xu, X.; Wang, Y.; Gedda, A.I. The effects of bush control methods on encroaching woody plants in terms of die-off and survival in Borana rangelands, southern Ethiopia. *Pastoralism* **2020**, *10*, 16. [CrossRef]

18. Graham, S.C.; Barrett, A.S.; Brown, L.R. Impact of *Seriphium plumosum* densification on Mesic Highveld Grassland biodiversity in South Africa. *R. Soc. Open Sci.* **2020**, *7*, 192025. [CrossRef]

19. Baltimore, M.M.; Jorrie, J.J.; Tieho, P.M.; Martin, J.P. The Effect of Root and Shoot Extracts of *Seriphium plumosum* as Allelopathic Agents. *Insights For. Res.* **2017**. [CrossRef]

20. Avenant, P. *Report on the National Bankrupt Bush (Seriphium plumosum) Survey (2010–2012)*; Department of Agriculture, Forestry & Fisheries: Pretoria, South Africa, 2015.

21. Huang, C.Y.; Archer, S.R.; McClaran, M.P.; Marsh, S.E. Shrub encroachment into grasslands: End of an era? *PeerJ* **2018**, *2018*. [CrossRef] [PubMed]

22. Urban, M.; Heckel, K.; Berger, C.; Schratz, P.; Smit, I.P.J.; Strydom, T.; Baade, J.; Schmullius, C. Woody cover mapping in the savanna ecosystem of the Kruger National Park using Sentinel-1 C-Band time series data. *KOEDOE Afr. Prot. Area Conserv. Sci.* **2020**, *62*, 6. [CrossRef]

23. Bucini, G.; Hanan, N.; Boone, R.; Smit, I.; Saatchi, S.; Lefsky, M.; Asner, G. Woody Fractional Cover in Kruger National Park, South Africa: Remote Sensing–Based Maps and Ecological Insights. In *Ecosystem Function in Savannas*; Hill, M.J., Hanan, N.P., Eds.; CRC Press: Boca Raton, FL, USA, 2010; pp. 219–237.

24. Higginbottom, T.P.; Symeonakis, E.; Meyer, H.; van der Linden, S. Mapping fractional woody cover in semi-arid savannahs using multi-seasonal composites from Landsat data. *ISPRS J. Photogramm. Remote Sens.* **2018**, *139*, 88–102. [CrossRef]

25. Urbazaev, M.; Thiel, C.; Mathieu, R.; Naidoo, L.; Levick, S.R.; Smit, I.P.J.; Asner, G.P.; Schmullius, C. Assessment of the mapping of fractional woody cover in southern African savannas using multi-temporal and polarimetric ALOS PALSAR L-band images. *Remote Sens. Environ.* **2015**, *166*, 138–153. [CrossRef]

26. Ludwig, M.; Morgenthal, T.; Detsch, F.; Higginbottom, T.P.; Lezama Valdes, M.; Nauß, T.; Meyer, H. Machine learning and multi-sensor based modelling of woody vegetation in the Molopo Area, South Africa. *Remote Sens. Environ.* **2019**, *222*, 195–203. [CrossRef]

27. Skowno, A.L.; Thompson, M.W.; Hiestermann, J.; Ripley, B.; West, A.G.; Bond, W.J. Woodland expansion in South African grassy biomes based on satellite observations (1990–2013): General patterns and potential drivers. *Glob. Chang. Biol.* **2017**, *23*, 2358–2369. [CrossRef]

28. Venter, Z.S.; Scott, S.L.; Desmet, P.G.; Hoffman, M.T. Application of Landsat-derived vegetation trends over South Africa: Potential for monitoring land degradation and restoration. *Ecol. Indic.* **2020**, *113*, 106206. [CrossRef]

29. Wessels, K.; Mathieu, R.; Knox, N.; Main, R.; Naidoo, L.; Steenkamp, K. Mapping and monitoring fractional woody vegetation cover in the Arid Savannas of Namibia Using LiDAR training data, machine learning, and ALOS PALSAR data. *Remote Sens.* **2019**, *11*, 2633. [CrossRef]

30. Graw, V.; Ghazaryan, G.; Dall, K.; Gómez, A.D.; Abdel-Hamid, A.; Jordaan, A.; Piroska, R.; Post, J.; Szarzynski, J.; Walz, Y.; et al. Drought dynamics and vegetation productivity in different land management systems of Eastern Cape, South Africa—A remote sensing perspective. *Sustainability* **2017**, *9*, 1728. [CrossRef]

31. Marston, C.G.; Aplin, P.; Wilkinson, D.M.; Field, R.; O'Regan, H.J. Scrubbing up: Multi-scale investigation of woody encroachment in a Southern African savannah. *Remote Sens.* **2017**, *9*, 419. [CrossRef]

32. Kiker, G.A.; Scholtz, R.; Smit, I.P.J.; Venter, F.J. Exploring an extensive dataset to establish woody vegetation cover and composition in Kruger National Park for the late 1980s. *Koedoe* **2014**, *56*, 1–10. [CrossRef]

33. Pettorelli, N.; Wegmann, M.; Skidmore, A.; Mücher, S.; Dawson, T.P.; Fernandez, M.; Lucas, R.; Schaepman, M.E.; Wang, T.; O'Connor, B.; et al. Framing the concept of satellite remote sensing essential biodiversity variables: Challenges and future directions. *Remote Sens. Ecol. Conserv.* **2016**, *2*, 122–131. [CrossRef]

34. International Land Resources (Pty) Ltd. *Automated Land Cover Classification South Africa*; Report No: L06572/180618/1; Final Report—SSC WC 03(2017/2018) DRDLR; International Land Resources (Pty) Ltd.: Pietermaritzburg, South Africa, 2018.

35. World Bank. Roads in South Africa—Shape File Dataset. 2017. Available online: https://datacatalog.worldbank.org/dataset/south-africa-roads (accessed on 23 August 2021).
36. Nel, J.L.; Richardson, D.M.; Rouget, M.; Mgidi, T.N.; Mdzeke, N.; Le Maitre, D.C.; Van Wilgen, B.W.; Schonegevel, L.; Henderson, L.; Neser, S. A proposed classification of invasive alien plant species in South Africa: Towards prioritizing species and areas for management action. *S. Afr. J. Sci.* **2004**, *100*, 53–64.
37. Geudtner, D.; Torres, R.; Snoeij, P.; Davidson, M.; Rommen, B. Sentinel-1 System capabilities and applications. In Proceedings of the International Geoscience and Remote Sensing Symposium (IGARSS), Quebec, QC, Canada, 13–18 July 2014; pp. 1457–1460.
38. ESA Sentinel-2 User Handbook. 2015. Available online: https://sentinel.esa.int/documents/247904/685211/Sentinel-2_User_Handbook (accessed on 23 August 2021).
39. Department of Agriculture, Land Reform and Rural Development (DALRRD). *Crop Type Classification for the Free State Province—Spatial Data Layers, 2014–2018*; DALRRD: Pretoria, South Africa, 2020.
40. Statistics South Africa. Stats SA Releases Census of Commercial Agriculture 2017 Report. 2020. Available online: http://www.statssa.gov.za/?p=13144 (accessed on 23 August 2021).
41. Google Map Data ©2021 Google 2021. Available online: earth.google.com/web/ (accessed on 23 August 2021).
42. National-Geo-Spatial-International (CDNGI). Geospatial Data Portal. Available online: http://www.cdngiportal.co.za/cdngiportal/ (accessed on 23 August 2021).
43. Truckenbrodt, J.; Freemantle, T.; Williams, C.; Jones, T.; Small, D.; Dubois, C.; Thiel, C.; Rossi, C.; Syriou, A.; Giuliani, G. Towards Sentinel-1 SAR Analysis-Ready Data: A Best Practices Assessment on Preparing Backscatter Data for the Cube. *Data* **2019**, *4*, 93. [CrossRef]
44. Wegmüller, U.; Werner, C.; Strozzi, T.; Wiesmann, A.; Frey, O.; Santoro, M. Sentinel-1 Support in the GAMMA Software. *Procedia Comput. Sci.* **2016**, *100*, 1305–1312. [CrossRef]
45. United States Geological Survey (USGS). Shuttle Radar Topography Mission (SRTM) 1 Arc-Second Global—Version 3. Available online: https://earthexplorer.usgs.gov/ (accessed on 23 August 2021).
46. Small, D. Flattening gamma: Radiometric terrain correction for SAR imagery. *IEEE Trans. Geosci. Remote Sens.* **2011**, *49*, 3081–3093. [CrossRef]
47. Rocca, F.; Ferretti, A.; Monti-Guarnieri, A.V.; Prati, C.M.; Massonnet, D. Part C InSAR processing: A mathematical approach. In *InSAR Principles: Guidelines for SAR Interferometry Processing and Interpretation*; ESA Publications ESTEC: Noordwijk, The Netherlands, 2007; p. 110. ISBN 9290922338.
48. Ferretti, A.; Monti-Guarnieri, A.; Prati, C.; Rocca, F. Part B InSAR processing: A practical approach. In *InSAR Principles: Guidelines for SAR Interferometry Processing and Interpretation*; ESA Publications ESTEC: Noordwijk, The Netherlands, 2007; p. 67. ISBN 9290922338.
49. Jacob, A.W.; Notarnicola, C.; Suresh, G.; Antropov, O.; Ge, S.; Praks, J.; Ban, Y.; Pottier, E.; Mallorqui Franquet, J.J.; Duro, J.; et al. Sentinel-1 InSAR Coherence for Land Cover Mapping: A Comparison of Multiple Feature-Based Classifiers. *IEEE J. Sel. Top. Appl. Earth Obs. Remote Sens.* **2020**, *13*, 535–552. [CrossRef]
50. Sica, F.; Pulella, A.; Nannini, M.; Pinheiro, M.; Rizzoli, P. Repeat-pass SAR interferometry for land cover classification: A methodology using Sentinel-1 Short-Time-Series. *Remote Sens. Environ.* **2019**, *232*, 111277. [CrossRef]
51. Santoro, M.; Cartus, O. Research Pathways of Forest Above-Ground Biomass Estimation Based on SAR Backscatter and Interferometric SAR Observations. *Remote Sens.* **2018**, *10*, 608. [CrossRef]
52. Stelmaszczuk-Górska, M.A.; Rodriguez-Veiga, P.; Ackermann, N.; Thiel, C.; Balzter, H.; Schmullius, C. Non-parametric retrieval of aboveground biomass in siberian boreal forests with ALOS PALSAR interferometric coherence and backscatter intensity. *J. Imaging* **2016**, *2*, 1. [CrossRef]
53. Martone, M.; Rizzoli, P.; Wecklich, C.; González, C.; Bueso-Bello, J.L.; Valdo, P.; Schulze, D.; Zink, M.; Krieger, G.; Moreira, A. The global forest/non-forest map from TanDEM-X interferometric SAR data. *Remote Sens. Environ.* **2018**, *205*, 352–373. [CrossRef]
54. Thiele, A.; Dubois, C.; Boldt, M.; Hinz, S. Using TanDEM data for forest height estimation and change detection. In *Proceedings Volume 10005, Earth Resources and Environmental Remote Sensing/GIS Applications VII*; Michel, U., Schulz, K., Ehlers, M., Nikolakopoulos, K.G., Civco, D., Eds.; SPIE: Edinburgh, UK, 2016. [CrossRef]
55. Main-Knorn, M.; Louis, J.; Hagolle, O.; Müller-Wilms, U.; Alonso-Gonzalez, K. The Sen2Cor and MAJA cloud masks and classification products. In Proceedings of the 2nd Sentinel-2 Validation Team Meeting, Frascati, Italy, 28–31 January 2018.
56. Flood, N.; Gillingham, S. Python Implementation of Fmask. 2015. Available online: http://www.pythonfmask.org/en/latest/ (accessed on 23 August 2021).
57. Zhu, Z.; Wang, S.; Woodcock, C.E. Improvement and expansion of the Fmask algorithm: Cloud, cloud shadow, and snow detection for Landsats 4-7, 8, and Sentinel 2 images. *Remote Sens. Environ.* **2015**, *159*, 269–277. [CrossRef]
58. Zhu, Z.; Woodcock, C.E. Object-based cloud and cloud shadow detection in Landsat imagery. *Remote Sens. Environ.* **2012**, *118*, 83–94. [CrossRef]
59. Qiu, S.; Zhu, Z.; He, B. Fmask 4.0: Improved cloud and cloud shadow detection in Landsats 4–8 and Sentinel-2 imagery. *Remote Sens. Environ.* **2019**, *231*, 111205. [CrossRef]
60. Tucker, C.J. Red and photographic infrared linear combinations for monitoring vegetation. *Remote Sens. Environ.* **1979**, *8*, 127–150. [CrossRef]
61. Huete, A.R. A soil-adjusted vegetation index (SAVI). *Remote Sens. Environ.* **1988**, *25*, 295–309. [CrossRef]

62. Friedman, J. A Variable Span Smoother. *J. Am. Stat. Assoc.* **1984**. [CrossRef]
63. Brenning, A. Spatial cross-validation and bootstrap for the assessment of prediction rules in remote sensing: The R package sperrorest. In Proceedings of the International Geoscience and Remote Sensing Symposium (IGARSS), Munich, Germany, 22–27 July 2012; pp. 5372–5375.
64. Breiman, L. Random forests. *Mach. Learn.* **2001**, *45*, 5–32. [CrossRef]
65. Belgiu, M.; Drăgu, L. Random forest in remote sensing: A review of applications and future directions. *ISPRS J. Photogramm. Remote Sens.* **2016**, *114*, 24–31. [CrossRef]
66. Meyer, H.; Reudenbach, C.; Wöllauer, S.; Nauss, T. Importance of spatial predictor variable selection in machine learning applications—Moving from data reproduction to spatial prediction. *Ecol. Modell.* **2019**, *411*, 108815. [CrossRef]
67. Schratz, P.; Muenchow, J.; Iturritxa, E.; Richter, J.; Brenning, A. Hyperparameter tuning and performance assessment of statistical and machine-learning algorithms using spatial data. *Ecol. Modell.* **2019**, *406*, 109–120. [CrossRef]
68. Probst, P.; Boulesteix, A.L.; Bischl, B. Tunability: Importance of hyperparameters of machine learning algorithms. *arXiv* **2018**, arXiv:1802.09596.
69. Probst, P.; Wright, M.N.; Boulesteix, A.L. Hyperparameters and tuning strategies for random forest. *Wiley Interdiscip. Rev. Data Min. Knowl. Discov.* **2019**, *9*, e1301. [CrossRef]
70. Bernard, S.; Heutte, L.; Adam, S. Influence of hyperparameters on random forest accuracy. In *Lecture Notes in Computer Science*; Subseries Lecture Notes in Artificial Intelligence and Lecture Notes in Bioinformatics; Springer: Berlin/Heidelberg, Germany, 2009; Volume 5519, pp. 171–180.
71. Heckel, K.; Urban, M.; Schratz, P.; Mahecha, M.D.; Schmullius, C. Predicting forest cover in distinct ecosystems: The potential of multi-source Sentinel-1 and -2 data fusion. *Remote Sens.* **2020**, *12*, 302. [CrossRef]
72. Huang, B.F.F.; Boutros, P.C. The parameter sensitivity of random forests. *BMC Bioinform.* **2016**, *17*. [CrossRef]
73. Strobl, C.; Boulesteix, A.L.; Zeileis, A.; Hothorn, T. Bias in random forest variable importance measures: Illustrations, sources and a solution. *BMC Bioinform.* **2007**, *8*, 25. [CrossRef]
74. Lang, M.; Binder, M.; Richter, J.; Schratz, P.; Pfisterer, F.; Coors, S.; Au, Q.; Casalicchio, G.; Kotthoff, L.; Bischl, B. mlr3: A modern object-oriented machine learning framework in R. *J. Open Source Softw.* **2019**, *4*, 1903. [CrossRef]
75. Funk, C.; Peterson, P.; Landsfeld, M.; Pedreros, D.; Verdin, J.; Shukla, S.; Husak, G.; Rowland, J.; Harrison, L.; Hoell, A.; et al. The climate hazards infrared precipitation with stations—A new environmental record for monitoring extremes. *Sci. Data* **2015**, *2*, 150066. [CrossRef]
76. Ainembabazi, J.H. The 2015-16 El Niño-induced drought crisis in Southern Africa: What do we learn from historical data? In Proceedings of the International Association of Agricultural Economists Conference, Vancouver, BC, Canada, 28 July–2 August 2016.
77. Archer, E.R.M.; Landman, W.A.; Tadross, M.A.; Malherbe, J.; Weepener, H.; Maluleke, P.; Marumbwa, F.M. Understanding the evolution of the 2014–2016 summer rainfall seasons in southern Africa: Key lessons. *Clim. Risk Manag.* **2017**, *16*, 22–28. [CrossRef]
78. Urban, M.; Berger, C.; Mudau, T.E.; Heckel, K.; Truckenbrodt, J.; Odipo, V.O.; Smit, I.P.J.; Schmullius, C. Surface moisture and vegetation cover analysis for drought monitoring in the southern Kruger National Park using Sentinel-1, Sentinel-2, and Landsat-8. *Remote Sens.* **2018**, *10*, 1482. [CrossRef]
79. Louppe, G.; Wehenkel, L.; Sutera, A.; Geurts, P. Understanding variable importances in Forests of randomized trees. In Proceedings of the Advances in Neural Information Processing Systems Workshop, Lake Tahoe, NV, USA, 5–10 December 2013.
80. NASA's Fire Information for Resource Management System (FIRMS), Part of NASA's Earth Observing System Data and Information System (EOSDIS). Available online: https://firms.modaps.eosdis.nasa.gov/ (accessed on 23 August 2021).
81. Blair, D.; Shackleton, C.M.; Mograbi, P.J. Cropland abandonment in South African smallholder communal lands: Land cover change (1950-2010) and farmer perceptions of contributing factors. *Land* **2018**, *7*, 121. [CrossRef]
82. Brandt, M.; Hiernaux, P.; Rasmussen, K.; Mbow, C.; Kergoat, L.; Tagesson, T.; Ibrahim, Y.Z.; Wélé, A.; Tucker, C.J.; Fensholt, R. Assessing woody vegetation trends in Sahelian drylands using MODIS based seasonal metrics. *Remote Sens. Environ.* **2016**, *183*, 215–225. [CrossRef]
83. Wilgen, V.B.; Maitre, L.D. Working for Water Trivial and political reasons for the failure of classical biological control of weeds: A personal view. *S. Afr. J. Sci.* **2004**, *100*, 189–230.
84. Oelofse, M.; Birch-Thomsen, T.; Magid, J.; de Neergaard, A.; van Deventer, R.; Bruun, S.; Hill, T. The impact of black wattle encroachment of indigenous grasslands on soil carbon, Eastern Cape, South Africa. *Biol. Invasions* **2016**, *18*, 445–456. [CrossRef]
85. Oldeland, J.; Dorigo, W.; Wesuls, D.; Jürgens, N. Mapping bush encroaching species by seasonal differences in hyperspectral imagery. *Remote Sens.* **2010**, *2*, 1416–1438. [CrossRef]
86. Montandon, L.M.; Small, E.E. The impact of soil reflectance on the quantification of the green vegetation fraction from NDVI. *Remote Sens. Environ.* **2008**, *112*, 1835–1845. [CrossRef]
87. Caldeira, M.C.; Lecomte, X.; David, T.S.; Pinto, J.G.; Bugalho, M.N.; Werner, C. Synergy of extreme drought and shrub invasion reduce ecosystem functioning and resilience in water-limited climates. *Sci. Rep.* **2015**, *5*, 15110. [CrossRef] [PubMed]
88. Roques, K.G.; O'Connor, T.G.; Watkinson, A.R. Dynamics of shrub encroachment in an African savanna: Relative influences of fire, herbivory, rainfall and density dependence. *J. Appl. Ecol.* **2001**, *38*, 268–280. [CrossRef]
89. Grandini, M.; Bagli, E.; Visani, G. Metrics for Multi-Class Classification: An Overview. *arXiv* **2020**, arXiv:2008.05756.

Article

Spatial Heterogeneity of Vegetation Response to Mining Activities in Resource Regions of Northwestern China

Hanting Li [1], Miaomiao Xie [1,2,*], Huihui Wang [1], Shaoling Li [1] and Meng Xu [1]

[1] School of Land Science and Technology, China University of Geosciences (Beijing), Xueyuan Road 29, Beijing 100083, China; lht_2014@cugb.edu.cn (H.L.); wanghh@cugb.edu.cn (H.W.); lishaoling@cugb.edu.cn (S.L.); xumeng@cugb.edu.cn (M.X.)

[2] Key Laboratory of Land Consolidation, Ministry of Natural Resources of the PRC, Guanying Yuan West 37, Beijing 100035, China

*   Correspondence: xiemiaomiao@cugb.edu.cn

Received: 4 September 2020; Accepted: 2 October 2020; Published: 6 October 2020

Abstract: Aggregated mining development has direct and indirect impacts on vegetation changes. This impact shows spatial differences due to the complex influence of multiple mines, which is a common issue in resource regions. To estimate the spatial heterogeneity of vegetation response to mining activities, we coupled vegetation changes and mining development through a geographically weighted regression (GWR) model for three cumulative periods between 1999 and 2018 in integrated resource regions of northwestern China. Vegetation changes were monitored by Sen's slope and the Mann–Kendall test according to a total of 72 Landsat images. Spatial distribution of mining development was quantified, due to four land-use maps in 2000, 2005, 2010, and 2017. The results showed that 80% of vegetation in the study area experienced different degrees of degradation, more serious in the overlapping areas of multiple mines and mining areas. The scope of influence for single mines on vegetation shrunk by about 48%, and the mean coefficients increased by 20%, closer to mining areas. The scope of influence for multiple mines on vegetation gradually expanded to 86% from the outer edge to the inner overlapping areas of mining areas, where the mean coefficients increased by 92%. The correlation between elevation and vegetation changes varied according to the average elevation of the total mining areas. Ultimately, the available ecological remediation should be systematically considered for local conditions and mining consequences.

Keywords: spatial heterogeneity; vegetation trends; mining development; geographically weighted regression (GWR); Sen's slope; Mann-Kendall; arid and semi-arid areas

1. Introduction

Vegetation dominating terrestrial ecosystems connects the material circulation and energy flow of the biosphere [1] and plays a critical role in supporting ecosystem services and functions [2,3]. Vegetation changes, thus, have increasingly become an inevitable indicator in global climate changes and regional eco-environmental assessment [4,5]. Changes in natural conditions and strong human activities involve ecological elements and ecological processes, and alter the regional environment [6]. As intensive human activities, mining activities have an impact on 11 out of the 17 United Nations Sustainable Development Goals (SDGs) [7], and are a constraint for achieving sustainable development [8]. Mining activities, especially extractive ones, directly destroy vegetation and indirectly lead to environmental problems, including air and water pollution [9], heavy-metal pollution [10], groundwater loss [11], soil erosion and degradation [12]. These problems profoundly change the environment of vegetation growth, and, in turn, disproportionately damage broader range of vegetation coverages and show spatial differences on vegetation changes. The vegetation changes representing local ecosystem health are severely disturbed by mining activities [13]. Research on the effect of mining activities on vegetation is essential for further ecological construction and achieving the SDGs.

Analyzing the mechanism of mining activities on vegetation growth in mining areas provides significant insight for constructing ecological coal mining [14]. Researchers have made many findings through field surveys and experiments focusing on soil parameters [15], microorganisms [16], root environments [17], toxicological effects [18,19], colony symbiosis and photosynthesis [20], heavy-metal pollution and enrichment in vegetation [21,22], the extinction of major dominant species [23], and biodiversity loss [24]. Related studies have revealed that mining approaches impacting vegetation growth are diversiform on a local scale and more complicated on a regional scale [25]. However, mining development in resource regions is not a single sporadic mine pit, but a complex and systematic industrial chain [26]. This chain involves a wholly integrated process and establishes diversified industrial branches from mining excavation, transportation, preprocessing, and deep processing, to material consumption and utilization [26,27]. The successive impacts are constantly accumulated by the aggregation of one or more activities on receptors [28,29]. The difference in the spatial accumulation degree over time and space causes different responses of various vegetation types to mining development on a regional scale, resulting in significant spatial heterogeneity. Understanding how mining impacts accumulate, and change over time is the key issue for assessing and monitoring vegetation response to mining activities.

The regional ecological impact of mining development could be revealed through large-scale observation [30]. Recent achievements include that coal mining is an important driving factor resulting in serious regional vegetation degradation, especially in China's Mongolia Plateau and alpine areas [29–31]. Vegetation disturbance caused by mining is evident on a large scale [30], and much more significant in arid and semiarid areas [25,31]. The combined effect of climate conditions and ecological restoration activities also make vegetation changes more volatile and show vast spatial differences [30,32]. In relation to the regional scale, the relationship between mining development and vegetation changes during the aggregation progress of mining development and the typical region where mining activities influence more significantly, are still not well-understood. Establishing a mathematical coupling model between vegetation trends and human activities is essential in a complex system under the coupling of natural conditions and human activities [33].

Spatial analysis provides an advantage in understanding the variation in the impact of mining on vegetation [34]. Traditional multivariate statistical analysis and simple spatial analysis, such as ordinary least squares (OLS) models, usually assume that spatial relationships between variables are stable in the entire study area and reflect any variation of spatial characteristics with difficulty [35]. Geographically weighted regression (GWR) constructs local regression equations from any given geographic location to represent accurate quantitative characteristics of spatial relationships, thereby avoiding the problems of spatial non-stationarity, heterogeneity, and autocorrelation [36]. Computed correlation coefficients in the GWR model quantitatively express the spatial relationships at each location. Geographically weighted regression models are widely used in urban landscape pattern analysis [35,37,38], $PM_{2.5}$ concentration estimation [39], carbon emissions [40], and ecosystem services [41,42]. Sawut et al. [43] also estimated the heavy metal arsenic (As) contents of an open-pit coal mine in soil on the basis of GWR.

Arid environments occupy more than 47% of Earth's landmass with constant expansion throughout the world [44]. Exploitation of mineral resources has had extensive environmental and social consequences [45]. China is the leading country in energy production and consumption [46]. More than 70% of coal reserves are distributed in arid, semiarid, and fragile ecological regions, with high-strength exploitation activities [23]. Analyzing the relationship of vegetation and mining development provides practical guidance and reference for the development of mineral resources and ecological construction in the Belt and Road Initiatives.

As a representative resource-based city of China, Wuhai is not only a city that has maintained coal exploitation for decades, but also an important ecological function zone in Inner Mongolia. In the context of simultaneous ecological destruction and construction, setting Wuhai and its surroundings as the research area was of great significance to regional ecological security and harmonious development. The purpose of this article is to determine the spatial variability of mining impact on vegetation changes. There were two detailed objectives: (1) To identify the mining development pattern and associated vegetation dynamics in different periods, and (2) to explore the spatial variability of vegetation response to mining development.

2. Study Area and Data Sources

2.1. Study Area

The study area (106.36°E–107.05°E, 39.15°N–39.52°N) mainly comprised the whole city of Wuhai, and parts of Alxa League and Ordos according to the planning (2010–2030) of Wuhai and its surrounding areas. The whole study area is located in the middle of Inner Mongolia with six districts (Figure 1a) and surrounded by three deserts—the Uulan Buh, the Kubuqi, and the Maowusu [47,48]. The north–south-oriented Yellow river runs through the whole city and forms irrigation districts of about 175 km^2 with a narrow river beach wetland and an agricultural oasis [47]. Topographically, the study area is low-lying in the northwest, and high-lying in the middle and east (Figure 1b). The study area belongs to the middle-latitude temperate continental climate zone, a region affected by the East Asian monsoon belt [49]. Annual precipitation is 160 mm, and annual evaporation is 20 times that of rainfall [47]. The main vegetation types in the study area are grassland and shrubland. The combination of the Yellow river and the complex natural environment gives the entire region a unique desertification ecosystem, including national wetland parks and an extremely precious plant, *Tetraena mongolica* [48].

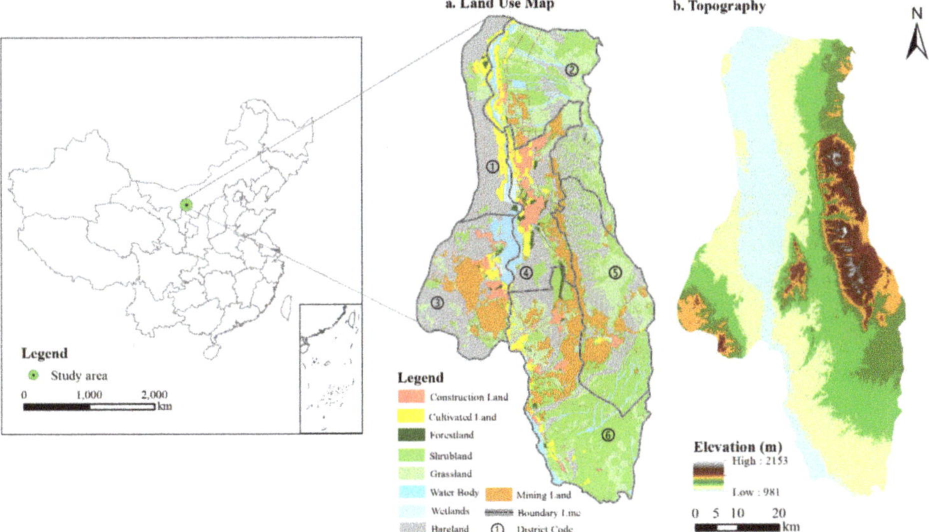

Figure 1. (**a–b**) Location, administrative divisions, land-use and land-cover map (a), and topography (b) of study area. Land-use and land-cover map was monitored at 2017, provided by the Institute of Geographic Sciences and Natural Resources Research in China. Note: districts of ①, Uulan Buh; ②, Mengxi; ③, Wuda; ④, Haibowan; ⑤, Etuoke Banner; and ⑥, Hainan.

Wuhai is a typical resource-based city where mining development comprises many mining activities, including industrial base, surface mining, and waste dumping, and they play a dominant role in social–economic development [49]. Industrial base mainly consists of coal washing, coal storage, and primary and deep processing sites, while surface mining and waste dumping are the main activity sites for mineral mining and disposal [48]. There are three industrial bases distributed across the study area: the Wuda industrial base in the northwest, the Hainan industrial base in the midland, and the Mengxi industrial base in the Mengxi district. Mining areas are attached to the Shendong coalfield in Inner Mongolia and adjacent to the Ningdong Energy and Industrial Base, one of China's largest coal bases [23]. As the development progresses, decline transformation, and deepening of social economic reforms, mining, coal, and chemical industries were introduced in this region over the course of 30 years by enterprises with severe pollution and an extensive development model from the developed eastern part of China [49]. Under the pressure of the inherent irreconcilable conflict between social-economic development and ecological protection, it is more and more urgent to recognize the internal relationship between ecological degradation and mining development

2.2. Data Sources

2.2.1. Landsat Data and Mining Maps

Normalized difference vegetation index (NDVI) values of all clear-sky Landsat images during the growing seasons from April to October of 1999–2018 were obtained to composite interannual maximal sequence to detect vegetation variation trends. Growing seasons included the vegetative and reproductive phases of vegetation growth [50]; the maximal value of NDVI in the arid and semiarid areas represented the best state of vegetation in a year. All Landsat data were obtained from the United States Geological Survey (Table 1). Land-use maps were used to present the spatial distribution of mining activities and calculate the distance from vegetation areas to mining areas. The maps were extracted from land-use maps monitoring at 2000, 2005, 2010, 2017, respectively, from the Institute of Geographic Sciences and Natural Resources Research. All maps were accurately interpreted on the corresponding historical Google Earth images. Topography data at 30 m spatial resolution from the digital elevation model in ASTER GDEM 2 (http://www.gscloud.cn/) were used to reveal the relationships between vegetation dynamics and terrain features. All data were converted into a common coordinate system (WGS1984, UTM Zone 49N), and raster data were resampled into 1000 × 1000 m.

Table 1. Sources of remote sensing data.

Theme	Data Type/Images Number	Resolution	Time	Source
Landsat 4-5 TM C1 Level-1	Satellite Imagery/ 41 Imageries	30 m	1999–2011	U.S. Geological Survey (USGS) (http://www.glovis.usgs.gov/)
Landsat 7 ETM+ C1 Level-1	Satellite Imagery/ 11 Imageries	30 m	1999–2003	U.S. Geological Survey (USGS) (http://www.glovis.usgs.gov/)
Landsat 8 OIL/TIRS C1 Level-1	Satellite Imagery/ 20 Imageries	30 m	2013–2018	U.S. Geological Survey (USGS) (http://www.glovis.usgs.gov/)
Historical Google Earth Image	Satellite imagery	17 m/4 m/2 m	2000, 2005, 2010, and 2017	Google Earth Pro (http://www.google.com/intl/en_uk/earth/versions/#earth-pro)

2.2.2. Boundary Data in Vector Format and Climate Dataset

The boundary of the research area was set according to the coal industry planning (2010–2030) of Wuhai and its surrounding areas, which was made by the government of the Inner Mongolia Autonomous Region. Basic geographic information was provided by the National Geomatics Center of China (http://218.244.250.94:9003/English/html/1/), including a set of regional boundaries, major roads, and river basins. The boundary of the conservation zone in the study area was drawn on the basis of the Western Ordos national nature reserve [51]. The observed annual precipitation and average temperature datasets were downloaded by the National Meteorological Information Center of China (http://data.cma.cn/en), to describe the impact of climate conditions on vegetation changes. This dataset, comprising monthly observations, was obtained from 5 meteorological reference stations around the research area in 1999–2018.

3. Methodology

The purpose of the article was to analyze the relationship between vegetation changes and mining development on the basis of remote sensing data and the GWR model. Vegetation changes were described by interannual NDVI trends (1999–2018), and the spatial distribution of mining activities were obtained via four land-use maps (2000, 2005, 2010, 2017, respectively). Considering the intensity of the potential influence of mining activities relying on distance [21,52], all data were divided into 1 km units to calculate the distance from vegetation units to mining units on the basis of Euclidean distance in ArcGIS 10.2. Minimal distance emphasizing the ecological impact of a single mine and summary distance emphasizing regional mining impact on vegetation were differently analyzed. Minimal distance was the shortest one of distances of central point between a vegetation unit and mining units. The summary distance was the sum of distances of central point between a vegetation unit and mining units. Topography was a limiting factor affecting vegetation changes in geographical conditions, such as water and radiation balance. Elevation was regarded as an important factor in the analysis of vegetation response to mining activities.

The methodology framework was divided into three steps (Figure 2). The first step was to identify vegetation dynamics. Vegetation changes were divided into three stages, 1999–2005, 1999–2010, and 1999–2018, to correspond to the cumulative effect of mining development in three periods, where the starting year was set to 1999 to ensure the initial stability of the NDVI sequence. The second was to present the spatial distribution of mining development, and calculate the minimal distance and summary distance from vegetation areas to mining areas in units in different stages. The third was to quantify the spatial relationships between two kinds of distances, elevation, the combination of distance and elevation, and vegetation changes in the GWR model. Removing the improvement areas of vegetation in the 1999–2018 period was to highlight the cumulative effects of mining development. Detailed descriptions are provided in the following sections.

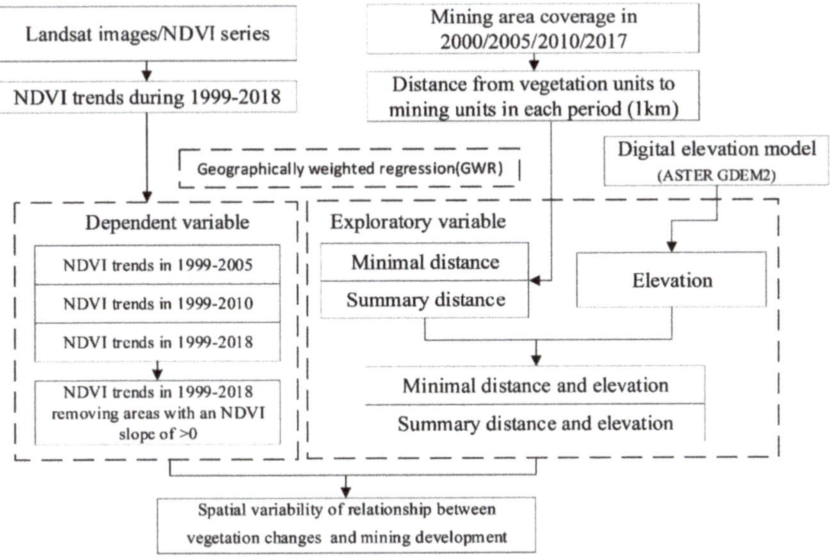

Figure 2. Framework of data processing flow.

3.1. Trend Analysis of Vegetation Changes

NDVI is an indispensable indicator for mapping green biomass to describe vegetation dynamics because they are closely related with biophysical and biochemical variables [53]. The calculation method was detailed in a study by Maneja et al. [54,55]. All NDVI series were synthesized according to the maximal value of the growing season in a year to eliminate interference caused by vegetation changes, clouds, and the atmosphere. Sen's slope is calculated by the median of the linear rate of change between any two points in the sequence, which accurately expresses the trend and relatively reduces noise interference [56,57]. The Mann–Kendall trend test is a quick and effective method for detecting significance level with the advantage of not requiring time distributions and being insensitive to outliers [58]. Sen's slope estimator was used to first detect the direction and magnitude of vegetation changes, and the Mann–Kendall trend test was then applied to quantify the significance level. Therefore, vegetation trends could be estimated by the combination of Sen's slope and the Mann–Kendall trend test in light of the NDVI series.

Sen's slope equation is shown in Equation (1) [57]:

$$\theta slope = Median\left[\left(NDVI_j - NDVI_i\right)/\left(j-i\right)\right], \forall j > i \tag{1}$$

The Mann–Kendall test is shown by test statistic S in Equation (2) [59,60]:

$$S = \sum_{i=1}^{n=1} \sum_{j=i+1}^{n} sign(NDVI_j - NDVI_i) \tag{2}$$

where signal $sign\left(NDVI_j - NDVI_i\right)$ is;

$$sign\left(NDVI_j - NDVI_i\right) = \begin{cases} 1 & \left(NDVI_j - NDVI_i > 0\right) \\ 0 & \left(NDVI_j - NDVI_i = 0\right) \\ -1 & \left(NDVI_j - NDVI_i < 0\right) \end{cases} \tag{3}$$

The test statistic Z is defined as;

$$Z = \begin{cases} (S-1)/\sqrt{V(S)} & S > 0 \\ 0 & S = 0 \\ (S+1)/\sqrt{V(S)} & S < 0 \end{cases} \tag{4}$$

where variance $V(S)$ is;

$$V(S) = n(n-1)(2n+5)/18 \tag{5}$$

where $\theta slope$ is the annual variation rate of the NDVI trend on a pixel scale, and $NDVI_i$ and $NDVI_j$ represent the maximal NDVI values of monitoring years j and i, respectively; $V(S)$ is the variance. A positive value of $\theta slope$ indicates an upward trend for vegetation, and a negative value means a downward trend. Moreover, the appropriate statistical test in the process of inferring significance is determined through the n values of the time-series lengths; when n < 10, the bilateral trend test was used to directly show a slight upward or downward trend by test statistic S. When n >= 10, test statistic S obeyed standardized normal distribution. Given confidence level $\alpha = 0.05$, whether the trend changed significantly depended on $|Z| \geq 1.96$. Four kinds of classification were obtained through trend and significance analysis: $\theta slope \geq 0\&|Z| \geq 1.96$ denoted significant improvement, and $\theta slope \geq 0\&|Z| \leq 1.96$ indicated slight improvement, whereas $\theta slope \leq 0\&|Z| \geq 1.96$ denoted significant degradation, and $\theta slope \leq 0\&|Z| \leq 1.96$ meant slight degradation.

3.2. Relationship between Vegetation Changes and Mining Development in GWR Model

The GWR model was explored to examine the relationship between mining development and vegetation changes, and their spatial variability. OLS is a global regression model, and parameter estimates are consistent throughout the study area. The GWR model makes important improvements in solving non-stationary spatial relationships and cross-space spatial autocorrelation by estimating local parameter characteristics and geographic map variability in the association between results and predictors [34,61]. This regional exploratory analysis technique can measure a set of local parameters that could be mapped, estimated, and analyzed in each unit to provide new insights about window movement and the global correlation of variables in a single modeling frame [62]. The GWR model is expressed in Equation (6) [34],

$$y = \beta_0\left(\mu_j, v_j\right) + \sum_{i=1}^{k} \beta_i(\mu_j, v_j)\chi_{ij} + \varepsilon_j \tag{6}$$

where μ_j and v_j denotes the spatial coordinates of sample points j and i; $\beta_0\left(\mu_j, v_j\right)$ indicates the intercept of location j; $\beta_i\left(\mu_j, v_j\right)$ denotes the local estimated coefficient of independent variable χ_{ij}; and ε_j is the error term.

Local parameter estimation was conducted through a spatial weight matrix by a distance decay weighted function in GWR modeling. The function was spatially modified by kernel function bandwidth. Kernel function bandwidth determines the scope of spatial dependence, which means the total numbers of neighborhood points. The Akaike information criterion (AIC) determined the optimal bandwidth. More details about GWR were shown in Alahmadi et al. [62]. GWR analysis was performed in the GWR tools of ArcGIS 10.2. All data were normalized by a standardized min–max method before regressions.

The multicollinearity of the explanatory variables was excluded by the variance inflation factor (VIF) value of the running OLS model [63]; all values were less than 7.5, which indicated that slight or no collinearity existed in the explanatory variables. The performance of the GWR and OLS models was compared using the values of AIC and R^2; these two values were used to determine the predictive capacity of the model. The higher the R^2 was, the more reliable the independent variable's explanation of the dependent variable. AIC estimated the accuracy of the estimated value, and lower values could better describe the observed data.

4. Results

Vegetation has been considerably degraded as the mining development rapidly expanded according to Sen's slope and the Mann–Kendall test. Spatial correlation of GWR expressed significant spatial differences between minimal distance, summary distance, elevation, and vegetation changes.

4.1. Temporal Trends and Spatial Distribution in NDVI

The tendency of vegetation changes to first rise and then quickly decline appeared in the whole study area. The proportion of the significant degradation area increased (Figure 3) according to vegetation trend analysis of Sen's slope. Clear improvements of vegetation changes accounted for the majority of the study area (74%) in the initial stages, especially a significant improvement gathering in the south of the study area (17%) with the NDVI value increasing nearly by 100% (Figure 3b). Initially degraded areas were mainly distributed in the mining areas and eastern mountainous areas, and 85% of them turned degradation into improvement during 1999–2010. Positive growth conditions drove an upward vegetation trend in 1999–2010. Nevertheless, in 1999–2018, the overall trend of vegetation had deteriorated, and degraded areas accounted for more than 80% of the total study area (Figure 3c). Severely degraded vegetation areas (27%) were distributed in the north of the study area and mining areas. Most significant improvements in the south of the study area were lost, and continuous degradation occurred in the western areas. The few improvements that were gathered in the central town may result from ecological construction.

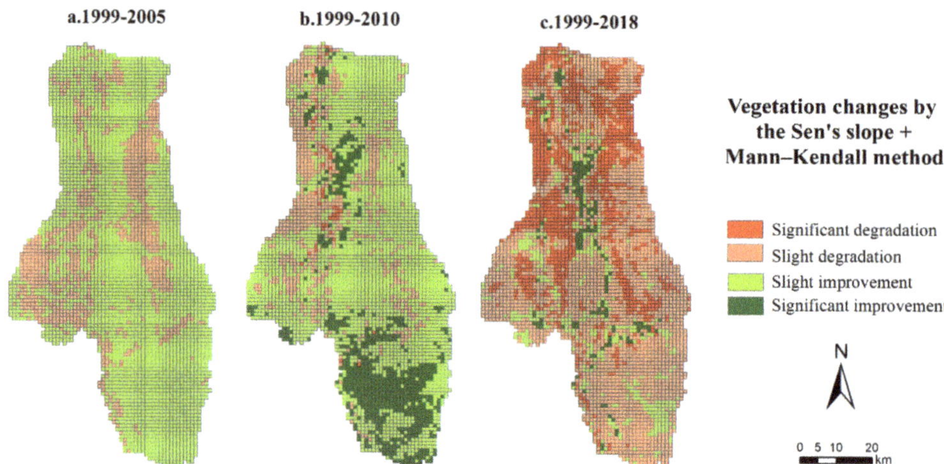

Figure 3. (**a–c**) Spatial distribution of vegetation changes by Sen's slope and Mann–Kendall method. (**a**) 1999–2005; (**b**) 1999–2010; (**c**)1999–2018.

4.2. Spatiotemporal Distribution of Mining Development

Mining development rapidly expanded over the past 20 years, and established a connected spatial pattern in three major industrial bases. As shown in Figure 4, the total area of the industrial base was 9.57 km^2 in 2000 and 184.98 km^2 in 2017. Open pits gradually expanded with a uniform growth rate of 9.17 km^2/a around the core industrial base, the distribution of which was in a narrow pattern along the terrain of the valley in the middle, and an aggregate pattern in the northwest of the study area. The waste dump was staggered with open pits and expanded from 2.85 km^2 in 2000 to 69.36 km^2 in 2017. Areas of mining activities expanded from 55.83 km^2 in 2000 to 453.78 km^2 in 2017, accounting for more than one-tenth of the research area and 9 times the production scale in 2000. The average expansion rate was 16%, 32%, and 7% per year in 2000–2005, 2005–2010 and 2010–2017, respectively, with the highest expansion in the period of 2005–2010, as the market price of coal was at historic highs.

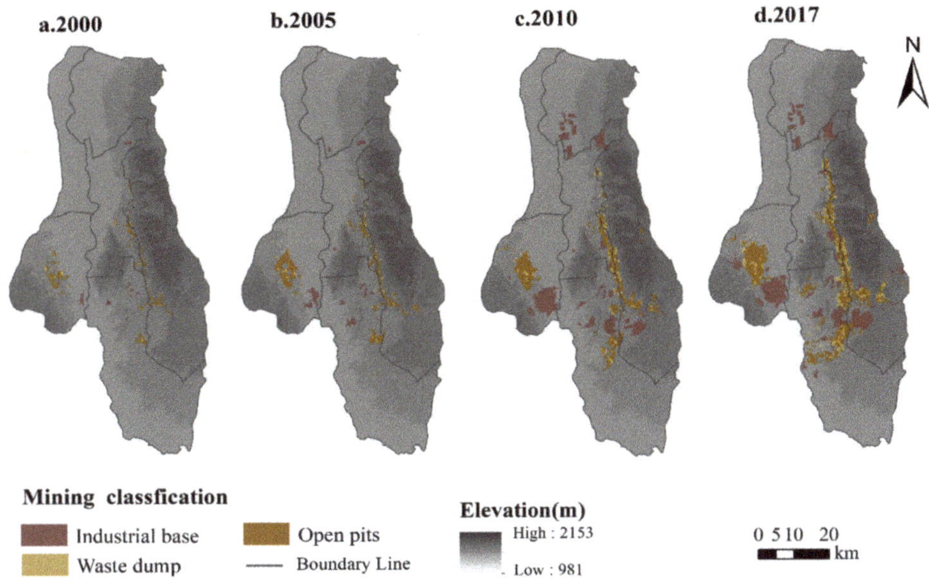

Figure 4. (a–d) Components and spatial distribution of mining activities in study area.

4.3. Mining Development and Elevation of Influencing Vegetation Changes

4.3.1. Correlation between Minimal Distance, Summary Distance, and Vegetation Changes

Spatial variation, mapping the relationship between distances and vegetation changes, was clearly shown in Figure 5. The positive coefficient indicated that vegetation changes moved towards an upward trend as the increase in distance to mining areas in the minimal and summary distance models. The depth of colors expressed the level of the correlation coefficient and fitness to match the variables.

For minimal distance, as distance increased, the impact of single mines on vegetation was shrunken, but dominated around areas of mining activities. Areas with positive coefficient (above 0.01) was 1418, 969, 866, and 733 km^2, respectively, with a continuous decline trend of 48.31%. The mean coefficient (above 0.02) continued growing by 20% (0.025, 0.026, 0.03, 0.033, respectively). Positive coefficients were significantly higher around mining areas than those of other areas, and a clear shift from negative to positive correction constantly occurred in the mining areas (Figure 5a–c). The spatial pattern of areas with higher positive correlation (above 0.01) was consistent with spatial pattern of mining development, especially in the middle of study area after removing the areas with an NDVI slope of >0.

For summary distance, in the agglomeration process of mining development, the impact of multiple mines gradually shifted from the outside to the inside and continued to be increasingly concentrated in overlapping areas of mining activities. Areas with positive correlation (above 0.01) were 1129, 704, 587, and 588 km^2 respectively, showing a downward trend. The higher ones (above 0.02) maintained a downward trend in the east but an increasing trend in the west, with an area of 186, 0, 393, and 373 km^2. The average coefficient continuously increased by 92% from 0.026 to 0.050. The increase in both the area and coefficient of positive correlation had emerged and gathered in the west overlapping areas of three industrial bases. This was more evident in the increase in areas with positive correlation after removing areas with an NDVI slope of >0, with mean coefficients of up to 0.056. The comprehensive impact of coal bases is more influential for regional vegetation changes.

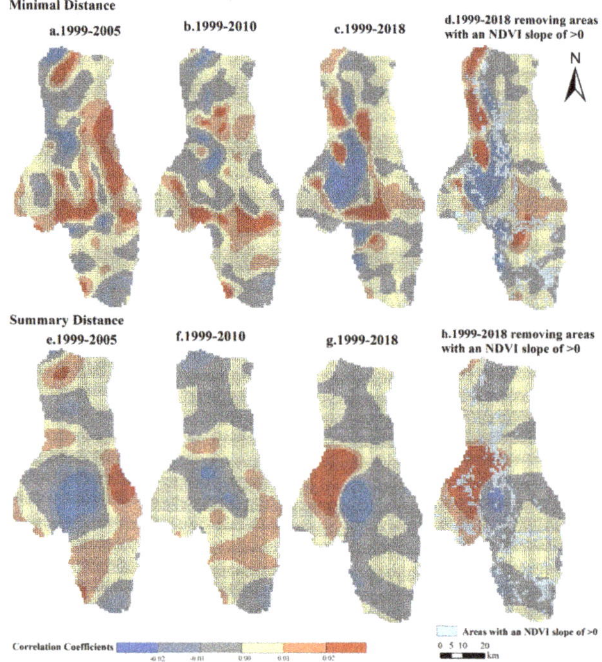

Figure 5. (**a–h**) Spatial patterns of correlation coefficients between distance and vegetation changes. (**a–d**) Minimal and (**e–h**) summary distance in 1999–2018.

4.3.2. Correlation between Elevation and Vegetation Changes

Significant spatial differences were shown in the relationship between vegetation changes and elevation (Figure 6). The positive coefficient indicated that vegetation degradation improved with the increase in elevations, and the negative coefficient meant that vegetation degradation was worsening with the elevation's increasing.

Spatial relationship between elevation and vegetation changes varied according to the average elevation of the total mining areas. Negative correlation in higher-elevation areas and positive correlation in low-altitude areas were expressed in the relationship between elevation and vegetation changes. The two were approximately bounded by the average elevation of the total mining areas. Most areas with negative correlation were distributed in the middle and southwest of the study area, with the proportion gradually decreasing from 56%, 62%, 47%, to 38%. Low-elevation areas in the south gradually changed from negative into positive correlation as mining activities continuously expanded, indicating that the degradation of low-elevation vegetation was more serious with the decrease in elevation. Furthermore, this may partly explain the disappearance of extremely significant improvement areas in southern areas as the Hainan industrial base agglomerating. Three positively correlated aggregation areas of about 362 km^2 were located in the middle, north, and south around the Yellow river and constructed areas. The boundary between positive and negative correlations (–0.03 to 0 and 0 to 0.03) was consistent with the 1200–1230 m contour line, shown in Figure 6c,d.

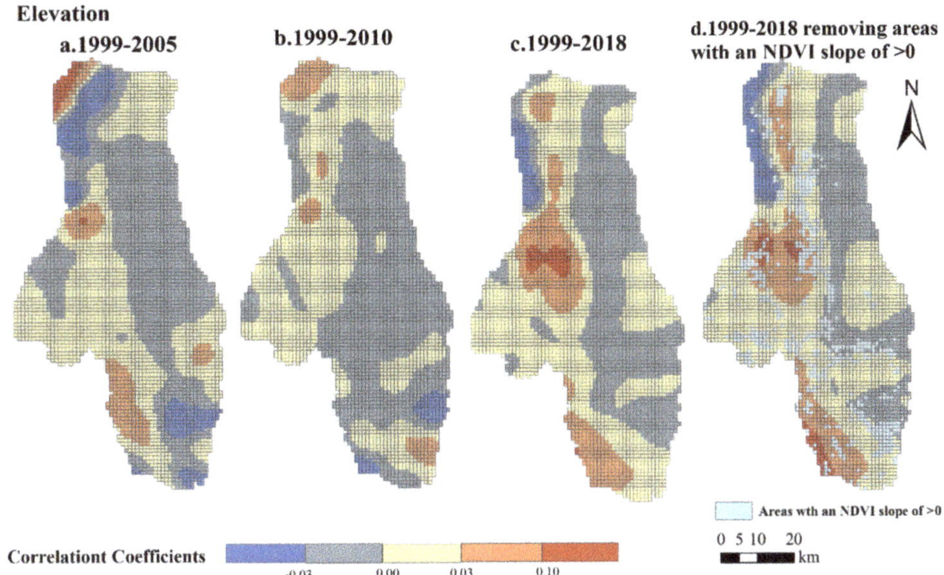

Figure 6. (**a–d**) Spatial patterns of correlation coefficients between elevation and vegetation changes in 1999–2018.

4.3.3. Correlation between Minimal Distance, Summary Distance, Elevation, and Vegetation Changes

Proper spatial stationarity was enormously maintained in maps of distance factors and vegetation changes after the combination of elevation and distances (Figure 7).

For minimal distance, quantity disappearance in positively correlated areas and expansion of negatively correlated areas, especially in the western areas, were clearly shown in the combination of minimal distance and elevation (Figure 7a–d). Such disappearance illustrated that the ecological impact of a single mine on vegetation and its action pathway were not closely related to elevation. As minimal distance to mining areas increased, elevation and distance had disproportionately opposite effects on vegetation changes at different elevation levels, as per Sections 4.3.1 and 4.3.2.

For summary distance, the spatial distribution of positively correlated areas was relatively stable and showed no significant difference in whether elevation was involved in the relationship of summary distance and vegetation changes. Areas of positive coefficients continued growing with a proportion of up to 86.36%, and the quantities of higher positive coefficients (above 0.02) in 1999–2018 had increased by 17% compared with Figure 5e–h. Consequently, the comprehensive impact of summary distance on vegetation changes was relatively stable and not determined by the elevation, but by the spatial pattern of mining development. Moreover, after removing areas with an NDVI slope of >0, significant improvements of the positive coefficients occupied the majority, and less than 14% of the negative areas were distributed in the middle of the study area. The comprehensive ecological influence of summary distance was constantly strengthened with the systemization of mining activities.

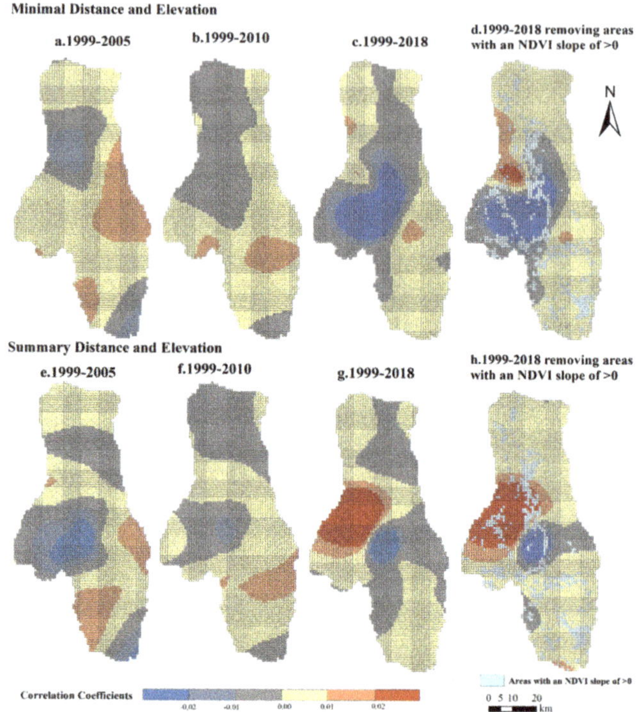

Figure 7. (**a–h**) Spatial patterns of correlation coefficients between distance and vegetation changes. (**a–d**) Minimal distance and elevation, and (**e–h**) summary distance and elevation in 1999–2018.

5. Discussion

5.1. Importance of Applying GWR in Studying Spatial Heterogeneity of Vegetation

Significant spatial heterogeneity of the relationship between mining development and vegetation changes was revealed by comparing the GWR and OLS models (Tables 2 and 3). The adjusted R^2 of the GWR model was in the range of 0.19 to 0.62, which was higher and better than that of the OLS model (all less than 0.1). This showed that the GWR model could greatly explain the impact of mining on vegetation changes. Therefore, it was concluded that the spatial relationship between mining development and vegetation changes was almost not linear, but showed great spatial heterogeneity. All R^2 of GWR also gradually increased with the expansion and aggregation of mining development, indicating that mining development had increasingly significant impact on vegetation changes.

However, the question was why vegetation changes in the resource regions showed significant spatial heterogeneity. Numerous studies suggested that vegetation greening rate was elevation-dependent by the different sensitivity levels to precipitation and temperature changes in arid and semiarid regions [64–66]. In mining areas, Liu et al. [67] found significant positive correlation in the relationship between NDVI and elevation factor, while Li et al. [68] discovered that as the elevation increased, the area covered by medium and high vegetation gradually decreased. We concluded that the vegetation changes were closely related with the average elevation of mining activities, and the higher degradation rate occurred away from the average elevation of mining activities due to the gravity convergence of the basin in low elevation, and fragile conditions of vegetation growth and high external sensitivity in high elevation.

Table 2. Comparison in adjusted R^2 of geographically weighted regression (GWR) and ordinary least squares (OLS) models.

	Adjusted R^2	1999–2005	1999–2010	1999–2018	1999–2018R
Elevation	Adjusted R^2_G	0.27	0.31	0.50	0.59
	Adjusted R^2_O	0.01	0.00	0.02	0.04
Minimal distance	Adjusted R^2_G	0.28	0.33	0.52	0.62
	Adjusted R^2_O	0.05	0.04	0.00	0.00
Summary distance	Adjusted R^2_G	0.20	0.28	0.27	0.41
	Adjusted R^2_O	0.01	0.01	0.00	0.01
Minimal distance and elevation	Adjusted R^2_G	0.20	0.27	0.29	0.41
	Adjusted R^2_O	0.05	0.03	0.02	0.05
Summary distance and elevation	Adjusted R^2_G	0.19	0.26	0.26	0.38
	Adjusted R^2_O	0.02	0.01	0.02	0.06

Adjusted R^2_G: adjusted R^2 of GWR; adjusted R^2_O: adjusted R^2 of OLS; 1999–2018R: 1999–2018 after removing areas with normalized difference vegetation index (NDVI) slope of > 0.

Table 3. Comparison of Akaike information criterion (AIC) from GWR and OLS models.

	AIC	1999–2005	1999–2010	1999–2018	1999–2018R
Elevation	AIC_G	−31717.3	−34396.9	−32869.7	−28356.0
	AIC_O	−30523.8	−32951.2	−30187.1	−25504.8
Minimal distance	AIC_G	−31739.0	−34488.6	−33030.7	−28700.3
	AIC_O	−30686.0	−33086.7	−30113.6	−25358.1
Summary distance	AIC_G	−31367.3	−34238.5	−31372.5	−27137.2
	AIC_O	−30517.9	−32968.3	−30118.3	−25381.6
Minimal distance and elevation	AIC_G	−31364.3	−34189.6	−31489.8	−27145.2
	AIC_O	−30692.2	−33084.9	−30186.6	−25519.6
Summary distance and elevation	AIC_G	−31311.1	−34138.6	−31329.3	−26979.4
	AIC_O	−30549.2	−32969.9	−30214.0	−25589.5

Extensive human activities, including ecological management, are other drivers to determine vegetation changes. Areas with negative correlation, where vegetation improved with the decline of summary distance to mining areas (Figure 7h), are mainly distributed in the natural protected areas with strong ecological management, though these are at the shortest distance in the summary distance model, and should be the region with the highest degradation according to the assumption. The effect of ecological protection could also be seen in the relationship between vegetation changes and elevation (Figure 6d), such as the eastern, southeastern, and northeastern areas of positive correlation. The significant improvement areas of vegetation were mainly distributed around the central town because of ecological-engineering activities. Ecological activities are important measures to maintain and improve vegetation growth, and become important factors to increase the spatial heterogeneity of vegetation changes. Furthermore, all-natural protection areas were disproportionately degraded, so existing protected areas must be strengthened, and the cumulative regionwide effects of mining activities must be mitigated [69–71].

5.2. Effect of Distance on Vegetation Disturbance in Mining Areas

The results suggested that the ecological impact of a single mine was continuously strengthened around the mining areas. A nationwide survey in China concluded that the distance of environmental impact from mining sites varied from a few hundred meters to 10 km [72]. Previous studies have shown that the influencing range of ecological disturbance in a single mine was mostly between 1000 and 3200 m [73,74]. In areas with poor vegetation growth conditions, the range of mining disturbance is much greater [74]. The average distance of vegetation disturbance in large mining sites is greater than that of small ones [75]. Empirical evidence from the Mongolian Plateau shows that this disturbance range of large coal exacting areas had increased to over 5 km [75], where the ecosystem could maintain stable, sustainable, and rich ecosystem services [76]. The aggregated development of mining activities enhances the range of ecological disturbance of individual mines.

For the comprehensive impact of mining activities, it is difficult to judge the distance threshold of mining disturbance. Our research results showed that its impact on vegetation changes was determined by the comprehensive spatial pattern of current and future mining development. Related scholars made some preliminary explorations. Cheng et al. [77] suggested that, in large coal bases, the source and distribution of heavy-metal pollution have significant spatial heterogeneity, there is a hot spot in the overlapping area of multiple coal mines, and the dispersion of pollutants is higher than that of single mines. Moreover, the distance of its heavy-metal pollution far exceeds the capacity of a single mine, reaching more than 15 km. In the aggregation process of mining development, the centralization effect of mining activities emerged, and was continuously enhanced and stabilized.

5.3. Response of Vegetation Changes to Climate Conditions

A warming–wetting trend occurred in the study area that provided excellent conditions for vegetation growth, and the increase in precipitation showed spatial differences during the last two decades (Figure 8). In the study area, the precipitation in the north was much lower than that in the south, in which the precipitation growth reached the peak of about 7 mm/a on average during 1999–2010. This strongly explained overall improvements of vegetation in the study area, especially in the south, where almost all significant improvements had emerged in 1999–2010. However, the degree of vegetation degradation was relatively high in 1999-2018, although there was a significant vegetation improvement in the previous period and a constant precipitation increase throughout the period. This properly meant that the impact that resulted from mining development was far greater than the impact of positive climate factors on vegetation during 2010–2018.

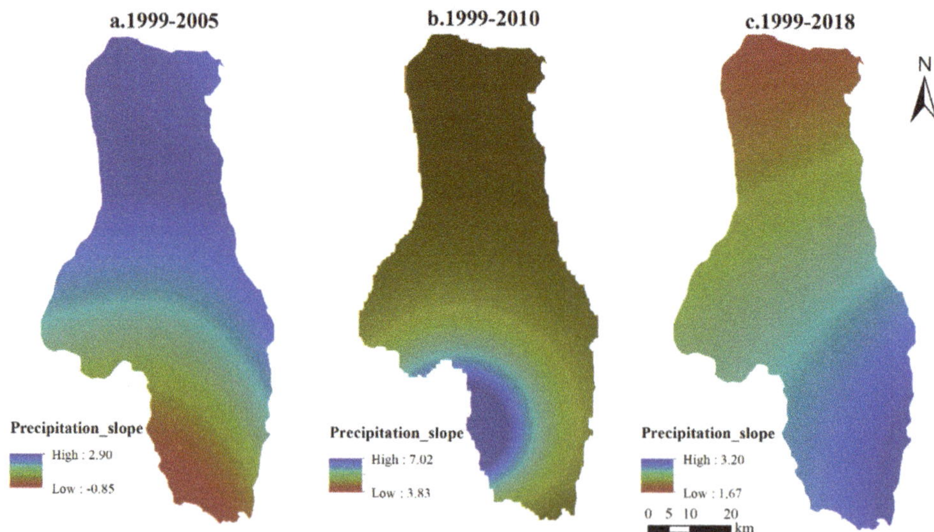

Figure 8. (**a–c**) Spatial distribution of precipitation slope by Kriging interpolation. Precipitation data in 1999–2018 were acquired from five meteorological reference stations: Hanjinqi, Linhe, Huinong, Etuokeqi, and Taole. Precipitation slope was obtained by linear regression based on interannual precipitation data, and then imported into ArcGIS to obtain changes in the study area by Kriging interpolation.

Under the context of global warming, the research area experienced notable climate change. Prior to 2010, temperature changes were relatively stable and even slightly decreased, and the average temperature then rose rapidly by about 1.5°C. Increased temperature promotes the germination of vegetation in spring and improves the growth of vegetation; on the other hand, it increases the transpiration and evaporation of plants in summer, which limits vegetation growth in arid and semiarid regions. Ma et al. [78] reported that evapotranspiration variation was consistent with changes in vegetation coverage, with a marginally increasing trend of about 0–5 mm/a during 2000–2010 and no significant increasing trend during 2011–2015 in the northwestern Loess Plateau. Numerous studies also showed that climate warming is one of the main driving factors of greening in northern China by enhancing photosynthesis and increasing vegetation activity [66,79–81].

In summary, climatic conditions were very favorable for the growth of vegetation during 1999–2018. It is almost impossible that climate changes led to vegetation degradation in such a water-scarce area with huge evapotranspiration, instead promoting vegetation growth.

5.4. Limitations

Although reliable and extensive data were used in this research, there were inevitable uncertainties or limitations. The interannual series of the maximal NDVI can characterize the dynamic changes of regional vegetation, but the shortage of data in some years and the image time deviation of the growing season had a specific effect on the accurate assessment of vegetation changes. More accurate time-series data and error analysis of calculation should be implemented in future studies. In addition, studies on the spatial differences of dominant areas by single mine and regional mining impacts, and their dominant areas conversion in resource regions should be strengthened in future research. Vegetation changes in resource regions are a comprehensive manifestation of indicators, such as vegetation types, climate conditions, groundwater depth, mining activities, and ecological management [30]. In relation to effect analysis of factors on vegetation, there may have been uncertainties brought about by other indicators. The contribution of other indicators to vegetation changes is notable and should be strengthened in future research.

6. Conclusions

Spatial heterogeneity is a great challenge in exploring the correlation between vegetation changes and mining development. Through spatial correlation based on the GWR model, three dominating factors were detected to quantify the correlation between vegetation changes and mining development across time and space. Our analysis indicated that incremental and combined mining activities could reverse the incremental trend of regional vegetation, leading to 86% degradation in the entire study area. Vegetation experienced a trend first of growth and then decline in the aggregation process of mining development. The scope of influence for single mines on vegetation had shrunk by about 48%, and the mean coefficients increased by 20%, closer to mining areas. The scope of influence for multiple mines on vegetation gradually expanded to 86% from the outer edge to the inner overlapping areas of mining areas, where the mean coefficients increased by 92%. Elevation dependence of vegetation changes varied according to the average elevation of total mining areas and played an important role in causing the spatial heterogeneity of mining impact on vegetation. Ecological measures should be implemented according to local conditions to achieve sustainable vegetation ecology.

Author Contributions: Conceptualization, H.L. and M.X. (Miaomiao Xie); methodology, H.L.; software, H.L., H.W. and S.L.; validation, H.L., H.W. and S.L.; formal analysis, H.L.; investigation, M.X. (Meng Xu); resources, H.W., S.L. and M.X. (Meng Xu); data curation, H.L.; writing—original draft, H.L.; writing—review editing, M.X. (Miaomiao Xie), H.W. and M.X. (Meng Xu) visualization, H.W. and S.L.; supervision, S.L. and M.X. (Meng Xu); project administration, M.X. (Miaomiao Xie); funding acquisition, M.X. (Miaomiao Xie); All authors have read and agreed to the published version of the manuscript.

Funding: This work is supported by the National Key R&D Program of the Ministry of Science and Technology of China (Grant number 2017YFC0504401).

Remote Sens. **2020**, *12*, 3247

Acknowledgments: The authors would thank the editors and four anonymous reviewers for their suggestions. We also appreciate professor Qing Chang and professor Yanxu Liu for improving the study.

Conflicts of Interest: The authors declare no conflict of interest.

References

1. Vereecken, H.; Kollet, S.; Simmer, C. Patterns in Soil-Vegetation-Atmosphere Systems: Monitoring, Modeling, and Data Assimilation. *Vadose Zone J.* **2010**, *9*, 821–827. [CrossRef]
2. Boyd, I.L.; Freer-Smith, P.H.; Gilligan, C.A.; Godfray, H.C.J. The Consequence of Tree Pests and Diseases for Ecosystem Services. *Science* **2013**, *342*, 823. [CrossRef] [PubMed]
3. Franklin, J.; Serra-Diaz, J.M.; Syphard, A.D.; Regan, H.M. Global change and terrestrial plant community dynamics. *Proc. Natl. Acad. Sci. USA* **2016**, *113*, 3725–3734. [CrossRef] [PubMed]
4. Fensholt, R.; Langanke, T.; Rasmussen, K.; Reenberg, A.; Prince, S.D.; Tucker, C.; Scholes, R.J.; Le, Q.B.; Bondeau, A.; Eastman, R.; et al. Greenness in semi-arid areas across the globe 1981–2007—An Earth Observing Satellite based analysis of trends and drivers. *Remote Sens. Environ.* **2012**, *121*, 144–158. [CrossRef]
5. Zhu, Z.C.; Piao, S.L.; Myneni, R.B.; Huang, M.T.; Zeng, Z.Z.; Canadell, J.G.; Ciais, P.; Sitch, S.; Friedlingstein, P.; Arneth, A.; et al. Greening of the Earth and its drivers. *Nat. Clim. Chang.* **2016**, *6*, 791–795. [CrossRef]
6. Li, F.; Wang, R.S.; Hu, D.; Ye, Y.P.; Yang, W.R.; Liu, H.X. Measurement methods and applications for beneficial and detrimental effects of ecological services. *Ecol. Indic.* **2014**, *47*, 102–111. [CrossRef]
7. United Nations(UN). Sustainable Development Goals: 17 Goals to Transform Our World. Available online: http://www.un.org/sustainabledevelopment/sustainable-development-goals/ (accessed on 3 September 2020).
8. United Nations(UN). Decade on Ecosystem Restoration 2021–2030. Available online: https://www.decadeon restoration.org/ (accessed on 3 September 2020).
9. Hendryx, M.; Zullig, K.J.; Luo, J.H. Impacts of Coal Use on Health. In *Annual Review of Public Health*; Fielding, J.E., Ed.; Annual Reviews: Palo Alto, CA, USA, 2020; Volume 41, pp. 397–415.
10. Qureshi, A.A.; Kazi, T.G.; Baig, J.A.; Arain, M.B.; Afridi, H.I. Exposure of heavy metals in coal gangue soil, in and outside the mining area using BCR conventional and vortex assisted and single step extraction methods. Impact on orchard grass. *Chemosphere* **2020**, *255*, 11. [CrossRef]
11. Liu, S.L.; Li, W.P. Zoning and management of phreatic water resource conservation impacted by underground coal mining: A case study in arid and semiarid areas. *J. Clean. Prod.* **2019**, *224*, 677–685. [CrossRef]
12. Ma, K.; Zhang, Y.X.; Ruan, M.Y.; Guo, J.; Chai, T.Y. Land Subsidence in a Coal Mining Area Reduced Soil Fertility and Led to Soil Degradation in Arid and Semi-Arid Regions. *Int. J. Environ. Res. Public Health* **2019**, *16*, 3929. [CrossRef]
13. Martins, W.B.R.; Lima, M.D.R.; Barros, U.D.; Amorim, L.; Oliveira, F.D.; Schwartz, G. Ecological methods and indicators for recovering and monitoring ecosystems after mining: A global literature review. *Ecol. Eng.* **2020**, *145*, 11. [CrossRef]
14. Xu, J.P.; Ma, N.; Xie, H.P. Ecological coal mining based dynamic equilibrium strategy to reduce pollution emissions and energy consumption. *J. Clean. Prod.* **2017**, *167*, 514–529. [CrossRef]
15. Kompala-Baba, A.; Bierza, W.; Blonska, A.; Sierka, E.; Magurno, F.; Chmura, D.; Besenyei, L.; Radosz, L.; Wozniak, G. Vegetation diversity on coal mine spoil heaps—How important is the texture of the soil substrate? *Biologia* **2019**, *74*, 419–436. [CrossRef]
16. Lefticariu, L.; Walters, E.R.; Pugh, C.W.; Bender, K.S. Sulfate reducing bioreactor dependence on organic substrates for remediation of coal-generated acid mine drainage: Field experiments. *Appl. Geochem.* **2015**, *63*, 70–82. [CrossRef]
17. Fiket, Z.; Medunic, G.; Vidakovic-Cifrek, Z.; Jezidzic, P.; Cvjetko, P. Effect of coal mining activities and related industry on composition, cytotoxicity and genotoxicity of surrounding soils. *Environ. Sci. Pollut. Res.* **2020**, *27*, 6613–6627. [CrossRef] [PubMed]
18. Artico, L.L.; Kommling, G.; Migita, N.A.; Menezes, A.P.S. Toxicological Effects of Surface Water Exposed to Coal Contamination on the Test System Allium cepa. *Water Air Soil Pollut.* **2018**, *229*, 12. [CrossRef]
19. Freitas, L.A.d.; Rambo, C.L.; Franscescon, F.; Barros, A.F.P.d.; Lucca, G.d.S.D.; Siebel, A.M.; Scapinello, J.; Lucas, E.M.; Magro, J.D. Coal extraction causes sediment toxicity in aquatic environments in Santa Catarina, Brazil. *Rev. Ambiente Água* **2017**, *12*, 591–604. [CrossRef]

20. Wang, Z.Y.; Hou, J.; Guo, J.Y.; Wang, C.J.; Wang, M.J. Coal Dust Reduce the Rate of Root Growth and Photosynthesis of Five Plant Species in Inner Mongolian Grassland. *J. Residuals Sci. Technol.* **2016**, *13*, S63–S73.

21. Shi, Y.K.; Mu, X.M.; Li, K.R.; Shao, H.B. Soil characterization and differential patterns of heavy metal accumulation in woody plants grown in coal gangue wastelands in Shaanxi, China. *Environ. Sci. Pollut. Res.* **2016**, *23*, 13489–13497. [CrossRef]

22. Sun, L.; Liao, X.Y.; Yan, X.L.; Zhu, G.H.; Ma, D. Evaluation of heavy metal and polycyclic aromatic hydrocarbons accumulation in plants from typical industrial sites: Potential candidate in phytoremediation for co-contamination. *Environ. Sci. Pollut. Res.* **2014**, *21*, 12494–12504. [CrossRef]

23. National Development and Reform Commission; PRC. 13th Five-Year Plan for Coal Industry Development. Available online: https://www.ndrc.gov.cn/xxgk/zcfb/ghwb/201612/t20161230_962216.html (accessed on 19 September 2020).

24. Giam, X.; Olden, J.D.; Simberloff, D. Impact of coal mining on stream biodiversity in the US and its regulatory implications. *Nat. Sustain.* **2018**, *1*, 176–183. [CrossRef]

25. Zeng, Q.; Shen, L.; Yang, J. Potential impacts of mining of super-thick coal seam on the local environment in arid Eastern Junggar coalfield, Xinjiang region, China. *Environ. Earth Sci.* **2020**, *79*, 15. [CrossRef]

26. Li, Y.M.; Zhang, B.; Wang, B.; Wang, Z.H. Evolutionary trend of the coal industry chain in China: Evidence from the analysis of I-O and APL model. *Resour. Conserv. Recycl.* **2019**, *145*, 399–410. [CrossRef]

27. Zhang, Y.Q.; Wu, D.; Wang, C.X.; Fu, X.; Wu, G. Impact of coal power generation on the characteristics and risk of heavy metal pollution in nearby soil. *Ecosyst. Health Sustain.* **2020**, *6*, 12. [CrossRef]

28. Franks, D.M.; Brereton, D.; Moran, C.J. The cumulative dimensions of impact in resource regions. *Resour. Policy* **2013**, *38*, 640–647. [CrossRef]

29. Porter, M.; Franks, D.M.; Everingham, J.A. Cultivating collaboration: Lessons from initiatives to understand and manage cumulative impacts in Australian resource regions. *Resour. Policy* **2013**, *38*, 657–669. [CrossRef]

30. Liu, S.L.; Li, W.P.; Qiao, W.; Wang, Q.Q.; Hu, Y.B.; Wang, Z.K. Effect of natural conditions and mining activities on vegetation variations in arid and semiarid mining regions. *Ecol. Indic.* **2019**, *103*, 331–345. [CrossRef]

31. Pei, H.; Fang, S.F.; Lin, L.; Qin, Z.H.; Wang, X.Y. Methods and applications for ecological vulnerability evaluation in a hyper-arid oasis: A case study of the Turpan Oasis, China. *Environ. Earth Sci.* **2015**, *74*, 1449–1461. [CrossRef]

32. Fang, A.M.; Dong, J.H.; Cao, Z.G.; Zhang, F.; Li, Y.F. Tempo-Spatial Variation of Vegetation Coverage and Influencing Factors of Large-Scale Mining Areas in Eastern Inner Mongolia, China. *Int. J. Environ. Res. Public Health* **2020**, *17*, 47. [CrossRef]

33. Li, Y.R.; Cao, Z.; Long, H.L.; Liu, Y.S.; Li, W.J. Dynamic analysis of ecological environment combined with land cover and NDVI changes and implications for sustainable urban-rural development: The case of Mu Us Sandy Land, China. *J. Clean. Prod.* **2017**, *142*, 697–715. [CrossRef]

34. McMillen, D.P. Geographically Weighted Regression: The Analysis of Spatially Varying Relationships. *Am. J. Agric. Econ.* **2004**, *86*, 554–556. [CrossRef]

35. Li, H.L.; Peng, J.; Liu, Y.X.; Hu, Y.N. Urbanization impact on landscape patterns in Beijing City, China: A spatial heterogeneity perspective. *Ecol. Indic.* **2017**, *82*, 50–60. [CrossRef]

36. Tenerelli, P.; Demsar, U.; Luque, S. Crowdsourcing indicators for cultural ecosystem services: A geographically weighted approach for mountain landscapes. *Ecol. Indic.* **2016**, *64*, 237–248. [CrossRef]

37. Dadashpoor, H.; Azizi, P.; Moghadasi, M. Land use change, urbanization, and change in landscape pattern in a metropolitan area. *Sci. Total Environ.* **2019**, *655*, 707–719. [CrossRef] [PubMed]

38. Shaker, R.R.; Altman, Y.; Deng, C.B.; Vaz, E.; Forsythe, K.W. Investigating urban heat island through spatial analysis of New York City streetscapes. *J. Clean. Prod.* **2019**, *233*, 972–992. [CrossRef]

39. Liu, Y.; Zhao, N.Z.; Vanos, J.K.; Cao, G.F. Revisiting the estimations of PM2.5-attributable mortality with advancements in PM2.5 mapping and mortality statistics. *Sci. Total Environ.* **2019**, *666*, 499–507. [CrossRef]

40. Xia, C.; Xiang, M.; Fang, K.; Li, Y.; Ye, Y.; Shi, Z.; Liu, J. Spatial-temporal distribution of carbon emissions by daily travel and its response to urban form: A case study of Hangzhou, China. *J. Clean. Prod.* **2020**, *257*, 120797. [CrossRef]

41. Sannigrahi, S.; Zhang, Q.; Pilla, F.; Joshi, P.K.; Basu, B.; Keesstra, S.; Roy, P.S.; Wang, Y.; Sutton, P.C.; Chakraborti, S.; et al. Responses of ecosystem services to natural and anthropogenic forcings: A spatial regression based assessment in the world's largest mangrove ecosystem. *Sci. Total Environ.* **2020**, *715*, 13. [CrossRef]

42. Sun, X.; Tang, H.J.; Yang, P.; Hu, G.; Liu, Z.H.; Wu, J.G. Spatiotemporal patterns and drivers of ecosystem service supply and demand across the conterminous United States: A multiscale analysis. *Sci. Total Environ.* **2020**, *703*, 17. [CrossRef] [PubMed]

43. Sawut, R.; Kasim, N.; Abliz, A.; Li, H.; Yalkun, A.; Maihemuti, B.; Shi, Q.D. Possibility of optimized indices for the assessment of heavy metal contents in soil around an open pit coal mine area. *Int. J. Appl. Earth Obs. Geoinf.* **2018**, *73*, 14–25. [CrossRef]

44. Intergovernmental Panel on Climate Change(IPCC). Special Report On Climate Change And Land: Desertification. Available online: https://www.ipcc.ch/srccl/chapter/chapter-3/ (accessed on 18 September 2020).

45. Mancini, L.; Sala, S. Social impact assessment in the mining sector: Review and comparison of indicators frameworks. *Resour. Policy* **2018**, *57*, 98–111. [CrossRef]

46. International Energy Agency (IEA). World Energy Outlook 2017: China. Available online: https://www.iea.org/reports/world-energy-outlook-2017-china (accessed on 18 September 2020).

47. Bu, Q.W.; Li, Q.S.; Zhang, H.D.; Cao, H.M.; Gong, W.W.; Zhang, X.; Ling, K.; Cao, Y.B. Concentrations, Spatial Distributions, and Sources of Heavy Metals in Surface Soils of the Coal Mining City Wuhai, China. *J. Chem.* **2020**, *2020*, 10. [CrossRef]

48. Wang, W.F.; Hao, W.D.; Sian, Z.F.; Lei, S.G.; Wang, X.S.; Sang, S.X.; Xu, S.C. Effect of coal mining activities on the environment of Tetraena mongolica in Wuhai, Inner Mongolia, China-A geochemical perspective. *Int. J. Coal Geol.* **2014**, *132*, 94–102. [CrossRef]

49. Guan, Q.Y.; Wang, L.; Pan, B.T.; Guan, W.Q.; Sun, X.Z.; Cai, A. Distribution features and controls of heavy metals in surface sediments from the riverbed of the Ningxia-Inner Mongolian reaches, Yellow River, China. *Chemosphere* **2016**, *144*, 29–42. [CrossRef] [PubMed]

50. Dai, S.W.; Shulski, M.D.; Hubbard, K.G.; Takle, E.S. A spatiotemporal analysis of Midwest US temperature and precipitation trends during the growing season from 1980 to 2013. *Int. J. Clim.* **2016**, *36*, 517–525. [CrossRef]

51. Ministry of Ecology and Environment; PRC. List of National Nature Reserves. Available online: http://www.gov.cn/guoqing/2019-04/09/content_5380702.htm (accessed on 19 September 2020).

52. Batunacun; Wieland, R.; Lakes, T.; Hu, Y.F.; Nendel, C. Identifying drivers of land degradation in Xilingol, China, between 1975 and 2015. *Land Use Policy* **2019**, *83*, 543–559. [CrossRef]

53. Xue, J.R.; Su, B.F. Significant Remote Sensing Vegetation Indices: A Review of Developments and Applications. *J. Sens.* **2017**, *2017*, 17. [CrossRef]

54. Maneja, R.H.; Miller, J.D.; Li, W.Z.; El-Askary, H.; Flandez, A.V.B.; Dagoy, J.J.; Alcaria, J.F.A.; Basali, A.U.; Al-Abdulkader, K.A.; Loughland, R.A.; et al. Long-term NDVI and recent vegetation cover profiles of major offshore island nesting sites of sea turtles in Saudi waters of the northern Arabian Gulf. *Ecol. Indic.* **2020**, *117*, 13. [CrossRef]

55. Schell, J.A. Monitoring vegetation systems in the great plains with ERTS. *Nasa Spec. Publ.* **1973**, *351*, 309.

56. Gocic, M.; Trajkovic, S. Analysis of changes in meteorological variables using Mann-Kendall and Sen's slope estimator statistical tests in Serbia. *Glob. Planet. Chang.* **2013**, *100*, 172–182. [CrossRef]

57. Sen, P.K. Estimates of the Regression Coefficient Based on Kendall's Tau. *J. Am. Stat. Assoc.* **1968**, *63*, 1379–1389. [CrossRef]

58. Pingale, S.M.; Khare, D.; Jat, M.K.; Adamowski, J. Spatial and temporal trends of mean and extreme rainfall and temperature for the 33 urban centers of the arid and semi-arid state of Rajasthan, India. *Atmos. Res.* **2014**, *138*, 73–90. [CrossRef]

59. Mann, H.B. Nonparametric test against trend. *Econometrica* **1945**, *13*, 245–259. [CrossRef]

60. Kendall, M.G. *Rank Correlation Methods*; [Oxford University Press, Biometrika Trust]; Griffin: London, UK, 1957; Volume 44, p. 298.

61. Brunsdon, C.; Fotheringham, S.; Charlton, M. Geographically Weighted Regression-Modelling Spatial Non-Stationarity. *J. R. Stat. Soc. Ser. D (Stat.)* **1998**, *47*, 431–443. [CrossRef]

62. Alahmadi, S.; Al-Ahmadi, K.; Almeshari, M. Spatial variation in the association between NO2 concentrations and shipping emissions in the Red Sea. *Sci. Total Environ.* **2019**, *676*, 131–143. [CrossRef]

63. Marquardt, D.W. Generalized Inverses, Ridge Regression, Biased Linear Estimation, and Nonlinear Estimation. *Technometrics* **1970**, *12*, 591–612. [CrossRef]

64. Qi, X.Z.; Jia, J.H.; Liu, H.Y.; Lin, Z.S. Relative importance of climate change and human activities for vegetation changes on China's silk road economic belt over multiple timescales. *Catena* **2019**, *180*, 224–237. [CrossRef]
65. Tai, X.L.; Epstein, H.E.; Li, B. Elevation and Climate Effects on Vegetation Greenness in an Arid Mountain-Basin System of Central Asia. *Remote Sens.* **2020**, *12*, 1665. [CrossRef]
66. Zhu, Y.K.; Zhang, J.T.; Zhang, Y.Q.; Qin, S.G.; Shao, Y.Y.; Gao, Y. Responses of vegetation to climatic variations in the desert region of northern China. *Catena* **2019**, *175*, 27–36. [CrossRef]
67. Liu, X.Y.; Zhou, W.; Bai, Z.K. Vegetation coverage change and stability in large open-pit coal mine dumps in China during 1990-2015. *Ecol. Eng.* **2016**, *95*, 447–451. [CrossRef]
68. Li, X.H.; Lei, S.G.; Cheng, W.; Liu, F.; Wang, W.Z. Spatio-temporal dynamics of vegetation in Jungar Banner of China during 2000–2017. *J. Arid Land* **2019**, *11*, 837–854. [CrossRef]
69. Duran, A.P.; Rauch, J.; Gaston, K.J. Global spatial coincidence between protected areas and metal mining activities. *Biol. Conserv.* **2013**, *160*, 272–278. [CrossRef]
70. Neri, A.C.; Dupin, P.; Sanchez, L.E. A pressure-state-response approach to cumulative impact assessment. *J. Clean. Prod.* **2016**, *126*, 288–298. [CrossRef]
71. Siqueira-Gay, J.; Sonter, L.J.; Sánchez, L.E. Exploring potential impacts of mining on forest loss and fragmentation within a biodiverse region of Brazil's northeastern Amazon. *Resour. Policy* **2020**, *67*, 101662. [CrossRef]
72. Ministry of Natural Resources; PRC. The Report on the National General Survey of Soil Contamination. Available online: http://g.mnr.gov.cn/201701/t20170123_1428712.html (accessed on 3 September 2020).
73. Liao; Liu, F. Identifying the Mining Impact Range on the Vegetation of Yangquan Coal Mining Region by Using 3S Technology. *J. Nat. Resour.* **2010**, *25*, 185–191.
74. Yao, F.; Guli, J.; Bao, A.; Zhang, J.; Li, C.; Liu, J. Damage assessment of the vegetable types based on remote sensing in the open coalmine of arid desert area. *China Environ. Sci.* **2013**, *33*, 707–713.
75. Ma, Q.; He, C.Y.; Fang, X.N. A rapid method for quantifying landscape-scale vegetation disturbances by surface coal mining in arid and semiarid regions. *Landsc. Ecol.* **2018**, *33*, 2061–2070. [CrossRef]
76. Lv, X.J.; Xiao, W.; Zhao, Y.L.; Zhang, W.K.; Li, S.C.; Sun, H.X. Drivers of spatio-temporal ecological vulnerability in an arid, coal mining region in Western China. *Ecol. Indic.* **2019**, *106*, 18. [CrossRef]
77. Cheng, W.; Lei, S.G.; Bian, Z.F.; Zhao, Y.B.; Li, Y.C.; Gan, Y.D. Geographic distribution of heavy metals and identification of their sources in soils near large, open-pit coal mines using positive matrix factorization. *J. Hazard. Mater.* **2020**, *387*, 11. [CrossRef]
78. Ma, Z.; Yan, N.; Wu, B.; Stein, A.; Zhu, W.; Zeng, H. Variation in actual evapotranspiration following changes in climate and vegetation cover during an ecological restoration period (2000–2015) in the Loess Plateau, China. *Sci Total Environ.* **2019**, *689*, 534–545. [CrossRef]
79. Forzieri, G.; Alkama, R.; Miralles, D.G.; Cescatti, A. Response to Comment on "Satellites reveal contrasting responses of regional climate to the widespread greening of Earth". *Science* **2018**, *360*, 3. [CrossRef]
80. Li, Z.Y.; Ma, W.H.; Liang, C.Z.; Liu, Z.L.; Wang, W.; Wang, L.X. Long-term vegetation dynamics driven by climatic variations in the Inner Mongolia grassland: Findings from 30-year monitoring. *Landsc. Ecol.* **2015**, *30*, 1701–1711. [CrossRef]
81. Piao, S.L.; Yin, G.D.; Tan, J.G.; Cheng, L.; Huang, M.T.; Li, Y.; Liu, R.G.; Mao, J.F.; Myneni, R.B.; Peng, S.S.; et al. Detection and attribution of vegetation greening trend in China over the last 30 years. *Glob. Chang. Biol.* **2015**, *21*, 1601–1609. [CrossRef] [PubMed]

 *remote sensing*

Article

Vegetation Resilience under Increasing Drought Conditions in Northern Tanzania

Steye L. Verhoeve [1,*], Tamara Keijzer [2], Rehema Kaitila [3], Juma Wickama [4] and Geert Sterk [1]

1 Faculty of Geosciences, Utrecht University, Princetonlaan 8A, 3584 CB Utrecht, The Netherlands; g.sterk@uu.nl
2 PBL Netherlands Environmental Assessment Agency, Bezuidenhoutseweg 30, 2594 AV The Hague, The Netherlands; tamara.keijzer@pbl.nl
3 Tanzania National Parks, Lake Manyara National Park, Arusha P.O. Box 3134, Tanzania; rehema.kaitila@tanzaniaparks.go.tz
4 Tanzania Agricultural Research Institution, Mlingano, Tanga P.O. Box 5088, Tanzania; wickama@yahoo.com
* Correspondence: s.l.verhoeve@uu.nl

Abstract: East Africa is comprised of many semi-arid lands that are characterized by insufficient rainfall and the frequent occurrence of droughts. Drought, overgrazing and other impacts due to human activity may cause a decline in vegetation cover, which may result in land degradation. This study aimed to assess drought occurrence, vegetation cover changes and vegetation resilience in the Monduli and Longido districts in northern Tanzania. Satellite-derived data of rainfall, temperature and vegetation cover were used. Monthly precipitation (CenTrends v1.0 extended with CHIRPS2.0) and monthly mean temperatures (CRU TS4.03) were collected for the period of 1940–2020. Eight-day maximum value composite data of the normalized difference vegetation index (NDVI) (NOAA CDR—AVHRR) were obtained for the period of 1981–2020. Based on the meteorological data, trends in rainfall, temperature and drought were determined. The NDVI data were used to determine changes in vegetation cover and vegetation resilience related to the occurrence of drought. Rainfall did not significantly change over the period of 1940–2020, but mean monthly temperatures increased by 1.06 °C. The higher temperatures resulted in more frequent and prolonged droughts due to higher potential evapotranspiration rates. Vegetation cover declined by 9.7% between 1981 and 2020, which is lower than reported in several other studies, and most likely caused by the enhanced droughts. Vegetation resilience on the other hand is still high, meaning that a dry season or year resulted in lower vegetation cover, but a quick recovery was observed during the next normal or above-normal rainy season. It is concluded that despite the overall decline in vegetation cover, the changes have not been as dramatic as earlier reported, and that vegetation resilience is good in the study area. However, climate change predictions for the area suggest the occurrence of more droughts, which might lead to further vegetation cover decline and possibly a shift in vegetation species to more drought-prone species.

Keywords: drought impacts; NDVI; drought adaptation; drought index; vegetation resilience; drought vulnerability; standardized precipitation evapotranspiration index; AVHRR; land degradation

Citation: Verhoeve, S.L.; Keijzer, T.; Kaitila, R.; Wickama, J.; Sterk, G. Vegetation Resilience under Increasing Drought Conditions in Northern Tanzania. *Remote Sens.* **2021**, *13*, 4592. https://doi.org/10.3390/rs13224592

Academic Editor: Elias Symeonakis

Received: 14 September 2021
Accepted: 22 October 2021
Published: 15 November 2021

Publisher's Note: MDPI stays neutral with regard to jurisdictional claims in published maps and institutional affiliations.

1. Introduction

The main component of terrestrial ecosystems is vegetation, which has a direct link to many ecosystem services, such as food production, soil retention, climate regulation, water purification and disease management [1]. The value of these services could decline or disappear with an increasing pressure on vegetation resources. Not only natural influences such as wildlife grazing and the weather, but anthropogenic pressures can also have a negative influence on the productivity of vegetation [2].

Land degradation is defined as changes in land use from productive to unproductive due to natural or human-made factors [3]. Land degradation is one of the world's major socio-economic and environmental problems, affecting two-fifths of humanity [4]. Agricultural expansion has led to severe land degradation all over the world, particularly when accompanied by high water consumption and the conversion of natural landscapes into cultivated lands [5–7]. Land degradation undermines the land's productivity and contributes to the degradation of ecosystem services. Land degradation disproportionately affects the poor and is sometimes the decisive component that causes poverty and social conflict [8,9]. The loss of productive land is part of a vicious circle for many rural people in developing countries in which land degradation can be both the cause and the effect of poverty [10].

Semi-arid ecosystems in particular are under pressure from the rising demand for natural resources and an increase in weather extremes. This is caused by an increasing human population and by climate change, respectively. More frequent and severe droughts have been forecasted in the 21st century, particularly in the mid-latitudes [11]. Increases in drought occurrences are driven by a decrease in precipitation and/or an increase in evapotranspiration due to higher daily temperatures [12]. As water availability acts as the main driver of vegetation distribution and productivity in arid and semi-arid regions [13], droughts impose a serious risk on the livelihood of many people [14].

Previous studies show that Eastern Africa has been suffering from an increase in temperature and more frequent droughts, which have continued in the 21st century [15]. In this region a browning trend of the vegetation has occurred in the past 40 years [16,17], which is among the most notable vegetation browning in the world. Some reports also show a decline in vegetation productivity and an increase in land degradation [1,3,18]. On the other hand, other remote-sensing-based studies have shown that areas in East Africa have experienced fluctuations in vegetation cover, which were largely driven by variations in soil moisture [19]. This can be explained by the quick response of vegetation in arid and semi-arid biomes to rainfall fluctuations. Plant species have adopted mechanisms that allow them to rapidly adapt to changing water availability and are also able to withstand water deficits [20]. These mechanisms suggest a strong revival of vegetation health during periods of water abundance [21]. However, the way vegetation responds to drought on different time scales remains largely unknown because of the different response times and vulnerability that species have to drought. By knowing this, the severity of degradation can be assessed, and an estimate can be made on the importance of applying measures.

The semi-arid zone in northern Tanzania is an example of an area that suffers from increasing droughts and enhanced soil degradation [22]. The area close to Lake Manyara, covered by the Monduli and Longido districts, is primarily comprised of savanna and rangeland, which is widely used for livestock grazing by the local Masai herders. According to Wynants et al. [23], 2.0% of this area is degraded while there has been a serious increase in the soil erosion risk from 1988 to 2016. Masai herders in the area complain about more frequent droughts and a lack of sufficient grazing resources. Part of the problem faced by the Masai is an increase in livestock numbers, which results in more pressure on grazing resources [18,24,25]. Using Landsat satellite imagery from the Google Earth Engine, Verhoeve [24] studied land use/cover changes in the Monduli and Longido districts over the period of 1985–2018. The results showed widely fluctuating land use/cover classes over time, and neither revealed any significant changes in land cover nor provided evidence for large-scale vegetation degradation. These results contradict previous studies that showed overall degrading vegetation cover in the study area [17,23,25,26]. According to Verhoeve [24], vegetation cover is largely influenced by the amount of precipitation and the occurrence of drought. However, it was also clear that vegetation resilience in the study area is high, with a good recovery of vegetation cover following a drought year. Similar observations were made for Sahelian West Africa, where the vegetation recovered following the devastating droughts of the late 1970s and early 1980s [21].

In this study we used satellite-derived hydrometeorological data for drought analysis and the normalized difference vegetation index (NDVI) as a proxy for vegetation response to drought. The main objective was to investigate the resilience of vegetation to drought over different time scales in northern Tanzania during 1981–2020. To achieve this objective, we first investigated the long-term hydrometeorological data and the occurrence of drought in the region. Next, we analyzed the long-term interannual NDVI trends in the region. Finally, we determined the short-term effects of drought on vegetation health and recovery at different time periods.

2. Materials and Methods

2.1. Study Area

The study area comprises the districts of Monduli and Longido in Arusha Region, northern Tanzania (Figure 1) [25,27]. The total area covers approximately 16,000 km^2 and has some 282,000 inhabitants [28]. The majority of the area lies in the East African Rift Valley and is bordered by a high (~1300 m a.s.l) escarpment in the west. The valley floor is about 300 m lower in the western part and gradually rises towards the east, where there is no clear escarpment. Several small mountains are scattered throughout the Rift Valley, which are mostly volcano remnants. It is an important area for wildlife conservation, including or bordering the Lake Manyara, Arusha, Tarangire, Mount Kilimanjaro and Serengeti National Parks, as well as the Ngorongoro Conservation Area.

The climate of the study area is semi-arid and has four climatic seasons [29]. In general, the short rains season from November to January (NDJ) is followed by a short dry season (Feb) until the long rains start, which typically occur from March to May (MAM). From June to October (JJASO), a long dry season can be identified with cooler temperatures. Because of high interannual variability, the NDJ season often continues into February. In wet years the NDJ and MAM seasons often overlap in an almost continuous rainy season. The annual rainfall is between 450 and 1200 mm, averaging around 750 mm. The lower lying areas receive a mean annual rainfall of about 650 mm, whereas in the higher parts the annual rainfall ranges from 1000 mm to 1200 mm on average [30,31]. The average temperature is between 20–25 °C, with a minimum of 11 °C in July to September and a maximum temperature of 31 °C in January and February [32,33].

The physical characteristics of the area, such as its morphology, geology and soils, are strongly influenced by tectonic activities and volcanism [3]. These characteristics have influenced the rainfall distribution, vegetation types and wildlife of the area. The Monduli district is part of the Lake Manyara catchment. Lake Manyara, part of an endorheic basin, is the southernmost lake within the eastern arm of the East African Rift System. The lake is shallow and saline, and is situated at 960 m a.s.l.

In the districts are multiple large volcanic mountains, both active and inactive. These mountains stand out in the dominantly flatter landscape, and often have higher rainfall on or near their slopes. Apart from the forests on the slopes of the mountains, savanna is the major land cover type. Savannas are generally on the transition area between tropical rainforests to deserts, which in this area is represented by the forests of the Monduli mountains, Mt. Meru and the Ngorongoro Conservation Area, and the drier, more arid regions of Simanjiro District and Dodoma Region in the south. The savannas have been managed extensively by the Masai through fire and grazing by their livestock, suppressing the growth of bushes and trees [34,35].

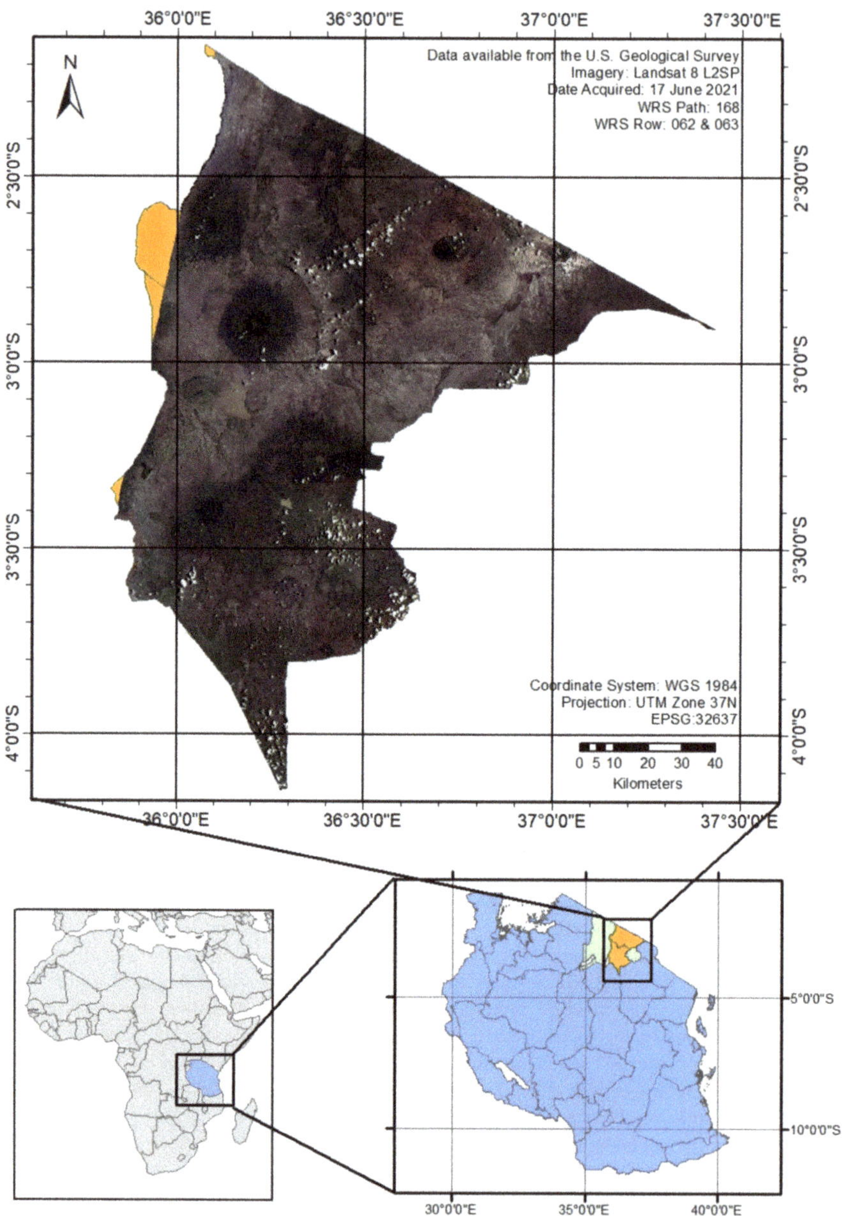

Figure 1. Study area: the Monduli and Longido districts within Arusha Region, northern Tanzania. Source: [27].

2.2. Data

The NDVI is an indicator of the vitality and density of vegetation of a remote sensing image pixel [36]. It is regarded as a reliable indicator for land cover conditions and variations, and over the years it has been widely used for vegetation monitoring [37]. The NDVI produced from historical satellite image archives captures long-term changes in vegetation health and density, enabling the measurement of responses to climate variability [38].

For this study we used the NOAA Climate Data Record (CDR) of AVHRR NDVI, Version 5. This dataset contains daily measurements of surface vegetation cover, gridded at a resolution of 0.05° and computed globally over land surfaces [39]. The AVHRR provides data on a long-term basis (1981–current day) with a moderate spatial resolution. Other datasets could have a higher spatial resolution, but they were not suitable because they start providing data of the study area at a later date. The online platform Google Earth Engine (GEE) was used to extract the daily NDVI values of the study area. GEE is a high-performance cloud-based platform that gives access to a vast and growing amount of earth observation data and provides the processing power necessary to analyze the data [40]. The daily NDVI was used to compute 8-day maximum value composites to filter out cloud irregularities. Of these maximum value composites, the mean NDVI of the study area was used in this study.

The NDVI time series runs from 24 June 1981, the start of the AVHRR mission, until 24 June 2020. In 1988 a series of 51 negative NDVI values were measured, which coincides with the service start of a new AVHRR satellite (NOAA-11), and were therefore left out from further analysis. From week 36 in 1994 to week three in 1995 no data were available due to sensor malfunctioning [41,42].

The hydrometeorological data included the monthly precipitation and temperature of the study area. Limited in situ data were available as the study area is poorly gauged. However, reanalysis and satellite-based techniques can provide continuous hydrometeorological data. Monthly precipitation was obtained from the CenTrends v1.0 extended with the CHIRPS-2.0 dataset. The CenTrends dataset was developed for East Africa in particular to overcome the precipitation data gaps and to enable the analysis of seasonal and decadal fluctuations within a centennial context [43]. The CenTrends dataset is available from 1900. CHIRPS (Climate Hazards Group InfraRed Precipitation with Station data) is the state-of-the-art observational daily precipitation dataset for East Africa. CHIRPS uses additional infrared satellite data, and is therefore available from 1981. CenTrends and CHIRPS are non-independent datasets as they are based on a similar assimilation technique and underlying observational data for their overlap period. They are highly correlated (0.95), justifying the extension of the CenTrends dataset with monthly averaged data from CHIRPS [44]. The combination provides information about both trends in the past and present. The combined dataset was obtained via the KNMI Climate Explorer and has a spatial resolution of 0.2°.

The temperature data used in this study were obtained from the CRU TS4.03 monthly mean temperature. This dataset uses observations interpolated into 0.5° latitude/longitude grid cells combined with existing climatology to obtain absolute monthly values [45]. The CRU was validated with the nearest available in situ data of a meteorological station. The in situ data were only available as mean monthly maximums. Therefore, the CRU TS4.03 mean monthly maximum temperature (CRU_{max}) was validated with the available in situ dataset. With the use of the Pearson's r, the correlation of the datasets was tested. This statistical test was used because the datasets were continuous and normally distributed (Shapiro–Wilk normality test: $W = 0.97$ and $W = 0.95$ for, respectively, the in situ and the CRU_{max} datasets, at $p < 0.05$). The station mean monthly maximum temperatures measured at Arusha and CRU_{max} are in close correspondence ($r = 0.96$, $R^2 = 0.91$), but show relatively high deviations in terms of their magnitude (RMSE = 1.84 °C) for 1979 to 2018. The strong correlation indicates a good representation of the annual temperature cycle. An overestimation of the CRU_{max} data was determined at T < 30.62 °C and an underestimation above this value was determined.

As both the CHIRPS and the CRU TS4.03 datasets are delivered grid-sized, the means of the monthly precipitation and the temperature of all grid cells were calculated for the area between coordinates ~2–4°S, ~35–37°E, which includes the study area. Drought and trends in precipitation and temperature were studied from 1940–2020, which is twice the temporal range of the NDVI data.

2.3. Trend Analysis

The health of vegetation and the corresponding NDVI value is dependent on many anthropogenic and natural factors. The most important natural factors are temperature and precipitation [46]. Therefore, the variations in NDVI and hydrometeorological data were determined over time for the two rainy seasons (NDJ and MAM) and the hydrological year (September–August). All the datasets showed a non-normal distribution over the research period (Shapiro-Wilk normality test, $p < 0.00$). Therefore, the Mann-Kenndall (MK) test was used to determine the direction and significance of trends. The MK test is a non-parametric rank-based test method which is widely used to assess the presence of trends in a time series of climatic, environmental or hydrological data [47–50]. The MK test results in a measure of the rank correlation of Kendall's τ (tau) and the significance (p-value). The magnitude of the trend determined by the MK test was computed with Sen's slope. This test calculates both the slope and the intercept of a linear rate of change [51].

2.4. Drought Analysis

Due to the lack of data on stream flow, groundwater and soil moisture in the study area, only the occurrence of meteorological drought was investigated. The standardized precipitation evaporation index (SPEI) was used to determine drought. The SPEI is an extension of the standardized precipitation index (SPI), which uses only precipitation anomalies to determine drought [52]. To determine hydrological anomalies the SPI uses only precipitation, while the SPEI uses both precipitation (P) and potential evapotranspiration (PET) to determine drought. It takes into account the impact of changing temperatures on water demand. This is important as evapotranspiration influences soil moisture variability and therefore vegetation water content [53]. PET was calculated in this study using the Thornthwaite equation [54], available in the SPEI package of the "R" language. It requires mean temperature and latitude as input values. Other equations (e.g., Hargreaves or Penman) require variables for which no data were available for the study area. The mean temperature data of CRU TS4.03 were used. The SPEI focusses on the anomalies, and therefore the CRU data are assumed to be useful because of the high correlation with in situ data despite the reported overestimation. As the study area lies between 2–4°S, a latitude of $-3°$N was used in the Thornthwaite equation.

The SPEI measures P-PET anomalies based on a comparison of observations for a period of interest (e.g., 1, 3, 6, 12 and 48 months) with the long-term historical record of that period. It requires monthly data, preferably continuous and for 30 years or longer. For each month a SPEI value is calculated using the month itself and a previous number of months, which are together equal to the period of interest. For instance: when calculating the SPEI of March with a period of interest of 3 months, the cumulative P-PET of January, February and March is used. This value is then compared with the long-term record of cumulative January–March P-PET. The period of interest of the SPEI represents typical time scales for water deficits to affect different types of water sources. For example, the 1- or 3-month SPEI represents short droughts and indicates immediate impacts, such as reduced soil moisture, while the 12- or 24-month SPEI represents long droughts, causing, for instance, changes in reservoir storages [52]. In this study the 3-month SPEI (SPEI-3) was used to represent short-term droughts, while the 12-month SPEI (SPEI-12) values were used to represent annual (medium-term) and multi-annual (long-term) droughts. Furthermore, the SPEI-3 was used to indicate dry/wet seasons and the SPEI-12 was used to indicate dry/wet years within the study period.

To calculate the SPEI, the P-PET record is fitted to a probability distribution (log-logistic) function. It is then transformed into a normal distribution with a mean of zero and a variance of one. The result is the SPEI, which represents the number of standard deviations from the mean. Positive SPEI values indicate anomalous wet periods, and negative values indicate dry periods [52,53,55]. The magnitude of the SPEI gives a probabilistic measure of drought/wetness intensity. For instance: an SPEI-3 equal to -2 in January–March of a certain year means that the cumulative January–March P-PET of that year is 2 standard deviations smaller than the long-term average of cumulative January–March P-PET. Events were defined according to the drought intensity classes of McKee et al. [52]: a drought event was classified when the index was below −1 and a wet event was classified when the index was higher than 1 (Table 1). Some studies suggest a denotation of −0.5 to 0.5 for normal conditions [56,57]; however, due to the adaptation of the vegetation to semi-arid conditions the effects of a mild drought on the vegetation are assumed to be neglectable.

Both seasons (NDJ and MAM) and years were classified as dry, wet or normal using the SPEI. Here, the SPEI-3 of January and May were considered for, respectively, the NDJ and MAM seasons. To determine whether a hydrological year was wet, dry or normal, the SPEI-12 of August was used as this value is based on the previous September–August P-PET values.

Table 1. SPEI-based classification of drought. Based on temperature and precipitation during a given moment at a specified latitude, the SPEI gives anomalies of that given moment compared to other years. Source: [52].

SPEI	Classification	Expected Probability (%)
0 to −0.99	Mild drought	34.1
−1.00 to −1.49	Moderate drought	9.2
−1.50 to −1.99	Severe drought	4.4
≤−2.00	Extreme drought	2.3

2.5. Vegetation Resilience

The resilience of vegetation was tested for multi-annual (long-term), annual (medium-term) and seasonal (short-term) responses to drought. The 8-day maximum value composited NDVI and the SPEI values were used to investigate the effects of drought on vegetation cover and resilience.

The long-term effect of changing climate conditions on vegetation cover is represented by changes in the NDVI values over the years. Such changes over time in the NDVI values represent a change in vegetation cover and vegetation health [58]. This was tested with the MK test of 8-day maximum value composited NDVI from 1981 to 2020 and compared to the long-term precipitation and temperature in the region. With the use of Sen's slope [51], the linear rate of change was calculated.

The annual (medium-term) dynamics in vegetation cover and its response to drought were quantified by separating the years into different classes based on hydrological conditions. Dry, normal and wet years were classified with the use of the SPEI-12. Additionally, the year following a dry year was classified as a "recovery-year", regardless of the hydrological conditions of that year. The data of each of those four classes were than fitted by using local polynomial regression (LOESS) to provide a smooth curve through a set of datapoints. The response of the NDVI throughout the seasons was then compared for these four different hydrological conditions. In a second step, vegetation resilience over time was evaluated. For this purpose, three time periods were selected: 1991–2000; 2001–2010; and 2011–2020. If the resilience was not affected over the long-term, similar intra-annual NDVI values were expected during dry years in each time period. The time periods were chosen to have an approximate equal number of years classified as dry (SPEI-12 in August < −1). The years 1981–1990 were left out because no drought had occurred during these years.

Finally, the seasonal (short-term) effect of drought on vegetation cover and resilience was evaluated. The dynamics of vegetation response to seasonal droughts were tested by comparing intra-annual NDVI trends during dry and non-dry periods. It was assumed that if the vegetation has adapted to the semi-dry environment, it will withstand droughts by reviving as soon as the conditions permit [59]. This would mean that the regrowth of the vegetation is not dependent on the severity of a drought, as indicated by the SPEI values. Hence, it was expected that the NDVI values of the subsequent non-dry season do not deviate substantially from other normal (non-dry) years. Seasonal resilience was tested by comparing the sinusoidal curve of the intra-annual NDVI pattern of dry years and the following normal or wet year. Moreover, the effect of the timing of a drought during a dry year was assessed. The timing of the drought was determined with the use of the SPEI-3. The moment of drought was characterized as an SPEI-3 smaller than -1 during the first or the second rainy season (NDJ or MAM, respectively). Four situations were compared, in which both seasons were dry (two occurrences), the first or the second season was dry (four occurrences for both situations) or both seasons were normal or wet (thirteen occurrences). The curves of the NDVI during these different timings of the drought show the response of vegetation to the seasonal drought, and thus provide a proxy of vegetation resilience.

3. Results

3.1. Rainfall and Temperature

Variations in precipitation during the NDJ and MAM rainy seasons as well as the hydrological year are shown in Figure 2. The rainfall in the study area is characterized by a high but mainly homogeneous variation in seasonal rainfall (Figure 2A,B) and annual rainfall (Figure 2C). Overall, the NDJ season has a higher variation than the MAM season ($CV_{NDJ} = 0.49$; $CV_{MAM} = 0.32$), but on average rainfall in the NDJ season is lower compared with the MAM season (226 mm vs. 327 mm) (Table 2).

The precipitation has a slightly decreasing trend in the MAM season and annual rainfall, and a small increasing trend in NDJ rainfall (Figure 2). However, all trends are insignificant according to the MK test (Table 3), which means it cannot be concluded that the amount of rainfall has changed in the study area over the period of 1940–2020. Temperature on the other hand does show a significant ($\alpha = 0.05$) increasing trend (Figure 3). This positive trend was observed both for the yearly averaged temperature as well as for the seasonally averaged temperatures (NDJ and MAM) (Table 3). Deviations from this trend in the form of relatively warm (e.g., 1951–1952) and cold (e.g., 1967) years are also visible. Overall, the yearly average temperature increased by 1.06 °C between 1940 and 2020.

Table 2. Statistics of seasonal and annual precipitation and temperature during the period of 1940–2020 in northern Tanzania. Trend based on Sen's slope. During the warmer NDJ season there is, on average, less precipitation with a higher level of variance compared to the MAM season.

Period	Precipitation			Temperature		
	Trend (mm/Decade)	Average (mm)	CV [1] (-)	Trend (°C/Decade)	Average (°C)	CV [1] (-)
NDJ	5.7	226	0.49	0.12	22.1	0.020
MAM	−3.7	327	0.32	0.12	21.8	0.022
Annual (Sept–Aug)	−4.5	656	0.27	0.13	21.3	0.019

[1], coefficient of variation.

Table 3. Seasonal and annual Mann–Kendall test trend results of rainfall and precipitation over the period of 1940–2020 in northern Tanzania. At $\alpha = 0.05$ the temperature is significantly increasing, but the precipitation is insignificant.

Period	Precipitation		Temperature	
	Tau	*p*-Value	Tau	*p*-Value
NDJ	0.117	0.12	0.448	$<2.2 \times 10^{-16}$
MAM	−0.0877	0.25	0.411	1.19×10^{-7}
Annual (Sept–Aug)	−0.0247	0.75	0.569	$<2.2 \times 10^{-16}$

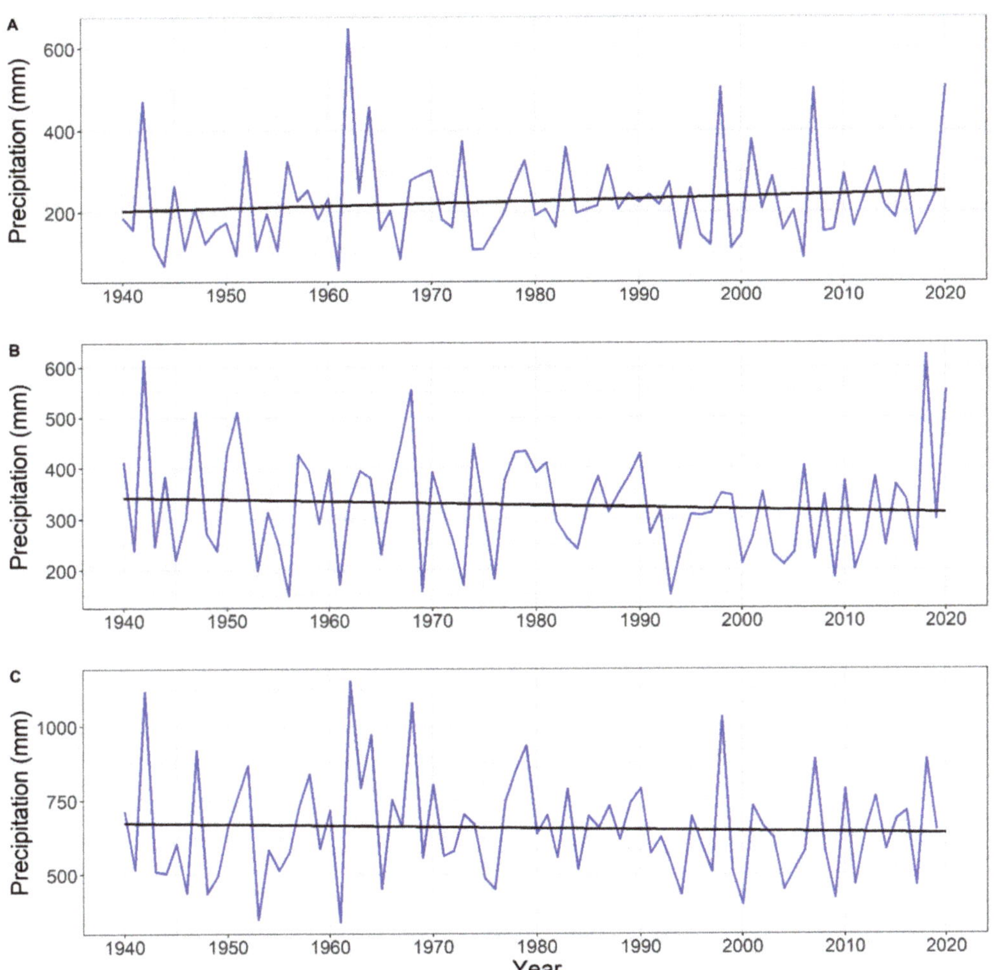

Figure 2. Total amounts of rainfall in northern Tanzania during the period of 1940–2020 based on the CenTrends v1.0 dataset extended with CHIRPS-2.0. (**A**) The NDJ (November, December and January) rainy season, (**B**) the MAM (March, April and May) rainy season and (**C**) the hydrological year (Sept–Aug). Linear trend lines (black) are based on Sen's slope. Insignificant at $\alpha = 0.05$. The NDJ-season has a higher variability compared to the MAM-season (CV = 0.49 and 0.32, respectively), but a lower seasonal mean (226 and 327 mm, respectively).

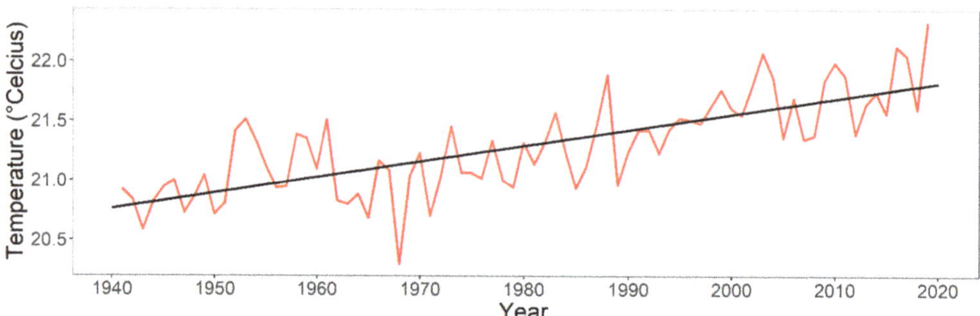

Figure 3. Average annual temperature in northern Tanzania during the period of 1940–2020 based on monthly mean data of the CRU TS4.03 dataset. Linear trend (black) based on Sen's slope. Significant at α = 0.05, increase of 0.13 °C/decade.

3.2. Drought Occurrence

The SPEI was calculated for periods of 3 months (SPEI-3), representing short-term dry or wet seasons, and 12 months (SPEI-12), representing medium-term drought/wet years. Figure 4 shows SPEI-3 and SPEI-12 time series over the time period of 1940–2020. As expected, fewer drought events are identified with the SPEI-12 compared to the SPEI-3 time series. Multiple smaller events identified by the SPEI-3 can either be flattened out or cumulated into one event of the SPEI-12. The latter effect is known as pooling [60]. Relatively wet and dry years can be distinguished, with either largely positive SPEI-12 values (e.g., the 1960s) or negative SPEI-12 values (e.g., the period of 1999–2006).

Both SPEI time series show a significant decreasing trend (SPEI-3: tau = −0.157, *p*-value = 4.1×10^{-13}; SPEI-12: tau = −0.148, *p*-value = 7.4×10^{-12}). This means that a drying trend is present in the study area, which is mainly the effect of the increasing temperature, given the non-significant changes in rainfall in the study area. Higher temperatures result in higher potential evapotranspiration values, which lead to overall more negative SPEI values. The drying trend since 1940 is also reflected by the relatively large area below zero (=dry) compared to the area above zero (=wet) in recent decades (1990–2020).

Zooming into the time period of the NDVI data (1981–2020), the 1980s was a decade which barely shows long or extreme dry and wet events. The 1990s and 2000s are characterized by more frequent and longer droughts. The period of 2001–2010 in particular suffered from extended and severe droughts, with some SPEI values going below −2 (extreme drought). The differences in drought severity between decades is reflected by the occurrence of low SPEI-3 and SPEI-12 values presented in Table 4. Assuming a normal distribution of the SPEI, the occurrence of SPEI < −1 or SPEI > 1 would occur 15.9% of the time and the mean would be 0. However, every decade since 1980 has fewer wet events, and since 1991 more dry events have occurred than expected. The 2001–2010 decade was the driest period, represented by a high occurrence of droughts and a low mean SPEI value over time, while the 2011–2020 decade experienced similar droughts as the 1991–2000 decade. During the 1981–1990 decade the study area experienced a low number of moderate to extreme events, which is also reflected by the steady rainfall values over this time period (Figure 2). Before 1980, several periods of serious drought occurred, for instance during 1953–1956 and 1975–1977 (Figure 4).

Table 4. Percentage of total SPEI values per decade and mean SPEI values per decade. Based on the definition of SPEI, the mean over the entire research period is 0, and the total of moderate to extreme wet/dry seasons ($-1 >$ SPEI > 1) should not exceed 15.9% of the time (Table 1). During the four most recent decades this is not the case.

Period	SPEI-3			SPEI-12		
	<-1	>1	Mean	<-1	>1	Mean
	%	%	-	%	%	-
1981–1990	5.0	11.7	0.138	0.8	7.5	0.247
1991–2000	25.0	6.7	−0.353	28.3	9.2	−0.470
2001–2010	34.2	8.3	−0.437	30.0	9.2	−0.482
2011–2020	19.2	10.0	−0.309	17.5	9.2	−0.285

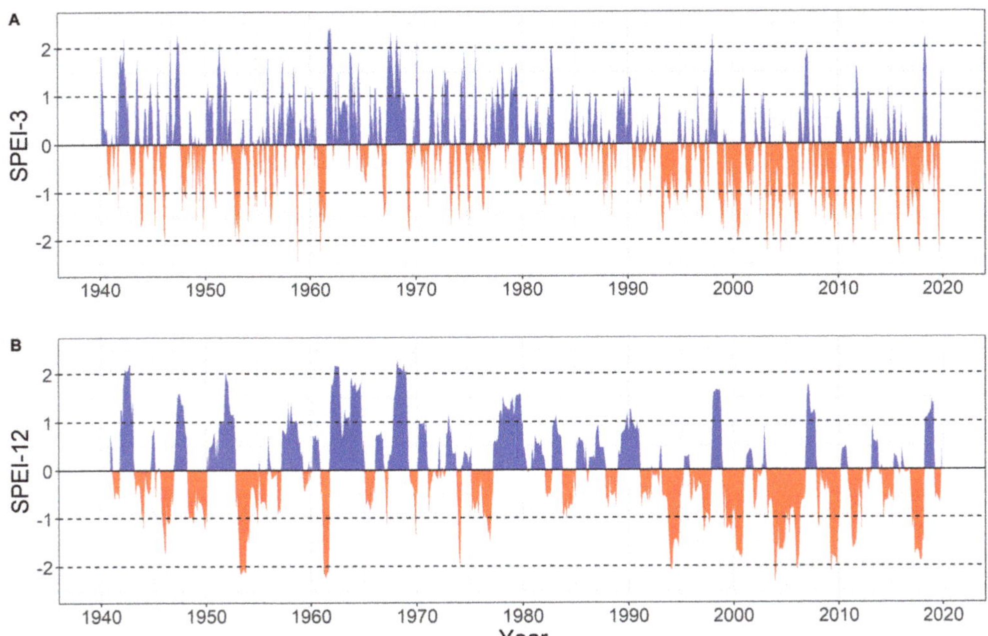

Figure 4. SPEI-3 (**A**) and SPEI-12 (**B**) values of 1940–2020. Blue indicates relatively wet conditions, while red indicates relatively dry conditions in the indicated (3 or 12) antecedent time period in months. Significant ($\alpha = 0.05$) negative trends were found in both the SPEI-3 and SPEI-12.

3.3. Vegetation Cover and Resilience

The effects of droughts and seasonal rainfall on the resilience of vegetation in the area were assessed by comparing the NDVI time series with the SPEI-3 and SPEI-12 values. Trend analysis was applied to the long-term NDVI time series and a visual comparison was applied to the intra-annual variations.

The NDVI values over the years (1981–2020) show a seasonal pattern, with relatively high NDVI values in the period of the two wet seasons, and low values during the dry season (Figure 5). The higher NDVI values indicate healthy vegetation with a high cover, while the lower values indicate bare soil or low vegetation cover. Figure 5 also shows that the NDVI values are generally lower during dry periods, as indicated by the negative SPEI-3 values.

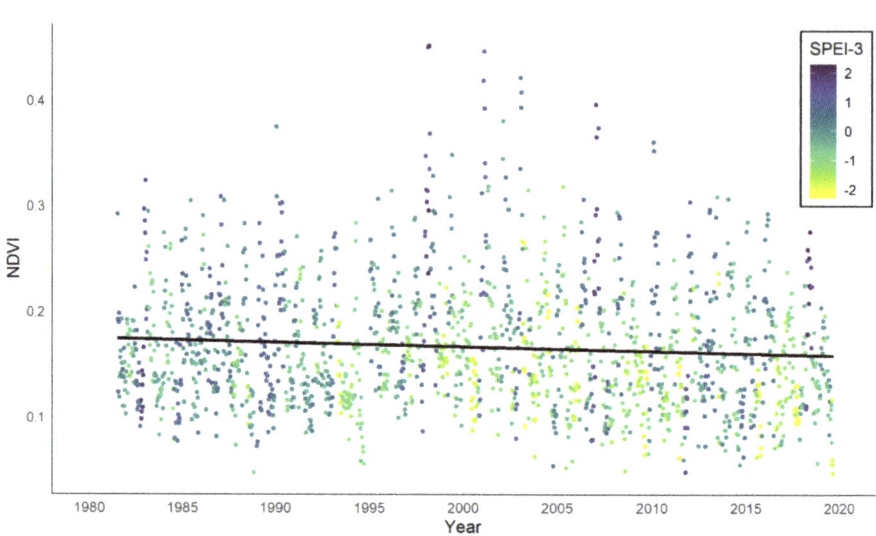

Figure 5. Interannual normalized difference vegetation index (NDVI) series, composed of 8-day maximum value composited NDVI. The NDVI values were classified according to SPEI-3 values, which indicates drought (strong negative SPEI-3 values) or wet (strong positive SPEI-3 values) conditions. A trend line (black) was fitted through all data based on Sen's slope (significant at $\alpha = 0.05$).

The long-term NDVI has a significant downward trend (tau = -0.0623, p-value 8.38×10^{-5}, Sen's slope = -1.02×10^{-5}). This downward trend results in a decrease of 0.017 NDVI points (from 0.175 to 0.158) between 1981 and 2020. Therefore, over the period of analysis, on average, vegetation cover in the study area has declined. Several possible reasons for this vegetation cover decline can be given. The first is the conversion of grazing land to arable land for crop production [23,24]. The second possibility is land degradation due to overgrazing in the area [23,25]. The last reason could be the increased temperatures and drought as exemplified by the SPEI values (Figure 4).

Figure 6 shows the same 8-day NDVI values as in Figure 5, but in this figure the values have been plotted versus the hydrological year (Sept–Aug). The polynomial regression lines of NDVI values for dry (red; SPEI-3 < -1), wet (blue; SPEI-3 > 1) and normal years (black; -1 < SPEI-3 < 1) are also plotted. An additional regression line (yellow color) shows the NDVI values for a year immediately following a drought year. Obviously, the wet years have higher NDVI values than normal years, and thus better vegetation cover. During dry years, the NDVI values are substantially lower than in normal years, indicating less vegetation cover or less healthy vegetation. However, in the years following a drought year, the NDVI values return to normal values, which indicates a high resilience of vegetation in the study area. Apparently, the drought is affecting the vegetation temporarily, but during the period of study (1981–2020) it has not led to dramatic vegetation degradation, apart from the slightly negative long-term trend that was detected (Figure 5).

The intra-annual NDVI time series has been split into three decades: 1991–2000, 2001–2010 and 2011–2020 (Figure 7). In each decade, two or three years were classified as dry years (SPEI-12 < -1). The polynomial regression curves show that in the first (Figure 7A) and last decade (Figure 7C) the NDVI values in a year following a drought year quickly return to normal values. In the decade 2001–2010, which experienced the most droughts, the NDVI values following a drought year also return to nearly similar levels as the long-term normal. The normal year NDVI values in this decade are higher than in the other two decades, which is surprising given the more severe drought conditions in this decade (Table 4). Apparently, the rainfall was well-distributed during the normal years in the period of 2001–2010, which resulted in good vegetation growth during those normal years.

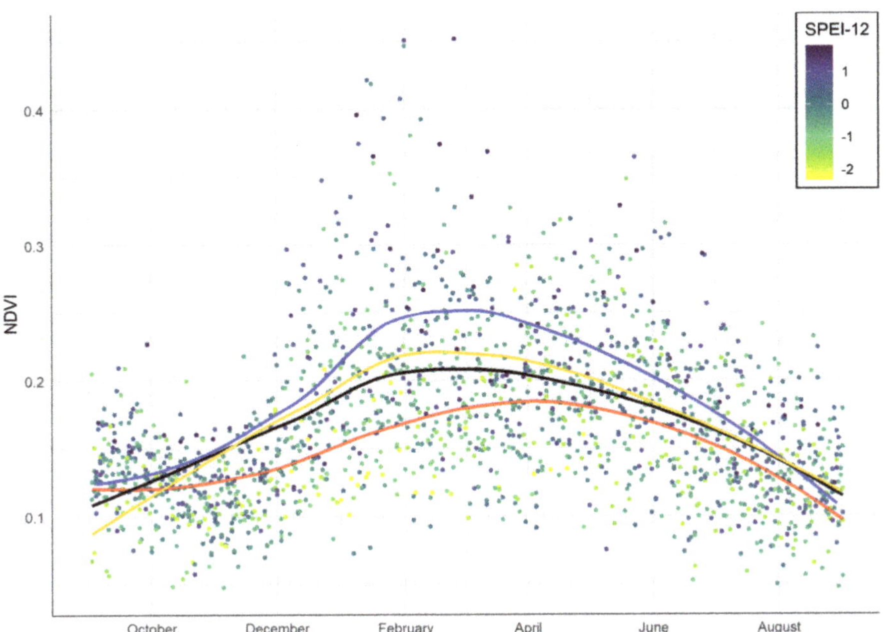

Figure 6. Intra-annual NDVI series, based on 8-day maximum value composited NDVI from 1981 to 2020. Based on the SPEI-12 of the hydrological year, the lines represent the NDVI values belonging to normal (black), dry (red) or the year subsequent to a dry hydrological year (yellow) with the use of locally estimated scatterplot smoothing (LOESS).

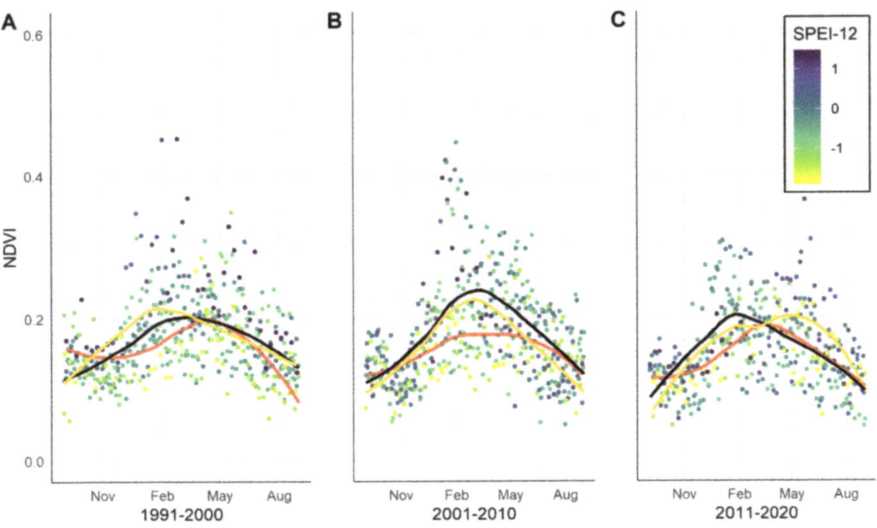

Figure 7. Intra-annual decadal NDVI series, composed of 8-day maximum value composited NDVI from (**A**) 1991–2000, (**B**) 2001–2010 and (**C**) 2011–2020. The lines represent the NDVI values belonging to normal (black), dry (red) or the year subsequent to a dry hydrological year (yellow) with the use of locally estimated scatterplot smoothing (LOESS).

In the last analysis, the impacts of seasonal droughts on NDVI development were evaluated. The short-term effects of drought were tested with the use of the SPEI-3 for the NDJ and MAM seasons. The NDJ season was classified as dry when the SPEI-3 of Jan was below −1. The MAM rainy season was considered dry when the SPEI-3 of May was below −1.

The timing of the drought during the hydrological year has an effect on the pattern and magnitude of the NDVI values (Figure 8). During a year in which both the short rainy season (NDJ) is normal and the long rainy season (MAM) is normal the NDVI reaches its peak of ~0.20 in early March (dark-green curve in Figure 8). On the other hand, the occurrence of drought during the NDJ, MAM or both seasons impacts the development of the NDVI over time. If the NDJ rainy season is normal the NDVI will pass the curve of two normal seasons at first but will decline more quickly during a dry MAM season (yellow curve). A year with a dry NDJ but normal MAM seasons (blue curve) shows a delay in the development in the NDVI but reaches similar peak values as in a normal year. The peak values of NDVI are reached about 1.5–2 months later than in a normal year, before subsequently declining again. This either indicates that the vegetation recovery from the dry NDJ season requires some time, or that vegetation cover in a normal year reduces more quickly due to heavy grazing, which can start much earlier in a good rainfall year.

The duration of increased NDVI levels is similar if one of the rainy seasons is a dry season. However, if both seasons are classified as dry (purple curve), this time period is shorter. The purple curve shows that despite both seasons being classified as dry, the MAM season still has enough rainfall to enable vegetation growth, albeit not as good as during a normal or wet MAM season. The minimum amount of rainfall in the MAM season is ~150 mm (Figure 2B), and in the two years that comprise the purple curve in Figure 8 the MAM precipitation was 210 mm (2004) and 235 mm (2017). This explains the relatively good vegetation cover in the MAM rainy season that on average has 78 mm more precipitation than the NDJ season (Table 2).

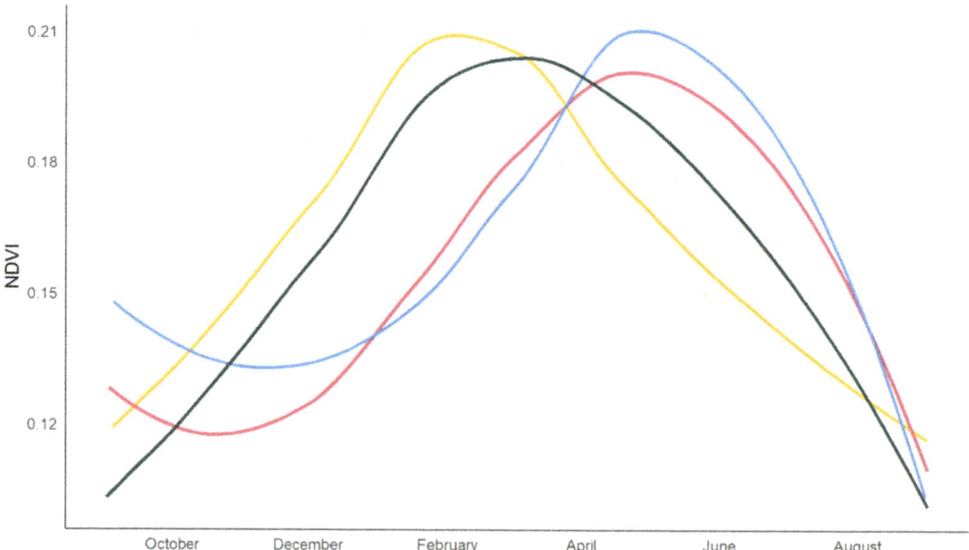

Figure 8. Seasonal NDVI response to different drought regimes between 1991 and 2020. The dark-green line represents a normal season, both during the NDJ and the MAM. The purple line represents two dry seasons, blue represents a dry NDJ followed by a normal MAM season and yellow represents a normal NDJ followed by a dry MAM season.

4. Discussion

4.1. Rainfall and Temperature

The total annual rainfall and the rainfall during the NDJ or MAM rain season in the study area did not change significantly ($\alpha = 0.05$) during the period of 1940–2020 (Table 3). These results contradict the results of earlier research which indicated decreasing East African rainfall due to lower MAM rainfall since the early 1980s [61–63]. The reason given for the declining rainfall is the rapid warming of the Indian Ocean, which leads to stronger convection and more rainfall over the Indian Ocean and less rainfall in East Africa. Our data for Monduli and Longido districts do not confirm those reported results, as the rainfall trends are all insignificant.

The rainfall data of the study area (Figure 2) are characterized by a high interannual variation in amounts of rainfall. As in other semi-arid regions, mean annual rainfall does not often occur; many years had much lower or much higher amounts of rainfall. This is also reflected by the seasonal amounts of rainfall. In most years the NDJ rainfall was below average, and in only a few years it was well above the average (e.g., 1962, 1998 and 2007), which leads to a positively skewed distribution (skew = 1.33). The MAM rainfall was less variable than the NDJ rainfall and more evenly distributed around the mean (skew = 0.56). The hydrological years (Sept–Aug) in which the amounts of rainfall were below average are usually caused by a lack of rainfall in one of the rainy seasons. During only 12 out of 80 years both rainy seasons were more than 25% below average. For the NDJ season this occurred in 28 of the 80 years, and for the MAM season this occurred in 21 years. For the rainfall in the hydrological years, the contribution of the MAM season varied from 24 to 76%, but on average it was 50%. The NDJ season contributed between 14 and 64% of rainfall to the hydrological year. On average this was 37%.

Unlike the amounts of rainfall, the temperature in the study area increased significantly ($\alpha = 0.05$) by 1.06 °C over the period of 1940–2020. A highly significant increasing trend was determined and can only be caused by global warming [64]. According to [65] the warming in East Africa started in the early 1980s, but our data series (Figure 3) shows that a more or less steady increase in temperature had already started since 1940. Only the 1960s were relatively cool, but since then the increase in temperature was again steady and approximately 0.12 °C per decade. This warming may have resulted in a more erratic rainfall pattern with higher rainfall intensities due to the stronger convection [66]. However, the rainfall data used in the study do not provide any information on the rainfall character, and thus it cannot be confirmed that the rainfall has actually become more extreme.

4.2. Drought

Occurrence of drought in the study area was analyzed using the SPEI-3 (short-term droughts) and SPEI-12 (long-term droughts). Both SPEI time series (Figure 4) show a significant ($\alpha = 0.05$) decreasing trend in SPEI values, indicating that drought has become more serious recently than it was in the past in the Monduli and Longido districts. As no significant changes in rainfall occurred, the enhanced drought can only be the result of the warming of the area. Increasing temperature will result in a higher potential evapotranspiration [54], which will lead to stronger desiccation of the land, and thus more drought stress in semi-arid areas such as the Monduli and Longido districts. Since 1993, six long-term droughts (SPEI-12 < −1) have occurred, while in the 53 years prior to 1993 only three of such droughts occurred. A similar pattern can be observed for short-term droughts (SPEI-3 < −1). A notable period without much drought was the period from the late 1970s until the early 1990s (Figure 4 and Table 4).

It is difficult to compare our results with other studies, as study designs may be different in timescales, datasets, research periods or study areas [67]. The importance of the latter is underlined by [68,69], which studied the Greater Horn of Africa and obtained large spatiotemporal variations in trends in precipitation and temperature between 1980 and 2010 [68], and trends in SPEIs between 1964 and 2015 [69]. A recent study on the entire Lake Manyara catchment, including the Monduli and Longido districts, showed the presence of a drying trend over the past century [22]. Furthermore, a general increase in decadal drought characteristics (duration, severity and frequency) from the 1930s to present, with the exception of the wet 1980s, was reported by [22].

The observed increasing drought is in agreement with local people that state that it is drier and warmer nowadays compared to 25–30 years ago [25]. The drying trend may raise concerns for the future of the Monduli and Longido districts. Conway et al. [70] analyzed the results of 34 climate models that simulated future precipitation and temperatures in Tanzania. These results showed wide spatiotemporal variations within Tanzania regarding future precipitation. The results indicate that the number of rainy days will decrease and the that the intensity of events will increase. This suggests more variable rainfall with a higher chance of droughts or floods in the future. In contrast to rainfall, the climate models predict increasing temperatures between 0.8 and 1.8 °C for 2040 in addition to between 1.6 and 5.0 °C for 2090 (relative to the period of 1976–2005). The change is evenly distributed across Tanzania. Thus, while future changes in precipitation are uncertain, it can be stated that temperatures will continue to rise, which will further increase the drought risk in the study area due to higher potential evapotranspiration rates.

4.3. Vegetation Trends and Resilience

Based on the NDVI timeseries a significant ($\alpha = 0.05$) decline in vegetation cover was observed between 1981 and 2020 (Figure 5). The fitted trendline indicates that the average NDVI declined by 9.7%, from 0.175 in 1981 to 0.157 in 2020. This observed decline in vegetation cover is less dramatic than previously reported numbers in other studies on East Africa [1,23,71], which generally show 1.5 to 3 times more vegetation reduction than our results. It is not clear why those other studies come to these higher vegetation cover decline values for the same study area. One reason could be that our analysis is based on near-continuous NDVI values while most other studies take NDVI values from fewer moments in time. As clearly visible in Figure 5, the timing of the satellite imagery for NDVI calculation can result in rather different NDVI values. This is equally true for the year (dry versus wet) which is chosen and the timing of the image within that year (dry season versus wet season).

The decline in vegetation cover observed here could be the result of different causes. Parts of the study area have been converted from grassland into arable land, which on average has lower vegetation cover [23,24]. Additionally, vegetation degradation due to overgrazing leading to bare soil might play a role in the contemporary lower vegetation cover [23,25,32]. Finally, the increase in drought severeness and frequency (Figure 4) might also result in generally lower vegetation cover. Drought will not only reduce the amount of green vegetation but will also affect the condition of the growing plants, which is reflected by a lower NDVI [72]. Based on the available data and analyses, it is not possible to conclude which of these reasons is the most important, and it could well be that all reasons play a role in the declining vegetation in the study area. However, given the significant increase in drought occurrence in the study area (Figure 4), it is believed that drought is the main cause of declining vegetation cover.

Vegetation resilience of the study area can be characterized as high. When a year was dry the NDVI values were lower than in a normal year, meaning less vegetation cover, but the next year the vegetation always recovered to normal year values (Figure 6). The recovery of the vegetation following a drought year did not change over time (Figure 7). In the 2011–2020 decade the resilience was not different from the 1991–2000 and 2001–2010 decades. At the seasonal scale, vegetation resilience appeared to be good. A dry season resulted in lower vegetation cover, but the vegetation came back quickly during the next normal or wet season and reached similar NDVI values (Figure 8).

These results show that, despite a general decline in vegetation cover over the period of 1981–2020, vegetation resilience was still good in the Munduli and Longido districts. Drought had an immediate impact on vegetation cover, but once rains came back at normal or above normal levels in a new season the vegetation quickly responded and returned to normal levels. However, what NDVI observations do not tell is possible changes in the species composition. Pressures on the grazing systems could be drought and overgrazing, which may lead to changes in the species that grow in the study area. Drought-tolerant species could replace other species, while continuous preferential grazing of grass species may result in the spread of less favorable plant species that are not eaten by the livestock, and therefore considered a negative change [73,74]. The Masai herders in the study did complain about the lower quality and less availability of grass resources in the area, and mentioned drought and high livestock numbers as the main causes for the decline [24,25].

5. Conclusions

This study used remote-sensing-based datasets of meteorology and vegetation cover to analyze vegetation resilience in the Monduli and Longido districts of North Tanzania. The results of meteorological analysis show that the amounts of rainfall in the Monduli and Longido districts did not change significantly during the study period (1940—2020), but that temperatures increased by 1.06 °C over the same period. The rising temperature resulted in higher potential evapotranspiration rates, which significantly increased drought occurrence and frequency. Since the early 1990s serious droughts became more frequent as well as longer, and it can be expected that in the future this trend will continue, given the climate projections for Tanzania.

Vegetation cover in the two districts declined significantly by 9.7% over the period of 1981–2020. This decline in cover could be due to several reasons, but the increase in drought most likely played an important role. Other reasons such as overgrazing by livestock, land use conversions and species changes may have played a role as well, which have all been indicated to occur according to local Masai herders. Despite the overall decline in vegetation cover and more severe drought conditions, the resilience of the vegetation was high. A drought year or season affected vegetation cover and health, as indicated by lower NDVI values, but the vegetation recovered quickly during the following rainy season when the amounts of rainfall were back to normal or above-normal levels.

Finally, it is concluded that despite the overall decline in vegetation cover, the changes have not been as dramatic as earlier reported, and vegetation resilience is still good in the study area. The climate change predictions for the area suggest a higher occurrence of drought, which could cause a further decline in vegetation cover. In addition, a shift in vegetation species to more drought-prone species could occur, which may lead to fewer grazing resources for the local Masai herdsmen.

Author Contributions: Conceptualization, S.L.V. and G.S.; methodology, S.L.V., T.K. and G.S.; formal analysis, S.L.V. and T.K.; resources, S.L.V., T.K., R.K. and J.W.; writing—original draft preparation, S.L.V., T.K. and G.S.; writing—review and editing, S.L.V., T.K., R.K., J.W. and G.S.; visualization, S.L.V. and T.K.; supervision, G.S. All authors have read and agreed to the published version of the manuscript.

Funding: This research received no external funding.

Institutional Review Board Statement: Not applicable.

Informed Consent Statement: Not applicable.

Data Availability Statement: The data presented in this study are openly available on GitHub/Zenodo at https://doi.org/10.5281/zenodo.4635298 accessed on 21 October 2021.

Conflicts of Interest: The authors declare no conflict of interest.

References

1. Zhu, Z.; Piao, S.; Myneni, R.B.; Huang, M.; Zeng, Z.; Canadell, J.G.; Ciais, P.; Sitch, S.; Friedlingstein, P.; Arneth, A.; et al. Greening of the Earth and its drivers. *Nat. Clim. Chang.* **2016**, *6*, 791–795. [CrossRef]
2. Newbold, T.; Hudson, L.N.; Hill, S.L.L.; Contu, S.; Lysenko, I.; Senior, R.A.; Börger, L.; Bennett, D.J.; Choimes, A.; Collen, B.; et al. Global effects of land use on local terrestrial biodiversity. *Nature* **2015**, *520*, 45–50. [CrossRef]

3. Kiunsi, R.B.; Meadows, M.E. Assessing land degradation in the Monduli District, northern Tanzania. *Land Degrad. Dev.* **2006**, *17*, 509–525. [CrossRef]
4. IPBES. *The IPBES Assessment Report on Land Degradation and Restoration*; Montanarella, L., Scholes, R., Brainich, A., Eds.; Secretariat of the Intergovernmental Science-Policy Platform on Biodiversity and Ecosystem Services: Bonn, Germany, 2018. [CrossRef]
5. Tangud, T.; Nasahara, K.; Borjigin, H.; Bagan, H. Land-cover change in the Wulagai grassland, Inner Mongolia of China between 1986 and 2014 analysed using multi-temporal Landsat images. *Geocarto Int.* **2018**, *6049*, 1–15. [CrossRef]
6. Pullanikkatil, D.; Palamuleni, L.G.; Ruhiiga, T.M. Land use/land cover change and implications for ecosystems services in the Likangala River Catchment, Malawi. *Phys. Chem. Earth* **2016**, *93*, 96–103. [CrossRef]
7. Wickama, J.; Okoba, B.; Sterk, G. Effectiveness of sustainable land management measures in West Usambara highlands, Tanzania. *Catena* **2014**, *118*, 91–102. [CrossRef]
8. Millenium Ecosystem Assessment. *Ecosystems and Human Well-Being, Synthesis*; Island Press: Washington, DC, USA, 2005.
9. United Nations. *Convention to Combat Desertification in Those Countries Experiencing Serious Drought and/or Desertification, Particularly in Africa*; Paris, 14 October 1994; United Nations Treaty Series; United Nations: Paris, France, 1994; Volume 1954, p. 3. Available online: https://treaties.un.org/doc/Publication/MTDSG/Volume%20II/Chapter%20XXVII/XXVII-10.en.pdf (accessed on 21 October 2021).
10. Tal, A.; Cohen, J.A. Bringing Top-down to Bottom-up: A New Role for Environemtal Legislation in Combating Desertification. *Harv. Environ. Law Rev.* **2007**, *31*, 163–271.
11. Sheffield, J.; Wood, E.F. Projected changes in drought occurrence under future global warming from multi-model, multi-scenario, IPCC AR4 simulations. *Clim. Dyn.* **2008**, *31*, 79–105. [CrossRef]
12. Trenberth, K.E.; Dai, A.; van der Schrier, G.; Jones, P.D.; Barichivich, J.; Briffa, K.R.; Sheffield, J. Global warming and changes in drought. *Nat. Clim. Chang.* **2014**, *4*, 17–22. [CrossRef]
13. Churkina, G.; Running, S.W. Contrasting Climatic Controls on the Estimated Productivity of Global Terrestrial Biomes. *Ecosystems* **1998**, *1*, 206–215. [CrossRef]
14. *Drought Risk Reduction Framework and Practices: Contributing to the Implementation of the Hyogo Framework for Action*; UN Office for Disaster Risk Reduction: Geneva, Switzerland, 2009.
15. Busby, J.W.; Smith, T.G.; Krishnan, N. Climate security vulnerability in Africa mapping 3.01. *Polit. Geogr.* **2014**, *43*, 51–67. [CrossRef]
16. Landmann, T.; Dubovyk, O. Spatial analysis of human-induced vegetation productivity decline over eastern Africa using a decade (2001–2011) of medium resolution MODIS time-series data. *Int. J. Appl. Earth Obs. Geoinf.* **2014**, *33*, 76–82. [CrossRef]
17. Wei, F.; Wang, S.; Fu, B.; Pan, N.; Feng, X.; Zhao, W.; Wang, C. Vegetation dynamic trends and the main drivers detected using the ensemble empirical mode decomposition method in East Africa. *Land Degrad. Dev.* **2018**, *29*, 2542–2553. [CrossRef]
18. De Jong, R.; Verbesselt, J.; Zeileis, A.; Schaepman, M.E. Shifts in global vegetation activity trends. *Remote Sens.* **2013**, *5*, 1117–1133. [CrossRef]
19. Wei, F.; Wang, S.; Fu, B.; Wang, L.; Liu, Y.Y.; Li, Y. African dryland ecosystem changes controlled by soil water. *Land Degrad. Dev.* **2019**, *30*, 1564–1573. [CrossRef]
20. Vicente-Serrano, S.M.; Gouveia, C.; Camarero, J.J.; Begueria, S.; Trigo, R.; Lopez-Moreno, J.I.; Azorin-Molina, C.; Pasho, E.; Lorenzo-Lacruz, J.; Revuelto, J.; et al. Response of vegetation to drought time-scales across global land biomes. *Proc. Natl. Acad. Sci. USA* **2013**, *110*, 52–57. [CrossRef]
21. Sterk, G.; Stoorvogel, J.J. Desertification–Scientific Versus Political Realities. *Land* **2020**, *9*, 156. [CrossRef]
22. Keijzer, T. Drought Analysis of the Lake Manyara Catchment: Meteorological Drought Occurence, Influence of Atmospheric Teleconnections and Impact on Lake Manyara. Master's Thesis, Utrecht University, Utrecht, The Netherlands, 2020.
23. Wynants, M.; Solomon, H.; Ndakidemi, P.; Blake, W.H. Pinpointing areas of increased soil erosion risk following land cover change in the Lake Manyara catchment, Tanzania. *Int. J. Appl. Earth Obs. Geoinf.* **2018**, *71*, 1–8. [CrossRef]
24. Verhoeve, S.L. Satellite Based Analysis of Environmental Changes in the Monduli and Longido Districts, Tanzania. Master's Thesis, Utrecht University, Utrecht, The Netherlands, 2019.
25. Van Den Bergh, H.A.J. The Impacts of Maasai Settlements on Land Cover, Meteorological Conditions and Wind Erosion Risk in Northern Tanzania. Master's Thesis, Utrecht University, Utrecht, The Netherlands, 2016.
26. Blake, W.H.; Rabinovich, A.; Wynants, M.; Kelly, C.; Nasseri, M.; Ngondya, I.; Patrick, A.; Mtei, K.; Munishi, L.; Boeckx, P.; et al. Soil erosion in East Africa: An interdisciplinary approach to realising pastoral land management change. *Environ. Res. Lett.* **2018**, *13*, 124014. [CrossRef]
27. UN OCHA ROSA Tanzania Administrative Level 0–3 Boundaries. Available online: https://data.humdata.org/dataset/tanzania-administrative-boundaries-level-1-to-3-regions-districts-and-wards-with-2012-population (accessed on 11 December 2018).
28. National Bureau of Statistics. *2012 Population and Housing Census*; Ministry of Finance: Dar es Salaam, Tanzania, 2013.
29. Deus, D.; Gloaguen, R.; Krause, P. Water balance modeling in a semi-arid environment with limited in situ data using remote sensing in lake manyara, east african rift, tanzania. *Remote Sens.* **2013**, *5*, 1651–1680. [CrossRef]
30. Prins, H.H.T. Nature conservation as an integral part of optimal land use in East Africa: The case of the Masai ecosystem of Northern Tanzania. *Biol. Conserv.* **1987**, *40*, 141–161. [CrossRef]
31. Quénéhervé, G.; Bachofer, F.; Maerker, M. Experimental assessment of runoff generation processes on hillslope scale in a semiarid region in Northern Tanzania. *Geogr. Fis. Din. Quat.* **2015**, *38*, 55–66. [CrossRef]

32. Conroy, A.B. *Maasai Oxen, Agriculture and Land Use Change in Monduli District, Tanzania*; University of New Hampshire: Durham, NH, USA, 2001.
33. *Tanzania Meteorological Authority, Daily Rainfall Data, Monthly Rainfall Data, Monthly Temperature Data*; Retrieved through personal communication; Tanzania Meteorology Agency: Arusha, Tanzania, 2019.
34. Butz, R.J. Traditional fire management: Historical fire regimes and land use change in pastoral East Africa. *Int. J. Wildl. Fire* **2009**, *18*, 442–450. [CrossRef]
35. Salvatori, V.; Egunyu, F.; Skidmore, A.K.; De Leeuw, J.; Van Gils, H.A.M. The effects of fire and grazing pressure on vegetation cover and small mammal populations in the Maasai Mara National Reserve. *Afr. J. Ecol.* **2001**, *39*, 200–204. [CrossRef]
36. Lillesand, T.; Kiefer, R.W.; Chipman, J. *Remote Sensing and Image Interpretation*; John Wiley & Sons: Hoboken, NJ, USA, 2015; ISBN 111834328X.
37. Lanorte, A.; Manzi, T.; Nolè, G.; Lasaponara, R. On the Use of the Principal Component Analysis (PCA) for Evaluating Vegetation Anomalies from LANDSAT-TM NDVI Temporal Series in the Basilicata Region (Italy). In *International Conference on Computational Science and Its Applications*; Gervasi, O., Murgante, B., Misra, S., Gavrilova, M.L., Rocha, A.M.A.C., Torre, C., Taniar, D., Apduhan, B.O., Eds.; Springer International Publishing: Cham, Switzerland, 2015; pp. 204–216.
38. Hawinkel, P.; Thiery, W.; Lhermitte, S.; Swinnen, E.; Verbist, B.; Van Orshoven, J.; Muys, B. Vegetation response to precipitation variability in East Africa controlled by biogeographical factors. *J. Geophys. Res. Biogeosciences* **2016**, *121*, 2422–2444. [CrossRef]
39. Vermote, E. NOAA CDR Program NOAA Climate Data Record (CDR) of AVHRR Normalized Difference Vegetation Index (NDVI), Version 5. Available online: https://doi.org/10.7289/V5ZG6QH9 (accessed on 22 September 2020). [CrossRef]
40. Gorelick, N.; Hancher, M.; Dixon, M.; Ilyushchenko, S.; Thau, D.; Moore, R. Google Earth Engine: Planetary-scale geospatial analysis for everyone. *Remote Sens. Environ.* **2017**, *202*, 18–27. [CrossRef]
41. Rojas, O.; Vrieling, A.; Rembold, F. Assessing drought probability for agricultural areas in Africa with coarse resolution remote sensing imagery. *Remote Sens. Environ.* **2011**, *115*, 343–352. [CrossRef]
42. Tucker, C.J.; Pinzon, J.E.; Brown, M.E.; Slayback, D.A.; Pak, E.W.; Mahoney, R.; Vermote, E.F.; El Saleous, N. An extended AVHRR 8-km NDVI dataset compatible with MODIS and SPOT vegetation NDVI data. *Int. J. Remote Sens.* **2005**, *26*, 4485–4498. [CrossRef]
43. Funk, C.; Nicholson, S.E.; Landsfeld, M.; Klotter, D.; Peterson, P.; Harrison, L. The Centennial Trends Greater Horn of Africa precipitation dataset. *Sci. Data* **2015**, *2*, 150050. [CrossRef]
44. Funk, C.; Peterson, P.; Landsfeld, M.; Pedreros, D.; Verdin, J.; Shukla, S.; Husak, G.; Rowland, J.; Harrison, L.; Hoell, A.; et al. The climate hazards infrared precipitation with stations—A new environmental record for monitoring extremes. *Sci. Data* **2015**, *2*, 150066. [CrossRef]
45. Harris, I.; Osborn, T.J.; Jones, P.; Lister, D. Version 4 of the CRU TS monthly high-resolution gridded multivariate climate dataset. *Sci. Data* **2020**, *7*, 1–18. [CrossRef]
46. Moulin, S.; Kergoat, L.; Viovy, N.; Dedieu, G. Global-Scale Assessment of Vegetation Phenology Using NOAA/AVHRR Satellite Measurements. *J. Clim.* **1997**, *10*, 1154–1170. [CrossRef]
47. Chamaille-Jammes, S.; Fritz, H.; Murindagomo, F. Spatial patterns of the NDVI-rainfall relationship at the seasonal and interannual time scales in an African savanna. *Int. J. Remote Sens.* **2006**, *27*, 5185–5200. [CrossRef]
48. Forkel, M.; Carvalhais, N.; Verbesselt, J.; Mahecha, M.D.; Neigh, C.S.R.; Reichstein, M. Trend Change Detection in NDVI Time Series: Effects of Inter-Annual Variability and Methodology. *Remote Sens.* **2013**, *5*, 2113–2144. [CrossRef]
49. Mbatha, N.; Xulu, S. Time Series Analysis of MODIS-Derived NDVI for the Hluhluwe-Imfolozi Park, South Africa: Impact of Recent Intense Drought. *Climate* **2018**, *6*, 95. [CrossRef]
50. Teferi, E.; Uhlenbrook, S.; Bewket, W. Inter-annual and seasonal trends of vegetation condition in the Upper Blue Nile (Abay) Basin: Dual-scale time series analysis. *Earth Syst. Dyn.* **2015**, *6*, 617–636. [CrossRef]
51. Sen, P.K. Estimates of the Regression Coefficient Based on Kendall's Tau. *J. Am. Stat. Assoc.* **1968**, *63*, 1379–1389. [CrossRef]
52. McKee, T.; Doesken, N.; Kleist, J. The relationship of drought frequency and duration to time scales. In Proceedings of the 8th Conference on Applied Climatology, Anaheim, CA, USA, 17–22 January 1993. [CrossRef]
53. Vicente-Serrano, S.M.; Beguería, S.; López-Moreno, J.I. A Multiscalar Drought Index Sensitive to Global Warming: The Standardized Precipitation Evapotranspiration Index. *J. Clim.* **2010**, *23*, 1696–1718. [CrossRef]
54. Thornthwaite, C.W. An Approach toward a Rational Classification of Climate. *Geogr. Rev.* **1948**, *38*, 55. [CrossRef]
55. Maggioni, V.; Massari, C. *Extreme Hydroclimatic Events and Multivariate Hazards in a Changing Environment: A Remote Sensing Approach*; Elsevier: Amsterdam, The Netherlands, 2019; ISBN 0128149000.
56. Masud, M.B.; Khaliq, M.N.; Wheater, H.S. Analysis of meteorological droughts for the Saskatchewan River Basin using univariate and bivariate approaches. *J. Hydrol.* **2015**, *522*, 452–466. [CrossRef]
57. Liu, Y.; Zhou, Y.; Ju, W.; Wang, S.; Wu, X.; He, M.; Zhu, G. Impacts of droughts on carbon sequestration by China's terrestrial ecosystems from 2000 to 2011. *Biogeosciences* **2014**, *11*, 2583–2599. [CrossRef]
58. Easdale, M.H.; Fariña, C.; Hara, S.; Pérez León, N.; Umaña, F.; Tittonell, P.; Bruzzone, O. Trend-cycles of vegetation dynamics as a tool for land degradation assessment and monitoring. *Ecol. Indic.* **2019**, *107*, 105545. [CrossRef]
59. Borgogno, F.; D'Odorico, P.; Laio, F.; Ridolfi, L. Effect of rainfall interannual variability on the stability and resilience of dryland plant ecosystems. *Water Resour. Res.* **2007**, *43*. [CrossRef]
60. Van Loon, A.F. Hydrological drought explained. *Wiley Interdiscip. Rev. Water* **2015**, *2*, 359–392. [CrossRef]

61. Funk, C.; Dettinger, M.D.; Michaelsen, J.C.; Verdin, J.P.; Brown, M.E.; Barlow, M.; Hoell, A. Warming of the Indian Ocean threatens eastern and southern African food security but could be mitigated by agricultural development. *Proc. Natl. Acad. Sci. USA* **2008**, *105*, 11081–11086. [CrossRef]
62. Williams, A.P.; Funk, C. A westward extension of the warm pool leads to a westward extension of the Walker circulation, drying eastern Africa. *Clim. Dyn.* **2011**, *37*, 2417–2435. [CrossRef]
63. Lyon, B.; DeWitt, D.G. A recent and abrupt decline in the East African long rains. *Geophys. Res. Lett.* **2012**, *39*. [CrossRef]
64. Niang, I.; Ruppel, O.C.; Abdrabo, M.A.; Essel, A.; Lennard, C.; Padgham, J.; Urquhart, P. Africa. In *Climate Change 2014: Impacts, Adaptation and Vulnerability. Part B: Regional Aspects. Contribution of the Working Group II to the IPCC Fifth Assessment Report*; Barros, V.R., Field, C.B., Dokken, D.J., Mastrandrea, M.D., Mach, K.J., Bilir, T.E., Chatterjee, M., Ebi, K.L., Estrada, Y., Genova, R.C., Eds.; Cambridge University Press: Cambridge, UK; New York, NY, USA, 2014; pp. 1199–1265.
65. Anyah, R.O.; Qiu, W. Characteristic 20th and 21st century precipitation and temperature patterns and changes over the Greater Horn of Africa. *Int. J. Climatol.* **2012**, *32*, 347–363. [CrossRef]
66. Shongwe, M.E.; van Oldenborgh, G.J.; van den Hurk, B.; van Aalst, M. Projected Changes in Mean and Extreme Precipitation in Africa under Global Warming. Part II: East Africa. *J. Clim.* **2011**, *24*, 3718–3733. [CrossRef]
67. Nicholson, S.E. Climate and climatic variability of rainfall over eastern Africa. *Rev. Geophys.* **2017**, *55*, 590–635. [CrossRef]
68. Gebrechorkos, S.H.; Hülsmann, S.; Bernhofer, C. Analysis of climate variability and droughts in East Africa using high-resolution climate data products. *Glob. Planet. Chang.* **2020**, *186*, 103130. [CrossRef]
69. Haile, G.G.; Tang, Q.; Leng, G.; Jia, G.; Wang, J.; Cai, D.; Sun, S.; Baniya, B.; Zhang, Q. Long-term spatiotemporal variation of drought patterns over the Greater Horn of Africa. *Sci. Total Environ.* **2020**, *704*, 135299. [CrossRef] [PubMed]
70. Conway, D.; Mittal, N.; Vincent, K. Future Climate Projections for Tanzania. Available online: https://www.africaportal.org/publications/future-climate-projections-tanzania/ (accessed on 10 June 2020).
71. Kalisa, W.; Igbawua, T.; Henchiri, M.; Ali, S.; Zhang, S.; Bai, Y.; Zhang, J. Assessment of climate impact on vegetation dynamics over East Africa from 1982 to 2015. *Nat. Sci. Rep.* **2019**, *9*, 1–20. [CrossRef] [PubMed]
72. Bento, V.A.; Gouveia, C.M.; DaCamara, C.C.; Trigo, I.F. A climatological assessment of drought impact on vegetation health index. *Agric. For. Meteorol.* **2018**, *259*, 286–295. [CrossRef]
73. Herrmann, S.M.; Tappan, G.G. Vegetation impoverishment despite greening: A case study from central Senegal. *J. Arid Environ.* **2013**, *90*, 55–66. [CrossRef]
74. Van Auken, O.W. Shrub Invasions of North American Semiarid Grasslands. *Annu. Rev. Ecol. Syst.* **2000**, *31*, 197–215. [CrossRef]

remote sensing

Article

Use of A MODIS Satellite-Based Aridity Index to Monitor Drought Conditions in Mongolia from 2001 to 2013

Reiji Kimura [1,*] and Masao Moriyama [2]

[1]   Arid Land Research Center, Tottori University, Tottori 680-0001, Japan
[2]   Graduate School of Engineering, Nagasaki University, Nagasaki 852-8521, Japan; matsu@nagasaki-u.ac.jp
*   Correspondence: rkimura@tottori-u.ac.jp; Tel.: +81-857-21-7031

Abstract: The 4D disasters (desertification, drought, dust, and *dzud*, a Mongolian term for severe winter weather) have recently been increasing in Mongolia, and their impacts on the livelihoods of humans has likewise increased. The combination of drought and *dzud* has caused the loss of livestock on which nomadic herdsmen depend for their well-being. Understanding the spatiotemporal patterns of drought and predicting drought conditions are important goals of scientific research in Mongolia. This study involved examining the trends of the normalized difference vegetation index (NDVI) and satellite-based aridity index (SbAI) to determine why the land surface of Mongolia has recently (2001–2013) become drier across a range of aridity indices (AIs). The main reasons were that the maximum NDVI ($NDVI_{max}$) was lower than the $NDVI_{max}$ typically found in other arid regions of the world, and the SbAI throughout the year was large (dry), although the SbAI in summer was comparatively small (wet). Under the current conditions, the capacity of the land surface to retain water throughout the year caused a large SbAI because rainfall in Mongolia is concentrated in the summer, and the conditions of grasslands reflect summer rainfall in addition to grazing pressure. We then proposed a method to monitor the land-surface dryness or drought using only satellite data. The correct identification of drought was higher for the SbAI. Drought is more strongly correlated with soil moisture anomalies, and thus the annual averaged SbAI might be appropriate for monitoring drought during seasons. Degraded land area, defined as annual $NDVI_{max} < 0.2$ and annual averaged $SbAI > 0.025$, has decreased. Degraded land area was large in the major drought years of Mongolia.

Keywords: aridity index; drought; land degradation; remote sensing; satellite-based aridity index

Citation: Kimura, R.; Moriyama, M. Use of A MODIS Satellite-Based Aridity Index to Monitor Drought Conditions in Mongolia from 2001 to 2013. *Remote Sens.* **2021**, *13*, 2561. https://doi.org/10.3390/rs13132561

Academic Editor: Elias Symeonakis

Received: 27 May 2021
Accepted: 28 June 2021
Published: 30 June 2021

Publisher's Note: MDPI stays neutral with regard to jurisdictional claims in published maps and institutional affiliations.

1. Introduction

In recent years, global warming has caused an increase in temperature and decrease in precipitation in drylands at high latitudes [1–4]. An increase in environmental stress associated with human activities concurrent with climate change may spread the damage caused by these three disasters [3,5].

In Mongolia, the damage caused by cold and snow is called "*dzud*" and is a natural disaster that can lead to significant livestock mortality and economic damage [6,7]. The authors in [7] have called desertification, drought, dust, and *dzud* the 4D-related hazards. The impact of *dzud* during the winter is strongly affected by the drought conditions (low pasture production) during the last summer. For example, *dzud* occurred from October 2009 to March 2010 due to the effect of drought during summer 2009 [7]. In Mongolia, about 30% of the workforce is engaged in raising livestock, and the 4D hazards, in addition to global warming, pose a serious threat to their livelihood. The authors in [8] have indicated that about 60% of the decline in vegetation in Mongolia from 1988 to 2008 can be attributed to decreases in rainfall and increases in temperature.

The Aridity Index (AI) is a useful metric of the dryness or wetness in arid regions. The AI is defined as the ratio of annual precipitation (Pr) to annual potential evaporation (Ep) and is a water-balance-based climatic index. The total area of arid regions, determined from meteorological data collected from 1951 to 1980, was 41% of the terrestrial land surface, including Antarctica [9,10]. The corresponding percentage was 37% from 1981 to 2010 [11], and [12] have estimated that percentage to have been 39.5% from 2001 to 2013. These estimates suggest that the total area of arid regions has not changed or even decreased slightly since 1950.

The most widely used method to estimate aridity is the Palmer Drought Severity Index (PDSI) [13], not the AI. The PDSI is calculated using meteorological data and takes into consideration runoff, water supply, and the water retention capacity of the soil [14]. Some studies have estimated the PDSI in Mongolia [15–17], and all of them have concluded that the PDSI decreased (i.e., drought became more severe) from 2000 to 2010. However, there are some disadvantages to use of the PDSI. For example, the PDSI does not take into account changes in the spatial distributions of soil, vegetation, and hydrological processes [18].

With high resolution and frequency, satellite data offer advantages in monitoring environmental conditions in arid regions [19,20]. For example, the Moderate Resolution Imaging Spectroradiometer (MODIS) and Copernicus Missions (specifically Sentinel 1 or 2) have provided data since 2000 and 2014, respectively. Since lengthy cloudless periods are common in arid regions, much of the MODIS or Sentinel data are usable for global analyses.

Some drought indices are based on remote sensing [21,22]. Spectral reflectance has been widely used to calculate indices like the Normalized Difference Vegetation Index (NDVI) and normalized difference water index (NDWI) because the calculation procedures are simple [23]. The authors in [24] indicated that NDVI performed best in assessing land degradation compared with other indices using spectral reflectance like a soil-adjusted vegetation index (SAVI). Additionally, they revealed that thermal indices using land surface temperature (LST) were identified as the most influential variable for land degradation assessment. The authors in [25] have also suggested that a thermal index that uses the difference of the land surface temperature (LST) between day and night is much more useful as an indicator of water deficit: MODIS has provided daytime and nighttime LST data observed over equivalent locations every day, which have enabled the calculation of a thermal index since 2000.

The authors in [26] proposed a satellite-based aridity index (SbAI) that uses the difference of the LST between day and night, and the SbAI has already been validated and applied [12,26–29]. For example, years of major droughts in China have corresponded to years in which large increases in degraded land area were identified [28]. In addition, a comparison of the SbAI with the AI, that is, within Turc space (which is based on the water balance concept indicated by water limited to energy-limited lines) identified 15 categories in five zones: a stable zone, a zone transitioning toward dryness, a zone transitioning toward wetness, a dry zone, and a moist zone [30]. The authors of [30] have shown that the actual aridity was intensifying in most of Mongolia, though the climatic AI ranged from arid to semi-arid. The authors in [31] have demonstrated the NDVI relationship with precipitation and temperature in semi-arid regions and showed that the majority of sites displayed seasonal reversal, associated with transitions from water-limited to energy-limited conditions during wet winters.

Considering these past findings, the goal of this study was to examine the trends of NDVI and SbAI to determine why the land surface of Mongolia has recently become drier while the AI has ranged from arid to semi-arid and to propose a method to monitor the land-surface dryness or drought directly, using only satellite data.

2. Methods

2.1. Target Area and Analysis Period

Mongolia is a landlocked country surrounded by Russia, China, and Kazakhstan. The total land area is 1564,116 km^2. Mongolia is surrounded by high mountains and is located on a highland over 1500 m above sea level. The country has four distinct seasons, large temperature variations and little rainfall. The climate changes widely, not only due to differences in altitude but also in latitude. The annual mean temperature is between −8 °C and 6 °C, and the annual mean precipitation is between 50 mm and 400 mm [6].

The AI, SbAI, and NDVI in Mongolia were calculated for latitudes of 41–53° N and longitudes of 87–120° E (Figure 1). With the exception of the woodlands in northern Mongolia, most of the land surface in Mongolia has been classified as typical grasslands and bare soil, including the Gobi Desert in southern Mongolia [32]. The red circles in Figure 1 indicate the SYNOP (surface synoptic observations) meteorological stations —Ulaanbaatar, Mandalgovi, and Tsogt-Ovoo from north to south—which are located in grassland, the boundary between grassland and bare soil, and bare soil, respectively.

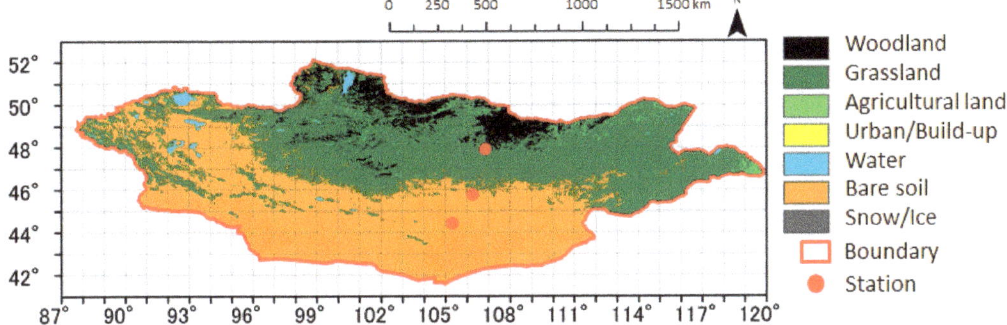

Figure 1. Land-use classification in Mongolia. Dots indicate the SYNOP meteorological stations (Ulaanbaatar, Mandalgovi, and Tsogt-Ovoo from north to south).

The time interval used to analyze the relationship between the AI and SbAI was 2001–2013. This period was chosen because precipitation data from the Global Precipitation Climatology Center's (GPCC) full data product (V7) are available throughout that time [12]. The GPCC has calculated precipitation for all global land areas during the target period through objective analysis of climatological average rainfall at the rain-gauge stations in its database.

The goal of this study is an examination of the trends of the NDVI and SbAI to determine why the land surface of Mongolia has recently (2001–2013) become drier. However, to establish the annual changes before and after 2001–2013, we calculated the NDVI from 1981 to 2000 using Advanced Very-High-Resolution Radiometer (AVHRR) data and from 2001 to 2020 using Moderate-Resolution Imaging Spectroradiometer (MODIS) data. We calculated the SbAI from 2001 to 2020 using MODIS data.

2.2. Data

The analysis of data regarding the relationship between the AI and SbAI from 2001 to 2013 was taken from [30], with a horizontal resolution of approximately 1° in both longitude and latitude.

We calculated the daily SbAI and NDVI from 2001 to 2020 using the Terra/MODIS data products MOD09CMG and MOD11C1 for surface reflectance and land surface temperature (LST) (https://modis-land.gsfc.nasa.gov/MODLAND_grid.html) (accessed on 27 May 2021). The spatial resolution of these two products was the same (0.05°). We used the "Collection 6" land product subsets web service to access and download the data [33].

We calculated daily NDVIs from 1981 to 2000 using the AVHRR surface reflectance (Channels 1 and 2) with 0.05° resolution (https://www.ncdc.noaa.gov/cdr/terrestrial/avhrr-surface-reflectance) (accessed on 27 May 2021). For long-term continuity of the NDVIs from the AVHRR and MODIS, we compared these two NDVIs in 2000 and obtained the following relationship with root mean squared errors (RMSE) = 0.09 (Figure 2):

$$\text{NDVI}_{\text{MODIS}} = 0.998 * \text{NDVI}_{\text{AVHRR}} + 1.975 * 10^{-5} \tag{1}$$

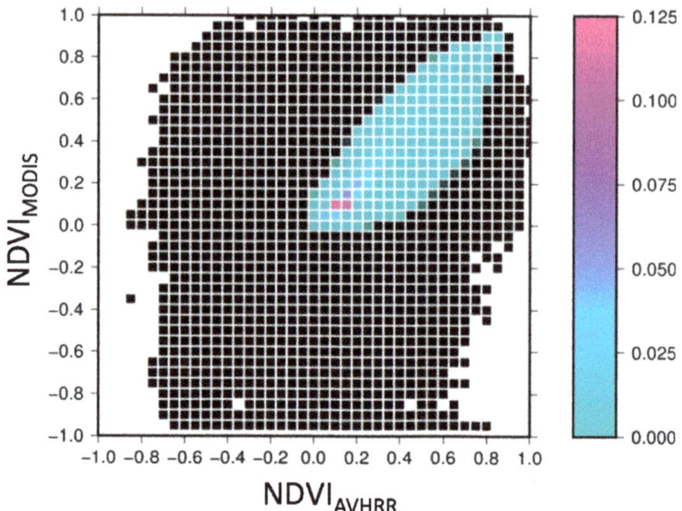

Figure 2. Relationship between the MODIS-NDVI and AVHRR-NDVI. Difference in color indicates the relative frequency.

In this study, Equation (1) was used to correct the NDVI$_{\text{AVHRR}}$ from 1981 to 2000.

A global land-cover map (GLCM) was used to characterize the distribution of land use in Mongolia (https://db.cger.nies.go.jp/dataset/landuse/en/) (accessed on 27 May 2021) (Figure 1). The GLCM is a raster image of Earth with a latitude-longitude resolution of 30 s, and it assigns the land cover of Earth into seven categories [34].

Annual rainfall data from 2001 to 2020 at Ulaanbaatar, Mandalgovi, and Tsogt-Ovoo (Figure 1) were downloaded from the Japan Meteorological Agency (JMA) website (http://www.data.jma.go.jp/gmd/cpd/monitor/climatview/frame.php) (accessed on 27 May 2021). When data were missing, the Climatic Resolution Unit Time Series monthly high-resolution gridded climate dataset was used to fill the gap (https://crudata.uea.ac.uk/cru/data/hrg/) (accessed on 27 May 2021).

2.3. Analytical Methods

The SbAI can be physically interpreted as a metric of the reciprocal heat capacity, which can be estimated from the ratio of the amplitude of the difference of the land surface temperature (LST) between day and night to the incident net solar radiation [26]. For a dry surface, the SbAI is large because the difference of the LST between day and night (ΔT_s) is large. The ΔT_s is large because the low water content of the land surface causes its heat capacity to be low. For the analysis in this study, we used an annual average SbAI, which we calculated by averaging daily SbAIs, because the AI that we used to examine the relationship between the AI and SbAI was also an annual average [27].

In the analysis, we used the maximum NDVI in each year, because the amounts of vegetation were low, and information from NDVIs is lost in arid regions like Mongolia when annual averages are used [27,35]. In this study, the maximum NDVI occurred in August, and we used that NDVI as a metric of the potential for vegetation growth because the vegetation was strongly affected by the amount of precipitation prior to August (85% of annual rainfall in Mongolia occurs from April to July) [8,36–38].

The authors in [12,30] have used the SbAI to classify arid regions according to their actual degree of aridity (Table 1).

Table 1. Classification of arid regions using the SbAI and AI.

Class	Range of SbAI	Range of AI
Hyper arid (HAr)	SbAI > 0.025	AI < 0.05
Arid (Ar)	$0.022 \leq$ SbAI ≤ 0.025	$0.05 \leq$ AI < 0.2
Semi-arid (SAr)	$0.017 \leq$ SbAI < 0.022	$0.2 \leq$ AI < 0.5
Dry sub-humid (DSH)	$0.015 \leq$ SbAI < 0.017	$0.5 \leq$ AI < 0.65

The range of the AI in each region, as defined by [9–11,39], is given in parentheses. The authors in [30] have subdivided these regions into 15 categories based on the SbAI and AI values of the four dryland regions (indicated by the double-headed red arrows along the Y and X axes) listed above (Figure 3). The stable zone (green), which includes points classified into the same dryland region by both the SbAI and AI, comprises categories 1, 2, 3, and 4. The transition zone, in which dryness is increasing (red), comprises categories 5 and 6, and the transition zone, in which wetness is increasing (blue), comprises categories 7, 8, and 9. In zones 10, 11, and 12, the magnitude of dryness is increasing in the dry zone (SbAI > 0.025), and zones 13, 14, and 15 are becoming wetter (SbAI < 0.015) (Figure 3).

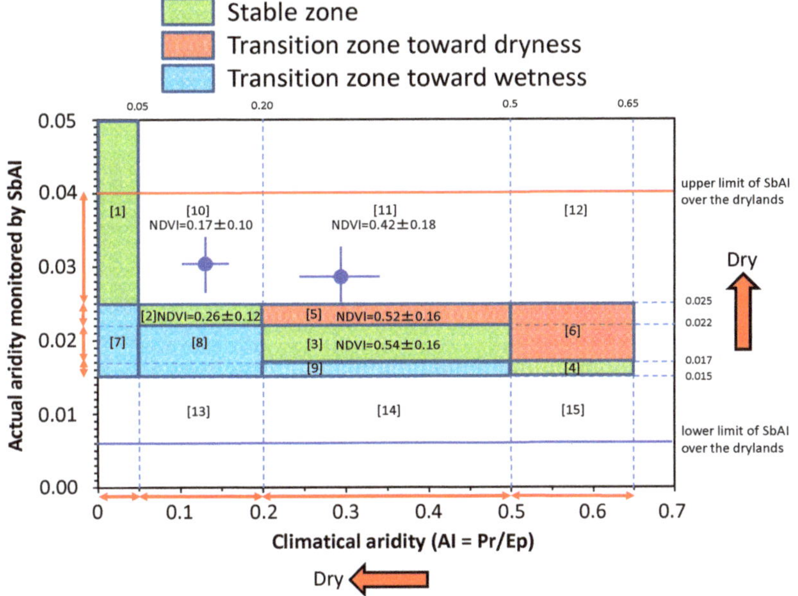

Figure 3. Relationship between the AI and SbAI, averaged over 2001–2013, and the 15 arid region categories. The red, double-headed arrows along the axes indicate the range of values of the indices that indicate hyper-arid, arid, semi-arid, and dry sub-humid regions. The blue dots with standard deviations indicate the actual ranges of the AIs and SbAIs in Mongolia (modified from [30]).

The authors in [27] examined the global distribution of annual maximum NDVI < 0.2 and annual averaged SbAI > 0.025 and defined areas that meet both these criteria as degraded land, which includes existing desert and land with both permanent and temporal dust erodibility. We examined yearly variation of degraded land area in Mongolia between 2001 and 2020 and discussed it in relation to drought.

3. Results

3.1. Distribution of Averaged AI in Mongolia from 2001 to 2013

The distribution of the averaged AI from 2001 to 2013 indicated that the northern region of Mongolia was dry sub-humid (DSH), the north central region was semi-arid (SAr), and the south region was arid (Ar) (Figure 4). A latitude of 47° N is the boundary between the SAr and Ar. The distribution of the AI corresponded to the distribution of land classification: northern Mongolia is woodland, north-central Mongolia is grassland, and southern Mongolia is bare soil (Figure 1).

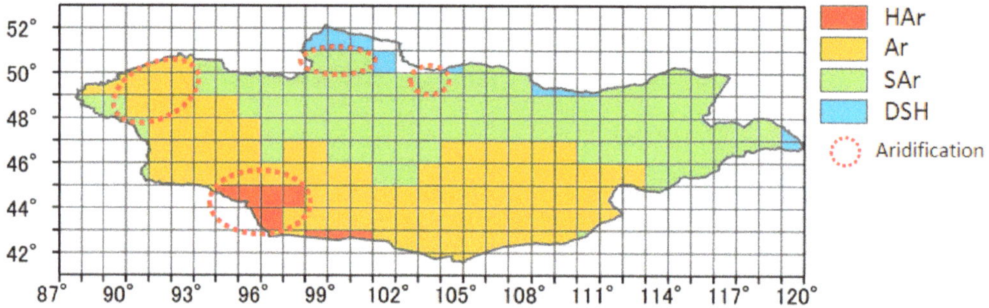

Figure 4. Spatial distribution of annual AI averaged during 2001–2013 in Mongolia with a resolution of 1° latitude × 1° longitude.

However, when the SbAIs were plotted against the AIs (blue dots with standard deviations in Figure 3), the areas that should have been in zones 2 and 3 were in zones 10 and 11 (Figure 5). This result indicates that the actual aridity in most of Mongolia was more severe than the climatic aridity. The authors in [40] have published a map of the distribution of aridity in Mongolia that shows a distribution of extremely strong to strong aridity, and middle to weak aridity similar to zones 10 and 11, respectively.

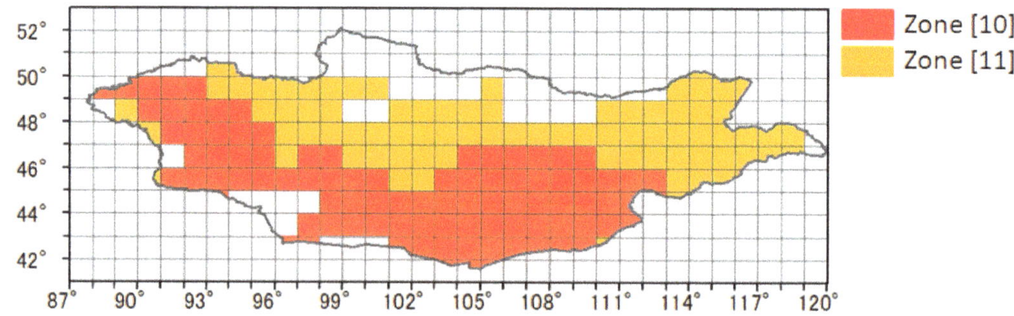

Figure 5. Spatial distribution of dry zone categories 10 and 11 during 2001–2013 in Mongolia with a resolution of 1° latitude × 1° longitude.

The authors in [30] have indicated that the yearly maximum of the NDVI ($NDVI_{max}$) in zones above the stable zone in Figure 3 has decreased, and the $NDVI_{max}$ in zones below the stable zone in Figure 3 has increased. The values of the $NDVI_{max}$ are therefore strongly related to the differences between the 15 zones in Figure 3. The values of the $NDVI_{max}$ in zones 10, 11, 2, 3, and 5 are shown in Figure 3, and the distribution of the annual $NDVI_{max}$, averaged over 2001 to 2013, is shown in Figure 6. The ranges of the $NDVI_{max}$ were 0.20 ± 0.11 and 0.51 ± 0.16 in zones 10 and 11, respectively (Figure 6), and were nearly consistent with the ranges of the $NDVI_{max}$ indicated in zones 10 and 5. That is, the smaller amount of vegetation may be one of the reasons why the land surface of Mongolia became drier from 2001 to 2013 across a range of AIs.

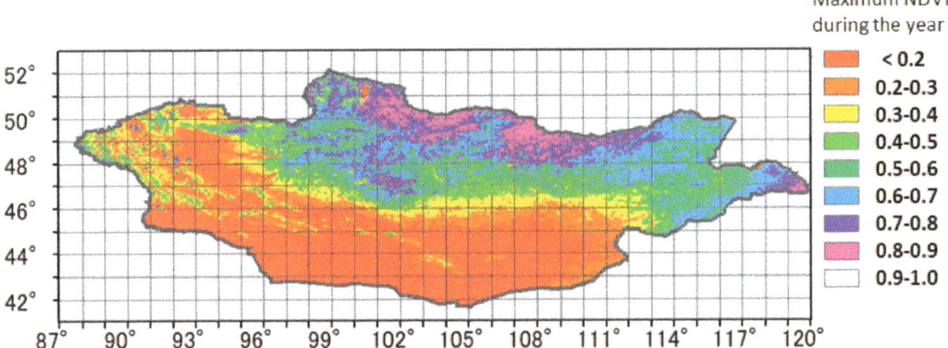

Figure 6. Spatial distribution of NDVI$_{max}$ values averaged during 2001–2013 in Mongolia with a resolution of 0.05° latitude × 0.05° longitude.

These results suggest the following characterizations of zones 10 and 11:

- Zone 10 is climatically an Ar region. From 2001 to 2013, however, this zone had less vegetation and was similar to an HAr region.
- Zone 11 is climatically a SAr region. The amount of vegetation in the summer is moderate. However, the actual water retention throughout the year in zone 11 inferred from its SbAI value was similar to that of an HAr region.

In Sections 3.2 and 3.3, we examine why the land surface of Mongolia became drier from 2001 to 2013 across a range of AIs by examining the trends of the NDVIs and SbAIs.

3.2. Difference of Climatic Conditions Using AI

We compared the AI distribution in Figure 4 to that from 1981–2010, calculated by [41], to identify the effects of climate change. Although a climatic trend toward aridity was apparent in some places (red circles), the distribution of AIs did not generally change (Figure 4).

Many studies have addressed trends of precipitation in Mongolia [7,8,17,42]. The authors in [43] examined the trend of annual precipitation from 1982 to 2010; they found that precipitation over Mongolia had been decreasing since 1993 (the trend was especially strong in northern and central Mongolia [17]) and the annual rainfall during 1994–2010 was about 30 mm lower than during 1982–1993. The authors in [7,8] found similar results. A decrease in annual rainfall by 30 mm has little effect on the classification of climates based on AI in Ar and SAr regions because the ranges of AIs in those categories are large: 0.05–0.2 and 0.2–0.5, respectively. The AI value itself may be reduced because of a decrease in rainfall and enhanced potential evaporation related to warmer temperatures [44]. The implication is that climatic effects are not revealed by a map of the distribution of AI values that reflect drier land surfaces in Mongolia because the range of AI values is large.

Monitoring the amount of vegetation will be an effective way to examine the effect of a decrease in rainfall by 30 mm over Mongolia. We therefore examined the trend of the NDVI$_{max}$ in August over Mongolia from 1981 to 2020 (Figure 7) because the NDVI$_{max}$ in August is sensitive to the amount of rainfall during the previous season [8,36,38]. In Mongolia, 85% of the annual rainfall is from April to July [37]. Figure 7 shows that the NDVI$_{max}$ decreased from 1994 to 2009. As previously mentioned, the annual rainfall during 1994–2010 decreased by about 30 mm compared with 1982–1993. The decreasing trend of the NDVI$_{max}$ up to 2009 was presumably due to decreased precipitation. However, the peak value of the NDVI$_{max}$ was only 0.39 in 1994, less than the average value of 0.4 in zones 2 and 3.

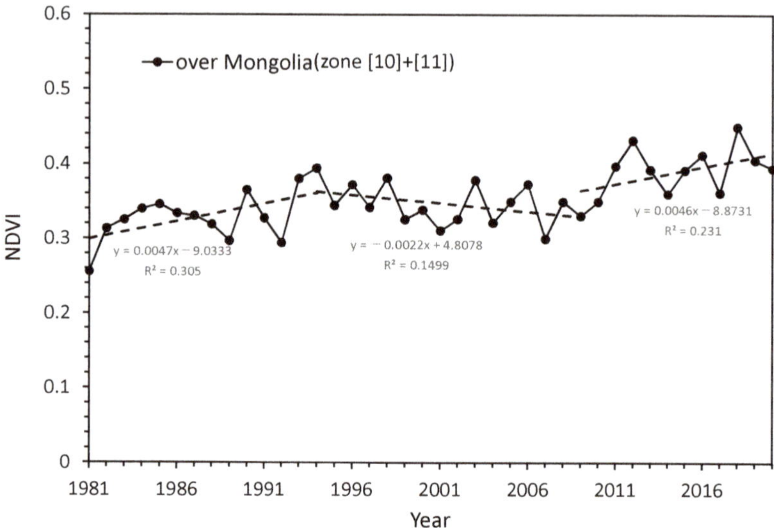

Figure 7. Annual changes (1981–2020) of the $NDVI_{max}$ in August over Mongolia. Dashed lines show the trends from 1981–1994, 1994–2009, and 2009–2020.

In contrast, an increasing trend of the $NDVI_{max}$ after 2009 can be found clearly. Since there have been few analyses of precipitation trends from 2001 to 2020 over Mongolia ([45] from 2000 to 2017; [23] from 2000 to 2016), we reexamined those trends at Ulaanbaatar, Mandalgovi, and Tsogt-Ovoo (Figure 8). At all three locations, annual rainfall increased after 2009, and those trends corresponded to an increase of the $NDVI_{max}$, which reached an averaged value of 0.4 in zones 2 and 3 (Figure 7). Based on an analysis of precipitation anomalies, the authors in [23] and [45] have also indicated that Mongolia became wetter from 2009 to 2017 compared with 2000–2008.

3.3. Trends of $NDVI_{max}$ and SbAI in Zones 10 and 11 during 2001–2020

We examined in detail the water retention of the land surface using the trends of the SbAI (averaged value in August and annual averaged value) and the $NDVI_{max}$ in August. In zone 10, the $NDVI_{max}$ decreased by a small amount from 2001 to 2009, and it increased very obviously after 2009 (Figure 9a). The average SbAI in August varied inversely with the $NDVI_{max}$, and the correlation between the two was high ($R^2 = 0.64$, $p < 0.001$). The averaged $NDVI_{max}$ from 2001 to 2010 was less than the limiting value of 0.2, below which land is considered to be degraded [27], but it increased after 2009. However, the $NDVI_{max}$ was smaller than the average value of 0.26 in Ar regions (zone 2). When the $NDVI_{max}$ exceeded 0.24, the averaged SbAI in August was lower than the limiting value of 0.025, below which land is considered to be degraded [27], but it was higher than 0.025 in many years. The annual averaged SbAI exceeded 0.03 in many years, and thus the environment of zone 10 appeared to be stressed in terms of water retention by the land surface through the year.

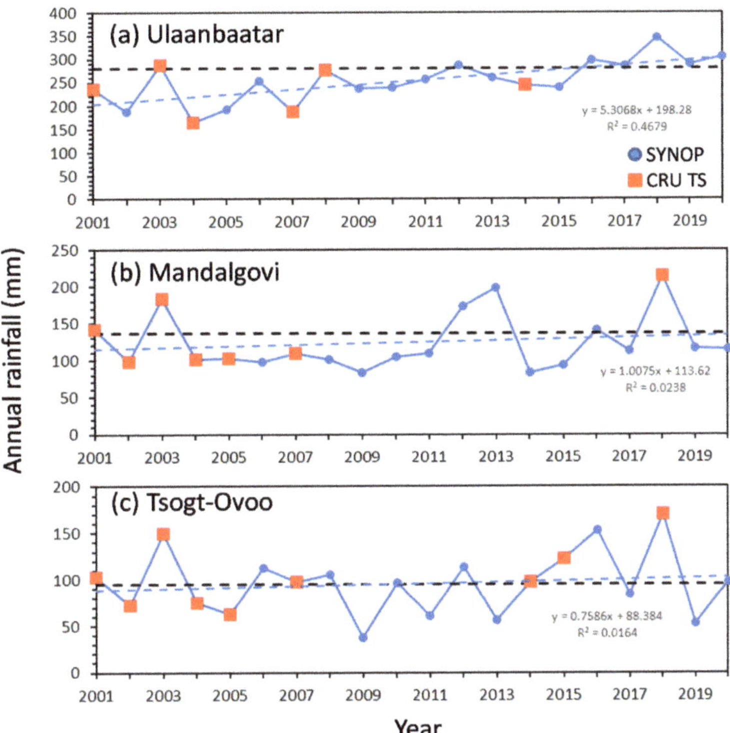

Figure 8. Annual changes (2001–2020) of annual rainfall in (**a**) Ulaanbaatar, (**b**) Mandalgovi, and (**c**) Tsogt-Ovoo. Black and blue dashed lines represent the normal values and trends during 1981–2010.

In zone 11, the $NDVI_{max}$ increased by a small amount from 2001 to 2009, and it increased very obviously after 2009 (Figure 9b). The averaged SbAI in August varied inversely with the $NDVI_{max}$, and the correlation between the two was high ($R^2 = 0.54$, $p < 0.001$). The $NDVI_{max}$ from 2001 to 2010 was lower than the general value of 0.54 in the SAr regions (zone 3), but it increased after 2009. The averaged SbAI in August from 2001 to 2009 was slightly lower than the limiting value of 0.025, below which land is considered to be degraded. However, the averaged SbAI in August after 2009 was substantially lower than 0.025, and it was even lower than 0.022, which is the upper bound for classification of a region as SAr (Figure 3). Although summers became wetter after 2009, water retention throughout the year was still low, because the annual averaged SbAI was 0.025–0.03.

The annual averaged SbAI was not necessarily correlated with the $NDVI_{max}$ ($R^2 = 0.31$, $p < 0.05$) (Figure 9b). For example, the $NDVI_{max}$ was lower in 2017 than in 2019, but the annual averaged SbAI was lower (wet) in 2017. There were similar relationships between the $NDVI_{max}$ and SbAI in 2010 and 2011. The authors in [45] indicated that water storage after summer in 2010 was higher than in 2011. It is inferred that precipitation in seasons other than summer affected water retention throughout the year. Numerical simulation results have shown that annual rainfall, especially rainfall during the winter, will increase over Mongolia from 2016 to 2035 and from 2081 to 2100 [37]. At the present time, it cannot definitively be concluded that the increasing trend of annual rainfall since 2009 (Figure 8) is in agreement with the simulation results. If the amount of precipitation increases enough that the annual averaged SbAI in zones 10 and 11 decreases to 0.025 and 0.022, respectively, the aridity in zones 10 and 11 will be close to climatically stable conditions.

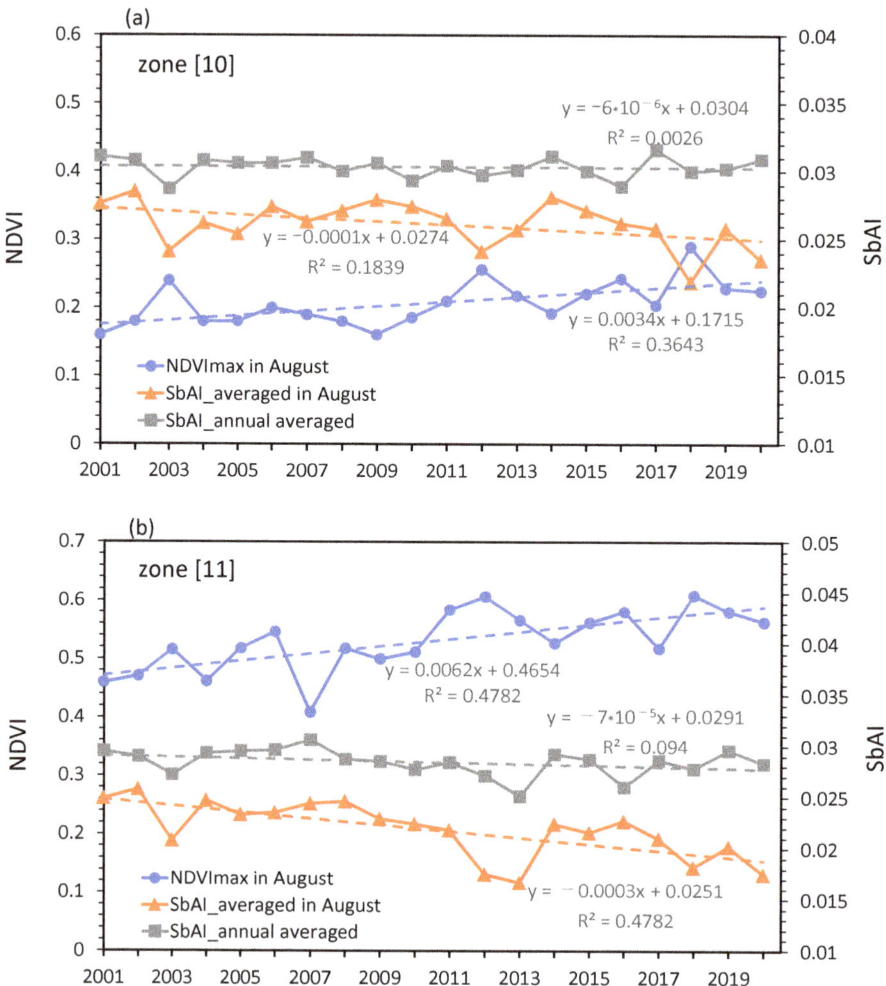

Figure 9. Annual changes (2001–2020) of the $NDVI_{max}$ in August, averaged SbAI in August, and annual averaged SbAI. (**a**) zone 10, (**b**) zone 11. Dashed lines show the trends of respective indices.

The annual averaged SbAI during the years shown in Figure 9b is the averaged value in zone 11, and thus there were wet regions with SbAIs below 0.022. For example, the distribution of annual averaged SbAIs in 2013 (the lowest SbAIs from 2001 to 2020) revealed wet regions with SbAIs below 0.022 (colored orange in Figure 10). The authors in [46] have indicated that although a large proportion of Mongolia's rangelands are not providing their potential ecosystem services, few have crossed an irreversible threshold of ecological change caused by current levels of grazing pressure. For the sustainable development of stock farming, continuous monitoring should be conducted to conserve the relatively wet regions (colored orange in Figure 10) and to prevent land degradation in nearly degraded regions (colored green in Figure 10).

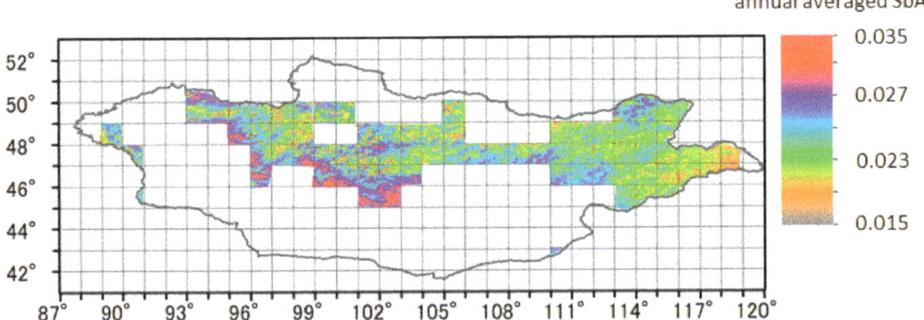

Figure 10. Spatial distribution of annual averaged SbAI values in zone 11 for 2003.

3.4. Detection of Drought Using SbAI

According to PDSI values, drought occurred frequently in all parts of Mongolia from 2000 to 2013 [7,16,37]. Since occurrences of *dzud* during the winter were strongly affected by drought conditions (low pasture production) during the preceding summer, understanding and predicting the characteristics of drought are of particular concern in Mongolia [15,37]. The authors in [15] have assessed drought frequency, duration, and severity over Mongolia from 2000 to 2010 using the PDSI and the standardized precipitation index (SPI). They have shown that droughts occurred in 2000, 2001, 2002, 2004, 2006, 2007, 2008, and 2009. Droughts therefore occurred in most years from 2000 to 2009 (red arrows in Figure 11). Figure 11 shows the yearly change in the $NDVI_{max}$, the averaged SbAI in August, and the annual averaged SbAI over Mongolia; the broken lines are the average values for the drought years. Both of the SbAI values equaled or exceeded these broken line averages during 2001–2009, but they have fallen below the broken lines in many years since 2009.

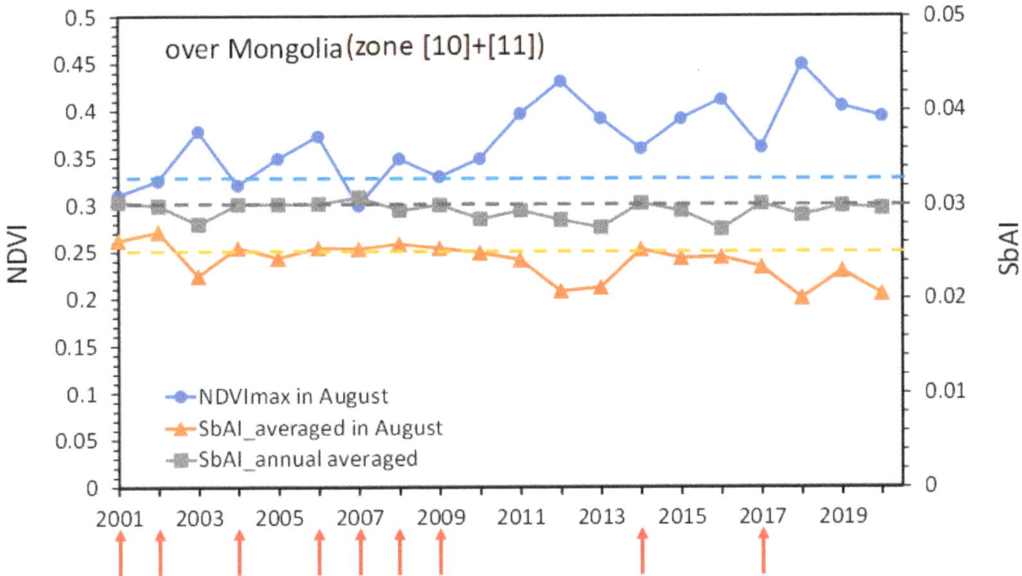

Figure 11. Annual changes (2001–2020) in the $NDVI_{max}$ in August, averaged SbAI in August, and annual averaged SbAI over Mongolia. Red arrows indicate drought years.

Drought years can be simply detected as follows:
$\mathrm{NDVI_{max}} \leq 0.33$
averaged SbAI in August ≥ 0.025
annual averaged SbAI ≥ 0.030.

The correct identification of drought (presence or absence) was higher for the SbAI than for the NDVI during 2001–2009. In particular, the identification accuracy was 100% for the averaged SbAI in August. The primary reason for this accuracy is that the droughts during 2001–2010 were summer droughts that led to a reduction in water retention [15].

The years 2014 and 2017 have recently been drought years [28,38,45,47], and they could be detected by the annual averaged SbAI. Drought is more strongly correlated with soil moisture anomalies [36,48], and thus the annual averaged SbAI might be appropriate for monitoring drought during seasons other than summer.

Degraded land area, defined as annual $\mathrm{NDVI_{max}} < 0.2$ and annual averaged SbAI > 0.025, has decreased ($R^2 = 0.24$, $p < 0.05$), especially since 2009 (Figure 12). Degraded land area was small in 2003, 2012, 2016, and 2018 but large in 2001, 2002, 2004, 2005, and 2009, which corresponded to the major drought years shown in Figure 11. Degraded land area can recover form one year to the next, as in 2017 to 2018. Since degraded land area was defined as areas including existing desert and land with both permanent and temporal dust erodibility [27], factors like an ecological processes and human impacts are also important in recovering degraded land [7].

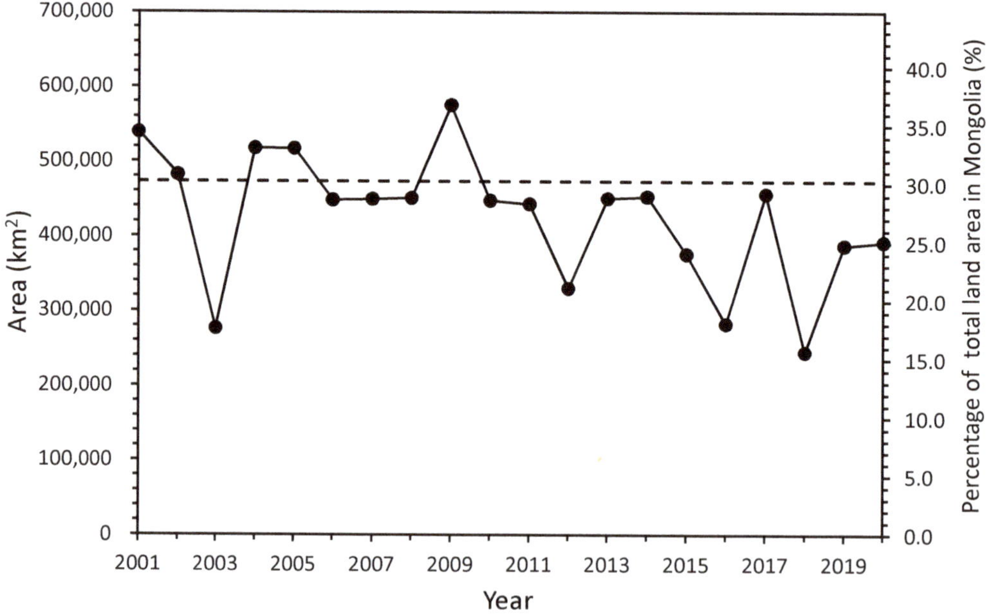

Figure 12. Annual changes of areas of degraded land and percentage of total land area in Mongolia. Dashed line represents the average extent of degraded land for 2001 to 2009.

The defined degraded land area should have been in zones 1 and 10 in Figure 3. Therefore, this method will be useful for general detection in very severe drought condition with a possibility of dust occurrences over Mongolia, particularly in zone 10 of Figure 5 (Figure 13). The spatial distribution of degraded land indicates that droughts have occurred frequently in southern-east Mongolia, that is, in Dundgovi, Omnogovi, and Dornogovi *aimags* (*aimag* is the first-level administrative subdivision). The authors in [49] also exhibited high risks of *dzud* in these provinces using social data from 1944 to 1993.

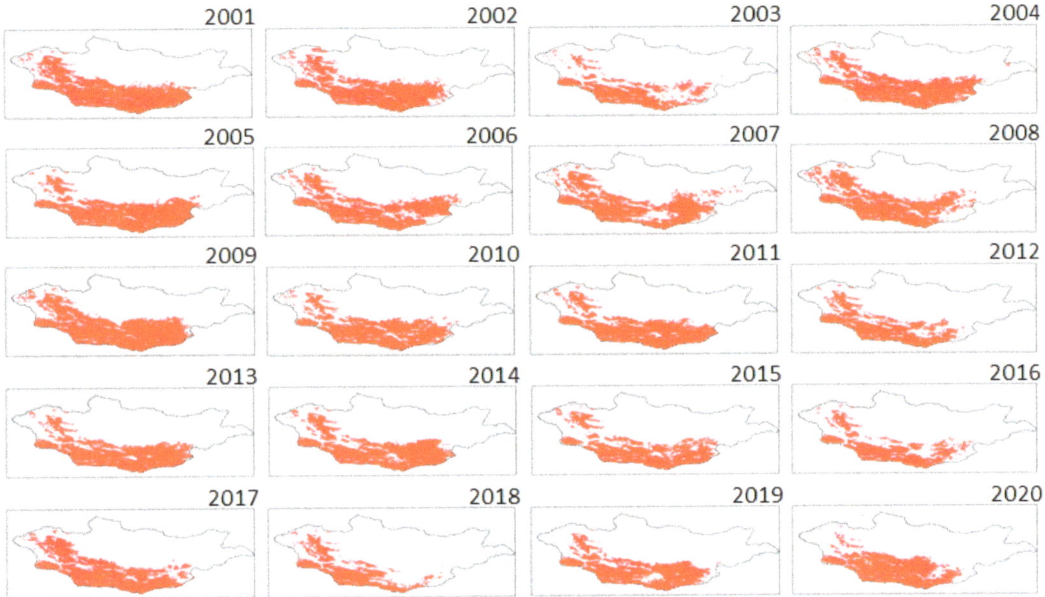

Figure 13. Spatial distribution of degraded land from 2001 to 2020.

4. Discussion

This study examined the reasons why the land surface in Mongolia has recently become drier across a range of AIs by examining the trends of the NDVI and SbAI. We then proposed a method to monitor drought conditions using only satellite data. Among explanations for why the SbAIs have been large within Mongolia are the following:

- The $NDVI_{max}$ was small compared with the $NDVI_{max}$ values in other Ar and SAr regions.
- Did Mongolia become drier climatically? Although the AI distribution was almost unchanged compared with 1981–2010, annual rainfall during 1994–2010 was about 30 mm less than during 1982–1993. There is a possibility that the amount of vegetation was sensitive to a rainfall decrease of 30 mm. In fact, the NDVImax had been decreasing up to 2010 after peaking in 1994. The NDVImax was small even at its peak value of 0.39 in 1994, and it did not reach its averaged value of 0.4 in zones 2 and 3.
- The SbAI during the summer was relatively small (wet). However, the SbAI through the year was large (dry). In Mongolia, most of the annual rainfall occurs from April to July, and that rainfall is reflected by the NDVImax in August. After August, vegetation is dried or eaten by livestock, and the land surface wetness decreases (large SbAI). At the same time, there is less rainfall during seasons other than summer.
- Under the current conditions, the capacity of the land surface to retain water leads to a large SbAI because the concentrated summer rainfall affects the growth of vegetation.

In contrast, the SbAI decreases when the annual rainfall and/or amount of rainfall increases during seasons other than summer. If the amount of precipitation, including precipitation during the winter, increases enough that the annual averaged SbAI decreases, the aridity of Mongolia will approach climatically stable conditions, and drought occurrences that are correlated with soil moisture anomalies will be less frequent. Since 2009, the $NDVI_{max}$ in August over Mongolia has tended to reach an average value of 0.4 in zones 2 and 3, and the frequency of drought years when SbAI values are over the threshold has also decreased.

For sustainable development in Mongolia, where 30% of the workforce is engaged in stock farming, continuous monitoring should be conducted to detect drought and prevent land degradation. The remote sensing techniques proposed in this study, in addition to other drought indices that make use of meteorological or satellite data, will facilitate this monitoring. We hope that the usefulness of our method will be confirmed by other researchers and in other arid countries, and that our method will serve as the basis for an improved system based on remote sensing techniques that will promote sustainable development in arid regions throughout the world.

5. Conclusions

The purpose of this study was to examine the trends of the NDVI and SbAI to determine why the land surface of Mongolia has recently become drier; that is, when the SbAIs were plotted against the AIs, actual aridity in most of Mongolia was more severe than climatic aridity. The main reasons were that the $NDVI_{max}$ was lower than the $NDVI_{max}$ found in the other drylands of the world, and the SbAI throughout the year was large. Under the current conditions, the capacity of the land surface to retain water throughout the year caused the SbAI to be large because rainfall in Mongolia is concentrated in the summer, and the conditions of grasslands reflect summer rainfall.

A method was proposed to monitor land-surface dryness or drought using satellite data. The correct identification of drought was higher for the SbAI than for the NDVI. Drought is more strongly correlated with soil moisture anomalies, and thus the annual averaged SbAI might be appropriate for monitoring drought during seasons other than summer. Degraded land area, defined as annual NDVImax <0.2 and annual averaged SbAI > 0.025, has decreased. Degraded land area was small in 2003, 2012, 2016, and 2018 but large in 2001, 2002, 2004, 2005, and 2009, which corresponded to the major drought years in Mongolia. However, it must be noted that degradation is not caused by not only drought events but also ecological processes and grazing pressure in Mongolia [7].

Author Contributions: Conceptualization, R.K.; methodology, R.K.; software, M.M.; validation, R.K. and M.M.; formal analysis, R.K.; resources, M.M.; data curation, R.K. and M.M.; writing—original draft preparation, R.K.; writing—review and editing, R.K.; visualization, R.K.; supervision, R.K.; project administration, R.K.; funding acquisition, R.K. All authors have read and agreed to the published version of the manuscript.

Funding: This research was funded by Grant-in-Aid for Scientific Research, grant Number KAKENHI 19H04239.

Institutional Review Board Statement: Not applicable.

Informed Consent Statement: Not applicable.

Acknowledgments: We are very grateful to the reviewers who significantly contributed to the im-provement of this paper.

Conflicts of Interest: The authors declare that there are no conflict of interest.

References

1. IPCC. Climate Change 2013: The Physical Science Basis. In *Contribution of Working Group I to the Fifth Assessment Report of the Intergovernmental Panel on Climate Change*; Stocker, T.F.D., Qin, G.K., Plattner, M., Tignor, S.K., Allen, J., Boschung, A., Nauels, Y., Xia, V.B., Midgley, P.M., Eds.; Cambridge University Press: Cambridge, UK, 2013; p. 1535.
2. Feng, S.; Fu, Q. Expansion of global drylands under a warming climate. *Atmos. Chem. Phys.* **2013**, *13*, 10081–10094. [CrossRef]
3. Huang, J.; Li, Y.; Fu, C.; Chen, F.; Fu, Q.; Dai, A.; Shinoda, M.; Ma, Z.; Guo, W.; Li, Z.; et al. Dryland climate change: Recent progress and challenges. *Rev. Geophys.* **2017**, *55*, 719–778. [CrossRef]
4. Koutroulis, A.G. Dryland changes under different levels of global warming. *Sci. Total Environ.* **2019**, *655*, 482–511. [CrossRef] [PubMed]
5. White, R.P.; Nackoney, J. *Drylands, People and Ecosystem Goods and Services*; World Resources Institute: Washington, DC, USA, 2003; p. 40.
6. Ministry of the Environment. Climate Change in Mongolia-Outputs from GCM. 2014. Available online: https://www.env.go.jp/earth/ondanka/pamph_gcm/gcm_mongolia_en.pdf (accessed on 17 February 2021).
7. Shinoda, M.; Nandintsetseg, G.U. *Climate Change and Hazards in Mongolia*; Scientific Report in Nagoya University: Nagoya, Japan, 2015; p. 109.

8. Liu, Y.Y.; Evans, J.P.; McCabe, M.F.; De Jeu, R.A.; van Dijk, A.I.; Dolman, A.J.; Saizen, I. Changing climate and overgrazing are decimating Mongolian steppes. *PLoS ONE* **2013**, *8*, e57599. [CrossRef] [PubMed]
9. UNEP. *World Atlas of Desertification*; Arnold: London, UK, 1997; p. 182.
10. Millennium Ecosystem Assessment (MA). *Ecosystems and Human Well-Being*; World Resources Institute: Washington, DC, USA, 2005; p. 25.
11. Cherlet, M.; Hutchinson, C.; Reynolds, J.; Hill, J.; Sommer, S.; von Maltitz, G. (Eds.) *World Atlas of Desertification*; Publication Office of the European Union: Luxembourg, 2018; p. 248.
12. Kimura, R.; Moriyama, M. Recent trends of annual aridity indices and classification of arid regions with satellite-based aridity indices. *Remote Sens. Earth Syst. Sci.* **2019**, *2*, 88–95. [CrossRef]
13. Palmer, W. *Meteorological Drought*; US Weather Bureau: Washington, DC, USA, 1965; p. 59.
14. Vicente-Serrano, S.M.; Begueria, S.; Lopez-Moreno, J.I. A multiscalar drought index sensitive to global warming: The standardized precipitation evapotranspiration. *Index J. Clim.* **2010**, *23*, 1696–1718. [CrossRef]
15. Nandintsetseg, B.; Shinoda, M. Assessment of drought frequency, duration, and severity and its impact on pasture production in Mongolia. *Nat. Hazards* **2013**, *66*, 995–1008. [CrossRef]
16. Hessl, A.E.; Anchukaitis, K.J.; Jelsema, C.; Cook, B.; Byambasuren, O.; Leland, C.; Nachin, B.; Pederson, N.; Tian, H.; Hayles, L.A. Past and future drought in Mongolia. *Sci. Adv.* **2018**, *4*, e1701832. [CrossRef]
17. Kakinuma, K.; Yanagawa, A.; Sasaki, T.; Rao, M.P.; Kanae, S. Socio-ecological interactions in a changing climate: A review of the Mongolian pastoral system. *Sustainability* **2019**, *11*, 5883. [CrossRef]
18. Tufaner, F.; Özbeyaz, A. Estimation and easy calculation of the Palmer drought severity index from the meteorological data by using the advanced machine learning algorithms. *Environ. Monit. Assess.* **2020**, *192*, 576. [CrossRef] [PubMed]
19. Srivastava, A.; Sahoo, B.; Raghuwanshi, N.S.; Singh, R. Evaluation of variable-infiltration capacity model and MODIS-terra satellite-derived grid-scale evapotranspiration estimates in a River Basin with Tropical Monsoon-Type climatology. *J. Irrig. Drain. Eng.* **2017**, *143*, 04017028. [CrossRef]
20. Pettorelli, N.; Vik, J.O.; Mysterud, A.; Gaillard, J.M.; Tucker, C.J.; Stenseth, N.C. Using the satellite-derived NDVI to assess ecological responses to environmental change. *Trends Ecol. Evol.* **2005**, *20*, 503–510. [CrossRef] [PubMed]
21. Bayarjargal, Y.; Karnieli, A.; Bayasgalan, M.; Khudulmur, S.; Gandush, C.; Tucker, C.J. A comparative study of NOAA-AVHRR derived drought indices using change vector analysis. *Remote Sens. Environ.* **2006**, *105*, 9–22. [CrossRef]
22. Chang, S.; Wu, B.; Yan, N.; Davdai, B.; Nasanbat, E. Suitability assessment of satellite-derived drought indices for Mongolian grassland. *Remote Sens.* **2017**, *9*, 650. [CrossRef]
23. Nanzad, L.; Zhang, J.; Tuvdendorj, B.; Nabil, M.; Zhang, S.; Bai, Y. NDVI anomaly for drought monitoring and its correlation with climate factors over Mongolia from 2000 to 2016. *J. Arid Environ.* **2019**, *164*, 69–77. [CrossRef]
24. Bakhtiari, M.; Boloorani, A.D.; Kakroodi, A.A.; Rangzan, K.; Mousivand, A. Land degradation modeling of dust storm sources using MODIS and meteorological time series data. *J. Arid Environ.* **2021**, *190*, 104507. [CrossRef]
25. Jones, H.G.; Vaughan, R.A. *Remote Sensing of Vegetation*; Oxford University Press: New York, NY, USA, 2010; p. 353.
26. Kimura, R.; Moriyama, M. Application of a satellite-based aridity index in dust source regions of northeast Asia. *J. Arid Environ.* **2014**, *109*, 31–38. [CrossRef]
27. Kimura, R.; Moriyama, M. Determination by MODIS satellite-based methods of recent global trends in land surface aridity and degradation. *J. Agric. Meteorol.* **2019**, *75*, 153–159. [CrossRef]
28. Kimura, R. Validation and application of the monitoring method for degraded land area based on a dust erodibility in eastern Asia. *Int. J. Remote Sens.* **2017**, *38*, 4553–4564. [CrossRef]
29. Kimura, R. Global distribution of degraded land area based on dust erodibility determined from satellite data. *Int. J. Remote Sens.* **2018**, *39*, 5859–5871. [CrossRef]
30. Kimura, R. Global detection of aridification or increasing wetness in arid regions from 2001 to 2013. *Nat. Hazards* **2020**, *103*, 2261–2276. [CrossRef]
31. Kumari, N.; Saco, P.M.; Rodriguez, J.F.; Johnstone, S.A.; Srivastava, A.; Chun, K.P.; Yetemen, O. The grass is not always greener on the other side: Seasonal reversal of vegetation greenness in aspect-driven semiarid ecosystems. *Geophys. Res. Lett.* **2020**, *47*, e2020GL088918. [CrossRef]
32. Nakano, T.; Shinoda, M. Spatial variability of photosynthetic production and ecosystem respiration on a hundred-kilometer scale within a Mongolian semiarid grassland. *J. Agric. Meteorol.* **2014**, *70*, 105–116. [CrossRef]
33. ORNL DAAC. *MODIS Collection 6 Land Product Subsets Web Service*; ORNL DAAC: Oak Ridge, TN, USA, 2017. [CrossRef]
34. Iwao, K.; Nishida, K.; Kinoshita, T.; Yamagata, Y. Validating land cover maps with Degree Confluence Project information. *Geophys. Res. Lett.* **2006**, *33*, L23404. [CrossRef]
35. Gamo, M.; Shinoda, M.; Maeda, T. Classification of arid lands, including soil degradation and irrigated areas, based on vegetation and aridity indices. *Int. J. Remote Sens.* **2013**, *34*, 6701–6722. [CrossRef]
36. Nandintsetseg, B.; Shinoda, M.; Kimura, R.; Ibaraki, Y. Relationship between soil moisture and vegetation activity in the Mongolian steppe. *Sola* **2010**, *6*, 29–32. [CrossRef]
37. Batjargar, Z. Climate change and use of renewable energy in Mongolia. *Erina Rep. Plus* **2018**, *145*, 25–34.
38. Kimura, R.; Moriyama, M. Use of a satellite-based aridity index to monitor decreased soil water content and grass growth in grasslands of north-east Asia. *Remote Sens.* **2020**, *12*, 3556. [CrossRef]

39. UNEP. *World Atlas of Desertification*; Arnold: London, UK, 1992; p. 69.
40. Fujita, N.; Kato, S.; Kusano, E.; Koda, R. *Mongolia*; Kyoto University Press: Kyoto, Japan, 2013; p. 685. (In Japanese)
41. Spinoni, J.; Vogt, J.; Naumann, G.; Carrao, H.; Barbosa, P. Towards identifying areas at climatological risk of desertification using the KÖppen-Geiger classification and FAO aridity index. *Int. J. Climatol.* **2015**, *35*, 2210–2222. [CrossRef]
42. Yu, W.; Wu, T.; Wang, W.; Li, R.; Wang, T.; Qin, Y.; Wang, W.; Zhu, X. Spatiotemporal changes of reference evapotranspiration in Mongolia during 1980–2006. *Adv. Meteorol.* **2016**, *2016*, 9586896. [CrossRef]
43. Bao, G.; Qin, Z.; Bao, Y.; Zhou, Y.; Li, W.; Sanjjav, A. NDVI-based long term vegetation dynamics and its response to climatic change in the Mongolian plateau. *Remote Sens.* **2014**, *6*, 8337–8358. [CrossRef]
44. Munkhtsetseg, E.; Kimura, R.; Wang, J.; Shinoda, M. Pasture yield response to precipitation and high temperature in Mongolia. *J. Arid Environ.* **2007**, *70*, 94–110. [CrossRef]
45. Yu, W.; Li, Y.; Cao, Y.; Schillerberg, T. Drought assessment using GRACE terrestrial water storage deficit in Mongolia from 2002 to 2017. *Water* **2019**, *11*, 1301. [CrossRef]
46. Fernández-Giménez, M.E.; Venable, N.H.; Angerer, J.; Fassnacht, S.R.; Reid, R.S.; Khishigbayar, J. Exploring linked ecological and cultural tipping points in Mongolia. *Anthropocene* **2017**, *17*, 46–69. [CrossRef]
47. FAO. *Special Report-FAO/WFP Crop and Livestock Assessment Mission to Mongolia*; FAO/WFP: Rome, Italy, 2017; p. 43.
48. Nandintsetseg, B.; Shinoda, M. Land surface memory effects on dust emission in a Mongolian temperate grassland. *J. Geophys. Res. Biogeosciences* **2015**, *120*, 414–427. [CrossRef]
49. Templer, G.; Swift, J.; Payne, P. The changing significance of risk in the Mongolian pastoral economy. *Nomadic Peoples* **1993**, *33*, 105–122.

remote sensing

MDPI

Article

Vegetation Trends, Drought Severity and Land Use-Land Cover Change during the Growing Season in Semi-Arid Contexts

Felicia O. Akinyemi [1,2]

1   Earth and Environmental Science Department, Botswana International University of Science and Technology, Private Bag 16, 10071 Palapye, Botswana; felicia.akinyemi@giub.unibe.ch or felicia.akinyemi@gmail.com
2   Institute of Geography, University of Bern, Hallerstrasse 12, CH-3012 Bern, Switzerland

Abstract: Drought severity and impact assessments are necessary to effectively monitor droughts in semi-arid contexts. However, little is known about the influence land use-land cover (LULC) has—in terms of the differences in annual sizes and configurations—on drought effects. Coupling remote sensing and Geographic Information System techniques, drought evolution was assessed and mapped. During the growing season, drought severity and the effects on LULC were examined and whether these differed between areas of land change and persistence. This study used areas of economic importance to Botswana as case studies. Vegetation Condition Index, derived from Normalised Difference Vegetation Index time series for the growing seasons (2000–2018 in comparison to 2020–2021), was used to assess droughts for 17 constituencies (Botswana's fourth administrative level) in the Central District of Botswana. Further analyses by LULC types and land change highlighted the vulnerability of both human and natural systems to drought. Identified drought periods in the time series correspond to declared drought years by the Botswana government. Drought severity (extreme, severe, moderate and mild) and the percentage of land areas affected varied in both space and time. The growing seasons of 2002–2003, 2003–2004 and 2015–2016 were the most drought-stricken in the entire time series, coinciding with the El Niño southern oscillation (ENSO). The lower-than-normal vegetation productivity during these growing seasons was evident from the analysis. With the above-normal vegetation productivity in the ongoing season (2020–2021), the results suggest the reversal of the negative vegetation trends observed in the preceding growing seasons. However, the extent of this reversal cannot be confidently ascertained with the season still ongoing. Relating drought severity and intensities to LULC and change in selected drought years revealed that most lands affected by extreme and severe drought (in descending order) were in tree-covered areas (forests and woodlands), grassland/rangelands and croplands. These LULC types were the most affected as extreme drought intersected vegetation productivity decline. The most impacted constituencies according to drought severity and the number of drought events were Mahalapye west (eight), Mahalapye east (seven) and Boteti west (seven). Other constituencies experienced between six and two drought events of varying durations throughout the time series. Since not all constituencies were affected similarly during declared droughts, studies such as this contribute to devising appropriate context-specific responses aimed at minimising drought impacts on social-ecological systems. The methodology utilised can apply to other drylands where climatic and socioeconomic contexts are similar to those of Botswana.

Keywords: Normalised Difference Vegetation Index (NDVI); Vegetation Condition Index (VCI); drought; land use-land cover; remote sensing; Botswana

Citation: Akinyemi, F.O. Vegetation Trends, Drought Severity and Land Use-Land Cover Change during the Growing Season in Semi-Arid Contexts. *Remote Sens.* **2021**, *13*, 836. https://doi.org/10.3390/rs13050836

Academic Editor: Elias Symeonakis

Received: 8 January 2021
Accepted: 18 February 2021
Published: 24 February 2021

1. Introduction

Drought as a slow-onset event is increasingly an environmental hazard due to its negative impacts on natural and human systems, including livelihoods [1–4]. Drought conditions are initiated by precipitation shortfall in comparison to the climatological normal in the focus context and amplified by concurrent heatwaves and extreme high-temperature events [5,6]. Impacts relate to insufficient availability of water to meet human and nature's needs, seasonal moisture deficit including soil moisture and extensive evaporation resulting in a decline in vegetation greenness and health, plant mortality, reduced water levels in dams and other adverse ecological and/or socioeconomic conditions [7–10]. Studies have found and are forecasting increasing drought duration, severity and frequency in different world regions [11–16].

Drylands, which by their very nature are water deficient due to limited rainfall and water supplies [17], are particularly vulnerable to droughts. In African drylands, the occurrence of droughts is not unusual. Some studies have found increasing drought events in African drylands [4,18], whereas other studies project future increases in droughts and other high-temperature events [19,20]. In the arid regions of South Africa, extreme high-temperature events in parks are increasing in frequency [18]. In Botswana, where this study was conducted, there is public awareness on issues relating to drought due to its negative impacts, particularly on agriculture, which is largely rainfed [21]. Studies have found observable changes of varying magnitudes in rainfall, temperature trends and drought over Botswana [21–24]. With future global warming, droughts are projected to increase in frequency and severity in this region based on regional climate model simulations [5,25]. In the face of climate variability and change, it is increasingly important to assess and monitor droughts because of the need to adapt and minimise impacts on the social-ecological systems in African drylands.

The need to monitor and assess droughts in drylands calls for the use of Remote Sensing (RS)-based data and methods as complementary to meteorological gauge data and climatological indices due to the drawbacks of using only ground-based data, such as inadequate spatial and temporal coverage of meteorological station data. RS image data and vegetation indices are widely used in drought monitoring [26–29]. With the increasing availability of free satellite image datasets of good temporal and spatial resolution, RS affords rapid and cost-effective assessment and monitoring of droughts. Drought effects on vegetation and ecosystem services were examined in the Bobirwa sub-district, Botswana using RS-based indices [30]. Although the drought situation is increasingly assessed in Botswana, the use of RS in examining droughts in Botswana is still very limited. Moreover, how drought severity differs between land systems to further exacerbate impacts in this region is not clear.

Using Normalised Difference Vegetation Index (NDVI) image time series datasets, this study contributes to the understanding of the spatial and temporal variations of droughts and the effects across LULC types and change. It integrated remote sensing with spatial statistics in Geographic Information System (GIS) for the analysis and mapping of drought. Variability and vegetation trends over 18 years for the growing seasons between 2000–2001 to 2017–2018 were assessed and afterwards compared to the ongoing growing season (2020–2021). Thus, this study better captured the seasonality of vegetation growth when considering drought severity as this relates to rainfall in dryland contexts. The spatial and temporal evolution of drought severity was assessed and mapped for each year in the time series. We provided an improved methodology for the examination of drought evolution and severity by incorporating indices of LULC change. Transitions in areas of LULC change and persistence, i.e., areas where no changes occurred, are useful indices to understand the processes, especially anthropogenic, driving the observed trends in vegetation productivity and how these relate to drought severity. Most RS-based studies on drought have not evaluated the effects according to LULC types and change. For the examination of drought severity by LULC and change to be meaningful, we further considered the differences in annual LULC configurations and sizes. Considering that both configuration and size differ from year to year, we utilised the annual LULC time series for analysis in each year identified as drought-stricken. Moreover, drought impacts on land-based resources upon which much of livelihoods are dependent would either be exacerbated or ameliorated depending on how the land is put to use and the land management practices. To better demonstrate RS capabilities for assessing drought severity at a finer, sub-national scale, the assessment was conducted in 17 constituencies (a constituency is the fourth administrative level in Botswana).

2. Materials and Methods

2.1. Description of the Study Location

Seventeen constituencies in the Central District of Botswana (CDB) in the eastern part of the country were used as case studies (Figure 1). The CDB is the largest amongst the nine districts of Botswana both in terms of population and geographic size (Table S1 details the population and geographic area of the 17 constituencies). The district has 576,064 inhabitants (29% of Botswana's population) as of the 2011 census. With 26% of Botswana's land area, it covers an area of approximately 147,730 km^2. With a semi-arid, hot steppe climate (Koppen's BSh classification), rains occur in the summer months with peak rainfall in January (71–142 mm). Annual average rainfall at the constituencies ranged from 321 mm to 430 mm. Temperature ranges from 32 to 39 °C and can occasionally exceed 40 °C [21]. As in most parts of Botswana, the annual evaporation rate of about 2000 mm year^{-1} far exceeds that of the rainfall (475–525 mm year^{-1}) [4].

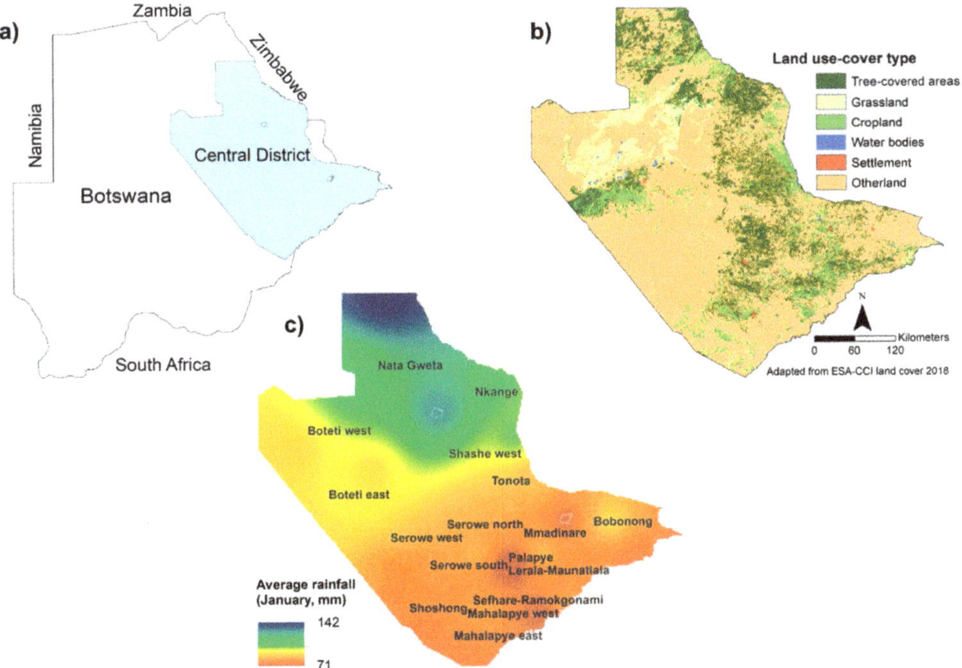

Figure 1. Study location: (**a**) Central District in eastern Botswana; (**b**) Land use-land cover for 2018; (**c**) Peak distribution of rainfall in the 17 constituencies examined in the study.

The district is of great economic importance to Botswana, with 23% (31,634 holdings) of all traditional agricultural holdings in Botswana [31]. Moreover, the majority of the mines in Botswana are located in the CDB, such as the Morupule coal mine and the diamond mines in Lerala, Orapa and Letlhakane.

2.2. Data Sources

Variability in vegetation condition and drought severity in the CDB were examined over 18 years using the 1 km Normalised Difference Vegetation Index (NDVI) decadal (i.e., 10-day composite) image time series from the Copernicus Land Monitoring Service (https://land.copernicus.eu/global/products/ndvi, accessed on 8 January 2021). These images were made available through the European Union–African Union-funded project on Monitoring for environment and security in Africa (MESA). The MESA was implemented for the Southern African Development Community (SADC) region comprising 15 countries and included Madagascar and the Democratic Republic of Congo. The dekadal NDVI datasets from October 2000 to 2014 were derived from SPOT VGT, and data from June 2014 to 2018 are from the PROBA-V [32]. These long-term, 1-km NDVI datasets from the two sensors have been pre-processed at the source to ensure compatibility and continuity [33].

The land cover datasets were from the European Space Agency (ESA-LC) Climate Change Initiative (CCI-LC v.2.0.7) ESA CCI and Copernicus Climate Change Service (C3S-LC Mv52 https://cds.climate.copernicus.eu/cdsapp#!/dataset/satellite-land-cover?tab=form, accessed on 8 January 2021). These datasets are a consistent series of multi-sensor annual maps from 1992 to 2018 [34]. The land categories in the ESA-LC were aggregated into six categories for compatibility with previous studies in Botswana [35]—tree-covered areas, grassland, cropland, water bodies, artificial surfaces (settlement including infrastructure) and otherland (Table S2). Tree-covered areas comprise forests and woodlands. In Botswana, forested areas are defined as comprising multi-layered tree canopies with over 10% cover, a minimum size of 0.5 ha and heights of more than 5 m [36]. Grassland incorporated shrubland and other sparse vegetation, whereas the otherland category combines bareland, rock outcrops and dunes [35].

2.3. Methods

2.3.1. Indicators of Vegetation Variability

Indices used for measuring vegetation variability and drought severity are based on the NDVI. NDVI is widely utilised for assessing and monitoring vegetation greenness, net primary productivity, plant phenology and land degradation in natural and human systems [35]. NDVI is calculated as in Equation (1), where NIR and R are the near infrared and visible red portions, respectively, of the electromagnetic spectrum. Photosynthetically active plants absorb more incident radiation in the visible red portion but reflect more in the near infrared portion [37].

$$\text{NDVI} = \frac{\text{NIR} - \text{R}}{\text{NIR} + \text{R}} \tag{1}$$

Three indicators of vegetation variability utilised in this study were derived from NDVI—NDVI difference, NDVI anomaly and NDVI trends. Other derived metrics to gauge seasonal vegetation productivity include the NDVI mean, maximum and cumulative values computed for each month in the growing season within the time series.

NDVI Difference

To analyse the variability of vegetation during the vegetation growing season over the 18-year study period, the NDVI Difference ($\text{NDVI}_{\text{diff}}$) function implemented in the MESA Drought Monitoring Services (DMS) software was utilised. $\text{NDVI}_{\text{diff}}$ is widely used to get an indication of vegetation state over a specific period by comparing vegetation productivity between two dekads or relative to the long-term average for the same period. This indicator highlights areas where vegetation is under stress as well as those performing well. For this study, seasonal $\text{NDVI}_{\text{diff}}$ was calculated for every growing season (i.e., annually) as the difference between the start dekad (i.e., first 10-day period) in October (D1) in a certain year i to the end dekad (i.e., last 10-day period) in March (D3) of the following year $i + 1$ in the time series data of 2000 to 2018 (Equation (2)):

$$\text{NDVI}_{\text{diff }(i,i+1)} = \text{NDVI}_{\text{D1}i} - \text{NDVI}_{\text{D3}i+1} \tag{2}$$

Standardised NDVI Anomalies

NDVI anomaly captures how vegetation productivity for a certain period deviates from the long-term average dynamics. It is calculated by subtracting the considered month NDVI from the month's long-term average and dividing it by the monthly standard deviation for that period. By distinguishing areas that are normal from those that are above or below normal vegetation productivity, NDVI anomaly is helpful to identify outliers, isolate the variability in the vegetation signal and consider the reviewed period within a meaningful historical context [26].

NDVI Trend

Using the NDVI time series of 2000 to 2020 as input, we computed vegetation change as trends and their significance based on the Mann-Kendall (MK) non-parametric test. Non-parametric approaches estimate trends in a time series by quantifying the rate of change in vegetation greenness for each pixel and characterises trends in the data using the median slope [38]. MK tau (τ) coefficient ranges from -1 to $+1$ with values greater than 0 indicating a continually increasing (greening) trend, and values less than 0 indicating a continually decreasing (browning) trend [39]. NDVI trends were reclassified into increase (>0), decrease (<0) and stable (0). The MK is useful to determine the significance of changes in vegetation productivity over time and is robust to outliers [40]. The reclassified NDVI trends were afterwards assessed by LULC type. In conjunction with LULC types, NDVI trends are useful in detecting land areas that are potentially degraded if significant negative trends are found over time [35].

2.3.2. Drought Indicators

Vegetation Condition Index

Drought severity during the growing season was measured using the NDVI-based Vegetation Condition Index (VCI) as in Equation (3) [27]. VCI compares the NDVI of a given period j ($NDVI_j$) with the long-term minimum NDVI ($NDVI_{ltmin}$) and long-term maximum NDVI ($NDVI_{ltmax}$) computed over a 10-year time series for the same period.

$$VCI_j = \left(\frac{NDVI_j - NDVI_{ltmin}}{NDVI_{ltmax} - NDVI_{ltmin}} \right) \times 100 \tag{3}$$

VCI is particularly useful for agriculture, as it assesses changes in NDVI through time since vegetation is water-stressed due to water deficiency such as during drought. VCI for the vegetation growing season was calculated between 2000–2001 and 2017–2018 starting with the first dekad of October in the previous year to the third dekad in March of the following year. VCI is measured as a percentage with values ranging between 0 (lowest) and 100 (highest), with values equal to or below 40% considered as drought to varying degrees of severity (Table 1). The two VCI-based indicators used for characterising drought are drought intensity and drought frequency.

Table 1. Vegetation Condition Index (VCI) -based drought severity classes (Adapted from [41]).

Drought Severity	VCI
Extreme	$0 \leq VCI < 10$
Severe	$10 \leq VCI < 20$
Moderate	$20 \leq VCI \leq 30$
Mild	$30 \leq VCI \leq 40$
No drought	$40 < VCI \leq 100$

Drought Intensity

Drought intensities for each growing season were calculated through the 18-year study period. These are percentages of pixels (a proxy for the surface land area) whose VCI values fell within the different drought severity and non-drought categories [41]. The evolution of the drought hazard was examined for each year in the time series based on the VCI.

Drought Frequency

To relate the pixel-level VCI to each of the 17 constituencies in the CDB, the zonal statistics function in GIS was used. The median VCI value within each constituency was utilised because there is the tendency for more years in drylands to have rainfall below the mean, with the median value deviating more from the mean. Due to the sensitivity of the mean to outliers, the median is a better data distribution measure to gain further insight regarding the frequency of droughts in each constituency. Drought severity was compared to the official declarations of drought years by the government. Drought frequency was computed as the count of the number of drought years in the entire time series (growing seasons 2000–2001 to 2017–2018). A year with an annual VCI value of 40 or less is identified as drought-stricken and the sum of such drought years per constituency in the entire time series amounts to the frequency of drought [26].

2.3.3. Land Use-Land Cover Change

In addition to vegetation trends, drought severity was examined according to LULC types. It was also examined at the constituency level, to take into cognisance the environmental and administrative basis of drought impacts, respectively. For each drought year, drought severity effects on each LULC type were assessed based on the intersection of VCI and LULC values for that particular year. Drought intensity is expressed as a percentage of the area under each LULC type affected by varying drought severity and non-drought conditions. Percentages of the surface land area affected were then normalised by the size of the LULC type for each year. Since LULC configurations and sizes vary from one year to another, the annual LULC map for each identified drought year was utilised. Only for the land change analysis aspect was the maps of the years 2000 and 2018 used for post-classification change detection.

3. Results

3.1. Vegetation Variability and Trend

The spatial and temporal variations in vegetation productivity were presented based on analyses during the growing seasons, i.e., October to March, of the 18 years. Providing insights regarding each growing season throughout the entire time series, NDVI$_{diff}$ maps (Figure 2) depict limited vegetation productivity in the CDB in years 2000–2001, 2003–2004, 2005–2006, 2007–2009, 2013–2014 and 2015–2018. In other years, such as 2002–2003 and 2010–2011, vegetation productivity was low mostly in the northern and the eastern parts of the district. Improvement in vegetation performance occurred during 2005–2006, 2009–2010, 2010–2011 and 2014–2015.

Figure 3a,b compared maximum, mean and cumulative NDVI in the entire time series (2000–2018) with those of the drought years (2002–2003 and 2003–2004) and non-drought year (2009–2012). Vegetation productivity for 2020–2021 as the ongoing growing season is also depicted. Vegetation productivity during the 2009–2010 non-drought growing season was higher than the mean of the entire time series for most months, except in mid-February to March, whereas, for this current growing season (2020–2021), vegetation productivity is well above the mean of the entire time series from mid-October 2020 to February 2021. NDVI anomalies (Figure 3c) captured lower or higher than normal vegetation productivity in the time series. There were some extended periods of very low vegetation productivity over multiple growing seasons. Examples are the period from the middle of the 2002–2004 growing seasons and 2004–2005 to the start of 2005–2006 growing seasons. The improved vegetation productivity during the growing seasons of 2005–2006, 2009–2010, 2010–2011 and 2014–2015 was further confirmed by the NDVI anomaly (Figure 3c). NDVI anomalies (Figure 3c) captured lower or higher than normal vegetation productivity in the time series.

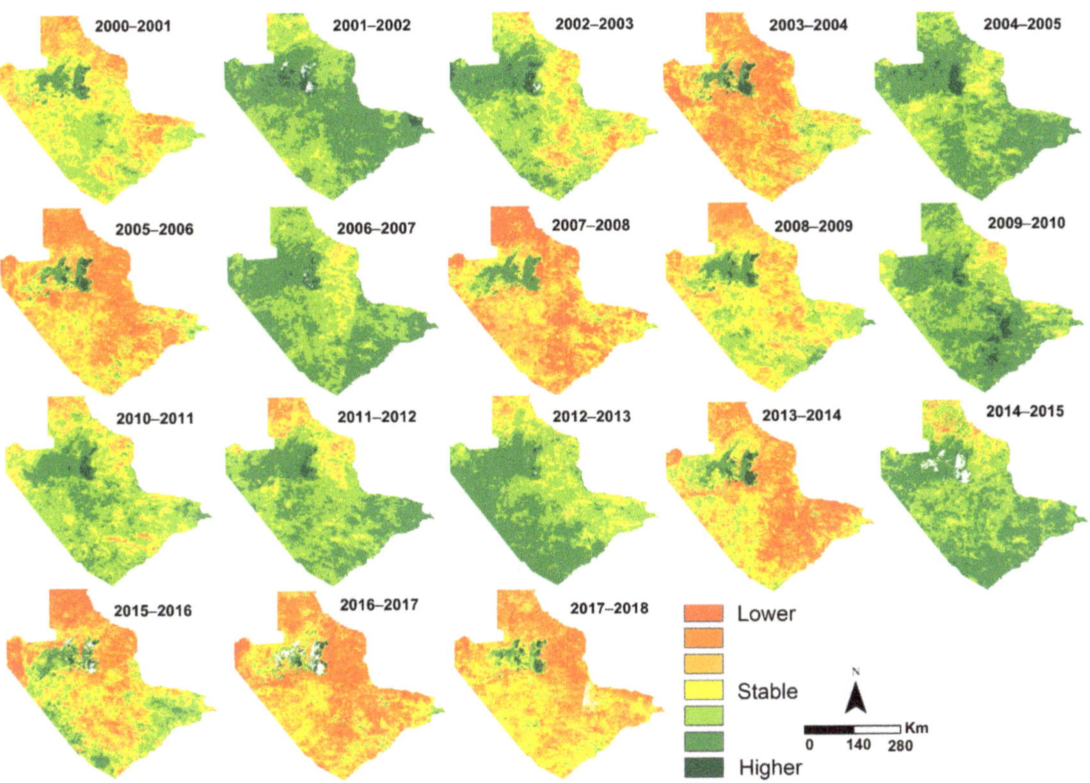

Figure 2. Variability in vegetation productivity based on the seasonal NDVI difference for the growing seasons (October to March) in the time series.

The NDVI trend was analysed as an indicator of vegetation productivity change. Areas of negative trends amounted to about 90% of the land area in the CDB (Figure 4). Decreasing NDVI trends mostly occurred in the north-east (Nkange, Shashe west, Tonota) and to the south (e.g., Sefhare-Ramokgonami, Mahalapye east). Decreasing vegetation trends imply that these areas experienced an overall decline in vegetation cover and biomass. Increasing NDVI trends (4% of CDB's land area), implying an improvement in vegetation productivity, were most pronounced in the north-west (e.g., Boteti west) and the eastern tip along the Motloutse River (Bobonong). Areas of stable vegetation productivity (6%) were mostly in the western parts around Serowe west, Shoshong, Palapye and Serowe north. The direction of these trends and the extent of land area affected are generally in line with the overall national vegetation productivity dynamics [35].

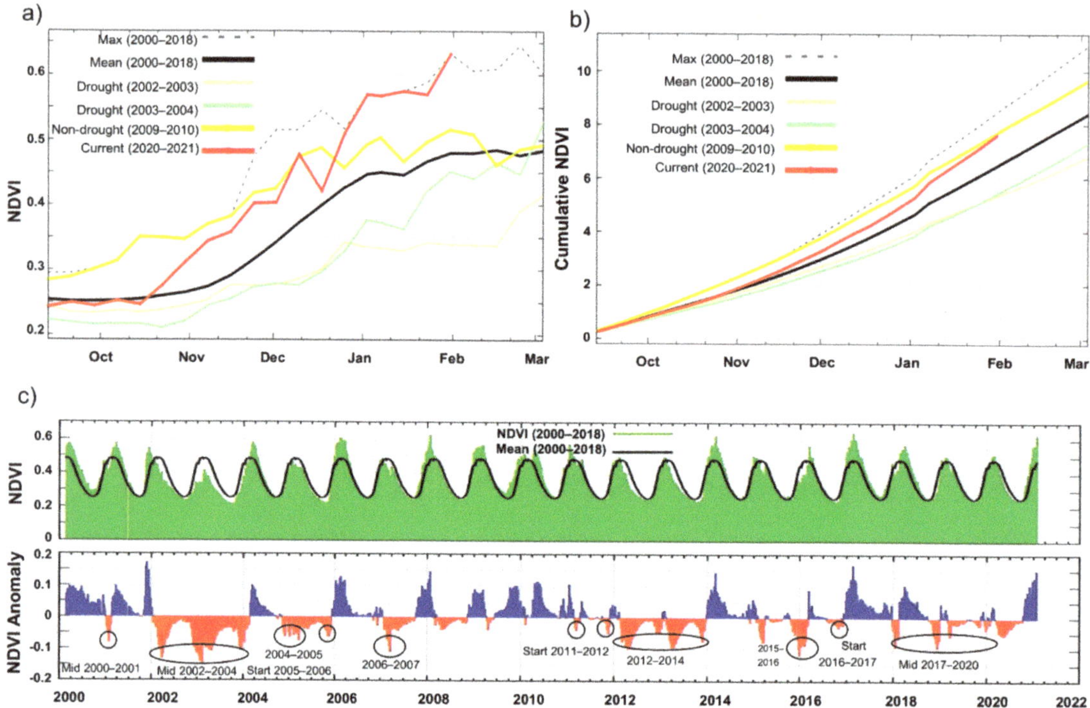

Figure 3. Profiles of NDVI metrics for the growing seasons: (**a**) Mean and maximum NDVI in the time series compared to multi-drought seasons (2002–2003 and 2003–2004), non-drought season (2009–2010) and 2020–2021 as the current growing season; (**b**) Seasonal cumulative NDVI for these metrics as in Figure 3a; (**c**) NDVI anomaly. Grey circles indicate when the anomaly occurred and sizes interpreted as follows: small circle = a part of the growing season was affected (indicated either as start or mid), medium circle = entire growing season was affected, large circle = multi-year growing seasons were affected.

3.2. Spatio-Temporal Evolution of Drought Severity during Vegetation Growing Seasons

Drought negatively impacts vegetation growth, the supply of water for nature's needs (e.g., to sustain wildlife) and human needs (e.g., for livelihoods and food security). Seasonal maps depicting spatial and temporal variations in drought severity for the CDB were produced (Figure 5a). The growing seasons of the years 2002–2004 and 2015–2016 were the worst drought periods in the entire series. The growing seasons of 2004–2007, 2012–2013 and 2016–2018 were also affected but to a lesser extent. During droughts, the most affected areas were towards the west, south and the eastern tip of the district except for 2002–2004, when the entire district was affected by drought. Land areas most affected by drought were in Mahalapye west and east, Boteti west, Shoshong, Shashe west, Palapye and Bobonong. The magnitude of the drought hazard varied between the years considered in the time series (Figure 5b). The highest percentages of extreme, severe, moderate and mild droughts were recorded during 2002–2005. These years corresponded to drought years declared by the government [31]. Drought severity ranged from extreme to mild as lower than normal, erratic rainfall amounts were recorded in the CDB, similar to most parts of Botswana during these years.

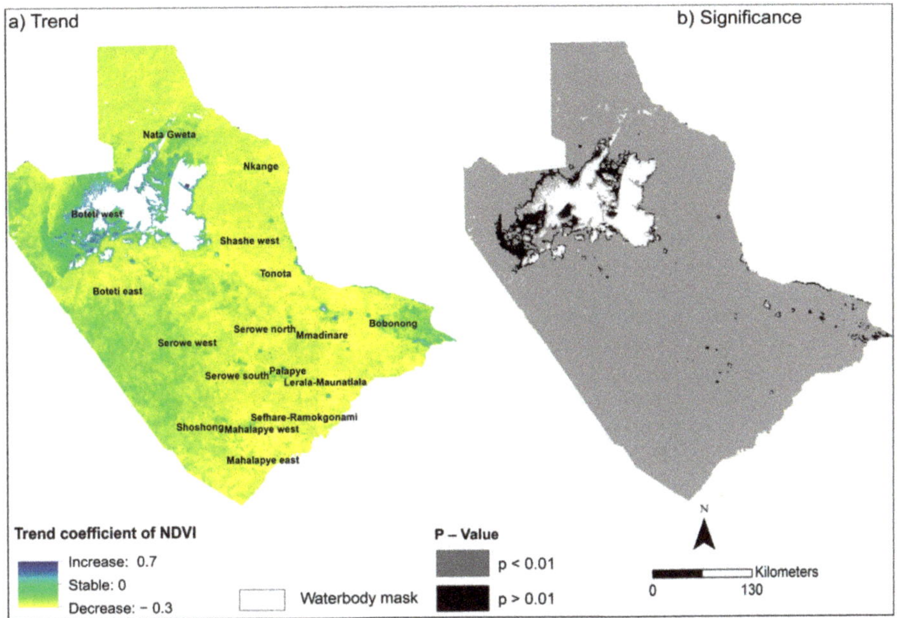

Figure 4. Spatial patterns of vegetation productivity change between 2000–2001 and 2020–2021 (Makgadikgadi salt pans in the north is indicated as white using the waterbody mask): (**a**) Vegetation trends; (**b**) Significance of trends.

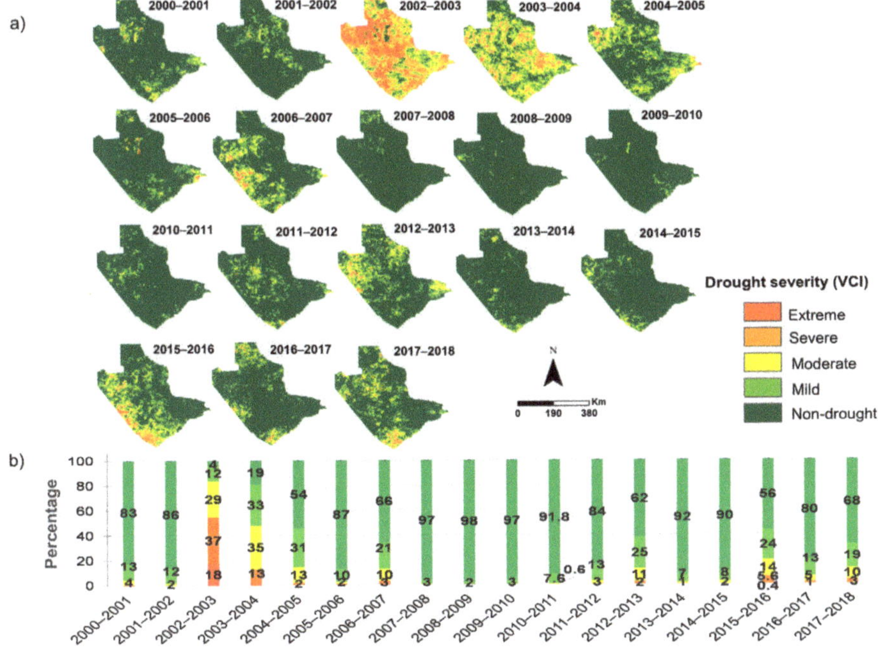

Figure 5. Evolution of VCI drought severity and intensity through the time series in the Central District: (**a**) Growing season VCI (October to March); (**b**) Percentage of land area affected by varying drought severity and non-drought conditions based on the VCI by years.

3.3. Area of Land Use-Land Cover Change and Persistence

In the CDB, areas of LULC persistence, i.e., unchanged, between 2000 to 2018 amounted to 88% (130,166 km^2). Figure 6a depicts the spatial distribution of areas of persistence as well as losses where LULC types have been displaced, i.e., transitioned to other land uses. Figure 6b shows the share of land under each persistent and transitioned land uses between the year 2000 and 2018 as a percentage of the surface land area (Table S3 provides the LULC transition matrix in km^2 between 2000–2018). Land transitions in the matrix are to be interpreted as 'from-to' changes, whereby a particular LULC type in 2000 (initial year) transitions to another LULC type in 2018 (target year). For example, 37% and 2% of tree-covered areas were derived from grasslands and croplands, respectively. Thus, tree-covered areas increased from 11.5% of the total land area of the CDB in 2000 to 14.5% in 2018. Other notable transitions are the expansion of artificial surfaces such as settlements, with 60% derived from grassland, 1.6% from tree-covered, 1.4% from cropland and 1.8% from otherland areas. Thus, artificial surfaces increased from 0.05% in 2000 to 0.13% in 2018. The main gains by grassland were from tree-covered areas (2.8%) and cropland (1%). The main gains by cropland were derived from tree-covered (8%) and grassland (~28%). Although cropland expanded over time (increased from 6.3% of the total land area in 2000 to 8% in 2018), it lost 2.3% through its conversion to tree-covered areas, 1% to grassland and 1.4% to artificial surfaces.

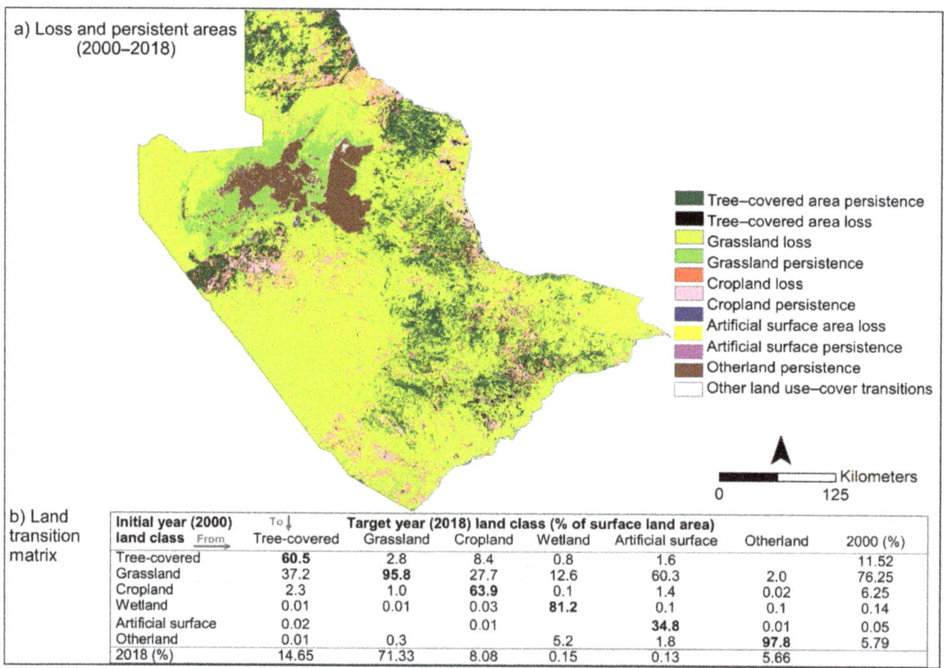

b) Land transition matrix

Initial year (2000) land class From	To ↓ Tree-covered	Grassland	Cropland	Wetland	Artificial surface	Otherland	2000 (%)
	Target year (2018) land class (% of surface land area)						
Tree-covered	**60.5**	2.8	8.4	0.8	1.6		11.52
Grassland	37.2	**95.8**	27.7	12.6	60.3	2.0	76.25
Cropland	2.3	1.0	**63.9**	0.1	1.4	0.02	6.25
Wetland	0.01	0.01	0.03	**81.2**	0.1	0.1	0.14
Artificial surface	0.02		0.01		**34.8**	0.01	0.05
Otherland	0.01	0.3		5.2	1.8	**97.8**	5.79
2018 (%)	14.65	71.33	8.08	0.15	0.13	5.66	

Figure 6. Changing land use-land cover conditions between 2000 and 2018: (**a**) Areas of land use-land cover loss and persistence; (**b**) Land use-land cover transitions from the year 2000 to 2018 in percentages (areas of persistence for each land use-land cover class are in bold, that is, areas of no change that remained in the same land class over the 18 years).

3.3.1. Land Change and Associated Vegetation Trends

With about 90% of the land area in the CDB experiencing negative vegetation trends, we investigated how changes in vegetation productivity (i.e., increasing, stable and decreasing) are associated with land change. The focus is on major LULC types (tree-covered area, cropland, otherland and grassland), as these made up over 95% of the study area as of 2018. Figure 7a–d depict the spatial distribution of vegetation trends found in these major LULC types. Figure 7e shows the percentage of land area by vegetation productivity change (direction and magnitude). By LULC categories as of 2018, vegetation productivity decreased in about 98% and 94% of tree-covered areas (such as forests and woodlands) and croplands, respectively, and 94% of grasslands. In other words, the majority of tree-covered, cropland and grassland areas experienced decreasing vegetation productivity as of 2018. Seventeen percent (17%) of wetlands and settlement areas, respectively, and 12% of otherland experienced increasing trends, signifying improved vegetation productivity.

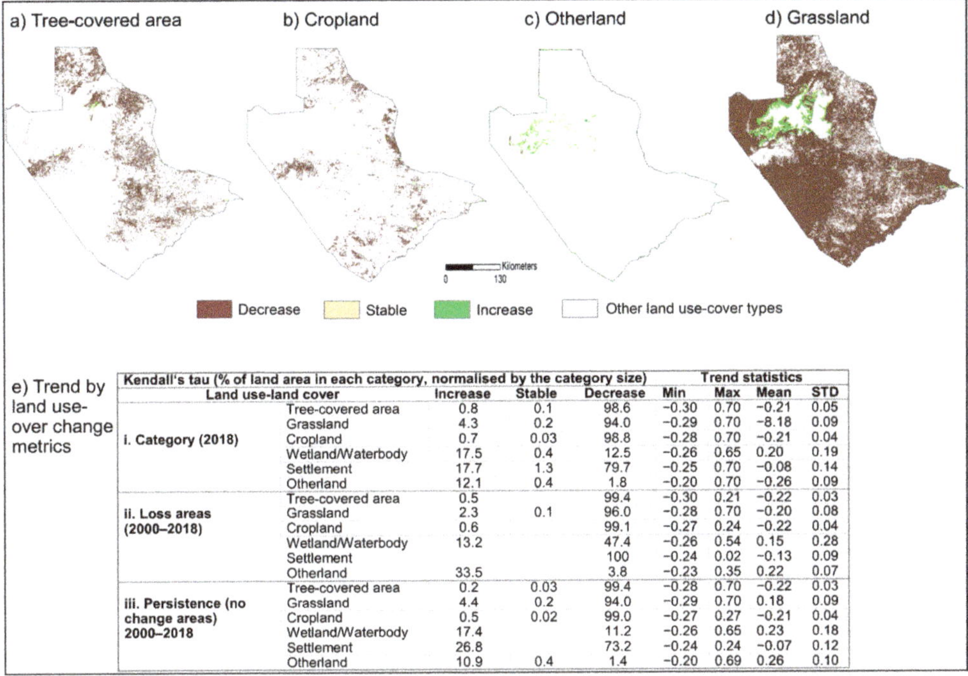

| Kendall's tau (% of land area in each category, normalised by the category size) | | | | Trend statistics | | | |
Land use-land cover	Increase	Stable	Decrease	Min	Max	Mean	STD
i. Category (2018)							
Tree-covered area	0.8	0.1	98.6	−0.30	0.70	−0.21	0.05
Grassland	4.3	0.2	94.0	−0.29	0.70	−8.18	0.09
Cropland	0.7	0.03	98.8	−0.28	0.70	−0.21	0.04
Wetland/Waterbody	17.5	0.4	12.5	−0.26	0.65	0.20	0.19
Settlement	17.7	1.3	79.7	−0.25	0.70	−0.08	0.14
Otherland	12.1	0.4	1.8	−0.20	0.70	−0.26	0.09
ii. Loss areas (2000–2018)							
Tree-covered area	0.5		99.4	−0.30	0.21	−0.22	0.03
Grassland	2.3	0.1	96.0	−0.28	0.70	−0.20	0.08
Cropland	0.6		99.1	−0.27	0.24	−0.22	0.04
Wetland/Waterbody	13.2		47.4	−0.26	0.54	0.15	0.28
Settlement			100	−0.24	0.02	−0.13	0.09
Otherland	33.5		3.8	−0.23	0.35	0.22	0.07
iii. Persistence (no change areas) 2000–2018							
Tree-covered area	0.2	0.03	99.4	−0.28	0.70	−0.22	0.03
Grassland	4.4	0.2	94.0	−0.29	0.70	0.18	0.09
Cropland	0.5	0.02	99.0	−0.27	0.27	−0.21	0.04
Wetland/Waterbody	17.4		11.2	−0.26	0.65	0.23	0.18
Settlement	26.8		73.2	−0.24	0.24	−0.07	0.12
Otherland	10.9	0.4	1.4	−0.20	0.69	0.26	0.10

Figure 7. Distribution of vegetation trends for major land use-land cover categories for the year 2018: (**a**) Tree−covered area; (**b**) Cropland; (**c**) Otherland; (**d**) Grassland; (**e**) Vegetation change as percent area of land in (i) 2018 land categories, (ii) areas of land change between 2000 and 2018 and (iii) areas of land persistence where no change occurred between 2000 and 2018.

Between years 2000 and 2018 in the loss areas, the greatest percentage of decreasing vegetation productivity (above 90%) were found in tree-covered, settlement and cropland areas. Areas with increasing trends in loss areas are otherland (34%), wetlands (137%) and grasslands (2%). In areas of persistence, vegetation productivity declined mostly in the same LULC types as in loss areas, whereas it improved in 27% of settlement areas, 17% of wetlands, 11% of otherlands and 4% of grasslands. Minimum and maximum values of NDVI trends in areas of land loss varied between forests (−0.3, 0.21), grassland (−0.28, 0.70), cropland (−0.27, 0.24), wetland (−0.26, 0.54), settlement (−0.24, 0.02) and otherland (−0.23, 0.35).

3.3.2. Drought Severity by Land Use-Land Cover Type

Focusing on the drought years identified earlier in the time series analysis of vegetation variability and drought severity (refer to Figures 2 and 5), we examined how drought severity differed between the major LULC types. The 2002–2003 drought-stricken growing season was used as an example because it was the worst drought experienced in the entire time series (Figure 8a–d). For example, land areas most impacted by extreme and severe droughts, respectively, in 2003–2004 are: over otherland (32%, 31%), grassland (19%, 39%) and cropland (12%, 37%). Drought intensities for areas under each LULC type for the identified drought years are shown in percentages alongside drought severity classes (Figure 8e).

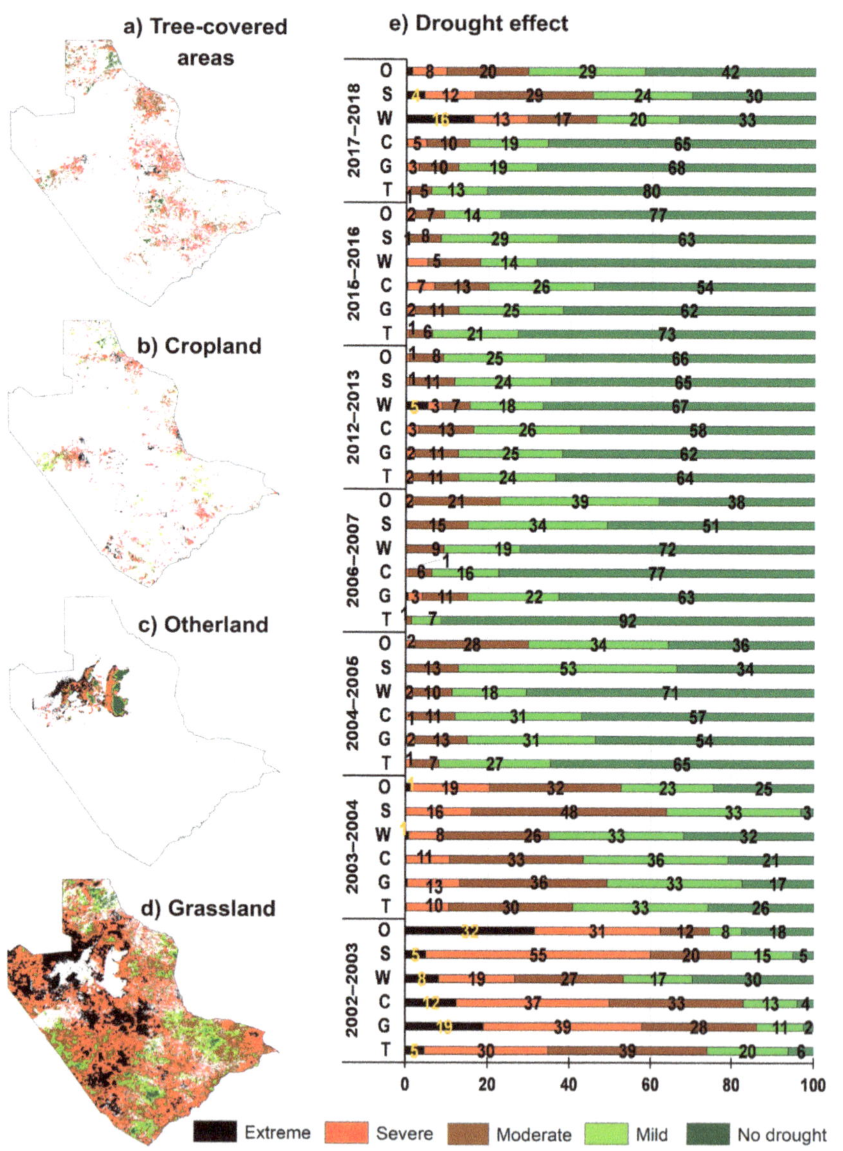

Figure 8. Distribution of drought severity for major land use-land cover types: (**a**) Tree-covered area; (**b**) Cropland; (**c**) Otherland; (**d**) Grassland, during the growing season of 2002–2003 (a drought year); (**e**) Percentage of land area under varying drought severity and non-drought conditions based on VCI for selected drought years by land use-land cover types (T = Tree-covered areas, G = Grassland, C = Cropland, W = Wetlands/Waterbodies, S = Settlement, O = Otherland).

4. Characterising Drought in the Constituencies

Drought Severity in Constituencies in Comparison with Drought Declaration

Drought and household food security vulnerability assessments are conducted annually during the mid-growing season in Botswana [42]. Since assessment and interventions are conducted at local levels, we further examined the severity of the drought in the 17 constituencies (Figure 9). The drought frequency and heatmap reveal how the constituencies were affected by droughts of differing magnitudes throughout the entire time series.

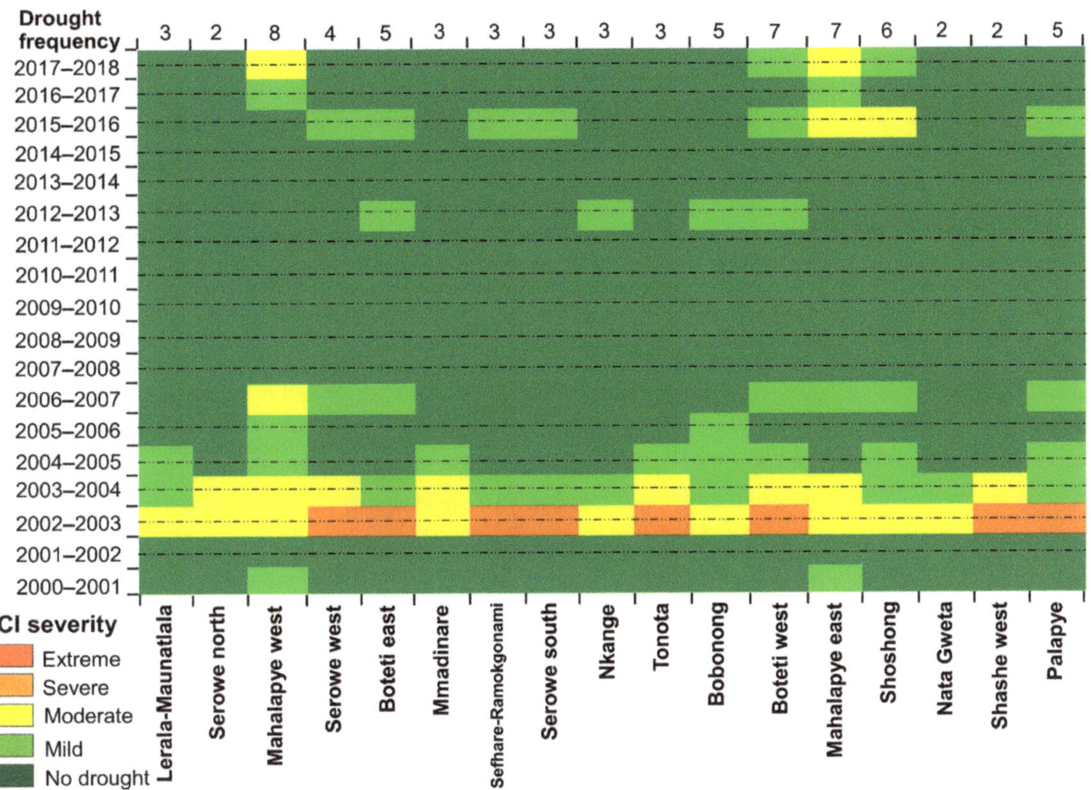

Figure 9. Heatmap of drought severity at constituency level during the growing seasons of 2000–2001 to 2017–2018. The years with dashed lines across were declared drought years by the Botswana government [31].

Figure 9 depicts drought intensities for each year's growing season and drought frequency per constituency in the entire time series. This heatmap is based on the median VCI value for each constituency (heatmaps of drought severity using the minimum and mean VCI values are shown in Figures S1 and S2, respectively). The prolonged drought during the multiple seasons of 2002–2004 is evident in the heatmap. Based on the count of drought occurrences, irrespective of severity classes between 2000–2018, constituencies with the most frequent droughts in descending order are Mahalapye west (eight), Mahalapye east (seven), Boteti west (seven), Shoshong (six), Bobonong (five), Boteti east (five) and Palapye (five). Other constituencies experienced between two and four drought occurrences with lesser severity. Of the 16 declared drought years by the government (the years with dashed lines in Figure 9), eight were evident in the time series, whereas the other years were favourable for the CDB.

5. Discussion

5.1. Vegetation Condition Change and Drought Severity

There was high spatial and temporal variability in vegetation productivity during the growing seasons in the 18-year study period. This is typical of dryland ecosystems which are often non-equilibrium and dynamic in response to both climatic and anthropogenic perturbations [43]. Recovery of vegetation during the start of the time series (2000–2001) after the prolonged droughts of the 1990s was evident. This finding is corroborated by [21], which noted above-normal rainfall in the CDB in the year 2000. For example, analysing rainfall amount between 1960–2015 for Palapye, the authors in [32] noted that vegetation condition improved in 1999–2000 after 649 mm rainfall was recorded that year, which was well above the long-term average of 351 mm.

Comparing vegetation productivity in the CDB during drought and non-drought years with the mean for the entire time series revealed the impacts droughts have on vegetation. For example, when compared to the mean, vegetation productivity was very limited in 2002–2003 (the worst drought episode in the time series). Drought occurrence was evident during the growing seasons of eight declared drought years in line with [21,30]. FAO special alert for Southern Africa in 2015 noted the retarded growth of early-planted crops as soil moisture was very low at the beginning of the growing season in most parts of the southern African region, including Botswana [44]. The drought and household food security outlook report of 2017 [45] attributed the decrease in vegetation productivity in most parts of Botswana to negative drought impacts on vegetation.

The lower than normal vegetation productivity during some of the drought-stricken growing seasons can be attributed to droughts linked to El Niño southern oscillation (ENSO). For example, the prolonged droughts in the growing seasons of 2002–2004 and 2015–2016 coincided with the El Niño years in the recent records [4]. Relating the association of ENSO to drought severity during the growing season as utilised in this study, the authors of [4] found the highest statistically significant correlations in January, February and March in Botswana, whereas they found negative non-significant correlations at the start of the season in October. At the regional level in southern Africa, the authors of [46] associated droughts with anomalies of negative Standardised Precipitation Evaporation Index and positive Sea Surface Temperature. At the global level, studies have also documented the effects of ENSO on drought severity, such as [46].

Comparing the ongoing growing season (2020–2021) with the mean for the entire time series, we found above-normal vegetation productivity after mid-October 2020 until February 2021. Thus, this suggests the full recovery of vegetation productivity during this season from the impacts of the prolonged droughts in the last couple of growing seasons. However, this observation is somewhat fraught with uncertainty judging from the below-normal vegetation productivity at the start of the season. Moreover, the growing season has not ended yet. The growing season spanning the first dekad in October to the third dekad in March was chosen to align the cropping and the raining season in Botswana, which enabled the exclusion of the dry season from the drought analyses.

5.2. Vegetation Trend and Drought Severity by Land Use-Land Cover and Change

Many studies on droughts have not examined how drought effects differ between LULC types. For those that have land use incorporated, little is known of the influence of drought severity on LULC—in terms of the differences in annual sizes and configurations—either in changing and/or persistent areas. Processes driving decreasing vegetation trends, either climatic or anthropogenic, are better identified when LULC and change are incorporated in the examination of drought severity. For example, in CDB between 2000 and 2018, vegetation productivity declined in most forests, woodlands, croplands and grasslands. In land change areas, the trend of declining vegetation was equally high. In areas of persistence, the greatest percentage of improved vegetation productivity was in wetlands, settlements, and otherlands. Minimal improvement of vegetation trends in forested areas can be attributed to the overall increase in tree-covered areas during the study period.

Previous studies in dryland contexts such as Botswana and elsewhere associate the improvement in vegetation productivity partly to bush-encroachment which remote sensing vegetation indices capture as vegetation greening but bush-encroachment is undesirable in cattle-based systems [32,47,48].

Relating LULC change to land degradation, conversion of tree-covered areas to grassland, otherland and cropland, is degrading. This is because these land transitions drive the removal of vegetation cover and contribute to land degradation processes. Similarly considered as degrading land transitions are those involving the conversion of grasslands into croplands, artificial surface areas and otherlands. For example, in Palapye, the authors of [32,49] found increases in barelands and rock outcrops with limited vegetation growth because of prolonged droughts. This bareland condition due to drought-induced vegetation decline is further exacerbated by human activities, such as overgrazing. In some instances, such as in Bobonong, researchers [30] found increases in bareland patches in communal grazing areas. Despite declining vegetation conditions during a four-year prolonged drought (2002–2005), livestock overgrazed and natural pastures degraded, as pastoralists had no incentive to destock or sell their cattle because of the slump in prices due to the prevalence of Foot-and-Mouth cattle disease during the drought. In other dryland instances, such as in the Sahel, bareland areas with minimal vegetation growth alternate with grasslands in response to rainfall variability and drought [50].

Drought impacts on grasslands, forests and wetlands imply negative impacts on the cattle system and biodiversity, including wildlife in savannas with the associated tourism and hospitality sector, whereas effects on croplands impact food production. For example, the water crisis of 2015–2016 resulting from relatively low, erratic rainfalls reduced water levels and water inflows into dams drastically across the country [51]. Regarding drought impacts on agriculture, for example, maize production in Botswana declined from 35,322 tons in 2011 to 13,911 tons in 2017 mainly due to drought constraints. This necessitated an expenditure of about P389 Million (~36 Million American Dollars) on maize import to meet cereal requirements not met through domestic crop production [31]. Other changes in LULC, such as the changes in the extent of settlements, as observed between 2000 and 2018 in this study, are not drought-related. Settlement expansion is more a reflection of the increasing land demand for human habitation and infrastructure due to the population growth experienced in the CDB.

5.3. Drought Severity in the Constituencies

Relating drought severity to the years declared as drought-stricken by the government, the results reveal that not all constituencies were equally affected by drought, as severity differed from severe to mild drought. Moreover, drought severities in some declared drought years were not as widespread in the CDB as in other parts of the country. For example, the growing season of the year 2009–2010 had improved vegetation productivity in response to above-normal rainfall recorded in previous months, which resulted in flooding events in five sub-districts in the CDB (Serowe/Palapye, Tutume, Boteti, Mahalapye and Bobirwa) [52]. In other drought-prone contexts in southern Africa, such as Zimbabwe, researchers [53] found that a drought's distribution and effects differed geographically and from season to season.

Drought severity was further gauged by the frequency at the constituency level. Over the study period, the constituencies experienced between eight to two drought events. Examples are Mahalapye west and east, with eight and seven drought occurrences, respectively, ranging from moderate to mild drought. Boteti west experienced seven drought events with severity ranging from severe to mild. Confirming these drought frequencies in the CDB, the authors of [30] found an average drought frequency of two to four (depending on the index) in the Bobonong region between 2000 and 2015. This is in line with our finding of five drought occurrences in Bobonong as our time series extending up to 2018 captured more drought events.

Drought effects on land-based resources and livelihoods will vary depending on the drought severity, the land use as well as the land management practices. For example, water use strategy adopted as part of land management practices was found to have impacted the responses of tree plantations to drought in China [54]. Droughts cannot be avoided, since these are an integral part of the climate cycle; however, the impacts on social-ecological systems are better minimised with monitoring as input for the use of both proactive and reactive measures [4,21].

6. Conclusions

This study proposed a spatial and temporal analysis of drought evolution in the Central District in eastern Botswana from 2000 to 2018. The results highlight the usefulness of incorporating land use-land cover and change in assessing the spatio-temporal variability of drought severity in drylands. Remote Sensing-based vegetation time series metrics were used, as complementary to climatological indices. Indicators characterising changes in vegetation conditions and drought severity during the growing seasons (October to March) from 2000–2001 to 2017–2018 were used. These indicators are NDVI difference—$NDVI_{diff}$, NDVI anomaly, NDVI trends, Vegetation Condition Index—VCI, Drought intensity and frequency. The use of these different NDVI-based indicators, which might seem redundant, are useful as complementary measures since they differ computation-wise. The NDVI difference as utilised in this study captured the in-season variability in vegetation productivity, whereas the VCI compared each growing season with the long-term minimum and maximum conditions. For example, limited vegetation productivity found during the growing seasons of 2002–2004 and 2015–2016, which was based on both NDVI difference and NDVI anomalies, agreed with the heightened levels of drought severity over the same periods as derived from the VCI.

Results further showed high temporal and spatial variability in vegetation productivity between drought and non-drought conditions in our case studies. The associated negative impact of droughts on vegetation resulting in limited vegetation productivity was further confirmed by results from this study. Drought effects on vegetation productivity during the study period were characterised by decreasing vegetation trends in most parts of the district. Although varying intensities of drought severity (severe, moderate and mild) occurred in the constituencies, the 2002–2003 and 2003–2004 growing seasons were found to be the worst drought periods in the entire series, as most parts of the district were affected. Assessing drought severity and intensities by LULC in selected drought years revealed varying drought effects. We found that drought effects differed between LULC types as well as whether these were areas of land change or persistence. Further examination of drought impacts in areas of no change is required, as our understanding of drought effects in areas with no change is still limited. More empirical studies in this regard will provide useful insights. Using the example of the 2002–2003 drought-stricken growing season, the highest percentage of land impacted by extreme and severe droughts were found in tree-covered areas, croplands and grasslands, whereas improved vegetation trends were found mostly in wetlands and some instances in otherland areas including barelands. Moreover, the results suggest that even in declared drought years, droughts severity varied, and the effects differed between constituencies. A further insight provided is that the magnitude of drought severity in some declared drought years was not as widespread in the CDB. For example, no other severe drought levels were recorded in the CDB after the extended drought which affected the growing seasons of 2002–2004.

Differences in spatial resolution of the datasets utilised and the coarse spatial resolution of the 1 km NDVI datasets compared to 300 m annual LULC are limitations identified in this study. With the increasing availability of images of higher spatial resolution, such as from SENTINEL-2, results from RS-based analysis of drought can be improved. However, methodological challenges ensue with the need to incorporate the newer images into the existing NDVI archives. For example, consideration ought to be given on parameterising across sensors and balancing the trade-offs between taking advantage of the superior spatial resolution of the newer satellite missions (e.g., Sentinel-1, 2 and 3) and the temporal resolution of images from the older missions (e.g., SPOT VGT and PROBA-V). As the 1 km, NDVI time series datasets extend way back to the 1990s, their use is indispensable for drought analysis at the current time. Moreover, as drought years were easily detected in the time series analyses, this is proof of the usefulness of the 1 km NDVI time series. For example, the lower than normal vegetation productivity during the prolonged drought periods that negatively impacted the 2002–2004 and 2015–2016 growing seasons coincided with strong El Niño years. With the above-normal vegetation productivity in the ongoing season (2020–2021), results suggest the reversal of the negative vegetation trends observed in the preceding growing seasons. How much these negative trends have been reversed remain uncertain, as the season is still ongoing. For clarity, future studies should examine the usefulness of RS-based indices for understanding the ongoing season's phenology in dryland contexts such as the CDB.

Remote Sensing-based time series enabled us to extend the analysis up to the ongoing season, demonstrating its usefulness for better characterisation of drought events. Remote Sensing-based results such as those obtained in this study, when provided at multiple administrative scales in a timely and cost-effective manner, have the potential to aid decision-makers to better plan and respond to drought situations. Scientific evidence is needed as input into the decision-making process to aid national resource mobilisation for drought management. Botswana requires both proactive and reactive approaches for drought management, for which remote sensing-based assessment and monitoring foster the implementation of drought early warning systems.

Supplementary Materials: The following are available online at https://www.mdpi.com/2072-4292/13/5/836/s1, Table S1: Geographic and socioeconomic characteristics of the constituencies in the Central District, Table S2: Land cover classification scheme, Table S3: Land use-land cover transition matrix in km^2 (2000–2018), Figure S1: VCI minimum value heatmap of drought severity at constituency level during the growing seasons of 2000–2001 to 2017–2018. The years in bold are declared drought years by the Botswana government (Source: [41]), Figure S2: VCI mean value heatmap of drought severity at constituency level during the growing seasons of 2000–2001 to 2017–2018. The years in bold are declared drought years by the Botswana government (Source: [41]).

Funding: This research received no external funding.

Data Availability Statement: Data supporting reported results can be found through the Copernicus Land Monitoring Service (https://land.copernicus.eu/global/products/ndvi, accessed on 8 January 2021) and the land cover datasets from the European Space Agency CDS (https://cds.climate.copernicus.eu/cdsapp#!/dataset/satellite-land-cover?tab=form, accessed on 8 January 2021).

Acknowledgments: The author is grateful for free access to all datasets used in this study including the provision of satellite image time series and support with the Drought Monitoring System software by the MESA-SADC Thema project. Four anonymous reviewers provided constructive feedback that improved the paper's quality.

Conflicts of Interest: The author declares no conflict of interest.

References

1. Vicente-serrano, S.M. Evaluating the impact of drought using Remote Sensing in a Mediterranean, semi-arid region. *Nat. Hazard.* **2007**, *40*, 173–208. [CrossRef]
2. Wilhite, D.A.; Svoboda, M.D.; Hayes, M.J. Understanding the complex impacts of drought: A key to enhancing drought mitigation and preparedness. *Water Resour. Manag.* **2007**, *21*, 763–774. [CrossRef]
3. Mogotsi, K.; Nyangito, M.M.; Nyariki, D.M. The role of drought among agro-pastoral communities in a semi-arid environment: The case of Botswana. *J. Arid Environ.* **2013**, *91*, 38–44. [CrossRef]
4. Byakatonda, J.; Parida, B.P.; Moalafhi, D.B.; Kenabatho, P.K.; Lesolle, D. Investigating relationship between drought severity in Botswana and ENSO. *Nat. Hazards* **2020**, *100*, 255–278. [CrossRef]
5. Akinyemi, F.O.; Abiodun, B.J. Potential impacts of global warming levels 1.5 °C and above on climate extremes in Botswana. *Clim. Chang.* **2019**, *154*, 387–400. [CrossRef]
6. Oikonomou, P.D.; Karavitis, C.A.; Tsesmelis, D.E.; Kolokytha, E.; Maia, R. Drought Characteristics Assessment in Europe over the Past 50 Years. *Water Resour. Manag.* **2020**, *34*, 4757–4772. [CrossRef]

7.   Madani, N.; Kimball, J.S.; Jones, L.A.; Parazoo, N.C.; Guan, K. Global analysis of bioclimatic controls on ecosystem productivity using satellite observations of solar-induced chlorophyll fluorescence. *Remote Sens.* **2017**, *9*, 530. [CrossRef]
8.   Anderegg, W.R.L.; Kane, J.M.; Anderegg, L.D.L. Consequences of widespread tree mortality triggered by drought and temperature stress. *Nat. Clim. Chang.* **2013**, *3*, 30–36. [CrossRef]
9.   Vicente-Serrano, S.M.; Begueria, S.; Camarero, J.J. Drought severity in a changing climate. In *Handbook of Drought and Water Scarcity: Principles of Drought and Water Scarcity*; Eslamian, S., Eslamian, F., Eds.; CRC Press: New York, NY, USA, 2017; pp. 279–303.
10. Xu, H.; Wang, X.; Zhao, C.; Yang, X. Assessing the response of vegetation photosynthesis to meteorological drought across northern China. *Land Degrad. Dev.* **2021**, *32*, 20–34. [CrossRef]
11. Friedlingstein, P.; Bopp, L.; Ciais, P.; Dufresne, J.-L.; Fairhead, L.; Letreut, H.; Monfray, P.; Orr, J. Positive feedback between future climate change and the carbon cycle. *Geophys. Res. Lett.* **2001**, *28*, 1543–1546. [CrossRef]
12. Trenberth, K.E.; Dai, A.; van der Schrier, G.; Jones, P.D.; Barichivich, J.; Briffa, K.R.; Sheffield, J. Global warming and changes in drought. *Nat. Clim. Chang.* **2014**, *4*, 17–22. [CrossRef]
13. Stagge, J.H.; Kingston, D.G.; Tallaksen, L.M.; Hannah, D.M. Observed drought indices show increasing divergence across Europe. *Sci. Rep.* **2017**, *7*, 14045. [CrossRef] [PubMed]
14. Spinoni, J.; Naumann, G.; Vogt, J.V. Pan-European seasonal trends and recent changes of drought frequency and severity. *Glob. Planet. Chang.* **2017**, *148*, 113–130. [CrossRef]
15. Máure, G.; Pinto, I.; Ndebele-Murisa, M.; Muthige, M.; Lennard, C.; Nikulin, G.; Dosio, A.; Meque, A. The southern African climate under 1.5 °C and 2 °C of global warming as simulated by CORDEX regional climate models. *Environ. Res. Lett.* **2018**, *13*, 065002. [CrossRef]
16. He, M.; Kimball, J.S.; Yi, Y.; Running, S.; Guan, K.; Jensco, K.; Maxwell, B.; Maneta, M.; Jencso, K. Impacts of the 2017 flash drought in the US Northern Plains informed by satellite-based evapotranspiration and solar-induced fluorescence. *Environ. Res. Lett.* **2019**, *14*, 074019. [CrossRef]
17. Christopherson, R.W. *Geosystems: An Introduction to Physical Geography*, 8th ed.; Pearson Prentice Hall: London, UK, 2013.
18. van Wilgen, N.J.; Goodall, V.; Holness, S.; Chown, S.L.; McGeoch, M.A. Rising temperatures and changing rainfall patterns in South Africa's national parks. *Int. J. Climatol.* **2016**, *36*, 706–721. [CrossRef]
19. Engelbrecht, F.; Adegoke, J.; Bopape, M.J.; Naidoo, M.; Garland, R.; Thatcher, M.; Gatebe, C. Projections of rapidly rising surface temperatures over Africa under low mitigation. *Environ. Res. Lett.* **2015**, *10*, 8. [CrossRef]
20. Nikulin, G.; Lennard, C.; Dosio, A.; Kjellstroem, E.; Chen, Y.; Haensler, A.; Kupiainen, M.; Laprise, R.; Mariotti, L.; Maule, C.F.; et al. The effects of 1.5 and 2 degrees of global warming on Africa in the CORDEX ensemble. *Environ. Res. Lett.* **2018**, *13*, 065003. [CrossRef]
21. Akinyemi, F.O. Climate change and variability in semi-arid Palapye; Eastern Botswana: An assessment from smallholder farmers' perspective. *Weather Clim. Soc.* **2017**, *9*, 349–365. [CrossRef]
22. Batisani, N.; Yarnal, B. Rainfall variability and trends in semi-arid Botswana: Implications for climate change adaptation policy. *Appl. Geogr.* **2010**, *30*, 483–489. [CrossRef]
23. Byakatonda, J.; Parida, B.P.; Kenabatho, P.K. Relating the dynamics of climatological and hydrological droughts in semiarid Botswana. *Phys. Chem. Earth.* **2018**, *105*, 12–24. [CrossRef]
24. Byakatonda, J.; Parida, B.P.; Kenabatho, P.K.; Moalafhi, D.B. Prediction of onset and cessation of austral summer rainfall and dry spell frequency analysis in semiarid Botswana. *Theor. Appl. Climatol.* **2018**, *135*, 101–117. [CrossRef]
25. Nkemelang, T.; New, M.; Zaroug, M.A.H. Temperature and precipitation extremes under current, 1.5 °C and 2.0 °C global warming above pre-industrial levels over Botswana, and implications for climate change vulnerability. *Environ. Res. Lett.* **2018**, *13*, 065016. [CrossRef]
26. Dubovyk, O.; Ghazaryan, G.; González, J.; Graw, V.; Löw, F.; Schreier, J. Drought hazard in Kazakhstan in 2000–2016: A remote sensing perspective. *Environ. Monit. Assess.* **2019**, *191*, 510. [CrossRef]
27. Kogan, F. Global Drought Watch from space. *Bull. Am. Meteorol. Soc.* **1997**, *78*, 621–636. [CrossRef]
28. AghaKouchak, A.; Farahmand, A.; Melton, F.S.; Teixeira, J.; Anderson, M.C.; Wardlow, B.D.; Hain, C.R. Remote Sensing of Drought: Progress, Challenges and Opportunities. *Rev. Geophys.* **2015**, *53*, 2014RG000456. [CrossRef]
29. Ahmadalipour, A.; Moradkhani, H.; Yan, H.; Zarekarizi, M. Remote Sensing of Drought: Vegetation, Soil Moisture, and Data Assimilation. In *Remote Sensing of Hydrological Extremes. Springer Remote Sensing/Photogrammetry*; Lakshmi, V., Ed.; Springer: Cham, Switzerland, 2017. [CrossRef]
30. Mugari, E.; Masundire, H.; Bolaane, M. Effects of Droughts on Vegetation Condition and Ecosystem Service Delivery in Data-Poor Areas: A Case of Bobirwa Sub-District; Limpopo Basin and Botswana. *Sustainability* **2020**, *12*, 8185. [CrossRef]
31. Statistics Botswana. *Annual Agricultural Survey Report 2014*; Statistics Botswana: Gaborone, Botswana, 2016; p. 155. Available online: http://www.statsbots.org.bw/sites/default/files/publications/Annual%20Agriculture%20Survey%20%202014.pdf (accessed on 31 December 2020).
32. Akinyemi, F.O.; Kgomo, M.O. Vegetation dynamics in African drylands: An assessment based on the Vegetation Degradation Index in an agro-pastoral region of Botswana. *Reg. Environ. Chang.* **2019**, *19*, 2027–2039. [CrossRef]
33. Vito. Operation of the Global Land Component: Algorithm Theoretical Basis. Issue I2.11 GIO-GL Lot1 Consortium. 2015. Available online: https://land.copernicus.eu/global/sites/cgls.vito.be/files/products/GIOGL1_ATBD_PROBA2VGT_I1.10.pdf (accessed on 31 December 2020).

34. Santoro, M.; Kirches, G.; Wevers, J.; Boettcher, M.; Brockmann, C.; Moreau, I.; Van Bogaert, E.; De Maet, T.; Bontemps, S.; Lamarche, C.; et al. Land Cover CCI Product User Guide Version 2.0. 2017. Available online: https://maps.elie.ucl.ac.be/CCI/viewer/download/ESACCI-LC-Ph2-PUGv2_2.0.pdf (accessed on 25 January 2020).
35. Akinyemi, F.O.; Ghazaryan, G.; Dubovyk, O. Assessing UN indicators of Land Degradation Neutrality and proportion of degraded land over Botswana using Remote Sensing based national level metrics. *Land Degrad. Dev.* **2021**, *32*, 158–172. [CrossRef]
36. DFRR. Enhancing National Forest Monitoring System for the Promotion of Sustainable Natural Resource Management Project Report. Gaborone: Ministry of Environment; Natural Resources Conservation and Tourism. 2017. Available online: http://open_jicareport.jica.go.jp/pdf/12301594_01.pdf (accessed on 1 January 2021).
37. Pu, R.L.; Gong, P.; Tian, Y.; Miao, X.; Carruthers, R.I.; Anderson, G.L. Using classification and NDVI differencing methods for monitoring sparse vegetation coverage: A case study of Saltcedar in Nevada; USA. *Int. J. Remote Sens.* **2008**, *29*, 3987–4011. [CrossRef]
38. Erasmi, S.; Schucknecht, A.; Barbosa, M.P.; Matschullat, J. Vegetation greenness in Northeastern Brazil and its relation to ENSO warm events. *Remote Sens.* **2014**, *6*, 3041–3058. [CrossRef]
39. De Jong, R.; de Bruin, S.; de Wit, A.; Schaepman, M.E.; Dent, D.L. Analysis of monotonic greening and browning trends from global NDVI time-series. *Remote Sens. Environ.* **2011**, *115*, 692–702. [CrossRef]
40. Neeti, N.; Eastman, J.R. A contextual Mann-Kendall approach for the assessment of trend significance in image time series. *Trans. GIS* **2011**, *15*, 599–611. [CrossRef]
41. Zambrano, F.; Lillo-Saavedra, M.; Verbist, K.; Lagos, O. Sixteen years of agricultural drought assessment of the BioBío region in Chile using a 250 m resolution Vegetation Condition Index (VCI). *Remote Sens.* **2016**, *8*, 530. [CrossRef]
42. Rural Development Council. Summary Report for the Drought and Household Food Security Vulnerability Assessment 2018/19. Gaborone: Botswana Vulnerability Assessment Committee; Ministry of Local Government and Rural Development. 2019. Available online: https://www.statsbots.org.bw/sites/default/files/Botswana%20Environment%20Statistics%20%20Natural%20and%20Technological%20Disasters%20Digest%202019%202%20(1).pdf (accessed on 7 January 2021).
43. Mogotsi, K.; Nyangito, M.M.; Nyariki, D.M. Drought management strategies among agro-pastoral communities in non-equilibrium Kalahari ecosystems. *Environ. Res. J.* **2011**, *5*, 156–162. [CrossRef]
44. FAO. Delayed Onset of Seasonal Rains in Parts of Southern Africa Raises Serious Concern for Crop and Livestock Production in 2016. FAO Special Alert for Southern Africa. 2015, Volume 336, p. 6. Available online: http://www.fao.org/3/a-i5258e.pdf (accessed on 25 December 2017).
45. Rural Development Council. *Drought and Household Food Security Outlook 2016/17*; Ministry of Local Government and Rural Development: Gaborone, Botswana, 2017.
46. Vicente-Serrano, S.M.; López-Moreno, J.I.; Gimeno, L.; Nieto, R.; Morán-Tejeda, E.; Lorenzo-Lacruz, J.; Beguería, S.; Azorin-Molina, C. A multiscalar global evaluation of the impact of ENSO on droughts. *J. Geophys. Res.* **2011**, *116*, 016039. [CrossRef]
47. Dardel, C.; Kergoat, L.; Hiernaux, P.; Mougin, E.; Grippa, M.; Tucker, C. Re-greening Sahel: 30 years of remote sensing data and field observations (Mali, Niger). *Remote Sens. Environ.* **2014**, *140*, 350–364. [CrossRef]
48. Brandt, M.; Tappan, G.; Diouf, A.A.; Beye, G.; Mbow, C.; Fensholt, R. Woody vegetation die off and regeneration in response to rainfall variability in the West African Sahel. *Remote Sens.* **2017**, *9*, 39. [CrossRef]
49. Akinyemi, F.O.; Mashame, G. Analysis of land change in the dryland agricultural landscapes of eastern Botswana. *Land Use Policy* **2018**, *76*, 798–811. [CrossRef]
50. UNEP 2012. Sahel Atlas of Changing Landscapes: Tracing Trends and Variations in Vegetation Cover and Soil Condition. UNEP: Nairobi. Available online: https://wle.cgiar.org/sahel-atlas-changing-landscapes-tracing-trends-and-variations-vegetation-cover-and-soil-condition (accessed on 7 February 2021).
51. Alemaw, B.F.; Keaitse, E.O.; Chaoka, T.R. Management of water supply reservoirs under uncertainties in arid and urbanized Environments. *J. Water Resour. Prot.* **2016**, *8*, 990–1009. [CrossRef]
52. Disaster Relief Emergency Fund. Botswana Floods. DREF Operation n° MDRBW001 GLIDE n° FL-2009-000120-BWA. 2009. Available online: https://reliefweb.int/sites/reliefweb.int/files/resources/D4C2CB5A61EB2BC6852575E500532FCC-Full_Report.pdf (accessed on 6 January 2021).
53. Mutowo, G.; Chikodzi, D. Remote sensing based drought monitoring in Zimbabwe. *Disaster Prev. Manag.* **2014**, *23*, 649–659. [CrossRef]
54. Gao, X.; Li, H.; Zhao, X. Impact of land management practices on water use strategy for a dryland tree plantation and subsequent responses to drought. *Land Degrad. Dev.* **2020**, *32*, 439–452. [CrossRef]

 *remote sensing*

Article

Classification Efficacy Using K-Fold Cross-Validation and Bootstrapping Resampling Techniques on the Example of Mapping Complex Gully Systems

Kwanele Phinzi [1,*], **Dávid Abriha** [1] and **Szilárd Szabó** [2]

[1] Department of Physical Geography and Geoinformatics, Faculty of Science and Technology, Doctoral School of Earth Sciences, University of Debrecen, Egyetem tér 1, 4032 Debrecen, Hungary; abriha.david@science.unideb.hu

[2] Department of Physical Geography and Geoinformatics, Faculty of Science and Technology, University of Debrecen, Egyetem tér 1, 4032 Debrecen, Hungary; szabo.szilard@science.unideb.hu

[*] Correspondence: phinzi.kwanele@science.unideb.hu

Abstract: The availability of aerial and satellite imageries has greatly reduced the costs and time associated with gully mapping, especially in remote locations. Regardless, accurate identification of gullies from satellite images remains an open issue despite the amount of literature addressing this problem. The main objective of this work was to investigate the performance of support vector machines (SVM) and random forest (RF) algorithms in extracting gullies based on two resampling methods: bootstrapping and k-fold cross-validation (CV). In order to achieve this objective, we used PlanetScope data, acquired during the wet and dry seasons. Using the Normalized Difference Vegetation Index (NDVI) and multispectral bands, we also explored the potential of the PlanetScope image in discriminating gullies from the surrounding land cover. Results revealed that gullies had significantly different ($p < 0.001$) spectral profiles from any other land cover class regarding all bands of the PlanetScope image, both in the wet and dry seasons. However, NDVI was not efficient in gully discrimination. Based on the overall accuracies, RF's performance was better with CV, particularly in the dry season, where its performance was up to 4% better than the SVM's. Nevertheless, class level metrics (omission error: 11.8%; commission error: 19%) showed that SVM combined with CV was more successful in gully extraction in the wet season. On the contrary, RF combined with bootstrapping had relatively low omission (16.4%) and commission errors (10.4%), making it the most efficient algorithm in the dry season. The estimated gully area was 88 ± 14.4 ha in the dry season and 57.2 ± 18.8 ha in the wet season. Based on the standard error (8.2 ha), the wet season was more appropriate in gully identification than the dry season, which had a slightly higher standard error (8.6 ha). For the first time, this study sheds light on the influence of these resampling techniques on the accuracy of satellite-based gully mapping. More importantly, this study provides the basis for further investigations into the accuracy of such resampling techniques, especially when using different satellite images other than the PlanetScope data.

Keywords: satellite imagery; gully mapping; machine learning; random forest; support vector machines; South Africa; semi-arid environment

Citation: Phinzi, K.; Abriha, D.; Szabó, S. Classification Efficacy Using K-Fold Cross-Validation and Bootstrapping Resampling Techniques on the Example of Mapping Complex Gully Systems. *Remote Sens.* **2021**, *13*, 2980. https://doi.org/10.3390/rs13152980

Academic Editor: Elias Symeonakis

Received: 6 June 2021
Accepted: 26 July 2021
Published: 28 July 2021

Publisher's Note: MDPI stays neutral with regard to jurisdictional claims in published maps and institutional affiliations.

1. Introduction

Defined as the detachment, transportation, and deposition of soil particles by the erosive forces of raindrop and runoff [1,2], soil erosion by water represents one of the most typical forms of land degradation affecting many countries around the world [3]. While soil erosion has many negative effects, the most concerning one include the decline in soil fertility, resulting in limited food production [4,5]. This, in turn, contributes to food insecurity in several developing countries, particularly in those ones where a considerable segment of their population strongly relies on agriculture for their survival [6]. South

Africa, with approximately six million people deriving a livelihood from agriculture [7], is extremely exposed to soil erosion. Formal agriculture provides employment to about 930,000 farm workers, including seasonal and contract workers [7]. Given the geomorphological conditions coupled with the strongly seasonal nature of rainfall across South Africa, it is not surprising that the country is predisposed to soil erosion, a serious threat to sustainable agriculture and natural environments [8]. Soil erosion in South Africa, especially in rural communities, has been further aggravated by human activities such as inappropriate agricultural practices and overstocking [9–12].

Although various types of water-borne erosion exist in the country, gully formation has been recognized as the major form of erosion in South Africa, accounting for considerable volumes of soil loss [13,14]. Accordingly, the Department of Agriculture, Forestry, and Fisheries (DAFF) in South Africa has identified the need to determine the spatial extent of gullies and their severity at a national scale [15]. Gullies occur when the soil and its parent material are scored and destroyed by surface runoff, resulting in the formation of v-shaped incised channels [16]. Gullies can either be classified as ephemeral or classical (also called permanent) based mainly on their depth. Unlike ephemeral gullies, classical gullies are deeper than 0.5 m and cannot be easily filled in by normal tillage [17], especially in highly dissected terrains [18]. Gullies also result from piping and tunneling due to the influence of soil chemistry on hydrological pathways [19]. The prevalence of erodible duplex and dispersive soils in certain parts of South Africa, especially the Eastern Cape where the subsurface (piping) erosion mostly occur, considerably facilities the formation and development of gullies [9,14]. Land use type and changes also trigger gully initiation [19]. In the context of South Africa, gullies are more prominent on gently sloping lands suitable for cultivation [15]. The spatial extent and severity of gully erosion vary from one province to another because of the differences in land use, soil types, vegetation, rainfall, and topography existing in different provinces. The Eastern Cape is one of the most gully-affected provinces in South Africa, with about 161,500 ha of land covered by gullies [15]. For this reason, most gully erosion studies in the country have been conducted in this province [9,14,20–22].

Accurate mapping of gullies is essential for monitoring gully erosion and understanding the associated environmental and socio-economic impacts [23], thereby supporting the implementation of practical erosion control measures [24,25]. Manual field-based assessments using tapes, rulers, and topographic profilers have been used for years to obtain gully information [26], but over the last few decades, rapid developments had been witnessed in digital aerial photography, and more recently, satellite images with different imaging capabilities [23]. Following the availability of such remotely sensed data, gully information has either been obtained through visual interpretation or automatic classification of remotely sensed data. Remote sensing related mapping, either based on visual interpretation or automatic method, is presently the only practical approach for mapping gully features over large areas, in arid or semi-arid regions, given the complexity of gully appearance (i.e., variability in size, shape, and occurrence) [27]. Although nowadays visual interpretation is regarded as the most traditional and time consuming method, some researchers still prefer it over the automatic method [15,28] because automatically-classified results are still subject to the characteristics of the selected training samples, algorithms, and satellite image, among other factors [29]. However, the low efficiency, uncertainty and high subjectivity associated with visual interpretation have made most researchers to investigate automatic methods [29].

The automatic extraction of gully information from satellite earth observation data takes two forms: pixel-based and object-based analysis [23,30]. The pixel-based analysis is relatively simple, and is the most frequently used and direct approach for image classification, using only the spectral information [31]. Such spectral information can be extracted using various image classification algorithms such as random forest (RF) and support vector machines (SVM), which thus far, are arguably the most commonly used algorithms due to their classification efficiency in relation to other algorithms, including k-nearest neighbor (kNN), maximum likelihood (ML), artificial Neural Network (ANN), convolutional neural networks (CNN), discriminant analysis (DA), and minimum distance (MD). One study mapped the areas susceptible to gully erosion using RF and ANN [32], and found that RF performed better than ANN. Noi and Kappas [33] compared SVM, RF, and kNN in land cover classification, and found that SVM, followed by RF, were better than kNN. Phinzi et al. [34] reported that both SVM and RF outperformed linear discriminant analyst (LDA) in a study on gully detection. Although deep learning methods such as CNNs have shown better performance over SVM and RF [35], like most deep learning methods, CNNs also strongly rely on the availability of abundant high-quality training/ground truth data [36]. While CNNs perform well in detecting and differentiating active gullies from other forms of surface erosion (e.g., sheet and rills), they have errors in detecting complex gully systems [37]. For these reasons, SVM and RF still attract most researchers' attention, because of their low computational complexity and higher interpretability capabilities compared to deep learning algorithms [36].

The wide usage of these machine learning algorithms in remote sensing proved that learning features from dataset is more efficient and practical than merely defining the features [38]. Although the application of machine learning in soil erosion research is not new, previous investigations commonly use coarser spatial imagery such as Landsat, ASTER and Sentinel/Sentinel-SAR (Synthetic Aperture Radar), which from an economic point of view makes sense, given that such images are obtainable at no cost. Besides, these sensors are good for wide area mapping of soil erosion. However, what has become apparent from previous studies, is that such sensors cannot identify individual gullies (especially small discontinuous gullies) with sufficient detail, this limitation is attributable to their low spatial resolution [15]. Whereas other optical sensors such as IKONOS, WorldView, and RapidEye with relatively higher spatial resolution exist for gully mapping, these sensors are not readily or freely available, as such, their high acquisition costs limit their application for gully mapping. Similarly, the use of LiDAR-derived elevation data from airborne surveys including Unmanned Aerial Vehicles (UAVs) is limited by a lack of financial resources. Depending on the availability of data and objective of a given study, multi-source and multi-sensor data fusion are common in remote sensing since this provides synthetic data that have the combined advantages of different sensors [39]. Multi-sensor or pixel level data fusion are mainly applied to optical images, for example, the fusion of high resolution panchromatic and low resolution multi-spectral images [40], was successfully applied in gully feature extraction [34]. Multi-source data fusion concerns feature level and decision level fusion of data from various sources such as SAR, optical images, LiDAR, geographic information system (GIS) data, and in-situ data [40]. In our case, we did not perform any data fusion due to lack of data (including the panchromatic band) with suitable spatial resolution necessary for detecting individual gullies.

Despite the unavailability of a higher spatial resolution panchromatic band, the 3 m PlanetScope image, which is available free of charge for research purposes, offers a great potential for detecting individual gullies. However, the capability of PlanetScope image in classifying gullies in different seasons (dry and wet) in an arid or semi-arid environment had been investigated only in areas of large forms (1–5 km length, 100–600 m width) [41]. While machine learning algorithms such as the SVM and RF have been frequently applied, little efforts have been made to investigate the influence of resampling techniques, particularly, bootstrapping and k-fold cross-validation (CV), on the accuracy relations. We identified gullies from PlanetScope images based on these resampling methods. Our aim was (i) to compare the satellite's bands reflectance values from the aspect of gullies, (ii) to reveal which classifier (RF or SVM) and resampling technique (CV or bootstrapping) perform better regarding the overall and class level accuracy metrics, and (iii) which season is more appropriate to identify the gullies.

2. Materials and Methods

2.1. Study Area

The study area was located in the rural part of eastern South Africa, characterized by extensive erosion where permanent gully erosion was the most prominent erosion type [42]. Geographically, the study area lies between 30°42′30″–30°43′55″S 28°46′22″–28°48′47″E, covering a surface of about 10 km² (Figure 1). Subsistence agriculture (e.g., crop farming and livestock rearing) and settlement were the main land use types. Grassland was the most common vegetation type throughout the area, with some forest patches found in the north-western section of the study area. The topography ranges from 1213 m–1658 m, with the north-western and south-western sections being steeper than other parts of the area. Steep mountain slopes with gently undulating footslopes characterize the geomorphology of the area [14]. The climate is semi-arid with temperatures ranging from 7–30 °C. Winters are cold and dry, with less vegetation due to limited rainfall. Rainfall mostly occurs during the summer season reaching approximately 670 mm on average per year. Although the study area has limited annual rainfall, it experiences high-intensity rainfall events. Gully development in the area was further fostered by the predominance of highly erodible soils such as duplex and dispersive soils [9,43], predominantly underlain by mudstone and sandstone of the Beaufort Group [44]. Although vegetation exists in the wet season, its effectiveness in protecting soil against erosion and inappropriate land-use practices such as overgrazing usually reduces vegetation cover, making the area susceptible to soil erosion. The study area features both continuous and discontinuous gully networks with distinct occurrences and appearances, i.e., narrow, wide, vegetated, shallow, deep with shadows, etc. [14,42]. Additionally, some gullies resemble the unpaved road network in appearance. Such complexity of gullies within the area makes the area particularly suitable for study.

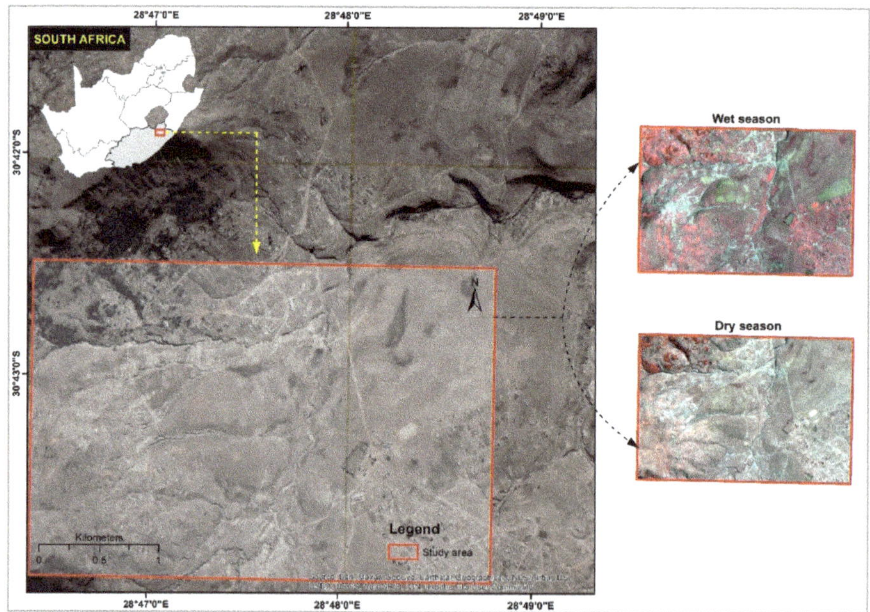

Figure 1. Location of the study area (PlanetScope false-color images).

2.2. Data Acquisition and Pre-Processing

Two cloud-free PlanetScope orthorectified products (Level 3B) for the wet and dry seasons acquired on 23 January 2017 and 25 June 2017, respectively, were used in this study. The images were downloaded from the Planet explorer website (https://www.

planet.com/explorer (accessed on 30 July 2020)). The orthorectified scenes had already been radiometrically and geometrically corrected and projected to the Universal Traverse Mercator (UTM) projection, referenced to the world geodetic system (WGS84) datum. With a spatial resolution of 3 m and temporal resolution of 1 day, the PlanetScope image is comprised of 4 spectral bands: red, green, blue (RGB), and near-infrared (NIR). The flowchart summarizing the workflow followed in this study is presented in Figure 2.

2.3. Gully Classification

Classification of gullies from the PlanetScope image was conducted in Python software using random forest (RF) and support vector machines (SVM). These were the most widely applied algorithms and their detailed description has been provided in the literature [10,34,36,45–47]. The RF, developed by Breiman [48], is a robust machine learning algorithm that is increasingly becoming more popular in remote sensing of soil erosion. The algorithm has several parameters that need to be tuned, amongst which the ntree (number of trees) and mtry (number of features in each split) are the most important that should be considered when training the algorithm [49]. The models were built using only 4 variables (e.g., four multispectral bands of the PlanetScope image), thus we tested all possible values of the mtry parameter. For the ntree parameter, we tested different values ranging from 50 to 1000. After ntree = 100, the accuracies stagnated while the computational time kept increasing [50]; thus, the final model was trained with 100 individual decision trees, selecting 2 random variables at each split.

The support vector machine (SVM) model was capable of overcoming both classification and regression problems [51,52]. To achieve this, SVM searched for the flat boundary (hyperplane) in some feature space that best separated the classes into homogeneous partitions where each partition contained only data points of a given class [34,49]. In reality, however, it was difficult to find a hyperplane that perfectly separated the classes using just the original features [49]. SVM overcomes this problem in 2 ways: first, loosen what is meant by "perfectly separates," and second, use the so-called kernel trick to expand the feature space to the extent that perfect separation of classes is more likely [49]. The radial basis function (RBF) was chosen for the kernel type. For RBF, a C penalty parameter against misclassifications and a kernel coefficient (γ) as a decision boundary have to be specified, which greatly affects the performance of the model [53]. Hyperparameter tuning was performed with the grid search method.

2.4. Reference Data Collection and Accuracy Assessment

The reference data were collected through field surveys and visual interpretation of high-resolution Google Earth images. We delineated the study area into 7 land cover classes, of which all were identifiable both in the field and in images (Google Earth and PanetScope): forest, built-up, agriculture, gully, bare soil, and mixed bare soil (i.e., exposed rocks, unpaved roads/dirty roads, and exposed soil mostly in ploughed fields). A total of 966 points were collected using stratified random sampling in ArcMap. Each land cover class was assigned a number of points proportional to its size.

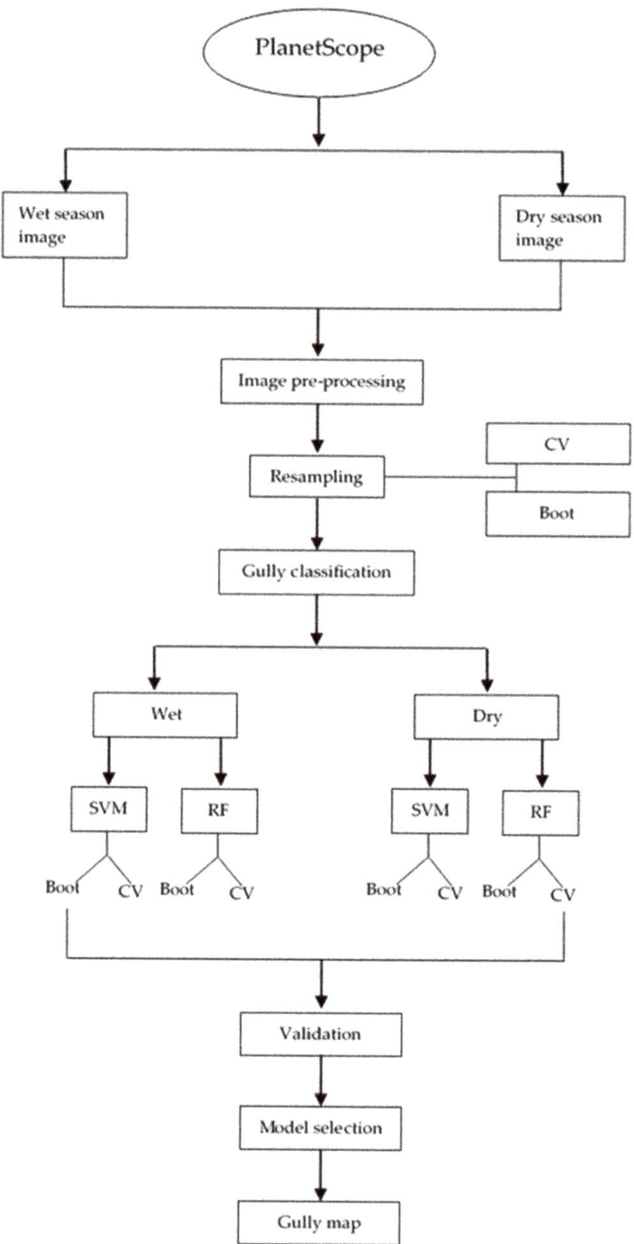

Figure 2. Workflow followed in this study (CV: cross-validation; boot: bootstrapping; SVM: support vector machines; RF: random forest).

We evaluated the overall performance of the RF and SVM algorithms using CV and bootstrapping. Kappa coefficients and overall accuracy (OA) were among the most commonly used metrics to evaluate classification accuracy [54]. However, the use of kappa in remote sensing classification accuracy is becoming less common [33,55]. Pontius and Millones [56], Flight and Julious [57], and more recently, Delgado and Tibau [58], recommend against using kappa because of its inherent limitations. A major limitation of kappa was that it is highly sensitive to the distribution of the marginal totals, potentially producing unreliable results [57]. Thus, we used OA to assess the overall performance of the models. Contrary to the conventional error matrix, which used all of the available data to test the model, CV splits the reference dataset into training and testing data. It used the majority of the data for training and the remainder, often called the holdout sample, was used to test the model, ensuring that the model was robust [49]. In total, we used 17,757 pixels for the wet season and 30,597 pixels for the dry season, generated from PlanetScope. We repeated the 5-fold CV 20 times, meaning that final accuracies were computed from 100 models. Before each repetition, the dataset was randomly shuffled and new folds were generated to increase the robustness of the models. Unlike CV, in bootstrapping, the original data were randomly sampled with replacement, meaning that, after a data point (bootstrap sample) was selected for inclusion in the subset, it was still available for further selection [49]. Two parameters must be chosen before running bootstrapping: sample size and the number of repetitions. In our case, the sample size was the same as the original dataset [59], and we applied 100 repetitions. The models were validated on the samples that were not included in the bootstrap sample.

We used the traditional error matrix to assess the model performance at class level as bootstrapping and CV do not provide class accuracies. An error matrix compared reference data to the classified map using various accuracy indices [54], but in this study, we only focused on class level accuracies/errors: producer's accuracy (PA) and user's accuracy (UA); PA was also known as sensitivity or recall while UA was sometimes referred to as precision. The difference of the possible 100% accuracy and the PA represented the omission error, which occurred when a pixel was excluded from the class to which it belonged. A difference of 100% and UA represented a commission error, which occurred when a pixel was incorrectly included in the class where it did not belong. We computed unbiased area-based PAs and UAs, following "good practice" recommendations for accuracy assessment [60]. The F1-score was also reported as the harmonic mean of UA and PA [61]. Additionally, we computed unbiased areal coverages (ha) of gullies along with their standard errors (ha) and associated $\pm$ 95% confidence intervals (ha). We generated 6 algorithms based on the combination of the classifiers: svm and rf, seasons: dry (d) and wet (w), and resampling methods: bootstrapping (b) and cross validation (cv), i.e., rf-d-b, rf-d-cv, rf-w-cv, rf-w-b, svm-d-cv, svm-d-b, svm-w-cv, and svm-w-b.

2.5. Statistical Analysis

NDVI values of the images, and specifically focusing on the gullies, were compared by the 2 seasons with the robust Mann–Whitney test using the Monte Carlo p (p_{MC}) with 9999 permutations. We applied the General Linear Model (GLM) to determine the effects of spectral bands (4 bands; RGB + NIR), seasons (wet and dry), and the LULC classes (7 classes). Furthermore, we also determined the statistical interactions to reveal if factorial variables had a common effect (e.g., effects of spectral bands differed by LULC classes or were different in the dry or wet seasons). Besides, we also determined the effect size (ω^2) as a standardized measure of the variables' contribution in the model (higher values indicate larger contribution, $\omega^2 > 0.14$ was considered as a large effect [62].

The Dunnett test [63] was used to determine if gullies had significant differences from other land cover types (H0: mean reflectance values of gullies was identical with the other land cover types). The Dunnett test was developed to perform multiple comparisons against 1 control group; in this case, gullies' land cover type was chosen as the control. As in the Dunnett test, the number of comparisons was limited (related to a full factorial approach; i.e., 6 instead of 21). Furthermore, the test compared the factor groups' means with the control group's mean (unlike other tests, which compare group means to the grand mean); thus, it can reveal small significant differences [64], and our intent was to find all overlaps in the reflectance with the gullies.

3. Results

3.1. Spectral Bands, Land Cover Classes and Seasons as Determinants of Reflectance

The difference of NDVI values was significant between the two seasons (U = 35303, z = 19.102, $p_{MC} < 0.0001$). The NDVI for the wet season had relatively higher values ranging from −0.36 to 0.81, while the values for the dry season lay in the range −0.41 to 0.59. The dry season had bimodal distribution while the wet season had multimodal distribution (Figure 3). Such bimodal distribution in the dry season represents non-vegetation (first mode) and vegetation pixels (second mode). Like in the dry season, the first mode in the wet season was indicative of non-vegetation pixels denoted by lower NDVI values compared to the last two modes, represented by relatively higher NDVI values. These last two modes represent vegetated areas: vegetation and forest pixels, respectively.

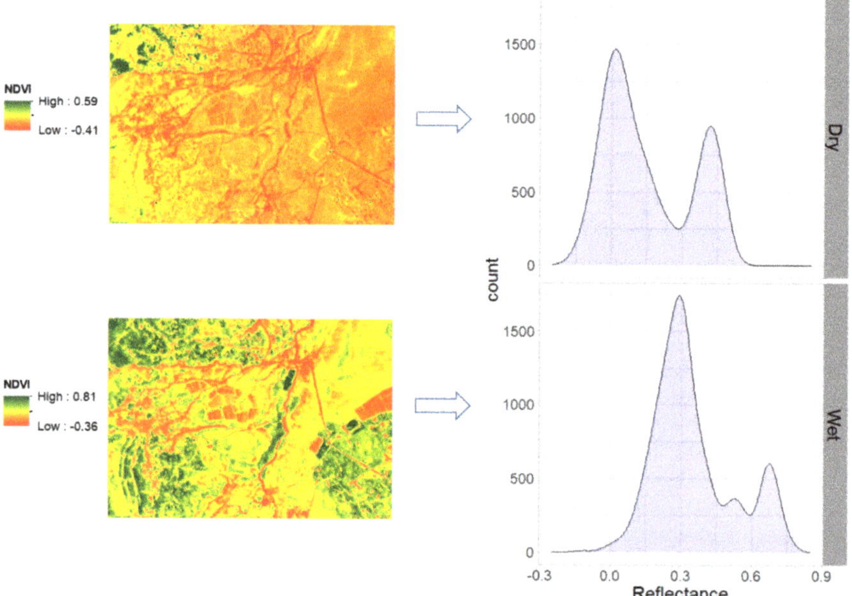

Figure 3. Distribution of NDVI reflectance values in the dry and wet season.

We also compared the NDVIs' of the gullies in the dry and wet seasons. Accordingly, the difference was significant (U = 162, z = 9.5534, $p < 0.0001$). The mean difference was 0.08 in the wet season, green vegetation was also present in gullies, thus, NDVI was larger. According to the results of the GLM we found that the spectral bands, LULC classes, and the seasons, as factorial variables and the interactions, were significant ($p < 0.001$) and explained 92.3% of the variance. Among the factors, the difference of dry and wet seasons had the largest effect on the reflectance (0.868). The bands and LULC classes had almost the same effect with a bit lower value (~0.6), however, also indicating a large effect. Regarding the interactions, we confirmed that reflectance by the band was different by LULC classes and seasons. The contribution of these interactions was large (Table 1). Furthermore, the effect size of the interaction between the seasons and the LULC classes was the lowest, being only third related to the interactions with the bands (0.141), but it still indicated a large effect. The interaction of all factors (spectral bands, seasons, LULC classes) also had a large effect but only with a smaller value (0.185).

Table 1. Results of General Linear Modelling (GLM) performed with reflectance as an independent variable (SS: Sum of Squares, df: degree of freedom, F: F-statistic, p: significance, ω^2p: effect size; $p < 0.05$: significance level).

Variables	SS	df	F	p	ω^2p
Model	6.99×10^9	55	860.4	<0.001	0.923
Bands	1.00×10^9	3	2256.1	<0.001	0.633
Season	3.80×10^9	1	25,715.0	<0.001	0.868
Class	9.79×10^8	6	1104.2	<0.001	0.629
Bands × Season	4.48×10^8	3	1010.0	<0.001	0.436
Bands × Class	5.30×10^8	18	199.3	<0.001	0.477
Season × Class	9.62×10^7	6	108.5	<0.001	0.141
Bands × Season × Class	1.34×10^8	18	50.3	<0.001	0.185
Residuals	5.70×10^8	3860			
Total	2.96×10^{10}	3916			

The post hoc test performed with the Dunnett test revealed significant differences ($p < 0.001$) between the gullies and other LULC classes in the dry season (Figure 4). The difference was not significant between the gullies and the agricultural areas (blue band), the vegetation and agricultural areas (green band and red band) in the wet season. Table 2 ranks the original band's importance in terms of discriminating gullies. We also studied the differences of NDVI and found that this spectral index was not as successful in discriminating the gullies as the original bands. It did not differ from the mixed bare soil and the vegetation in the dry season. Although NDVI performed better in the wet season, the difference was not significant with the built-up class.

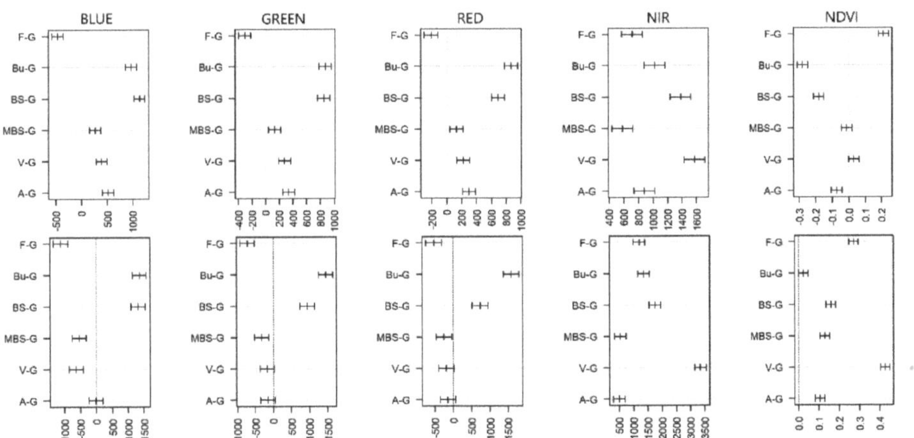

Figure 4. Differences of gullies and other land cover types' reflectance by bands and seasons (G: gully; F: forest; Bu: built-up; BS: bare soil; MBS: mixed bare soil; V: vegetation; A: agriculture; mean ± 95% confidence intervals; the difference was not significant if confidence range intersects the dashed line).

Table 2. PlanetScope bands ranking in discriminating gullies against the surrounding land cover.

Dry Season		Wet Season	
Band	Importance Ranking (%)	Band	Importance Ranking (%)
NIR	31	NIR	35
Red	26	Red	32
Green	25	Green	21
Blue	17	Blue	12

3.2. Accuracy Assessment of Gully Mapping

Using machine learning algorithms (RF and SVM), the cross-validation (CV) resampling method yielded better OA compared to bootstrapping for both the wet and dry seasons (Figure 5). Two apparent trends can be observed from these results based on OA: (i) RF consistently performed better than SVM irrespective of the season or resampling methods: bootstrapping and cross-validation; (ii) dry season had better OAs than the wet season, but this was not reflected in class level accuracy indices for gully classification. Based on the unbiased UA, all algorithms showed good performance in gully classification, recording UA above 70% (Figure 6). In particular, the best performance belonged to the svm-d-b (93.4%), whereas the worst UA belonged to the rf-w-b model (77%). For most models, PA was generally low relative to UA. Only half of the models recorded a PA greater than 70%, with the best performance belonging to svm-w-cv (89.2%), while the other half fell below 70%, with the svm-d-b model recording the lowest PA (32.5%).

An unbiased area estimate of gullies (ha) is presented in Table 3. With the highest PA (89.2%) and lowest standard error (3.7 ha), svm-w-cv provided the most accurate gully areal coverage (57.2ha). The highest standard error (11.5 ha) belonged to rf-w-b model, which had a gully area of 55.2 ± 25ha. However, in the F1-score ranking, rf-d-b and rf-d-cv algorithms achieved the best results (>0.90), but RF algorithms belonging to the wet season had relatively low score (0.82). On the other hand, all SVM algorithms (svm-d-cv, svm-d-b, svm-w-cv, and svm-w-b) recorded lower F1-scores, ranging 0.85–0.88. The two resampling techniques recorded the same omission error (85.1%), but slightly different commission errors, e.g., bootstrapping had 40.8% error of commission compared to 37.8% error for k-fold CV (Table 4).

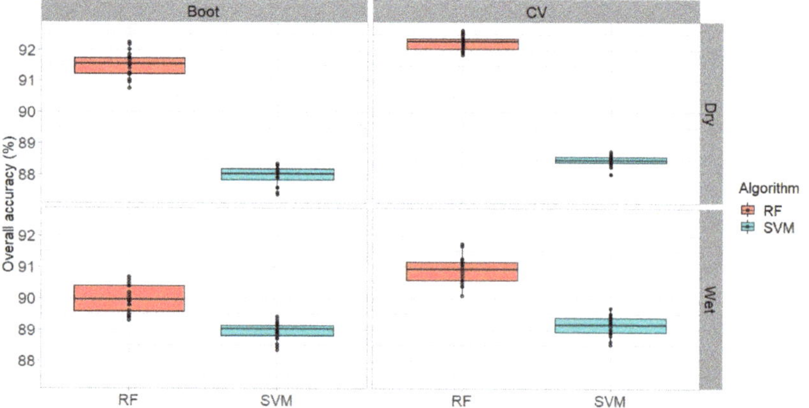

Figure 5. Accuracy assessment based on overall accuracy (OA) by the classification algorithm (RF: random forest, SVM: support vector machine), resampling method (boot: bootstrapping, CV: cross-validation), and season (wet and dry).

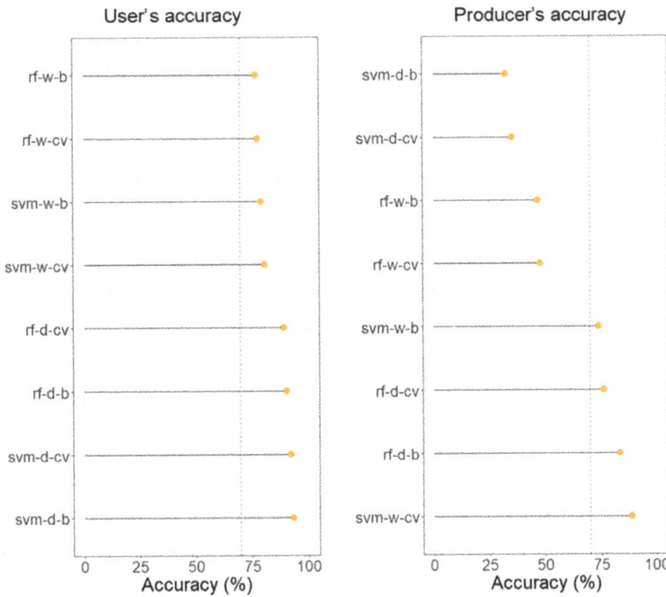

Figure 6. Unbiased user's accuracy and producer's accuracy (rf: random forest, svm: support vector machine, w: wet season, d: dry season, cv: cross-validation, b: bootstrapping, blue dashed line is 70% accuracy benchmark).

Table 3. Estimated gully area (ha) with associated standard error (ha) at ± 95% CI (ha) for each algorithm (rf: random forest, svm: support vector machine, d: dry, w: wet, b: bootstrapping, cv: cross-validation, CI: confidence interval).

Algorithm	Area (ha)	Standard Error (ha)	± 95% CI (ha)	PA (%)	UA (%)	F1-Score
rf-d-b	88	6.1	14.4	83.6	90.6	0.92
rf-d-cv	91.3	7.6	17.1	76.3	89.3	0.91
rf-w-cv	54.6	11.3	24.3	47.9	77.9	0.82
rf-w-b	55.2	11.5	25.0	46.8	77	0.82
svm-d-cv	32.6	10.1	21.1	35.4	92.3	0.86
svm-d-b	31.1	10.5	21.8	32.5	93.4	0.85
svm-w-cv	57.2	3.7	18.8	89.2	81	0.88
svm-w-b	57.4	6.4	19.3	74.1	79.4	0.86

The two resampling techniques recorded the same omission error (85.1%), but slightly different commission errors, e.g., bootstrapping had 40.8% error of commission compared to 37.8% error for k-fold CV.

Table 4. Summary of average error for resampling techniques, classifier, and season (RF: random forest, svm: support vector machine, CV: cross-validation).

	Resampling Technique		Classifier		Season	
Error	Bootstrap	k-Fold CV	RF	SVM	Dry	Wet
Commission (%)	40.8	37.8	36.4	42.2	43.1	35.5
Omission (%)	14.9	14.9	16.3	13.5	8.6	21.2
Standard error (ha)	8.6	8.2	9.1	7.7	8.6	8.2

3.3. Gully Distribution

Results indicated that gullies can be spectrally discriminated from other land cover classes, both in the dry and wet season; although there were observable differences in the distribution of the extracted gullies in these two seasons (Figure 7). In the wet season, there seem to be more gullies than there are in the dry season. This difference in gully areal coverage between the two seasons is more pronounced in Figure 7a, corresponding to rf-d-b and Figure 7b, representing the svm-w-cv model.

Differences in gully reflectance among the two seasons also had a bearing on gully classification. The underlying statistical test revealed that the difference was significant ($U = 162$, $z = 9.5534$, $p < 0.0001$), and the mean difference was 0.08. The wet season had more vegetation covering bare surfaces, and because of this, spectral differences were more pronounced during the wet season (Figure 8). On the contrary, in the dry season, most gullies spectrally resembled the bare surfaces they dissect. Consequently, the algorithms were less efficient in extracting gullies occurring on bare soil surfaces in the dry season. This probably explains the high commission error (43.1%) and standard error (8.6 ha) in the dry season.

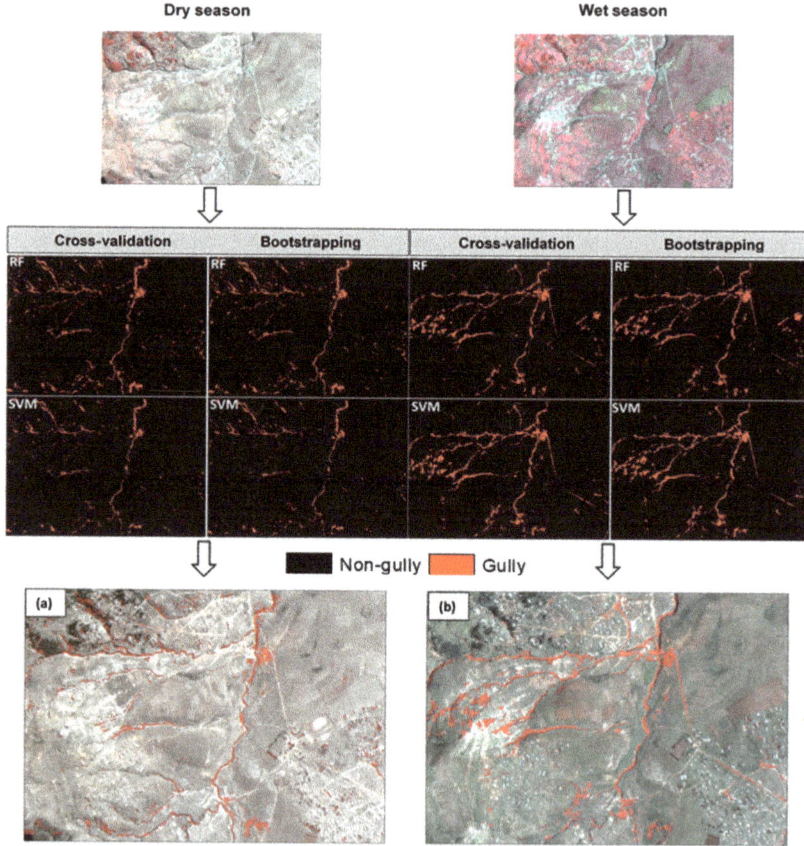

Figure 7. Spatial distribution of gullies: (**a**) rf-d-b and (**b**) svm-w-cv correspond to the best models for gully mapping in the dry and wet seasons, respectively (rf: random forest, svm: support vector machine, w: wet season, d: dry season, cv: cross-validation, b: bootstrapping).

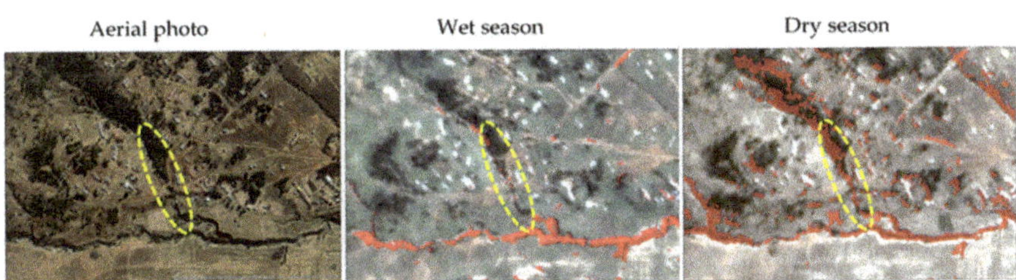

Figure 8. An example of a vegetated gully (dashed yellow ellipse) in the dry and wet seasons.

4. Discussion

Remotely sensed data are inherently subject to errors, hence, error assessment is essential for data assimilation, one of the primary uses of satellite data products [65]. In this section, we discuss errors associated with the derived gully maps, offering a possible explanation for such error sources. Different resampling methods undoubtedly play an important role in classification accuracy, hence, the final model selection. Specifically, we explored the influence of bootstrapping and k-fold cross-validation techniques in gully classification, considering different seasons (dry and wet) and classifiers (SVM and RF). Results revealed that k-fold CV performs slightly better than bootstrapping in terms of commission error. Kohavi et al. [66], in his study of CV and bootstrap for accuracy estimation and model selection, also reported k-fold CV as the best method to use over bootstrapping. Kim [67] estimated classification error rate, comparing repeated k-fold CV, repeated hold-out and bootstrap, and found that the repeated k-fold CV was better than bootstrap. The author further reported that bootstrapping had bias problems for both large and small samples, despite its small variance, hence, the expectation for better performance for small samples.

Although the results of our study are generally in agreement with previous studies, it is worth noting that the performance of the bootstrapping and k-fold CV varied considerably at class level with algorithm and season. There are instances where bootstrapping performed better than k-fold CV in gully classification. For instance, the best model, namely, svm-d-b, based on UA, belonged to bootstrapping. Such results are important because most studies using either bootstrapping or k-fold CV rarely focus on class level accuracy when evaluating the performance of these resampling techniques. More importantly, even at the class level, different accuracy metrics ought to be considered. This increases the robustness and reliability of the accuracy results, making it possible for researchers to draw correct deductions on the behavior of the algorithms under investigation [61]. However, various class accuracy metrics (UA, PA, standard error, and F1-score) used in the current study, all derived from the confusion matrix, disagreed with one another in some instances. For example, some algorithms that obtained high PA values had low corresponding UA values or vice-versa. This is also true with F1-score vs. either PA or UA. Based on the F1-score, the best algorithms belonged to RF (e.g., rf-d-b and rf-d-cv). Given the disagreement amongst various accuracy metrics, we relied on the standard error as a reliable measure to judge the accuracy of the algorithms.

In the wet season, the algorithms proved to be more efficient in gully classification on bare soil surfaces due to the existence of vegetation cover in bare soil surfaces, making it possible to discriminate gullies. Such findings are comparable or similar to those of previous studies. For example, one study automatically identified gullies based on ASTER images acquired during the dry and wet seasons [68]. The study concluded that the wet season-acquired image performed better than the dry season one. It is worth noting that the wet season is not always appropriate for gully identification in all situations. The success of gully identification depends on the complexity of gully appearance as influenced by their morphological characteristics (shape, size, length, depth, etc.) [42], sensor type and/or resolution, and classification algorithms [69], amongst other factors. For example,

Sentinel and Landsat images performed relatively well in the dry season than in the wet season [70]. Although gully classification was successful in the wet season relative to the dry season, there were few locations where gullies were filled up with vegetation. Such gullies could not be automatically classified, in which case we relied on visual interpretation of high-resolution aerial photographs and/or dry season PlanetScope images.

Gully appearance also played an important role in gully classification. Consistent with previous studies [42,71], the classification algorithms were efficient in detecting continuous gullies mostly in linear shape. Conversely, the algorithms proved to be less efficient in areas with high gully density, often surrounded by transitional zones to non-gully [71], but these areas form a relatively small portion of the study area and had negligible influence on the accuracy. The SVM combined with CV (e.g., svm-w-cv) reflected the best performance in the wet season with the least standard error (3.7 ha) and highest PA (89.2%), followed by a RF model (rf-d-b), recording slightly different standard error (6.1 ha) and PA (83.6%). Nevertheless, 50% of the models obtained a PA that is below 70%. Despite this discrepancy, the estimated gully areas (ha), based on area-weighted metrics, are unbiased and can be relied upon.

From a practical point of view, the identification of gullies from satellite images with reasonable accuracies is of paramount importance to gully rehabilitation. Like all remote sensing-derived products, gully maps are subject to errors, and hence, accuracy assessment is a prerequisite [54]. However, most remote sensing-based gully studies tend to rely on accuracy indices, such as PA and UA, without taking into account the uncertainty of the estimated gully areas. Although it is not a requirement, it is often recommended to provide not only PA and UA but also unbiased quantitative area estimates such as the area-weighted metrics and confidence intervals [60]. In this study, we quantified gullied areas (ha) together with their associated levels of uncertainties, such as standard errors (ha) and confidence intervals (ha).

RF combined with bootstrapping resampling provided the best gully area (88 ± 14.4 ha) estimate with the least standard error (6.1 ha) in the dry season. In the wet season, SVM combined with CV resampling estimated gully area (57.2 ± 18.8 ha) with the lowest standard error (3.7 ha). These findings shed light on the influence of these resampling techniques on the accuracy of satellite-based gully mapping but also provides the basis for further investigations into the accuracy of such resampling techniques, especially when using different satellite images other than the PlanetScope data, preferable, freely available ones, with higher spatial resolution. Initially, we planned to use both PlanetScope and SPOT-7 images, also obtainable free of charge for the test area, but SPOT-7 image scenes acquired in the wet and dry season months were not available for the test area. Nevertheless, given that we only mapped gullies in a small part of the problem area, we are planning to test the method in other areas with wider spatial coverage. However, mapping gullies over large areas, particularly using automatic methods, is still a challenge due to the complexity of gullies over such large areas [14]. Thus far, even advanced methods such as CNNs have errors in detecting complex gully systems [37]. It is worth noting that the detection of gullies mainly depends on the spatial resolution of the image used. For example, at larger scales, gullies have only been mapped at a spatial resolution of up to 2.5 m in South Africa [14,15]. To overcome this challenge, the future implementation of our method, will in part, require the use of a high spatial resolution (<2 m) image, for instance, pansharpened SPOT-7 image (1.5 m) or WorldView (0.5 m), which can detect individual gullies. Another limitation of this method relates to climate. The method is suitable for application in arid/semi-arid regions where gullies are often not covered by trees [42]. Our study demonstrated that gullies could be better identified in the dry season with RF combined with bootstrapping, whereas SVM combined with k-fold CV is best for identifying gullies in the wet season. Therefore, we recommend the use of RF and SVM for mapping gullies in the dry and wet seasons, respectively. Provided that PlanetScope provides global spatial coverage with daily revisit time, we particularly recommend it for continuous monitoring of gullies at any location.

5. Conclusions

The aim of this study was to assess the efficacy of cross-validation and bootstrapping in gully classification and also to reveal how well the PlanetScope images perform in gully extraction in the dry and wet seasons of a semi-arid climate. We found the following outcomes.

- Gullies were spectrally different in all bands of the PlanetScope images, both in the dry and the wet seasons.
- NDVI values did not differ from all land cover classes regarding the reflectance values; thus, it was not involved in gully classification.
- Dry and wet seasons ensured different classification accuracy, but gully extraction was successful. RF outperformed the SVM algorithm in terms of OA, but the differences of the OAs were < 4%. Differences were larger in the dry season (3.5%) and smaller in the wet season (~1%).
- Generally, based on the OAs, CV performed better with the RF algorithm than the bootstrapping (with ~1.0–1.5% differences), but on a class level, bootstrapping provided the most accurate gully extraction with the RF in the dry season, whereas CV was efficient with SVM in the wet season.

Accordingly, both resampling techniques were efficient, but RF with bootstrapping resampling technique in the dry season can be suggested to map gullies. In the future, we plan to extend the mapping in larger areas to help landowners and managers to fight against erosion and to plan the interventions at the hot spot areas.

Author Contributions: Conceptualization, K.P. and S.S.; methodology, K.P.; software, K.P. and D.A.; validation, K.P. and D.A., formal analysis, K.P.; investigation, K.P.; resources, S.S.; data curation, K.P. and D.A.; writing—original draft preparation, K.P.; writing—review and editing, K.P. and S.S.; visualization, K.P., D.A., and S.S.; supervision, S.S. All authors have read and agreed to the published version of the manuscript.

Funding: This research was funded by the Thematic Excellence Programme (TKP2020-NKA-04) of the Ministry for Innovation and Technology in Hungary projects and Department of Higher Education and Training (DHET) of South Africa.

Data Availability Statement: PlanetScope images can be purchased from the PlanetLabs Inc. Limited, non-commercial access to PlanetScope imagery can also be gained through the Education and Research Program (https://www.planet.com/markets/education-and-research/ (accessed on 30 July 2020)). Reference data can be provided by the authors on demand.

Acknowledgments: The first author (K.P.) greatly acknowledges the Tempus Public Foundation for funding his Ph.D. studies through the Stipendium Hungaricum Scholarship Programme. The author is equally grateful to the Department of Higher Education and Training (DHET) of South Africa for the supplementary support.

Conflicts of Interest: The authors declare no conflict of interest. The funders had no role in the design of the study; in the collection, analyses, or interpretation of data; in the writing of the manuscript, or in the decision to publish the results.

References

1. Meyer, L.D.; Wischmeier, W.H. Mathematical simulation of the process of soil erosion by water. *Trans. ASAE* **1969**, *12*, 754–758.
2. Morgan, R.P.C. *Soil Erosion and Conservation*; John Wiley & Sons: Hoboken, NJ, USA, 2009; ISBN 140514467X.
3. Borrelli, P.; Robinson, D.A.; Fleischer, L.R.; Lugato, E.; Ballabio, C.; Alewell, C.; Meusburger, K.; Modugno, S.; Schütt, B.; Ferro, V.; et al. An assessment of the global impact of 21st century land use change on soil erosion. *Nat. Commun.* **2017**, *8*, 1–13. [CrossRef]
4. Omuto, C.; Nachtergaele, F.; Rojas, R.V. *State of the Art Report on Global and Regional Soil Information: Where Are We? Where To Go?* Food and Agriculture Organization of the United Nations: Rome, Italy, 2013; ISBN 9251074496.
5. Kertész, A.; Křeček, J. Landscape degradation in the world and in Hungary. *Hung. Geogr. Bull.* **2019**, *68*, 201–221. [CrossRef]
6. Phinzi, K.; Ngetar, N.S.; Ebhuoma, O. Soil erosion risk assessment in the Umzintlava catchment (T32E), Eastern Cape, South Africa, using RUSLE and random forest algorithm. *S. Afr. Geogr. J.* **2020**, *103*, 139–162. [CrossRef]
7. Strategic Plan for the Department of Agriculture, Pretoria, South Africa. 2007. Available online: https://www.gov.za/sites/default/files/gcis_document/201409/agricstratplan2007.pdf (accessed on 16 July 2020).

8. Meadows, M.E.; Hoffman, M.T. The nature, extent and causes of land degradation in South Africa: Legacy of the past, lessons for the future? *Area* **2002**, *34*, 428–437. [CrossRef]

9. Beckedahl, H.R.; de Villiers, A.B. Accelerated erosion by piping in the Eastern Cape Province, South Africa. *S. Afr. Geogr. J.* **2000**, *82*, 157–162. [CrossRef]

10. Kakembo, V.; Rowntree, K.M. The relationship between land use and soil erosion in the communal lands near Peddie town, Eastern Cape, South Africa. *Land Degrad. Dev.* **2003**, *14*, 39–49. [CrossRef]

11. Mhangara, P.; Kakembo, V.; Lim, K.J. Soil erosion risk assessment of the Keiskamma catchment, South Africa using GIS and remote sensing. *Environ. Earth Sci.* **2012**, *65*, 2087–2102. [CrossRef]

12. Phinzi, K.; Ngetar, N.S. Land use/land cover dynamics and soil erosion in the Umzintlava catchment (T32E), Eastern Cape, South Africa. *Trans. R. Soc. S. Afr.* **2019**, *74*, 223–237. [CrossRef]

13. Kakembo, V.; Xanga, W.W.; Rowntree, K. Topographic thresholds in gully development on the hillslopes of communal areas in Ngqushwa Local Municipality, Eastern Cape, South Africa. *Geomorphology* **2009**, *110*, 188–194. [CrossRef]

14. Le Roux, J.J.; Sumner, P.D. Factors controlling gully development: Comparing continuous and discontinuous gullies. *Land Degrad. Dev.* **2012**, *23*, 440–449. [CrossRef]

15. Mararakanye, N.; Le Roux, J.J. Gully location mapping at a national scale for South Africa. *S. Afr. Geogr. J.* **2012**, *94*, 208–218. [CrossRef]

16. Poesen, J.; Nachtergaele, J.; Verstraeten, G.; Valentin, C. Gully erosion and environmental change: Importance and research needs. *Catena* **2003**, *50*, 91–133. [CrossRef]

17. Zhang, T.; Liu, G.; Duan, X.; Wilson, G.V. Spatial distribution and morphologic characteristics of gullies in the Black Soil Region of Northeast China: Hebei watershed. *Phys. Geogr.* **2016**, *37*, 228–250. [CrossRef]

18. Zgłobicki, W.; Poesen, J.; Cohen, M.; Del Monte, M.; García-Ruiz, J.M.; Ionita, I.; Niacsu, L.; Machová, Z.; Martín-Duque, J.F.; Nadal-Romero, E.; et al. The potential of permanent gullies in Europe as geomorphosites. *Geoheritage* **2019**, *11*, 217–239. [CrossRef]

19. Valentin, C.; Poesen, J.; Li, Y. Gully erosion: Impacts, factors and control. *Catena* **2005**, *63*, 132–153. [CrossRef]

20. Phinzi, K.; Ngetar, N.S. Mapping soil erosion in a quaternary catchment in Eastern Cape using geographic information system and remote sensing. *S. Afr. J. Geomat.* **2017**, *6*, 11. [CrossRef]

21. Seutloali, K.E.; Dube, T.; Mutanga, O. Assessing and mapping the severity of soil erosion using the 30-m Landsat multispectral satellite data in the former South African homelands of Transkei. *Phys. Chem. Earth* **2017**, *100*, 296–304. [CrossRef]

22. Phinzi, K.; Ngetar, N.S.; Ebhuoma, O.; Szabó, S. Comparison of rusle and supervised classification algorithms for identifying erosion-prone areas in a mountainous rural landscape. *Carpathian J. Earth Environ. Sci.* **2020**, *15*, 405–413. [CrossRef]

23. Shruthi, R.B.V.; Kerle, N.; Jetten, V. Object-based gully feature extraction using high spatial resolution imagery. *Geomorphology* **2011**, *134*, 260–268. [CrossRef]

24. Seutloali, K.E.; Beckedahl, H.R.; Dube, T.; Sibanda, M. An assessment of gully erosion along major armoured roads in south-eastern region of South Africa: A remote sensing and GIS approach. *Geocarto Int.* **2016**, *31*, 225–239. [CrossRef]

25. Phinzi, K.; Ngetar, N.S. The assessment of water-borne erosion at catchment level using GIS-based RUSLE and remote sensing: A review. *Int. Soil Water Conserv. Res.* **2019**, *7*, 27–46. [CrossRef]

26. Casalí, J.; López, J.J.; Giráldez, J.V. Ephemeral gully erosion in southern Navarra (Spain). *Catena* **1999**, *36*, 65–84. [CrossRef]

27. Knight, J.; Spencer, J.; Brooks, A.; Phinn, S. Large-area, high-resolution remote sensing based mapping of alluvial gully erosion in Australia's tropical rivers. In Proceedings of the 5th Australian Stream Management Conference, Thurgoona, Australia, 21–25 May 2007; Institute for Land, Water and Society, Charles Sturt University: Bathurst, Australia, 2007; Volume 2, pp. 199–204. [CrossRef]

28. Karydas, C.; Panagos, P. Towards an assessment of the ephemeral gully erosion potential in Greece using google earth. *Water* **2020**, *12*, 603. [CrossRef]

29. Liu, K.; Ding, H.; Tang, G.; Zhu, A.X.; Yang, X.; Jiang, S.; Cao, J. An object-based approach for two-level gully feature mapping using high-resolution DEM and imagery: A case study on hilly loess plateau region, China. *Chin. Geogr. Sci.* **2017**, *27*, 415–430. [CrossRef]

30. Duro, D.C.; Franklin, S.E.; Dubé, M.G. A comparison of pixel-based and object-based image analysis with selected machine learning algorithms for the classification of agricultural landscapes using SPOT-5 HRG imagery. *Remote Sens. Environ.* **2012**, *118*, 259–272. [CrossRef]

31. Zhang, L.; Zhang, L.; Du, B. Deep learning for remote sensing data: A technical tutorial on the state of the art. *IEEE Geosci. Remote Sens. Mag.* **2016**, *4*, 22–40. [CrossRef]

32. Ghorbanzadeh, O.; Shahabi, H.; Mirchooli, F.; Valizadeh Kamran, K.; Lim, S.; Aryal, J.; Jarihani, B.; Blaschke, T. Gully erosion susceptibility mapping (GESM) using machine learning methods optimized by the multi-collinearity analysis and K-fold cross-validation. *Geomat. Nat. Hazards Risk* **2020**, *11*, 1653–1678. [CrossRef]

33. Thanh Noi, P.; Kappas, M. Comparison of Random Forest, k-Nearest Neighbor, and Support Vector Machine Classifiers for Land Cover Classification Using Sentinel-2 Imagery. *Sensors* **2017**, *18*, 18. [CrossRef]

34. Phinzi, K.; Abriha, D.; Bertalan, L.; Holb, I.; Szabó, S. Machine learning for gully feature extraction based on a pan-sharpened multispectral image: Multiclass vs. Binary approach. *ISPRS Int. J. Geo Inf.* **2020**, *9*, 252. [CrossRef]

35. Heydari, S.S.; Mountrakis, G. Meta-analysis of deep neural networks in remote sensing: A comparative study of mono-temporal classification to support vector machines. *ISPRS J. Photogramm. Remote Sens.* **2019**, *152*, 192–210. [CrossRef]

36. Sheykhmousa, M.; Mahdianpari, M.; Ghanbari, H.; Mohammadimanesh, F.; Ghamisi, P.; Homayouni, S. Support Vector Machine Versus Random Forest for Remote Sensing Image Classification: A Meta-Analysis and Systematic Review. *IEEE J. Sel. Top. Appl. Earth Obs. Remote Sens.* **2020**, *13*, 6308–6325. [CrossRef]
37. Gafurov, A.M.; Yermolayev, O.P. Automatic gully detection: Neural networks and computer vision. *Remote Sens.* **2020**, *12*, 1743. [CrossRef]
38. Dong, L.; Xing, L.; Liu, T.; Du, H.; Mao, F.; Han, N.; Li, X.; Zhou, G.; Zhu, D.; Zheng, J.; et al. Very High Resolution Remote Sensing Imagery Classification Using a Fusion of Random Forest and Deep Learning Technique-Subtropical Area for Example. *IEEE J. Sel. Top. Appl. Earth Obs. Remote Sens.* **2020**, *13*, 113–128. [CrossRef]
39. Ghamisi, P.; Rasti, B.; Yokoya, N.; Wang, Q.; Hofle, B.; Bruzzone, L.; Bovolo, F.; Chi, M.; Anders, K.; Gloaguen, R. Multisource and multitemporal data fusion in remote sensing: A comprehensive review of the state of the art. *IEEE Geosci. Remote Sens. Mag.* **2019**, *7*, 6–39. [CrossRef]
40. Zhang, J. Multi-source remote sensing data fusion: Status and trends. *Int. J. Image Data Fusion* **2010**, *1*, 5–24. [CrossRef]
41. Shahabi, H.; Jarihani, B.; Tavakkoli Piralilou, S.; Chittleborough, D.; Avand, M.; Ghorbanzadeh, O. A Semi-Automated Object-Based Gully Networks Detection Using Different Machine Learning Models: A Case Study of Bowen Catchment, Queensland, Australia. *Sensors* **2019**, *19*, 4893. [CrossRef] [PubMed]
42. Phinzi, K.; Holb, I.; Szabó, S. Mapping Permanent Gullies in an Agricultural Area Using Satellite Images: Efficacy of Machine Learning Algorithms. *Agronomy* **2021**, *11*, 333. [CrossRef]
43. van Breda Weaver, A. The distribution of soil erosion as a function of slope aspect and parent material in Ciskei, Southern Africa. *GeoJournal* **1991**, *23*, 29–34. [CrossRef]
44. Hilbich, C.; Daut, G.; Mäusbacher, R.; Helmschrot, J. A landscape-based model to characterize the evolution and recent dynamics of wetlands in the Umzimvubu headwaters, Eastern Cape, South Africa. In *Wetlands: Modelling, Monitoring, Management*; Kotkowski, W., Maltby, E., Miroslaw–Swiatek, D., Okruszko, T., Szatylowicz, J., Eds.; Taylor & Francis: Abingdon, UK, 2007; pp. 61–69.
45. Adam, E.; Mutanga, O.; Odindi, J.; Abdel-Rahman, E.M. Land-use/cover classification in a heterogeneous coastal landscape using RapidEye imagery: Evaluating the performance of random forest and support vector machines classifiers. *Int. J. Remote Sens.* **2014**, *35*, 3440–3458. [CrossRef]
46. Sabat-Tomala, A.; Raczko, E. Comparison of Support Vector Machine and Random Forest Algorithms for Invasive and Expansive Species Classification Using Airborne Hyperspectral Data. *Remote Sens.* **2020**, *12*, 516. [CrossRef]
47. Papp, L.; van Leeuwen, B.; Szilassi, P.; Tobak, Z.; Szatmári, J.; Árvai, M.; Mészáros, J.; Pásztor, L. Monitoring invasive plant species using hyperspectral remote sensing data. *Land* **2021**, *10*, 29. [CrossRef]
48. Breiman, L. Random forests. *Mach. Learn.* **2001**, *45*, 5–32. [CrossRef]
49. Boehmke, B.; Greenwell, B.M. *Hands-On Machine Learning with R*; CRC Press: Boca Raton, FL, USA, 2019; ISBN 1000730190.
50. Oshiro, T.M.; Perez, P.S.; Baranauskas, J.A. How many trees in a random forest? In Proceedings of the 8th International Workshop on Machine Learning and Data Mining in Pattern Recognition, Berlin, Germany, 13–20 July 2012; Springer: Berlin/Heidelberg, Germany, 2012; pp. 154–168.
51. Vapnik, V. *The Nature of Statistical Learning Theory*; Springer Science & Business Media: Berlin/Heidelberg, Germany, 2013; ISBN 1475732643.
52. Brenning, A. Spatial prediction models for landslide hazards: Review, comparison and evaluation. *Nat. Hazards Earth Syst. Sci.* **2005**, *5*, 853–862. [CrossRef]
53. Pedregosa, F.; Varoquaux, G.; Gramfort, A.; Michel, V.; Thirion, B.; Grisel, O.; Blondel, M.; Prettenhofer, P.; Weiss, R.; Dubourg, V. Scikit-learn: Machine learning in Python. *J. Mach. Learn. Res.* **2011**, *12*, 2825–2830.
54. Congalton, R.G. A review of assessing the accuracy of classifications of remotely sensed data. *Remote Sens. Environ.* **1991**, *37*, 35–46. [CrossRef]
55. Heydari, S.S.; Mountrakis, G. Effect of classifier selection, reference sample size, reference class distribution and scene heterogeneity in per-pixel classification accuracy using 26 Landsat sites. *Remote Sens. Environ.* **2018**, *204*, 648–658. [CrossRef]
56. Pontius, R.G.; Millones, M. Death to Kappa: Birth of quantity disagreement and allocation disagreement for accuracy assessment. *Int. J. Remote Sens.* **2011**, *32*, 4407–4429. [CrossRef]
57. Flight, L.; Julious, S.A. The disagreeable behaviour of the kappa statistic. *Pharm. Stat.* **2015**, *14*, 74–78. [CrossRef] [PubMed]
58. Delgado, R.; Tibau, X.-A. Why Cohen's Kappa should be avoided as performance measure in classification. *PLoS ONE* **2019**, *14*, e0222916. [CrossRef] [PubMed]
59. Kuhn, M.; Johnson, K. *Applied Predictive Modeling*; Springer: Berlin/Heidelberg, Germany, 2013; Volume 26.
60. Olofsson, P.; Foody, G.M.; Herold, M.; Stehman, S.V.; Woodcock, C.E.; Wulder, M.A. Good practices for estimating area and assessing accuracy of land change. *Remote Sens. Environ.* **2014**, *148*, 42–57. [CrossRef]
61. Chicco, D.; Jurman, G. The advantages of the Matthews correlation coefficient (MCC) over F1 score and accuracy in binary classification evaluation. *BMC Genom.* **2020**, *21*, 1–13. [CrossRef] [PubMed]
62. Field, A. *Discovering Statistics Using IBM SPSS Statistics*; Sage: Newcastle upon Tyne, UK, 2013; ISBN 1446274586.
63. Lee, S.; Lee, D.K. What is the proper way to apply the multiple comparison test? *Korean J. Anesth.* **2018**, *71*, 353. [CrossRef] [PubMed]
64. McHugh, M.L. Multiple comparison analysis testing in ANOVA. *Biochem. Med.* **2011**, *21*, 203–209. [CrossRef] [PubMed]

65. Povey, A.C.; Grainger, R.G. Known and unknown unknowns: Uncertainty estimation in satellite remote sensing. *Atmos. Meas. Tech.* **2015**, *8*, 4699–4718. [CrossRef]
66. Kohavi, R. A study of cross-validation and bootstrap for accuracy estimation and model selection. In Proceedings of the 14th International Joint Conference on Artificial Intelligence (IJCAI), Montreal, QC, Canada, 20 August 1995; Volume 14, pp. 1137–1145.
67. Kim, J.-H. Estimating classification error rate: Repeated cross-validation, repeated hold-out and bootstrap. *Comput. Stat. Data Anal.* **2009**, *53*, 3735–3745. [CrossRef]
68. Vrieling, A.; Rodrigues, S.C.; Bartholomeus, H.; Sterk, G. Automatic identification of erosion gullies with ASTER imagery in the Brazilian Cerrados. *Int. J. Remote Sens.* **2007**, *28*, 2723–2738. [CrossRef]
69. Lu, D.; Weng, Q. A survey of image classification methods and techniques for improving classification performance. *Int. J. Remote Sens.* **2007**, *28*, 823–870. [CrossRef]
70. Sepuru, T.K.; Dube, T. Understanding the spatial distribution of eroded areas in the former rural homelands of South Africa: Comparative evidence from two new non-commercial multispectral sensors. *Int. J. Appl. Earth Obs. Geoinf.* **2018**, *69*, 119–132. [CrossRef]
71. Orti, M.V.; Winiwarter, L.; Corral-Pazos-de-Provens, E.; Williams, J.G.; Bubenzer, O.; Höfle, B. Use of TanDEM-X and Sentinel products to derive gully activity maps in Kunene Region (Namibia) based on automatic iterative Random Forest approach. *IEEE J. Sel. Top. Appl. Earth Obs. Remote Sens.* **2020**, *14*, 607–623. [CrossRef]

remote sensing

Article

Quantitative Soil Wind Erosion Potential Mapping for Central Asia Using the Google Earth Engine Platform

Wei Wang [1,2,3], Alim Samat [1,2,3], Yongxiao Ge [1,2], Long Ma [1,2,3], Abula Tuheti [1,4], Shan Zou [1,2,3] and Jilili Abuduwaili [1,2,3,*]

[1] State Key Laboratory of Desert and Oasis Ecology, Xinjiang Institute of Ecology and Geography, Chinese Academy of Sciences, Urumqi 830011, China; wangwei177@mails.ucas.ac.cn (W.W.); alim_smt@ms.xjb.ac.cn (A.S.); geyx@ms.xjb.ac.cn (Y.G.); malong@ms.xjb.ac.cn (L.M.); abltht@st.xju.edu.cn (A.T.); zoushan17@mails.ucas.ac.cn (S.Z.)
[2] Research Center for Ecology and Environment of Central Asia, Chinese Academy of Sciences, Urumqi 830011, China
[3] University of Chinese Academy of Sciences, Beijing 100049, China
[4] College of Resources and Environment Sciences, Xinjiang University, Urumqi 830046, China
* Correspondence: jilil@ms.xjb.ac.cn; Tel.: +86-0991-782-7353

Received: 1 September 2020; Accepted: 16 October 2020; Published: 19 October 2020

Abstract: A lack of long-term soil wind erosion data impedes sustainable land management in developing regions, especially in Central Asia (CA). Compared with large-scale field measurements, wind erosion modeling based on geospatial data is an efficient and effective method for quantitative soil wind erosion mapping. However, conventional local-based wind erosion modeling is time-consuming and labor-intensive, especially when processing large amounts of geospatial data. To address this issue, we developed a Google Earth Engine-based Revised Wind Erosion Equation (RWEQ) model, named GEE-RWEQ, to delineate the Soil Wind Erosion Potential (SWEP). Based on the GEE-RWEQ model, terabytes of Remote Sensing (RS) data, climate assimilation data, and some other geospatial data were applied to produce monthly SWEP with a high spatial resolution (500 m) across CA between 2000 and 2019. The results show that the mean SWEP is in good agreement with the ground observation-based dust storm index (DSI), satellite-based Aerosol Optical Depth (AOD), and Absorbing Aerosol Index (AAI), confirming that GEE-RWEQ is a robust wind erosion prediction model. Wind speed factors primarily determined the wind erosion in CA ($r = 0.7$, $p < 0.001$), and the SWEP has significantly increased since 2011 because of the reversal of global terrestrial stilling in recent years. The Aral Sea Dry Lakebed (ASDLB), formed by shrinkage of the Aral Sea, is the most severe wind erosion area in CA (47.29 kg/m^2/y). Temporally, the wind erosion dominated by wind speed has the largest spatial extent of wind erosion in Spring (MAM). Meanwhile, affected by the spatial difference of the snowmelt period in CA, the wind erosion hazard center moved from the southwest (Karakum Desert) to the middle of CA (Kyzylkum Desert and Muyunkum Desert) during spring. According to the impacts of land cover change on the spatial dynamic of wind erosion, the SWEP of bareland was the highest, while that of forestland was the lowest.

Keywords: wind erosion modeling; RWEQ; GEE; central Asia; spatial-temporal variation; land degradation

---

1. Introduction

During the past few decades, global climate change and human disturbance have meant that land degradation has become one of the most serious environmental problems of the 21st century [1]. Despite the lack of strong political will, the land degradation problem has attracted much attention throughout the world [2]. As of today, over 120 countries have committed to the Land Degradation

Neutrality (LDN) Target Setting Programme, which strives to achieve a land degradation neutral world before 2030 [2]. Soil erosion is the major global soil degradation threat to land, freshwater, and oceans [3]. More than 83% of the global extent of land degradation is caused by soil erosion [4]. Because of human activities and climatic variations, the soil erosion can cause topsoil loss, which leads to land degradation requiring centuries to recover, such as soil productivity loss and the thinning out of vegetative cover [5,6].

In the arid land, wind plays a more important role than water in removing the fertile topsoil, which requires centuries to build up [1]. The escalating loss of topsoil by wind erosion is a potential threat to sustainable agriculture, which is closely related to food security [7]. The soil organic matter and several other soil nutrients in the topsoil easily blow away due to strong near-surface wind, and in turn, the soil fertility and plant productivity are decreased. According to a report from the European Soil Data Center (ESDAC), about 28% of the global land degradation area suffers from the wind-driven soil erosion process [8]. Therefore, wind erosion is considered a significant threat to food security and human health, especially in arid and semi-arid regions of the world [9,10]. In addition, the hazard of sand and dust storms is one of the most severe consequences of wind erosion. Wind-blown soil particles and chemicals that lead to air pollution can affect the human respiratory system [11]. Therefore, the ability to accurately simulate and predict soil wind erosion is essential for land degradation control, suitable agricultural management, and sandstorm prevention, especially in arid regions.

As one of the largest land-locked arid regions, Central Asia (CA) has a critical need to combat desertification [12]. Moreover, CA has suffered from the most frequent sandstorms due to the frequent strong wind, limited rainfall, low vegetation coverage, and intense human disturbance [13]. Among those factors, it is generally accepted that the near-surface wind speed plays a vital role in wind erosion dynamics [14]. The need to investigate the soil wind erosion potential over CA is of great importance and brings unique challenges of large-scale wind erosion modeling and insufficient ground measurements [15,16]. Researches on the near-surface wind speed have revealed that there has been a global declining trend of the near-surface wind speed since the 1970s, which is known as global terrestrial stilling [14,17–19]. Affected by global terrestrial stilling, Li, et al. [20] found that the soil wind erosion modulus exhibited a declining trend across CA between 1986 and 2005. However, recent studies found a reversal in global terrestrial stilling around 2010 [21–23]. Furthermore, little is known about the wind speed variability across CA after 2010, let alone the impact of the reversed stilling on the wind erosion dynamic [20,24]. Additionally, existing researches cannot produce continuous high spatial-temporal resolution near-real-time (NRT) wind erosion products of the entire CA, especially for recent years [20,25]. These kinds of wind erosion maps are critical for ecological protection and land use practice in CA.

Wind erosion is a complex physical process controlled by both natural factors and human activities, and normally includes the wind speed, soil characteristics, surface roughness, vegetation cover, agricultural activities, and so on [14,26–30]. It is well-established that three conditions including strong enough wind, susceptible soil surface, and no surface protection by vegetation cover or snow cover, are required for soil wind erosion to occur [31]. The measurement of wind erosion has always been a major obstacle in wind erosion research. According to the literature, two categories of prominent methods, including the ^{137}Cs tracing technique and wind tunnel experiment, can estimate wind erosion more precisely [32,33]. However, they have limitations in that the measurement involves labor-intensive work and hardly describes the spatial variation of wind erosion. Additionally, due to the complex physical processes and driving mechanisms of wind erosion, it is still difficult to monitor the process and conduct quantitative measurements on wind erosion on a large-scale. Over the past few decades, substantial efforts have been made in terms of investigating the mechanism and driving factors of wind erosion. Based on small-scale regional field studies and wind tunnel experiments, several quantitative assessment models of wind erosion have been developed [34–38]. Since the scientific investigations of Bagnold [38] on the wind erosion prediction technology in 1941, soil wind erosion models ranging from empirical-based to physics-based models, have been put forward.

The most accepted models developed to quantify soil wind erosion include Wind Erosion Equation (WEQ) [39], Revised Wind Erosion Equation (RWEQ) [37], Wind Erosion Prediction System (WEPS) [40], Single-event Wind Erosion Evaluation Program (SWEEP) [41], Erosion Productivity Impact Calculator (EPIC) [42], Agricultural Policy/Environmental eXtender (APEX) [43], Texas Erosion Analysis Model (TEAM) [44], and Wind Erosion on European Light Soils (WEELS) [45]. Due to the limited parameters and data that can be obtained, it is difficult for this kind of wind erosion model to simulate soil loss by wind erosion at a larger geographic scale. In contrast with others, the RWEQ model proposed by Fryrear, et al. [37] employs a set of mathematical equations to input weather, soils, crops, and tillage data. Additionally, the RWEQ model has been validated with filed erosion data from 45 site years in several US states [37]. Due to the limitations of RWEQ input data acquisition, the original RWEQ was designed to calculate wind erosion loss at a field scale [36]. Zobeck, et al. [46] evaluated the feasibility of scaling up from fields to regions to estimate the soil wind erosion potential by a geographic information system (GIS)-based field scale wind erosion model in Texas, US. Chi, et al. [47] used the RWEQ model to calculate the soil wind erosion modulus based on field sampling point data regression and remote sensing data over China. Borrelli, et al. [48] developed the GIS-RWEQ model to evaluate the soil loss potential due to wind erosion in the European Union (EU). Although the RWEQ model has achieved some success in large-scale applications, a large amount of detailed local geodata and field work are still required [46–50]. With the development of Remote Sensing (RS) technology and cloud computing, Near Real-Time (NRT) wind erosion data have become more valuable for guiding agricultural production in specific areas. The challenge is to integrate global climate reanalysis data and remote sensing data into the RWEQ model so that it can provide essential knowledge about where and when wind erosion occurs. Another challenge is to consider processing terabyte geospatial data in continent-wide wind erosion quantitative mapping. Moreover, considering the limited computing resources and big data scenarios, it is difficult to use conventional software or programming languages to conduct computation.

The Google Earth Engine (GEE) platform, which has cloud computing capabilities and a multi-petabyte catalog of geospatial data, is a perfect tool for executing wind erosion models [51]. In this platform, the open-source geospatial data include RS data, ground observation data, model simulation data, assimilation data, and so on [52]. GEE's public data archive includes more than 40 years of historical imagery and scientific datasets, which almost cover the geospatial data needed to build the RWEQ model; for example, climate data (wind speed, snow depth, soil moisture, and so on), vegetation cover, soil characters, elevation, etc. These datasets are easily accessible and can be processed and computed in the Cloud, which means that it is not necessary to download data locally. In fact, GEE has shown great potential in change detection, mapping trends, and quantifying differences over the past few years [52]. To date, several studies have been conducted on the GEE platform from regional scales to global scales, such as large-scale land cover classification, vegetation monitoring, soil salinity mapping, disaster management, and so on [53–56].

As we discussed above, GEE is a novel and powerful tool for the quantitative mapping of wind erosion. However, to the best of our knowledge, almost no research has been done to simulate the soil wind erosion potential by using GEE, especially in CA. Additionally, in the context of the global terrestrial stilling reversal, it is important to figure out the wind speed variability in CA for the study of wind erosion in recent years. In view of this, the purposes of this study are (1) to evaluate the near-surface wind speed trend in CA from 2000–2019, based on multiple source climate data; (2) to quantify mapping the soil wind erosion potential (SWEP) in CA based on the RWEQ model by using the GEE platform; and (3) to analyze the monthly and seasonally change of soil wind erosion and the response of soil wind erosion dynamics to land cover change (LCC). This is the first study to execute the wind erosion model on the GEE platform. This provides new ideas for the construction and use of empirical models based on batch geospatial data and high-performance computing. The main conclusions could be beneficial for desertification control and land resource management in CA.

2. Study Area and Dataset

2.1. Study Area

The most common definition of CA is the official one of the Soviet Union, which includes the five former Soviet republics of Kazakhstan (KZ), Uzbekistan (UZ), Turkmenistan (TK), Kyrgyzstan (KG), and Tajikistan (TJ). The total area of CA is nearly 4×10^5 km^2, which is mainly covered with bareland and sparse vegetation. The landform types of CA are mainly plains and hills. Additionally, the mountains (Tianshan Mountain, Pamir Mountain, and Altai Mountain), which are known as the "Water Tower of Central Asia", are mainly distributed in the southeast [57]. As the Tianshan Mountain and Pamir Mountain block rain clouds that should enter CA from the east and south, CA is one of the largest land-locked arid regions in the world [58]. Most of CA lies in an arid climatic zone, which has low annual precipitation (less than 300 mm), a high air temperature, and strong evaporation. Five large temperate deserts (Karakum Desert, Kyzykum Desert, Muyunkum Desert, Sarresi-Atyray Desert, and Aralkum Desert) are distributed from the southwest to middle east (Figure 1). Additionally, desertification caused by large-scale agriculture practices has been an issue since 1960 and enhanced climate change presents many economic, social, and environmental problems in CA [15,24,59,60]. The most notorious example is the Aral Sea Crisis, which has been considered to be one of the planet's worst environmental disasters of the 21st century [59]. The large-scale construction of irrigation canals has reduced runoff from Syr Darya river and Amu Darya river into the Aral Sea, which in turn reduced the Aral Sea surface area from 68,000 square kilometers in 1960 to less than 7000 square kilometers in 2016 [60]. Meanwhile, a new anthropogenic desert known as Aralkum Desert in the eastern dry basin appeared in 1960. Salt and dust storms, which are caused by wind erosion occurring in Aralkum Desert, represent one of the most serious problems for human health and agricultural activities in CA [16].

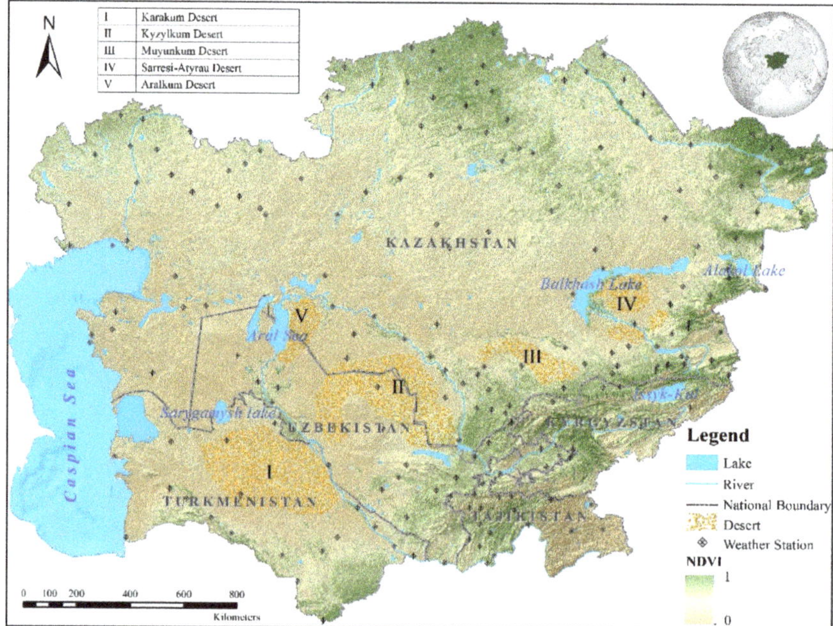

Figure 1. The study area (background image: Moderate-Resolution Imaging Spectroradiometer (MODIS) NDVI in 2019). It consists of the five former Soviet republics of Kazakhstan (KZ), Uzbekistan (UZ), Turkmenistan (TK), Kyrgyzstan (KG), and Tajikistan (TJ).

2.2. Data Collection and Source

The meteorological data included the wind speed, soil moisture, and snow depth, which were derived from Global Land Data Assimilation System 2.1 (GLDAS2.1) integrating satellite and ground-based observational products [61]. Three other sets of climate assimilation data, including The Fifth Generation ECMWF Atmospheric Reanalysis Data (ERA5), NCEP Climate Forecast System Reanalysis (CFSR), and the Famine Early Warning Systems Network (FEWS NET) Land Data Assimilation System (FLDAS), were used to investigate the wind speed variability across CA (Table 1). The soil mechanical composition, soil organic matter, and several other soil properties were obtained from the Harmonized World Soil Database (HWSD) and OpenLandMap (OLM), which are based on machine learning predictions from a global compilation of soil profiles and samples. A total of six standard depths (0, 10, 30, 60, 100, and 200 cm) were divided in the OLM dataset (Table 1), due to the lack of soil calcium carbonate content data in GEE datasets and based on the finding that a nonlinear positive correlation exists between the soil pH and soil calcium carbonate [62,63]. Huang, et al. [62] found that the relationship between the calcium carbonate content and pH value of surface soil in East Central Asia has the highest R^2 when they simulated the factors with an exponential equation. Liu, et al. [63] found that the soil pH and $CaCO_3$ content have a non-linear positive correlation during a study conducted in China. Based on more than 15,000 different soil mapping units, we proposed an exponential equation (Equation (1)) to quantify the relationship between the soil pH and soil calcium carbonate ($CaCO_3$).

$$pH = 4.576 \times CaCO_3{}^{0.09089} + 2.378, \tag{1}$$

where pH is the soil pH of different soil types in HWSD and $CaCO_3$ is the soil calcium carbonate content in HWSD (%).

Table 1. Data collection and sources.

Data	Source	Time	Spatial Resolution	Temporal Resolution
Wind Speed	NOAA GSOD ground measurement wind speed (GMWS)	2000–2019	-	Daily
	GLDAS2.1 *	2000–2019	0.25 degrees	3 h
	ERA5 *	2000–2019	0.25 degrees	Daily
	CFSR *	2000–2019	0.2 degrees	Monthly
	FLDAS *	2000–2019	0.1 degrees	Monthly
Visibility	NOAA GSOD	2000–2019	-	Daily
Soil Properties	OLM *	-	250 m	-
	HWSD	-	30 arc seconds	-
NDVI	MODIS Vegetation Indices (MOD13Q1) *	2000–2019	250 m	16 days
AOD	MODIS MAIAC Land Aerosol Optical Depth (MCD19A2) *	2000–2019	1000 m	Daily
AAI	Sentinel-5 Precursor NRTI/L3_AER_AI *	2019	0.01 arc degrees	Daily
DEM	NASA-SRTM *	-	90 m	-
Land Cover	ESA_CCI	2000–2018	300 m	Yearly

Note: * means that the dataset can be accessed on GEE.

NDVI was derived from the NASA Terra Moderate-Resolution Imaging Spectroradiometer (MODIS) Vegetation Indices (MOD13Q1). The National Aeronautics and Space Administration Shuttle Radar Topographic Mission (NASA-SRTM) provided Digital Elevation Model (DEM) data, which were used to calculate the slope data. The land cover data were provided by the European Space Agency (ESA)-based Climate Change Initiative (CCI) 300 m global land cover data products developed using the GlobCover unsupervised classification chain and by merging multiple available Earth observation products.

Ground observation wind speed data (2 m height) were derived from NOAA Global Surface Summary of Day (GSOD), which includes global data obtained from the United States Air Force (USAF) Climatology Center. We collected the daily ground measurement wind speed data from more than 400 weather stations in CA. Due to political or other reasons, weather stations in the former Soviet Union were abandoned and several new weather stations were established between 1990 and 2010 [60]. Most weather stations currently working were established in the 1960s. Therefore, we integrated ground observations of the wind speed based on 204 weather stations from 2000 to 2019 (Figure 1). The average wind speed data of all-weather stations were used to study the temporal variation characteristics of the wind speed in CA. The visibility, data which could be employed to calculate the Dust Storm Index (DSI), were derived from GSOD. Additionally, the RS technique has provided a new perspective on the validation of soil wind erosion. Aerosol data included the Aerosol Optical Depth (AOD) and Absorbing Aerosol Index (AAI), which were used to compare and validate the wind erosion in this research. AOD was derived from MODIS based on the Multi-angle Implementation of Atmospheric Correction (MAIAC). Moreover, AAI was derived from the Sentinel-5 Precursor, which launched on 13 October 2017. Details on the research data are listed in Table 1.

3. Methodology

3.1. GEE-RWEQ

The comprehensive assessment of wind erosion in a large-scale region like CA is complex and challenging. Dozens of parameters are employed to calculate the soil wind erosion modulus by using field-scale models, such as WEPS. GEE is a cloud computing platform specially designed to process raster data, including satellite images, climate assimilation grids, and other geospatial data. The advantage of the GEE platform lies in the instant access, manipulation, visualization, and real-time analysis of large amounts of geospatial data [52]. Therefore, the advent of GEE made it possible to launch global-scale environmental mapping and monitoring programs [53]. This is of great potential for integrating an environmental model on the GEE platform to build a GEE-based production framework. Furthermore, most developing countries where resources are limited have suffered from various environmental problems, including droughts, flooding, deforestation, soil degradation, and dust storms caused by wind erosion [16,53,64,65]. These countries often lack monitoring sites and networks for environmental problems, making these problems more serious [64–67]. In this study, by using multisource geospatial data, we present a fully automated algorithm for mapping NRT monthly wind erosion dynamics at a global scale using the GEE platform.

Although the computational efficiency should not be a concern in the GEE platform, the limited ground observation data present a big challenge for simulating soil wind erosion. Therefore, it is necessary to build a simplified and more practical model that can estimate the SWEP at a large scale on the GEE platform. Although it has a lower accuracy than mechanistic wind erosion models, this relatively simple model is not limited by the input data, location, and scale of the study area. RWEQ has been extensively tested, and good agreements between model results and field measurements were found in previous studies. In this study, a GEE cloud computing-based RWEQ model was developed to conduct quantitative mapping of soil wind erosion in a ground measurement limited area. As mentioned above, based on the progress of Earth observation and numerical modeling, several parameters that used to be filed, measured, or calculated can be easily acquired on the GEE platform. Although most of the input parameters are retained in GEE-RWEQ, the soil roughness factor (K') is difficult to estimate during farming production on a regional scale. Ouyang, et al. [68] replaced the soil ridge roughness with the roughness caused by topography, and it was calculated by the Smith–Carson equation. Because this equation has been widely used in many regions [68–71], it can applied when the study area is scaled up from a field to a region. Due to the limitations of wind erosion estimation based on RS on regional scales, the combined crop factor (C) was simplified based on previous findings [48–50,68,72].

The GEE-RWEQ involved basic equations, as follows [37]:

$$SWEP = \frac{2x}{s^2} Q_{max} e^{-\left(\frac{x}{s}\right)^2},$$ (2)

$$Q_{max} = 109.8 \times WF \times EF \times SCF \times K' \times C,$$ (3)

$$s = 150.71 \times (WF \times EF \times SCF \times K' \times C)^{-0.3711},$$ (4)

where $SWEP$ is the amount of soil wind erosion potential per unit area (kg/m^2); Q_{max} is the maximum transport capacity (kg/m); x is the distance from the upwind edge of the field (m), set to 55 m for the study area; s is the critical field length (m); WF is the weather factor (kg/m); EF is the erodible fraction (dimensionless); SCF is the soil crust factor (dimensionless); K' is the soil roughness factor (dimensionless); and C represents combined crop factors (dimensionless).

The weather factor can be calculated as

$$WF = \frac{SW \times SD \times \sum_{i-1}^{N} u_2 (u_2 - u_t)^2 \times N_d \times \frac{\rho}{g}}{500},$$ (5)

where SW is the soil wetness (dimensionless), SD is the snow cover factor (dimensionless), u_2 is the wind speed at 2 m (m/s), u_t is the threshold wind speed at 2 m (assumed 5 m/s), N is the number of wind speed observations ($u_2 > u_t$) in the period, N_d is the number of days in the period, ρ is the air density (kg/m^3), and g is the acceleration due to gravity (m/s^2).

The erodible fraction (EF) and soil crust factor (SCF) can be calculated as

$$EF = \frac{29.09 + 0.31Sa + 0.17Si + 0.33\frac{Sa}{Cl} - 2.59OM - 0.95CaCO_3}{100},$$ (6)

$$SCF = \frac{1}{1 + 0.0066(Cl)^2 + 0.021(OM)^2},$$ (7)

where Sa is the sand content (%), Si is the silt content (%), Sa/Cl is the sand to clay ratio (%), OM is the organic matter (%), $CaCO_3$ is the calcium carbonate (%), and Cl is the clay content (%).

The soil roughness factor (K') can be calculated as [70]

$$K' = \cos\alpha,$$ (8)

where α is the slope gradient (degree), which can be calculated by the Digital Elevation Model (DEM).

The combined crop factor (C) can be calculated as [70,73]

$$C = e^{-0.0483(SC)},$$ (9)

$$SC = (NDVI - NDVI_{soil}) / (NDVI_{max} - NDVI_{soil}),$$ (10)

where SC is the vegetation coverage (%), $NDVI_{soil}$ is the NDVI value of a bare soil pixel, and $NDVI_{max}$ is the maximum NDVI value of the study area.

3.2. Model Performance Evaluation

Because of the diverse land cover types and large area, it is extremely difficult to measure wind erosion for a whole region. Additionally, in the past two decades, almost no research has conducted field measurements on wind erosion in CA. Therefore, validation methods that can evaluate the reliability of wind erosion model results need to be proposed. Considering that the ground observed dust storm can indicate the frequency and intensity of wind erosion events, DSI was used to validate the spatial variation of SWEP.

DSI was calculated based on the meteorological record-visibility, which can represent the frequency and intensity of wind erosion events. DSI was first proposed by McTainsh [74] in the National Collaborative Project on Indicators for Sustainable Agriculture (NCPISA). Based on the relationship between meteorological records and DSI, O'Loingsigh, et al. [75] used daily visibility data acquired from 180 long-term meteorological stations to investigate a long-term national wind erosion record (1965–2011) in Australia. DSI is a methodology employed for monitoring wind erosion based on long-term daily meteorological observations. At present, it is generally accepted as an indicator of broad-scale wind erosion rates in Australia, Iran, and Northeast Asia [75,76]. Based on weather codes relating to wind erosion or visibility, wind erosion events were divided into three categories: (a) Severe Dust Storms (SDS); (b) Moderate Dust Storms (MDS); (c) Local Dust Events (LDE). The DSI was calculated using the following equation [75]:

$$DSI = \sum_{i=1}^{n} \left[(5 \times SDS) + MDS + (0.05 \times LDE) \right]_i, \tag{11}$$

where i is ith value of n stations for $i = 1$ to n, SDS is a severe dust storm (visibility < 200 m), MDS is a moderate dust storm (200 m < visibility < 1000 m), and LDE is a local dust event (1000 m < visibility < 20,000 m).

Due to the lower population and urban density, soil mineral particles produced by wind erosion are the main source of atmospheric aerosols in CA [24]. Therefore, the satellite-derived AOD data were used to evaluate the reliability of SWEP simulated by the RWEQ model. There are several satellite-based aerosol products, which have different spatial and temporal resolutions, such as CALIPSO Lidar Tropospheric Aerosol Profiles All sky data, VIIRS/SNPP Deep Blue L3 daily aerosol data, OMI/Aqua Multi-wavelength AOD Daily data, MODIS MO(Y)D08_M3 Terra (Aqua) Atmosphere Monthly data, MODIS MCD19A2 Terra & Aqua MAIAC Land Aerosol Optical Depth Daily data, and Sentinel-5P NRTI AER AI. However, most satellite-based aerosol products have a low spatial resolution and cannot meet the requirements of quantitative spatial comparisons [77–79]. In this study, we used the MODIS MCD19A2 dataset at the 0.47 μm blue band, along with the parameter Optical_Depth_047, which has a spatial resolution of 1 km [79]. Another aerosol dataset named the Absorbing Aerosol Index (AAI), with a 0.01-degree spatial resolution, was extracted from the Sentinel-5P NRTI AER AI product. Because the Sentinel-5P was launched on 13 October 2017, the aerosol dataset was released in 10 July 2018 [80]. Therefore, we used the 2019 annual average AAI to compare with the 2019 SWEP in this study.

3.3. Technical Flowchart of this Study

The Land Cover Change (LCC), which is influenced by both climate change and human activity, usually affects wind erosion on surface roughness and soil physical and chemical characteristics. Therefore, we studied the SWEP of different land cover types and SWEP changes caused by the conversion of different land cover types. In this study, we chose ESA-CCI 300 m global land use land cover data products developed using the GlobCover unsupervised classification chain and by merging multiple available Earth observation products. Based on the United Nation Land Cover Classification System's (LCCS) plant functional types (PET), the CCI-LC map is classified into 22 land types. According to the study area land characteristics, the land cover types are reclassified into nine categories based on the look-up table-conversion of CCI-LC classes to PET in the product user guide [81].

Based on the objective of this study, this manuscript is organized as presented in the technical flowchart (Figure 2). The research consists of four main steps: First, based on a time-series decomposition model, the wind speed variability of ground measurement data and reanalysis data was explored; second, by using multi-source geospatial data, the monthly SWEP across CA was generated based on GEE-RWEQ, and we explored the spatiotemporal variation of SWEP between 2000 and 2019; third, based on DSI and satellite-based AOD, validation was conducted to test the reliability of annual SWEP;

and finally, we investigated the responses of wind erosion to ground measurement wind speed change and land cover change.

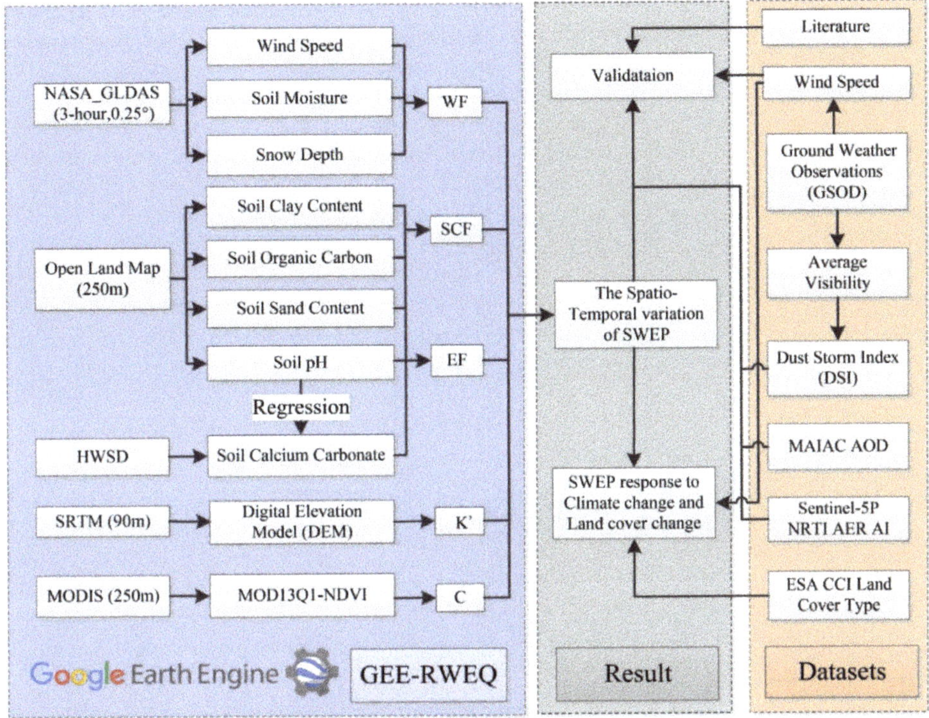

Figure 2. The technical flowchart of this study.

The ArcGIS10.6 software was implemented for land cover type reclassification in this research. The Pearson correlation coefficient (r) was calculated in R 3.6.3. The time-series decomposition model based on the "tseries" package was also run in R 3.6.3. The exponential fitting of the soil pH and soil calcium carbonate was performed in MATLAB 2018a.

4. Results, Analysis, and Validation

4.1. Variability of the Daily Average Wind Speed across CA

As the key factor of wind erosion, wind speed variability plays a vital role in wind erosion dynamics. A host of studies have reported that there was a declining trend in the global near-surface wind speed from 1970 to 2010 [14,17–19]. However, a recent study described an increase in the global wind speed during a particular year. Zeng, et al. [22] found that, after several decades of global terrestrial stilling, the wind speed has increased rapidly across the globe since 2010. Although Zeng, et al. [22] have investigated the global temporal variation of the wind speed, further studies are required because of the sensitivity of CA to global climate change [60].

To better understand the temporal variations of wind speed, it is possible to decompose wind speed time series data into sub-components by a time-series decomposition model. In this study, we used a multiplicative decomposition model, which is more effective when a seasonal value changes over time [82]. The calculation of this model included the following three steps [83]. The trend component was first determined and removed from time series by using the moving averages method. Secondly, the seasonal component was calculated and centered by averaging all periods for each time unit. In this study, a time-series decomposition multiplicative model was applied to Ground Measurement Wind Speed (GMWS) data. Figure 3 shows that the time series data were decomposed into various sub-components (the trend component, seasonal component, and random component). According to the trend component of the GMWS time series, we found that there was a significant decrease during the time period of 2000–2009 and a significant increase trend in the time period of 2009–2014. Moreover, GMWS exhibited steady fluctuations or a slight upward trend after 2014. From the perspective of quantitative analysis, we calculated the decade change rate of GMWS in these three time periods based on a linear regression analysis of ordinary least squares (OLS). The analysis shows that the daily average GMWS decreased significantly at a rate of -0.16 m s^{-1} decade^{-1} during the period of 2000–2009 ($p < 0.001$). After the turning point of 2009, the increasing rate of 0.42 m s^{-1} decade^{-1} was significantly higher than the decreasing rate during the period of 2009–2014 ($p < 0.001$). Although the GMWS shows a slight trend in recent years, the result is not statistically significant ($p > 0.05$). The time series of seasonal components indicated that the highest values for the daily average wind speed occurred during the spring, while the lowest values occurred during the autumn. The relationship between the daily average wind speed in different seasons is spring > winter > summer > autumn.

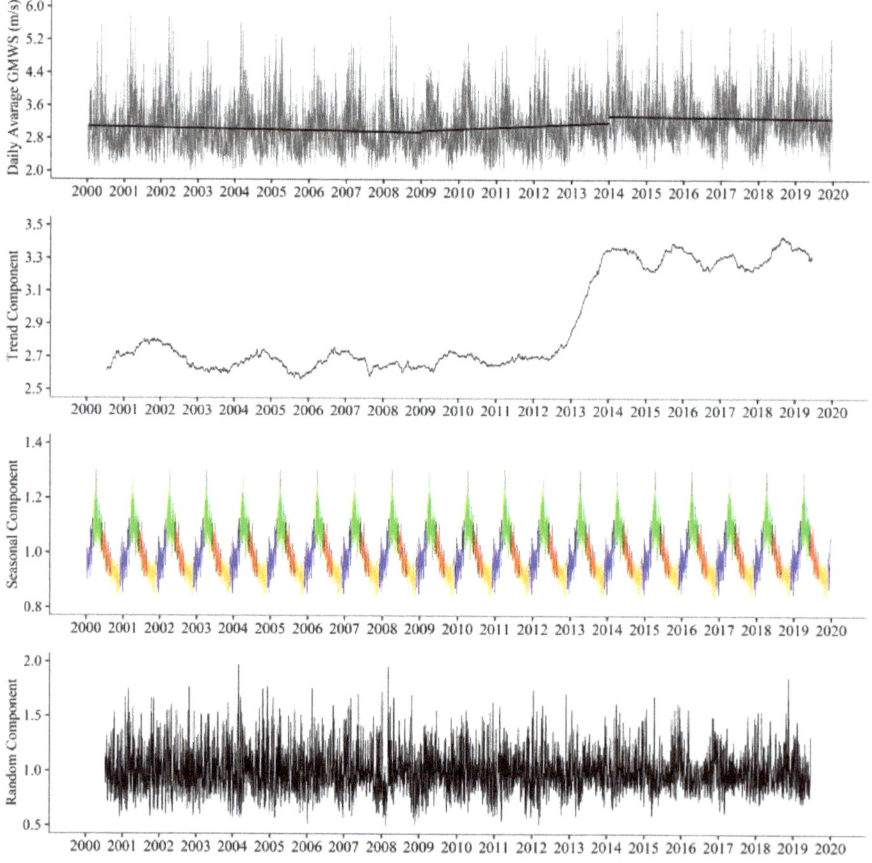

Figure 3. The time series decomposition of the average Ground Measurement Wind Speed (GMWS) in Central Asia (CA) from 2000 to 2019. Note: The different colors in the graph on the seasonal component indicate different seasons (blue: Winter (Dec., Jan., Feb.), green: Spring (Mar., Apr., May), red: Summer (Jun., Jul., Aug.), and yellow: Autumn (Sept., Oct., Nov.)).

In order to ensure the consistency of the reanalysis data input by the model and the actual observation data in CA, the trends of four reanalysis data (GLDAS2.1, ERA5, CFSR, and FLDAS) were introduced to conduct comparisons with GMWS. The comparison result showed that the daily average derived from GLDAS2.1 has the highest correlation with the ground measurement wind speed (Supplementary Materials Figure S1). Furthermore, in order to better compare the relationship between the wind speed in trend and seasonal components, we decomposed the GLDAS2.1 wind speed time series into trend component, seasonal component, and random component. Additionally, we calculated the correlation coefficient (r) of the daily average wind speed and components between ground measurement data and reanalysis data. The trend component had the highest correlation coefficient ($r = 0.829$), followed by the seasonal component ($r = 0.552$), daily average wind speed ($r = 0.125$), and random component ($r = -0.007$). On the other hand, according to the trend component, the turning point for the GLDAS wind speed (GLDASWS) time series occurred around 2011 (Figure 4). Moreover, the changing rate of the daily average wind speed is more significant. The result of linear regression shows that the daily average GLDASWS decreased significantly at a rate of -0.34 m s^{-1} decade^{-1} during the period of 2000–2010 ($p < 0.001$). The increasing rate of 1 m s^{-1} decade^{-1} is significantly higher than the decreasing rate during the period of 2010–2014 ($p < 0.001$) after the turning point of 2010 (Figure 4). The seasonal components of two wind speed data have a broadly similar pattern.

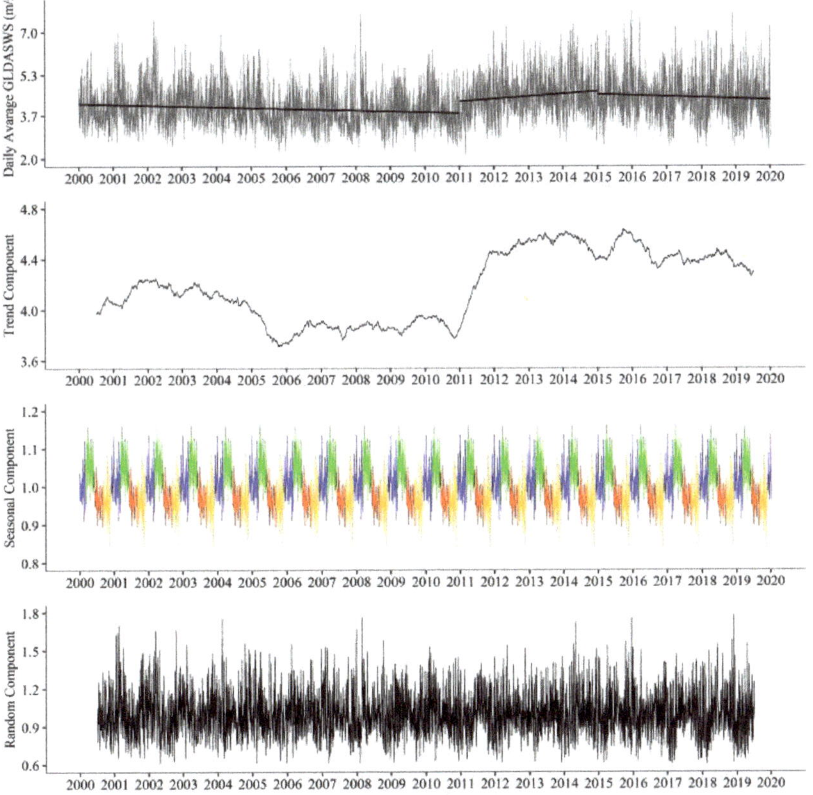

Figure 4. The time series decomposition of the average Global Land Data Assimilation System wind speed (GLDASWS) in CA from 2000 to 2019. Note: The different colors in the graph on the seasonal component indicate different seasons (blue: Winter (Dec., Jan., Feb.), green: Spring (Mar., Apr., May), red: Summer (Jun., Jul., Aug.), and yellow: Autumn (Sept., Oct., Nov.)).

4.2. The Spatiotemporal Variation of Wind Erosion across CA

Figure 5 shows the distribution of the annual mean SWEP in CA over the most recent 20 years (2000–2019). SWEP exhibits significant spatial variation, which has a range of 0–256 kg/m^2. The SWEP in the southwest CA is higher than that in the southeast and north CA, where the vegetation coverage and soil moisture are higher. During the past 20 years, the Aral Sea dry lake bed (ASDLB), which is one of the most active dust sources, was the most severe wind erosion area in CA (47.29 kg/m^2/y), followed by Kyzylkum Desert (10.64 kg/m^2/y), Karakum Desert (10.58 kg/m^2/y), and Muyunkum Desert (6.81 kg/m^2/y). Due to the dramatic shrinkage of Aral Sea from the second half of the 20th century, the ASDLB, also known as the Aral Sea Desert, was covered with the original salts and chemicals of the water [24]. The toxic sediments of the Aral Sea were blown away by strong winds and formed white sandstorms. These toxic particles from the dry Aral Sea lake bed had been found in Japan, Norway, Greenland, and even in the South Pole [84].

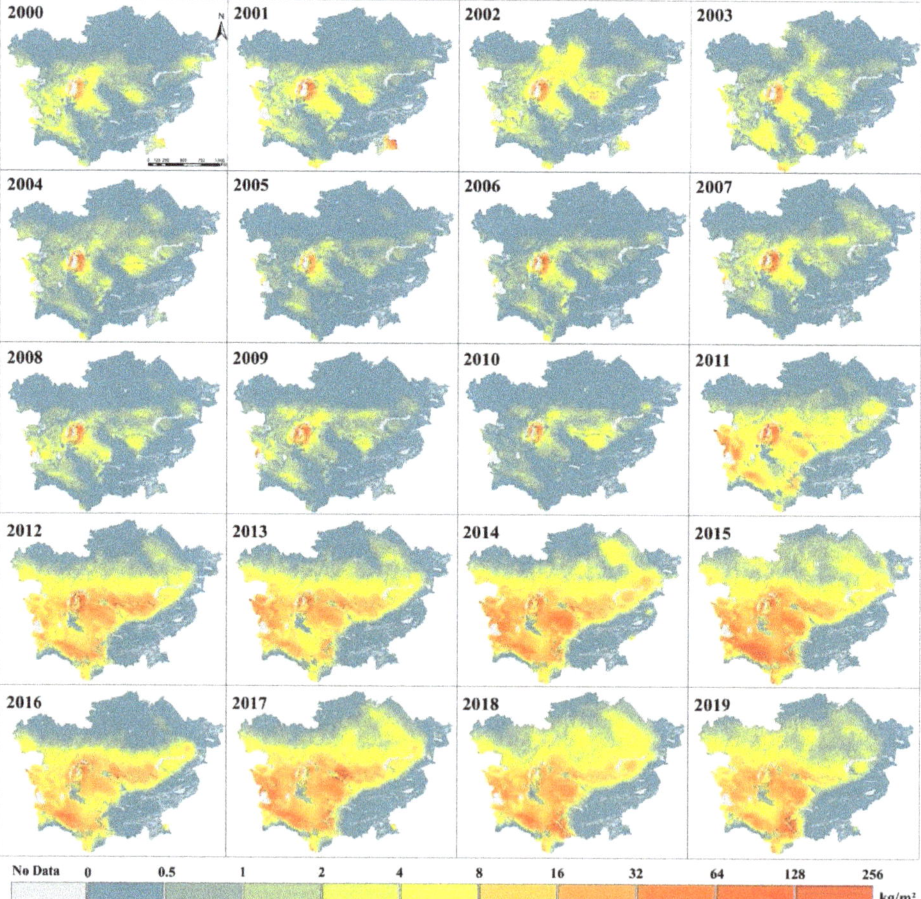

Figure 5. The spatial variation of the Soil Wind Erosion Potential (SWEP) over CA between 2000 and 2019.

From 2000 to 2019, the annual mean value of SWEP was 3.45 kg/m^2, with the lowest value occurring in 2010 and the highest value occurring in 2015. As mentioned above, the wind speed exhibited a significant increase around 2011, and the area and intensity of wind erosion have also increased significantly since 2011. However, the SWEP gradually decreased by -6.85 kg/m^2/y, over ASDLB, from 2011 (Figure 5). This reversal may have been caused by the recovery of the water body of the Aral Sea since 2010 [85]. This means that more dry lake beds are covered by water bodies and less bareland suffers from wind erosion. The Amu Darya River (ADR) and Syr Darya River (SDR), which are the two largest rivers across CA, are the principal water suppliers of the Aral Sea. Since the 1960s, many irrigation canals have been constructed in the middle and lower reaches of ADR and SDR [60]. Therefore, according to ESA land cover data, more than 43% of the irrigation cropland of CA survived on these canals, especially in Amu Darya Delta (ADD) and Syr Darya Delta (SDD). The main land cover type of these two deltas is irrigated cropland, which accounts for more than 20% of the total irrigated cropland in CA. Because of the high vegetation coverage, according to Figure 5, the ADD and SDD regions have a lower SWEP than the surrounded area. Similarly, on the edge of the Kyzylkum Desert, oasis agriculture that relies on irrigation also greatly reduces SWEP.

The seasonal variation characteristics of SWEP in CA are shown in Figure 6. Although it is usually consistent with the spatial distribution of the annual mean SWEP, the spatial pattern of SWEP varies in different seasons. As we can see in Figure 6, more land suffered from severe wind erosion in CA during the spring than in other seasons (spring: 1.48 kg/m^2 > summer: 0.70 kg/m^2 > winter: 0.69 kg/m^2 > autumn: 0.59 kg/m^2). Additionally, obvious wind erosion exists in several famous desert regions, such as the Karakum Desert, Kyzylkum Desert, and Aralkum Desert during spring. Due to the impact of snow cover, less wind erosion exists in north CA during the winter. However, the wind speed of the Aral region in winter markedly exceeds that in other seasons, especially in December. Therefore, the ASDLB region has higher SWEP in the winter. Figure 6 shows that the north Kazakhstan region, Kyrgyz, Tajikistan, has been slightly affected by wind erosion. Additionally, in summer, SWEP shows a significant high value in the middle reaches of Amu Darya (Figure 6). As shown in Figure 7, the SWEP displays significant monthly temporal variability, especially in ASDLB. March is the most severe month of wind erosion in the ASDLB. Alternatively, due to strong winds and dry surface soil in December, January, and February, sandstorms frequently occurred in ASDLB. We can confirm this from the true color image of MODIS (Supplementary Figure S2). Due to the difference in the solar radiation energy received by different latitudes, the melting time of snow cover varies in different regions. Figure 6 shows that the center of wind erosion moves from southwest to northeast during the spring (March, April, and May). These factors can explain why significant wind erosion in other famous deserts occurs in different months. The most severe month of wind erosion in Karakum is April, and that in Kyzylkum Desert and Muyunkum Desert is May. However, in general, Central Asia has suffered from the most severe wind erosion in April and the most widespread wind erosion in May.

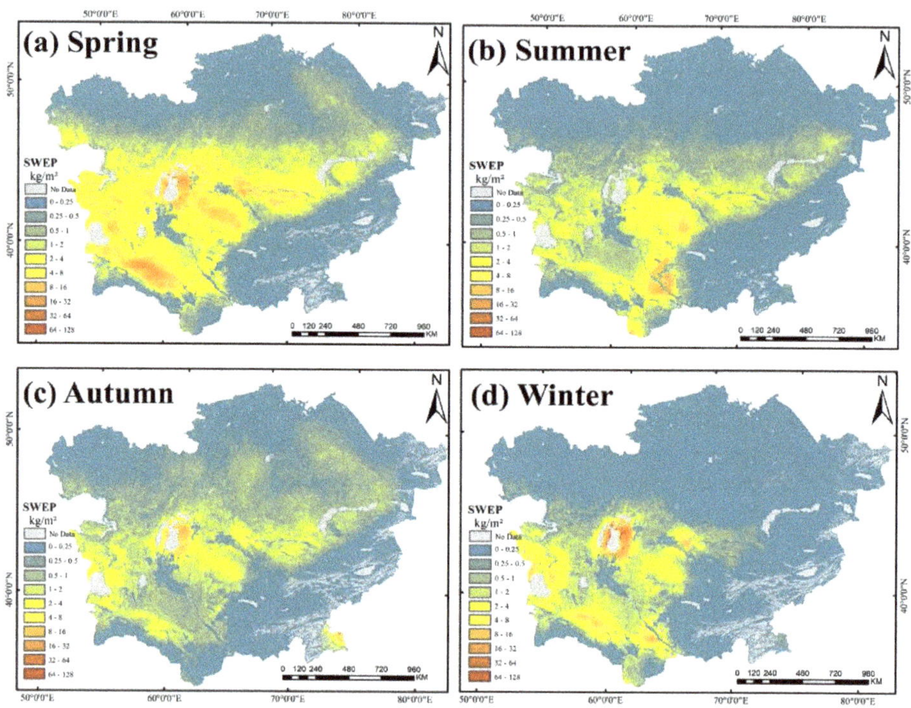

Figure 6. The seasonal variation of SWEP in (**a**) spring (Mar., Apr., May), (**b**) summer (Jun., Jul., Aug.), (**c**) autumn (Sept., Oct., Nov.), and (**d**) winter (Dec., Jan., Feb.) during the period of 2000–2019.

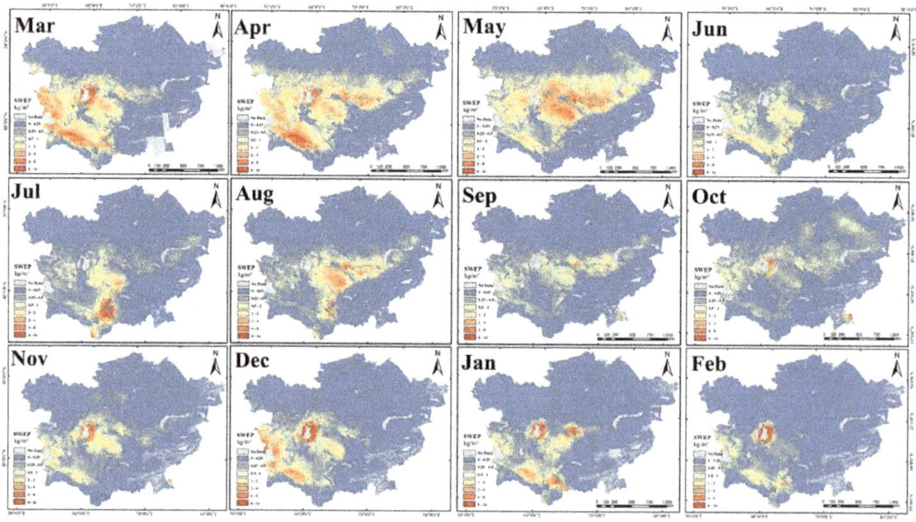

Figure 7. The monthly average variation of SWEP during the period of 2000–2019.

4.3. Responses to Wind Speed Change and Land Cover Change

4.3.1. Impacts of Ground Measurement Wind Speed Changes on the SEWP

Based on the research results mentioned above, the wind speed was found to be the dominant factor of wind erosion. Therefore, in this study, we analyzed the influence of wind speed as a factor of climate change on wind erosion. Based on the ground measurement wind speed data, we investigated the influence of wind speed changes on wind erosion. Figure 8 shows the strong and significant positive correlation ($r = 0.700$, $p < 0.001$) between the average GMWS and average ground SWEP. According to the trend line of these two sets of time-series data, the turning point roughly occurred between 2010 and 2011. The average GMWS showed a slowly decreasing trend (-0.07 ms^{-1} decade^{-1}) before the turning point in 2011, while it displayed a significant increasing trend during the period of 2011–2019 ($+0.6$ ms^{-1} decade^{-1}, $p < 0.001$). The recent increasing rate is almost tenfold the decreasing rate in the first decade. The average ground SWEP also showed a slightly decreasing trend during the period of 2000–2010 (-0.027 kgm^{-2} decade^{-1}), while it displayed a significant increasing trend ($+0.37$ kgm^{-2} decade^{-1}, $p < 0.001$). Shao, et al. [86] found that the global monthly mean dust concentration decreased from 2000 to 2012. This shows that the end of the quiet period of dust activities in Central Asia or globally marks the beginning of an active period of dust activities. Based on current research, it seems reasonable to relate the dust trend to the climate trend, especially the reversal in global terrestrial stilling [22].

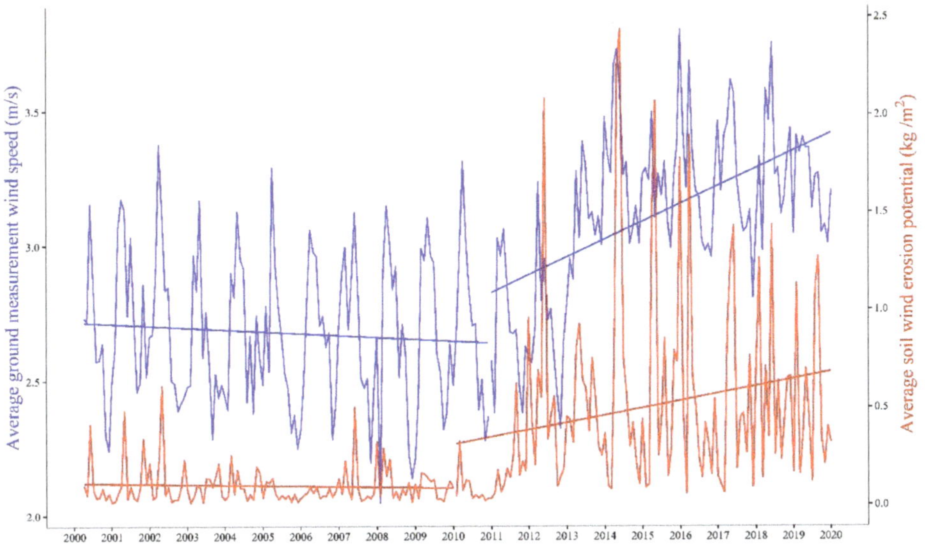

Figure 8. Average GMWS and average SWEP across CA from 2000 to 2019.

Additionally, the highest monthly average SWEP, which appeared in May 2014, was more than 2.4 kg/m^2. The seven months with the strongest wind erosion ($SWEP > 1.5$ kg/m^2) were March (2), April (2), May (2), and December (1). There were nine months (May: 3, March: 3, April: 2, Decembeer: 1) with an average wind speed greater than 3.5 m/s. Moreover, the highest monthly average wind speed, with a value of 3.79 m/s, appeared in Decembr 2015. Overall, a similar tendency of wind speed and wind erosion was observed for the past two decades. Both show a distinguishable declining trend, and then a sudden remarkable increase, and before slowly declining or finally stabilizing. During the study period, the wind speed ($+0.38$ ms^{-1} decade^{-1}, $p < 0.001$) and SWEP ($+0.34$ kgm^{-2} decade^{-1}, $p < 0.001$) increased very quickly from 2000 to 2019, indicating a more serious soil degradation and air pollution problem in CA.

4.3.2. Divergence of SWEP from Different Land Cover Types

According to the land characteristics of the study area, the ESA CCI land cover types can be reclassified into nine categories (cropland irrigated, cropland rain-fed, forestland, shrubland, grassland, sparse vegetation land, bareland, urbanland, and waterbody). The last subplot of Figure 9 shows the areas of different land cover types across CA in 2018. Figure 9 shows that the monthly average SWEP and its change rate of different land cover types were substantially different. The monthly average SWEP of bareland was more than 0.836 kg/m^2, followed by shrubland (0.572 kg/m^2), sparse vegetation land (0.203 kg/m^2), grassland (0.073 kg/m^2), irrigated cropland (0.043 kg/m^2), rainfed cropland (0.033 kg/m^2), and forestland (0.017 kg/m^2). This relationship is basically consistent with previous research conducted in CA and surrounding regions [20,87]. We also compared the soil wind erosion modulus with respect to regions with similar conditions or the use of different methodologies (Table 2). Li, et al. [20] assessed the soil wind erosion modulus variation in CA (including Xinjiang, China) between 1986 and 2005. Zhang, et al. [87] investigated the RWEQ-based soil wind erosion, which was validated by ^{137}Cs in Inner Mongolia (IM), during the time period of 1990–2015. Compared to other arid or semiarid regions, CA has relatively higher rates of soil wind erosion, which may be the result of the widespread distribution of deserts and wind speed increase in the past decade. Grassland has a relatively lower soil wind erosion rate, because it is mainly distributed in northern CA, which has a more humid climate and less erodible underlying surface condition [12]. Although the time periods and dataset sources were different, from the perspective of the wind erosion diversity of land cover, the wind erosion result of our research is reliable.

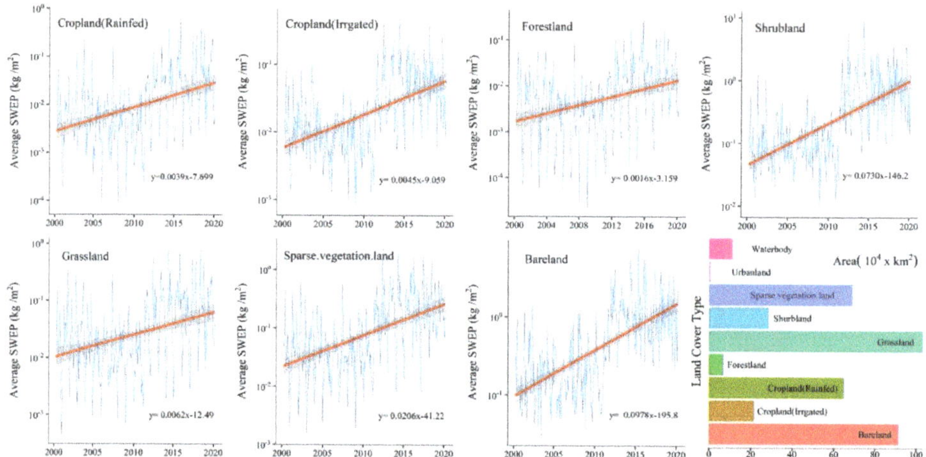

Figure 9. Monthly change of the average SWEP of different land cover types and the area of different land cover types across CA in 2018.

There was a significant increase in SWEP in CA in the past two decades. However, the change rate of wind erosion varied among regions with various types of land cover. Specifically, the change rate of wind erosion was the highest for bareland (0.0978 kg/m^2/y), followed by shrubland (0.0730 kg/m^2/y), sparse vegetation land (0.0206 kg/m^2/y), grassland (0.0062 kg/m^2/y), irrigated cropland (0.0045 kg/m^2/y), rainfed cropland (0.0038 kg/m^2/y), and forestland (0.0016 kg/m^2/y). Combined with the areas of different land covers, we could calculate the total amount of soil wind loss for different land covers in the period of 2000–2019. More than 2.8255×10^{11} t soil was eroded by the wind across CA during the past few decades. The soil wind erosion of bareland (1.838×10^{11} t) contributed more than 65% soil loss by the wind in CA, followed by shrubland (0.3907×10^{11} t), sparse vegetation land (0.3380×10^{11} t), grassland (0.1812×10^{11} t), rainfed cropland (0.0517×10^{11} t), irrigated cropland (0.0223×10^{11} t), and forestland (0.0028×10^{11} t).

Table 2. Soil wind erosion rate studies in CA and other regions with similar conditions.

Authors	Locations	Method	Study Period	Soil Wind Erosion Rate ($\times 10^{-1}$ kg/m^2/y)			
				Bareland	Grassland	Forestland	Cropland
This Study	CA	RWEQ	2000–2019	103.56	8.76	2.04	5.16(3.96)
Li, et al. [20]	CA (Included Xinjiang, China)	RWEQ	1986–2005	45.08	15.56	3.44	4.74
Zhang, et al. [87]	IM, China	RWEQ	1990–2015	101.96	24.21	2.96	11.31
Lin, et al. [50]	Hexi, China	RWEQ	1982–2015	85.19	40.07	9.48	21.43
Chi, et al. [47]	Arid land, China	RWEQ	2000–2010	57.61	6.73–28.07	16.03	17.66
Hu, et al. [32]	IM, China	137CS	2003	NA	18.08–42.7	NA	79.90
W. Cole, et al. [42]	New Mexico, USA	WEE/EPIC	50-years	NA	NA	NA	0.13–71.3
Hagen [88]	Arid land, USA	WEPS	1989–1997	NA	NA	NA	0–39.8

According to the continuous changes in the land cover area in CA during the period of 2000–2018, the cropland, forestland, urbanland, and shrubland showed an increased trend, while bareland and grassland showed a decreased trend (Supplementary Table S1). More than 2.67×10 km^2 land had undergone land cover change, including from bareland to grassland, sparse vegetation land to grassland, grassland to cropland (rain-fed), sparse vegetation land to cropland (rain-fed), and waterbody to bareland. In order to remove wind speed variability effects on SWEP, we calculated the annual SWEP of 2018 based on the wind speed data of 2000 to compare it with the annual SWEP of 2000. We found that the LCCs with the strongest inhibitions of wind erosion activity were bareland into shrubland (-0.782 kg/m^2), sparse vegetation land into forestland (-0.106 kg/m^2), and grassland into forestland (-0.073 kg/m^2). In comparison, the conversions of land cover which accelerated wind erosion the most were waterbody into bareland ($+3.784$ kg/m^2), sparse vegetation land into bareland ($+1.124$ kg/m^2), and grassland into bareland ($+0.490$ kg/m^2).

4.4. Validation of the GEE-RWEQ Model

Due to lack of long time series and wide range of ground-measured wind erosion data in CA, validation of the GEE-RWEQ model is challenging. Furthermore, due to the large area and complex terrain conditions, almost no previous research has conducted wind erosion field measurements in CA. Considering that most of the local dust storms are caused by surface wind erosion in CA [24], the dust storm index (DSI) can be used as a proxy to evaluate the wind erosion model performance [75]. Therefore, we used the DSI based on weather station visibility records to evaluate the reliability of the SWEP spatial distribution. Figure 10 displays the annual mean DSI across CA from 2000 to 2019. This map was interpolated by the annual average DSI of more than 200 weather stations based on Natural Neighbor Interpolation method. From Figure 10, we can see that the southwest desert region of CA has a high DSI, which means very frequent dust storms. However, there are some high values in the southeastern parts of CA, where there is a lower wind erosion risk. According to the research of Liu, et al. [89], affected by the strong southwest winds, the dust particles were transported from the western desert to eastern mountains and valley. Moreover, dust episodes were observed in these regions. Additionally, the southeastern parts of CA are the most densely populated areas in CA. Most of the weather stations across CA are located around densely populated cities. The anthropogenic pollutants will also be recognized as dusty weather due to reduced visibility. The spatial distribution of DSI is generally consistent with the spatial pattern of the annual SWEP. Although visibility records based on weather stations are a valuable and useful data resource for wind erosion monitoring, several limitations still exist. The low spatial density of weather stations is a challenge for conducting highly accurate wind erosion mapping, especially in the southeastern part of CA. Therefore, we need to obtain higher spatial resolution and more continuous SWEP verification data. It should be pointed out that satellite-based atmospheric aerosols, which refer to solid and liquid particles suspended in the atmosphere, have strong spatial correlations with wind erosion.

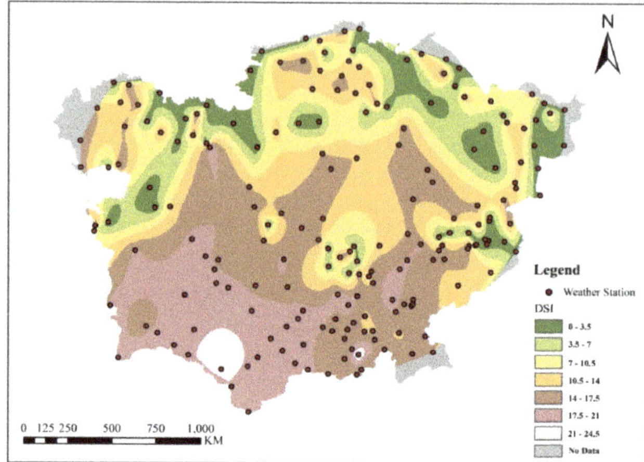

Figure 10. The location of weather stations that provide visibility records and the annual mean Dust Storm Index (DSI) for the period 2000 to 2019 across CA.

Figure 11 shows the spatial pattern of the annual AOD (a) and average AAI in 2019 (b), as well as the comparisons with SWEP in the Aral Sea region (ASR). From Figure 11, we can see that the Aral Sea region and its southwest surrounding area has the highest value in CA. This is because the dust of the Aral Sea is transported to the southwest under the action of the dominant wind-northeast wind [90]. Therefore, we chose ASDLB as a research hotspot area to compare with SWEP. Linear relationships between SWEP and aerosol parameters (AOD and AAI) were found, as shown in Figure 11. The results show that they had moderate positive linear relationships, with r values of 0.5623 ($p < 0.001$) and 0.5660 ($p < 0.001$), respectively. Ultimately, although it is difficult to verify the SWEP value, the comparison results obtained from the perspective of spatial and temporal distribution patterns showed that the RWEQ-based SWEP data in CA were reliable.

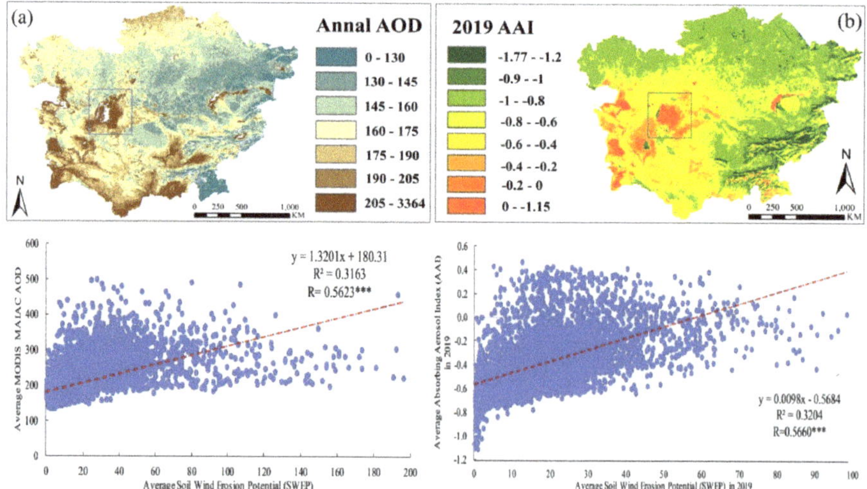

Figure 11. The spatial distribution of the annual average MODIS Aerosol Optical Depth (AOD) (**a**) and Sentinel-5P Absorbing Aerosol Index (AAI) of 2019 (**b**), and the comparisons of the average SWEP and AOD and AAI in the Aral Sea region.

5. Discussion

In this study, our results show that there are significant spatial and temporal differences in the wind erosion in CA. Controlled by the latitude zonality and vertical zonality, higher SWEP values are primarily distributed in the southwestern part of CA, which has low vegetation coverage and more fragile surface soil [91], while lower SWEP is mainly concentrated in the northern part of CA. Based on the land cover map of 2018, more than 89% of rain-fed cropland and 78% of forestland are distributed in these regions. Furthermore, due to latitude zonality and vertical zonality, the precipitation in these areas is higher than in other places of CA, and the temperatures in these areas are lower. Therefore, the higher soil moisture caused by lower evapotranspiration will reduce the wind erosion to a certain degree. Affected by the restoration of the Aral Sea in recent years, the vegetation coverage and other underlying surface factors of ASDLB are getting better [85]. Although the SWEP showed a decreasing trend in ASDLB (-6.85 kg/m^2/y), this region was still the most severe wind erosion area in CA. In addition, our results show that the SWEP has a clear seasonal and monthly variation. The land threatened by wind erosion has the largest range in spring, especially in May. Due to the difference in solar radiation heat at different latitudes and altitudes, the snowmelt period varies in different regions of CA [92]. The higher soil moisture caused by snowmelt and the snow cover both affected the movement of the wind erosion center across CA during spring [37]. Therefore, considering the major deserts of CA are located in different latitudes and altitudes, the severe wind erosion regions in CA will also migrate over time. The severe wind erosion regions move from the southwest CA (Karakum Desert) to the middle CA (Kyzylkum Desert and Muyunkum Desert) during the spring (MAM). However, due to the special meteorological conditions in certain areas of CA, such as the middle reaches of the ADR, severe wind erosion occurs in summer, but not spring, respectively [24]. We calculated the monthly average wind speed of four weather stations in this region. A comparison was made for the monthly average wind speed of all the weather stations in CA and the four weather stations in this region (Supplementary Figure S3). The most extreme value of the surface wind speed in CA appeared in March, while the most extreme value of the wind speed in the middle reaches of the ADR appeared in July. This result demonstrates that wind speed plays the key role in the spatial distribution of SWEP.

As the most dominant factors of wind erosion, wind speed variability, such as wind stilling or wind stilling reversal, account for the majority of the spatial-temporal variation of wind erosion in CA. Although global terrestrial stilling has been confirmed by many pieces of research, most studies have only looked at global or regional wind speed changes from the 1980s to 2010, and few have involved recent (after 2010) wind speed changes [14,17–19]. According to several climate assimilation datasets (GLDAS, ERA5, CFSR, and FLDAS) and a ground measurement dataset (GSOD), we found a turning point of wind speed stilling during the period of 2009–2012 in CA. This finding is supported by other studies that have reported that global terrestrial stilling has rebounded over the past few decades and has increased rapidly since 2010 [21–23]. Our research proves that the increase rate of the average wind speed in CA (0.6 m s^{-1} decade^{-1}) is higher than the increase rate of the average global wind speed (0.24 m s^{-1} decade^{-1}) over the same period, which means that stronger wind erosion occurred in CA. Indeed, the result shows a strong and significant positive correlation ($r = 0.7$) between the average GMWS and average SWEP ($p < 0.001$). A number of studies have demonstrated that CA is more sensitive to climate change compared to the global average [12,20,57,60,91,93]. While it is widely acknowledged that the global wind speed rebound is beneficial to the wind power industry for the near future [22], this study suggests that more severe wind erosion activity happened in CA.

According to the significant differences in natural conditions, such as the air temperature and precipitation, and the disturbance of human activities such as irrigation, the land cover in CA exhibited strong spatial differentiation. We found that SWEP differs greatly in different land cover types. This result is roughly consistent with a previous study on wind erosion in CA and surrounding regions [20,87]. Most of the shrubland in CA is made up of deserts and xeric shrublands in which *Haloxylon ammodendron*, *Calligonum aphyllum*, and *Ephedra lomatolepis*, as well as grasses such as

Agropyron fragile, grow [94–96]. These kinds of vegetation have good windproof and sand fixing functions in CA. In addition, LCCs are closely related to wind erosion activity and affect each other. From 2000 to 2018, more than 2.6×10^5 km^2 of land has changed land cover types (Supplementary Table S1). We compared the SWEP for the land with the LCC from 2000 to 2018, in which the effect of the wind speed variability was removed. The conversion of bareland to shrubland helped reduce wind erosion by -0.782 kg/m^2/y. Furthermore, due to the shrinkage of the Aral Sea, the conversion of waterbody to bareland increased wind erosion by $+3.784$ kg/m^2/y. The restoration of the Aral Sea has not only superficially reduced the possibility of wind erosion in ASDLB, but also increased the increased vegetation coverage of ASDLB caused by the higher groundwater level, making the long-distance transport of dust difficult. Although the wind speed has shown an increasing trend in the past 10 years, the wind erosion risk in the Aral Sea area is gradually decreasing due to the continuous recovery of the Aral Sea area. In the past 30 years, a large number of engineering projects have attempted to improve the Aral Sea environment, directly or indirectly. Although Aral sea restoration is the most effective way to restrain wind erosion in ASDLB, the complex political relations among countries in the Aral Sea basin make cross-border water management difficult [84]. Besides, in other parts of CA, more effective measures should be taken for wind erosion artificial control, for example, increasing the grassland area in regions with a suitable temperature and precipitation, developing cropland by using limited water resources, and planting cold- and drought-resistant shrub vegetation in bareland.

In this study, the RWEQ model was adopted as a wind erosion model in the GEE cloud computing platform. Compared with the local computing platform, GEE can process large amounts of geospatial data in a short time, which means that its processing power is completely unconstrained by time and space. Therefore, we do not need to spend a lot of time on downloading, preprocessing, and model running of a large amount geo-spatial data, which can greatly shorten the time required for long-term wind erosion mapping. As mentioned above, GEE-RWEQ provides the possibility of wind erosion monitoring in developing regions lacking on-site monitoring data. Meanwhile, the GEE platform makes it easier for researchers to publish their results for decision makers, and even the public [55]. Therefore, our research has a broader application value for decision makers than previous studies on wind erosion. In the future, we will interactively develop Earth Engine App to explore our result, which can then be used by experts and non-experts alike.

The RWEQ is a process-based, field-scale, empirical model that can quantitatively estimate wind erosion. However, the RWEQ was initially developed for the middle western area of the United States [37]. Therefore, it still presents some limitations in other regions [37,50,72,97]. Although the most important input parameters were retained in this study, the dataset required by some parameters was unavailable on GEE platforms. Therefore, several factors were simplified to simulate the global-scale wind erosion more effectively and more accurately. The soil moisture data were used to simulate how the surface wetness influences the wind speed required to erode the soil. Additionally, the cosine of the slope gradient which was calculated by DEM represented the soil roughness factor. On the other hand, we only used wind erosion-related data such as visibility data and remote sensing data for the verification of wind erosion in CA, but this still has uncertainties on a global scale. Central Asia, which is one of the most severe wind erosion regions, is restricted in terms of wind erosion modeling studies due to the lack of wind erosion measurement data for this region. Therefore, more ground soil loss measurement data on a global scale should be acquired to conduct more verification studies.

6. Conclusions

In this study, we developed a fully automated algorithm for quantitatively mapping wind erosion based on the Google Earth Engine, processed terabytes of geo-spatial data, and retrieved spatial and temporal patterns of monthly SWEP in CA, over 20 years (2000 to 2019). Several conclusions were reached in our study, as follows:

(1) With respect to the conventional methods, GEE-RWEQ does not require any ground measurement data, which need lots of manpower and resources, especially in developing countries

or sparsely populated regions. However, based on the Cloud computing platform, GEE-RWEQ uses climate assimilation data, soil property data, vegetation data, terrain data, and other underlying data to automatically generate high spatial resolution NRT soil wind erosion potential products. After verification using ground observation-based DSI and satellite-based AOD, the results still reach an acceptable accuracy and can be used for quantitative wind erosion mapping. This methodology provides new ideas for the construction and use of empirical models based on batch geospatial data and high-performance computing;

(2) According to the comparison of GMMS and SWEP, the wind speed is the main driving factor of wind erosion ($r = 0.7, p < 0.001$). Affected by the wind speed variability, the SWEP decreased first and increased remarkably during 2011. From the perspective of the temporal and spatial distribution, due to the sparse vegetation distribution and special meteorological conditions, the deserts in southwestern Central Asia are most affected by wind erosion, especially in ASDLB (47.29 kg/m^2/y). The severe wind erosion period of CA occurred in spring (MAM), especially in May. We also found that the SWEP distribution has obvious latitude zonality due to the distribution of snow cover and the start time of snow melt, and the wind erosion hot spot in spring moves from the southwest to central area across CA;

(3) Land cover change has strong effects on the soil wind erosion in CA, with the most obvious being the conversion of bareland into the water body in ASDLB. Affected by the restoration of the Aral Sea, the SWEP in this area has shown a downward trend (-6.85 kg/m^2/y) since 2011. Additionally, the conversion of bareland to shrubland helped reduce wind erosion by -0.782 kg/m^2/y. According to the SWEP variation based on LCC, more effective measures should be taken to maintain wind erosion artificial control, for example, restoring the Aral Sea water area to prevent more bareland from being exposed to wind erosion, increasing the grassland area in regions with a suitable temperature and precipitation, developing cropland by using limited water resources, and planting cold- and drought-resistant shrub vegetation in bareland.

Supplementary Materials: The following are available online at http://www.mdpi.com/2072-4292/12/20/3430/s1, Figure S1: Comparisons of the wind speed trend for different datasets (FLDAS, CFSR, ERA5, GLDAS2.1, and GMMS); Figure S2: Dust events in the Aral Sea during winter captured by MODIS; Figure S3: Monthly average wind speed of CA and the other four weather stations in the middle reaches of Amu Darya; and Table S1: The areas of different land cover types in Central Asia during the period of 2000–2018.

Author Contributions: Conceptualization, W.W.; data curation, W.W., Y.G., and A.T.; formal analysis, W.W. and Y.G.; funding acquisition, J.A.; methodology, W.W. and A.S.; project administration, A.S. and L.M.; resources, A.S. and J.A.; software, W.W. and S.Z.; supervision, J.A.; writing—original draft, W.W.; writing—review and editing, W.W. and A.S. All authors have read and agree to the published version of the manuscript.

Funding: This research was funded by the Strategic Priority Research Program of the Chinese Academy of Sciences, grant number XDA20060301; by the Youth Innovation Promotion Association Foundation of the Chinese Academy of Sciences under Grant 2018476; and in part by the National Natural Science Foundation of China, grant number 42071424.

Acknowledgments: The authors would like to give special thanks to the Google Earth Engine team for their support and allowing access. We appreciate the FLDAS, CFSR, ERA5, GLDAS, MODIS, Sentinel5P, and OLM soil properties data made available via the GEE. We are grateful for meteorological station data provided by the National Oceanic and Atmospheric Administration (NOAA). The GEE JavaScript code for retrieving SWEP can be made available by contacting the first author.

Conflicts of Interest: The authors declare no conflict of interest.

References

1. Borrelli, P.; Robinson, D.A.; Fleischer, L.R.; Lugato, E.; Ballabio, C.; Alewell, C.; Meusburger, K.; Modugno, S.; Schütt, B.; Ferro, V.; et al. An assessment of the global impact of 21st century land use change on soil erosion. *Nat. Commun.* **2017**, *8*, 2013. [CrossRef]

2. Chasek, P.; Akhtar-Schuster, M.; Orr, B.J.; Luise, A.; Rakoto Ratsimba, H.; Safriel, U. Land degradation neutrality: The science-policy interface from the UNCCD to national implementation. *Environ. Sci. Policy* **2019**, *92*, 182–190. [CrossRef]

3. Borrelli, P.; Robinson, D.A.; Panagos, P.; Lugato, E.; Yang, J.E.; Alewell, C.; Wuepper, D.; Montanarella, L.; Ballabio, C. Land use and climate change impacts on global soil erosion by water (2015–2070). *Proc. Natl. Acad. Sci. USA* **2020**, *117*, 21994. [CrossRef] [PubMed]

4. Oldeman, L. *The Global Extent of Soil Degradation*; ISRIC: Wageningen, The Netherlands, 1994; pp. 19–36.

5. Lal, R. Soil degradation and global food security: A soil science perspective. In *Land Quality, Agricultural Productivity, and Food Security: Biophysical Processes and Economic Choices at Local, Regional, and Global Levels*; Wiebe, K., Ed.; Edward Elgar Publishing: London, UK, 2003; pp. 16–35.

6. UNCCD. What is Desertification? Available online: https://www.unccd.int/frequently-asked-questions-faq (accessed on 21 July 2020).

7. Pimentel, D.; Burgess, M. Soil Erosion Threatens Food Production. *Agriculture* **2013**, *3*, 443–463. [CrossRef]

8. Joint Research Centre; European Soil Data Centre. Wind Erosion. Available online: https://esdac.jrc.ec.europa.eu/themes/wind-erosion (accessed on 21 July 2020).

9. Duniway, M.C.; Pfennigwerth, A.A.; Fick, S.E.; Nauman, T.W.; Belnap, J.; Barger, N.N. Wind erosion and dust from US drylands: A review of causes, consequences, and solutions in a changing world. *Ecosphere* **2019**, *10*, e02650. [CrossRef]

10. Jiang, C.; Liu, J.; Zhang, H.; Zhang, Z.; Wang, D. China's progress towards sustainable land degradation control: Insights from the northwest arid regions. *Ecol. Eng.* **2019**, *127*, 75–87. [CrossRef]

11. Rashki, A.; Eriksson, P.G.; Rautenbach, C.J.d.W.; Kaskaoutis, D.G.; Grote, W.; Dykstra, J. Assessment of chemical and mineralogical characteristics of airborne dust in the Sistan region, Iran. *Chemosphere* **2013**, *90*, 227–236. [CrossRef]

12. Jiang, L.; Jiapaer, G.; Bao, A.; Kurban, A.; Guo, H.; Zheng, G.; De Maeyer, P. Monitoring the long-term desertification process and assessing the relative roles of its drivers in Central Asia. *Ecol. Indic.* **2019**, *104*, 195–208. [CrossRef]

13. Shen, H.; Abuduwaili, J.; Samat, A.; Ma, L. A review on the research of modern aeolian dust in Central Asia. *Arab. J. Geosci.* **2016**, *9*, 625. [CrossRef]

14. Zhang, G.; Azorin-Molina, C.; Shi, P.; Lin, D.; Guijarro, J.A.; Kong, F.; Chen, D. Impact of near-surface wind speed variability on wind erosion in the eastern agro-pastoral transitional zone of Northern China, 1982–2016. *Agric. For. Meteorol.* **2019**, *271*, 102–115. [CrossRef]

15. Qi, J.; Kulmatov, R. An Overview of Environmental Issues in Central Asia. In *Proceedings of Environmental Problems of Central Asia and their Economic, Social and Security Impacts*; Springer: Dordrecht, The Netherlands, 2008; pp. 3–14.

16. Ge, Y.; Abuduwaili, J.; Ma, L. Lakes in Arid Land and Saline Dust Storms. *E3s Web Conf.* **2019**, *99*, 01007. [CrossRef]

17. Guo, H.; Xu, M.; Hu, Q. Changes in near-surface wind speed in China: 1969–2005. *Int. J. Climatol.* **2011**, *31*, 349–358. [CrossRef]

18. Azorin-Molina, C.; Rehman, S.; Guijarro, J.A.; McVicar, T.R.; Minola, L.; Chen, D.; Vicente-Serrano, S.M. Recent trends in wind speed across Saudi Arabia, 1978–2013: A break in the stilling. *Int. J. Climatol.* **2018**, *38*, e966–e984. [CrossRef]

19. McVicar, T.R.; Van Niel, T.G.; Li, L.T.; Roderick, M.L.; Rayner, D.P.; Ricciardulli, L.; Donohue, R.J. Wind speed climatology and trends for Australia, 1975–2006: Capturing the stilling phenomenon and comparison with near-surface reanalysis output. *Geophys. Res. Lett.* **2008**, *35*. [CrossRef]

20. Li, J.; Ma, X.; Zhang, C. Predicting the spatiotemporal variation in soil wind erosion across Central Asia in response to climate change in the 21st century. *Sci. Total Environ.* **2020**, *709*, 136060. [CrossRef] [PubMed]

21. Kim, J.; Paik, K. Recent recovery of surface wind speed after decadal decrease: A focus on South Korea. *Clim. Dyn.* **2015**, *45*, 1699–1712. [CrossRef]

22. Zeng, Z.; Ziegler, A.D.; Searchinger, T.; Yang, L.; Chen, A.; Ju, K.; Piao, S.; Li, L.Z.X.; Ciais, P.; Chen, D.; et al. A reversal in global terrestrial stilling and its implications for wind energy production. *Nat. Clim. Chang.* **2019**, *9*, 979–985. [CrossRef]

23. Azorin-Molina, C.; Menendez, M.; McVicar, T.R.; Acevedo, A.; Vicente-Serrano, S.M.; Cuevas, E.; Minola, L.; Chen, D. Wind speed variability over the Canary Islands, 1948–2014: Focusing on trend differences at the land-ocean interface and below–above the trade-wind inversion layer. *Clim. Dyn.* **2018**, *50*, 4061–4081. [CrossRef]

24. Issanova, G.; Abuduwaili, J. Relationship between Storms and Land Degradation. In *Aeolian Proceses as Dust Storms in the Deserts of Central Asia and Kazakhstan*; Issanova, G., Abuduwaili, J., Eds.; Springer: Singapore, 2017; pp. 71–86. [CrossRef]
25. Pi, H.; Sharratt, B.; Lei, J. Wind erosion and dust emissions in central Asia: Spatiotemporal simulations in a typical dust year. *Earth Surf. Process. Landf.* **2019**, *44*, 521–534. [CrossRef]
26. Chappell, A.; Webb, N.P.; Guerschman, J.P.; Thomas, D.T.; Mata, G.; Handcock, R.N.; Leys, J.F.; Butler, H.J. Improving ground cover monitoring for wind erosion assessment using MODIS BRDF parameters. *Remote Sens. Environ.* **2018**, *204*, 756–768. [CrossRef]
27. Chappell, A.; Webb, N.P.; Leys, J.F.; Waters, C.M.; Orgill, S.; Eyres, M.J. Minimising soil organic carbon erosion by wind is critical for land degradation neutrality. *Environ. Sci. Policy* **2019**, *93*, 43–52. [CrossRef]
28. Okin, G.S.; Gillette, D.A. Distribution of vegetation in wind-dominated landscapes: Implications for wind erosion modeling and landscape processes. *J. Geophys. Res. Atmos.* **2001**, *106*, 9673–9683. [CrossRef]
29. Van Pelt, R.S.; Hushmurodov, S.X.; Baumhardt, R.L.; Chappell, A.; Nearing, M.A.; Polyakov, V.O.; Strack, J.E. The reduction of partitioned wind and water erosion by conservation agriculture. *Catena* **2017**, *148*, 160–167. [CrossRef]
30. Du, H.; Zuo, X.; Li, S.; Wang, T.; Xue, X. Wind erosion changes induced by different grazing intensities in the desert steppe, Northern China. *Agric. Ecosyst. Environ.* **2019**, *274*, 1–13. [CrossRef]
31. Shao, Y.; Leslie, L.M. Wind erosion prediction over the Australian continent. *J. Geophys. Res. Atmos.* **1997**, *102*, 30091–30105. [CrossRef]
32. Hu, Y.; Liu, J.; Zhuang, D.; Cao, H.; Yan, H.; Yang, F. Distribution characteristics of 137Cs in wind-eroded soil profile and its use in estimating wind erosion modulus. *Chin. Sci. Bull.* **2005**, *50*, 1155–1159. [CrossRef]
33. Zhang, C.-L.; Zou, X.-Y.; Yang, P.; Dong, Y.-X.; Li, S.; Wei, X.-H.; Yang, S.; Pan, X.-H. Wind tunnel test and 137Cs tracing study on wind erosion of several soils in Tibet. *Soil Tillage Res.* **2007**, *94*, 269–282. [CrossRef]
34. Shao, Y. Integrated Wind-Erosion Modelling. In *Physics and Modelling of Wind Erosion*; Springer: Dordrecht, The Netherlands, 2008; pp. 303–360. [CrossRef]
35. Williams, J.; Nearing, M.; Nicks, A.; Skidmore, E.; Valentin, C.; King, K.; Savabi, R. Using soil erosion models for global change studies. *J. Soil Water Conserv.* **1996**, *51*, 381–385.
36. Fryrear, D.; Sutherland, P.; Davis, G.; Hardee, G.; Dollar, M. Wind erosion estimates with RWEQ and WEQ. In *Proceedings of the 10th International Soil Conservation Organization Meeting*; Purdue University and USDA-ARS National Soil Erosion Research Laboratory: West Lafayette, IN, USA, 2001; pp. 760–765.
37. Fryrear, D.W.; Bilbro, J.D.; Saleh, A.; Schomberg, H.; Stout, J.E.; Zobeck, T.M. RWEQ: Improved wind erosion technology. *J. Soil Water Conserv.* **2000**, *55*, 183.
38. Bagnold, R.A. *The Physics of Blown Sand and Desert Dunes*; William Morrow & Company: New York, NY, USA, 1941.
39. Woodruff, N.P.; Siddoway, F.H. A Wind Erosion Equation. *Soil Sci. Soc. Am. J.* **1965**, *29*, 602–608. [CrossRef]
40. Tatarko, J.; Sporcic, M.A.; Skidmore, E.L. A history of wind erosion prediction models in the United States Department of Agriculture prior to the Wind Erosion Prediction System. *Aeolian Res.* **2013**, *10*, 3–8. [CrossRef]
41. Feng, G.; Sharratt, B. Evaluation of the SWEEP model during high winds on the Columbia Plateau. *Earth Surf. Process. Landf.* **2009**, *34*, 1461–1468. [CrossRef]
42. Cole, G.W.; Lyles, L.; Hagen, L.J. A Simulation Model of Daily Wind Erosion Soil Loss. *Trans. ASAE* **1983**, *26*, 1758–1765. [CrossRef]
43. Pi, H.; Sharratt, B.; Feng, G.; Lei, J. Evaluation of two empirical wind erosion models in arid and semi-arid regions of China and the USA. *Environ. Model. Softw.* **2017**, *91*, 28–46. [CrossRef]
44. Gregory, J.M.; Wilson, G.R.; Singh, U.B.; Darwish, M.M. TEAM: Integrated, process-based wind-erosion model. *Environ. Model. Softw.* **2004**, *19*, 205–215. [CrossRef]
45. Böhner, J.; Schäfer, W.; Conrad, O.; Gross, J.; Ringeler, A. The WEELS model: Methods, results and limitations. *Catena* **2003**, *52*, 289–308. [CrossRef]
46. Zobeck, T.M.; Parker, N.C.; Haskell, S.; Guoding, K. Scaling up from field to region for wind erosion prediction using a field-scale wind erosion model and GIS. *Agric. Ecosyst. Environ.* **2000**, *82*, 247–259. [CrossRef]
47. Chi, W.; Zhao, Y.; Kuang, W.; He, H. Impacts of anthropogenic land use/cover changes on soil wind erosion in China. *Sci. Total Environ.* **2019**, *668*, 204–215. [CrossRef]

48. Borrelli, P.; Lugato, E.; Montanarella, L.; Panagos, P. A New Assessment of Soil Loss due to Wind Erosion in European Agricultural Soils Using a Quantitative Spatially Distributed Modelling Approach. *Land Degrad. Dev.* **2017**, *28*, 335–344. [CrossRef]

49. Borrelli, P.; Panagos, P.; Ballabio, C.; Lugato, E.; Weynants, M.; Montanarella, L. Towards a Pan-European Assessment of Land Susceptibility to Wind Erosion. *Land Degrad. Dev.* **2016**, *27*, 1093–1105. [CrossRef]

50. Lin, J.; Guan, Q.; Pan, N.; Zhao, R.; Yang, L.; Xu, C. Spatiotemporal Variations and Driving Factors of the Potential Wind Erosion Rate in the Hexi Region. *Land Degrad. Dev.* **2020**. [CrossRef]

51. Gorelick, N.; Hancher, M.; Dixon, M.; Ilyushchenko, S.; Thau, D.; Moore, R. Google Earth Engine: Planetary-scale geospatial analysis for everyone. *Remote Sens. Environ.* **2017**, *202*, 18–27. [CrossRef]

52. Mutanga, O.; Kumar, L. Google Earth Engine Applications. *Remote Sens.* **2019**, *11*, 591. [CrossRef]

53. Ivushkin, K.; Bartholomeus, H.; Bregt, A.K.; Pulatov, A.; Kempen, B.; de Sousa, L. Global mapping of soil salinity change. *Remote Sens. Environ.* **2019**, *231*, 111260. [CrossRef]

54. Pradhan, B.; Moneir, A.A.A.; Jena, R. Sand dune risk assessment in Sabha region, Libya using Landsat 8, MODIS, and Google Earth Engine images. *Geomat. Nat. Hazards Risk* **2018**, *9*, 1280–1305. [CrossRef]

55. Jin, Y.; Liu, X.; Yao, J.; Zhang, X.; Zhang, H. Mapping the annual dynamics of cultivated land in typical area of the Middle-lower Yangtze plain using long time-series of Landsat images based on Google Earth Engine. *Int. J. Remote Sens.* **2020**, *41*, 1625–1644. [CrossRef]

56. Xie, Z.; Phinn, S.R.; Game, E.T.; Pannell, D.J.; Hobbs, R.J.; Briggs, P.R.; Beutel, T.S.; Holloway, C.; McDonald-Madden, E. Using Landsat observations (1988–2017) and Google Earth Engine to detect vegetation cover changes in rangelands—A first step towards identifying degraded lands for conservation. *Remote Sens. Environ.* **2019**, *232*, 111317. [CrossRef]

57. Chen, Y.; Li, W.; Deng, H.; Fang, G.; Li, Z. Changes in Central Asia's Water Tower: Past, Present and Future. *Sci. Rep.* **2016**, *6*, 39364. [CrossRef] [PubMed]

58. Groll, M.; Opp, C.; Issanova, G.; Vereshagina, N.; Semenov, O. Physical and Chemical Characterization of Dust Deposited in the Turan Lowland (Central Asia). *Proc. E3S Web Conf.* **2019**, *99*, 03005. [CrossRef]

59. Micklin, P. The Aral Sea Crisis. In *Dying and Dead Seas Climatic Versus Anthropic Causes*; Springer: Dordrecht, The Netherlands, 2004; pp. 99–123.

60. Abuduwaili, J.; Issanova, G.; Saparov, G. Water Resources and Impact of Climate Change on Water Resources in Central Asia. In *Hydrology and Limnology of Central Asia*; Abuduwaili, J., Issanova, G., Saparov, G., Eds.; Springer: Singapore, 2019; pp. 1–9. [CrossRef]

61. Rodell, M.; Houser, P.; Jambor, U.E.A.; Gottschalck, J.; Mitchell, K.; Meng, J.; Arsenault, K.; Brian, C.; Radakovich, J.; Mg, B.; et al. The Global Land Data Assimilation System. *BAMS* **2004**, *85*, 381–394. [CrossRef]

62. Huang, D.K.; Xin-Qing, L.; Wei, J.; Zhao, Y.L.; Wen, C.; Ding, W. Geographic variation of carbonate content and pH in surface soil in East Central Asia: Significance as climate proxies. *Geochimica* **2008**, *37*, 129–138.

63. Liu, S.; Zhang, S.; Wu, J.; Pang, X.; Yuan, D. The relationship between soil pH and calcium carbonate content. *Soil* **2002**, *5*. [CrossRef]

64. Symeonakis, E.; Drake, N. Monitoring desertification and land degradation over sub-Saharan Africa. *Int. J. Remote Sens.* **2004**, *25*, 573–592. [CrossRef]

65. Fenta, A.A.; Tsunekawa, A.; Haregeweyn, N.; Poesen, J.; Tsubo, M.; Borrelli, P.; Panagos, P.; Vanmaercke, M.; Broeckx, J.; Yasuda, H.; et al. Land susceptibility to water and wind erosion risks in the East Africa region. *Sci. Total Environ.* **2020**, *703*, 135016. [CrossRef]

66. Naumann, G.; Barbosa, P.; Carrao, H.; Singleton, A.; Vogt, J. Monitoring Drought Conditions and Their Uncertainties in Africa Using TRMM Data. *J. Appl. Meteorol. Climatol.* **2012**, *51*, 1867–1874. [CrossRef]

67. Kalantari, Z.; Ferreira, C.S.S.; Keesstra, S.; Destouni, G. Nature-based solutions for flood-drought risk mitigation in vulnerable urbanizing parts of East-Africa. *Curr. Opin. Environ. Sci. Health* **2018**, *5*, 73–78. [CrossRef]

68. Ouyang, Z.; Zheng, H.; Xiao, Y.; Polasky, S.; Liu, J.; Xu, W.; Wang, Q.; Zhang, L.; Xiao, Y.; Rao, E.; et al. Improvements in ecosystem services from investments in natural capital. *Science* **2016**, *352*, 1455. [CrossRef] [PubMed]

69. Wang, L.-Y.; Xiao, Y.; Rao, E.-M.; Jiang, L.; Xiao, Y.; Ouyang, Z.-Y. An Assessment of the Impact of Urbanization on Soil Erosion in Inner Mongolia. *Int. J. Environ. Res. Public Health* **2018**, *15*, 550. [CrossRef]

70. Xu, J.; Xiao, Y.; Xie, G.; Wang, Y.; Jiang, Y. Computing payments for wind erosion prevention service incorporating ecosystem services flow and regional disparity in Yanchi County. *Sci. Total Environ.* **2019**, *674*, 563–579. [CrossRef] [PubMed]

71. Zhang, J.; Li, X.; Buyantuev, A.; Bao, T.; Zhang, X. How Do Trade-Offs and Synergies between Ecosystem Services Change in the Long Period? The Case Study of Uxin, Inner Mongolia, China. *Sustainability* **2019**, *11*, 6041. [CrossRef]

72. Jiang, L.; Xiao, Y.; Zheng, H.; Ouyang, Z. Spatio-temporal variation of wind erosion in Inner Mongolia of China between 2001 and 2010. *Chin. Geogr. Sci.* **2016**, *26*, 155–164. [CrossRef]

73. Wu, D.; Zou, C.; Cao, W.; Xiao, T.; Gong, G. Ecosystem services changes between 2000 and 2015 in the Loess Plateau, China: A response to ecological restoration. *PLoS ONE* **2019**, *14*, e0209483. [CrossRef] [PubMed]

74. McTainsh, G.H. *Sustainable Agriculture: Assessing Australia's Recent Performance. A Report of the National Collaborative Project on Indicators for Sustainable Agriculture*; CSIRO: Canberra, Australia, 1998; pp. 65–72.

75. O'Loingsigh, T.; McTainsh, G.H.; Tews, E.K.; Strong, C.L.; Leys, J.F.; Shinkfield, P.; Tapper, N.J. The Dust Storm Index (DSI): A method for monitoring broadscale wind erosion using meteorological records. *Aeolian Res.* **2014**, *12*, 29–40. [CrossRef]

76. Lim, J.-Y.; Chun, Y. The characteristics of Asian dust events in Northeast Asia during the springtime from 1993 to 2004. *Glob. Planet. Chang.* **2006**, *52*, 231–247. [CrossRef]

77. Wang, D.; Zhang, F.; Yang, S.; Xia, N.; Ariken, M. Exploring the spatial-temporal characteristics of the aerosol optical depth (AOD) in Central Asia based on the moderate resolution imaging spectroradiometer (MODIS). *Environ. Monit. Assess.* **2020**, *192*, 383. [CrossRef]

78. Qin, W.; Wang, L.; Lin, A.; Zhang, M.; Bilal, M. Improving the Estimation of Daily Aerosol Optical Depth and Aerosol Radiative Effect Using an Optimized Artificial Neural Network. *Remote Sens.* **2018**, *10*, 1022. [CrossRef]

79. Lee, S.; Pinhas, A.; Alexandra, C.A. Aerosol pattern changes over the dead sea from west to east—Using high-resolution satellite data. *Atmos. Environ.* **2020**, 117737. [CrossRef]

80. De Vries, J.; Voors, R.; Barend, O.; Jos, D.; Pepijn, V.; Antje, L.; Quintus, K.; Ruud, H.; Ilse, A. TROPOMI on ESA's Sentinel 5p ready for launch and use. *Proc. SPIE* **2016**. [CrossRef]

81. Kirches, G.; Brockmann, C.; Boettcher, M.; Peters, M.; Bontemps, S.; Lamarche, C.; Schlerf, M.; Santoro, M.; Defourny, P. Land Cover CCI-Product User Guide-Version 2. ESA Public Document CCI-LC-PUG. 2014. Available online: http://maps.elie.ucl.ac.be/CCI/viewer/download/ESACCI-LC-PUG-v2.5.pdf (accessed on 10 June 2019).

82. Prema, V.; Rao, K.U. Time series decomposition model for accurate wind speed forecast. *Renew. Wind Water Sol.* **2015**, *2*, 18. [CrossRef]

83. Kendall, M.K.; Stuart, A. The Advanced Theory of Statistics. *J. R. Stat. Soc.* **1983**, *3*, 410–414.

84. Grove, A.N. The Aral Sea: Going, Going, Gone. Available online: https://www.angelanealworld.com/the-pulse/climate/aral-sea-one-of-planets-most-shocking-disasters/ (accessed on 24 July 2020).

85. Yang, X.; Wang, N.; Chen, A.; He, J.; Hua, T.; Qie, Y. Changes in area and water volume of the Aral Sea in the arid Central Asia over the period of 1960–2018 and their causes. *Catena* **2020**, *191*, 104566. [CrossRef]

86. Shao, Y.; Klose, M.; Wyrwoll, K.-H. Recent global dust trend and connections to climate forcing. *J. Geophys. Res. Atmos.* **2013**, *118*, 107–111. [CrossRef]

87. Zhang, H.; Fan, J.; Cao, W.; Harris, W.; Li, Y.; Chi, W.; Wang, S. Response of wind erosion dynamics to climate change and human activity in Inner Mongolia, China during 1990 to 2015. *Sci. Total Environ.* **2018**, *639*, 1038–1050. [CrossRef] [PubMed]

88. Hagen, L. Evaluation of the Wind Erosion Prediction System (WEPS) erosion submodel on cropland fields. *Environ. Model. Softw.* **2004**, *19*, 171–176. [CrossRef]

89. Liu, Y.; Zhu, Q.; Wang, R.; Xiao, K.; Cha, P. Distribution, source and transport of the aerosols over Central Asia. *Atmos. Environ.* **2019**, *210*, 120–131. [CrossRef]

90. Ge, Y.; Abuduwaili, J.; Ma, L.; Liu, D. Temporal Variability and Potential Diffusion Characteristics of Dust Aerosol Originating from the Aral Sea Basin, Central Asia. *Water Air Soil Pollut.* **2016**, *227*, 63. [CrossRef]

91. Wang, W.; Alim, S.; Jilili, A. Geo-detector based spatio-temporal variation characteristics and driving factors analysis of NDVI in Central Asia. *Remote Sens. Land Resour.* **2019**, *31*, 32–40.

92. Dietz, A.J.; Conrad, C.; Kuenzer, C.; Gesell, G.; Dech, S. Identifying Changing Snow Cover Characteristics in Central Asia between 1986 and 2014 from Remote Sensing Data. *Remote Sens.* **2014**, *6*, 12752–12775. [CrossRef]

93. Li, Z.; Fang, G.; Chen, Y.; Duan, W.; Mukanov, Y. Agricultural water demands in Central Asia under 1.5 °C and 2.0 °C global warming. *Agric. Water Manag.* **2020**, *231*, 106020. [CrossRef]

94. Yu, F.; Price, K.P.; Ellis, J.; Feddema, J.J.; Shi, P. Interannual variations of the grassland boundaries bordering the eastern edges of the Gobi Desert in central Asia. *Int. J. Remote Sens.* **2004**, *25*, 327–346. [CrossRef]

95. Alina, A.Z.; Jiri, C.; Niels, T.; Anar, B.M.; Saule, S.A. Natural Regeneration Potential of the Black Saxaul Shrubforests in Semi-Deserts of Central Asia—The Ili River Delta Area, SE Kazakhstan. *Pol. J. Ecol.* **2017**, *65*, 352–368. [CrossRef]

96. Zheng, C.; Wang, Q. Water-use response to climate factors at whole tree and branch scale for a dominant desert species in central Asia: Haloxylon ammodendron. *Ecohydrology* **2014**, *7*, 56–63. [CrossRef]

97. Visser, S.M.; Sterk, G.; Karssenberg, D. Wind erosion modelling in a Sahelian environment. *Environ. Model. Softw.* **2005**, *20*, 69–84. [CrossRef]

Publisher's Note: MDPI stays neutral with regard to jurisdictional claims in published maps and institutional affiliations.

Article

Using Synthetic Remote Sensing Indicators to Monitor the Land Degradation in a Salinized Area

Tao Yu [1,2], **Guli Jiapaer** [1,3,*], **Anming Bao** [1,3], **Guoxiong Zheng** [1,2], **Liangliang Jiang** [4], **Ye Yuan** [1,2,5] and **Xiaoran Huang** [1,2]

[1] State Key Laboratory of Desert and Oasis Ecology, Xinjiang Institute of Ecology and Geography, Chinese Academy of Sciences, Urumqi 830011, China; yutao171@mails.ucas.ac.cn (T.Y.); baoam@ms.xjb.ac.cn (A.B.); zhengguoxiong17@mails.ucas.edu.cn (G.Z.); yeyuan.rs@gmail.com (Y.Y.); huangxiaoran14@mails.ucas.edu.cn (X.H.)

[2] University of Chinese Academy of Sciences, Beijing 100049, China

[3] Research Center for Ecology and Environment of Central Asia, Chinese Academy of Sciences, Urumqi 830011, China

[4] School of Geography and Tourism, Chongqing Normal University, Chongqing 401331, China; jiang@cqnu.edu.cn

[5] Sino-Belgian Joint Laboratory of Geo-information, 9000 Ghent, Belgium

* Correspondence: glmr@ms.xjb.ac.cn

Citation: Yu, T.; Jiapaer, G.; Bao, A.; Zheng, G.; Jiang, L.; Yuan, Y.; Huang, X. Using Synthetic Remote Sensing Indicators to Monitor the Land Degradation in a Salinized Area. *Remote Sens.* **2021**, *13*, 2851. https://doi.org/10.3390/rs13152851

Academic Editor: Bas van Wesemael

Received: 29 May 2021
Accepted: 16 July 2021
Published: 21 July 2021

Publisher's Note: MDPI stays neutral with regard to jurisdictional claims in published maps and institutional affiliations.

Abstract: Land degradation poses a critical threat to the stability and security of ecosystems, especially in salinized areas. Monitoring the land degradation of salinized areas facilitates land management and ecological restoration. In this research, we integrated the salinization index (SI), albedo, normalized difference vegetation index (NDVI) and land surface soil moisture index (LSM) through the principal component analysis (PCA) method to establish a salinized land degradation index (SDI). Based on the SDI, the land degradation of a typical salinized area in the Central Asia Amu Darya delta (ADD) was analysed for the period 1990–2019. The results showed that the proposed SDI had a high positive correlation ($R^2 = 0.89$, $p < 0.001$) with the soil salt content based on field sampling, indicating that the SDI can reveal the land degradation characteristics of the ADD. The SDI indicated that the extreme and strong land degradation areas increased from 1990 to 2019, mainly in the downstream and peripheral regions of the ADD. From 1990 to 2000, land degradation improvement over a larger area than developed, conversely, from 2000 to 2019, and especially, from 2000 to 2010, the proportion of land degradation developed was 32%, which was mainly concentrated in the downstream region of the ADD. The spatial autocorrelation analysis indicated that the SDI values of Moran's I in 1990, 2000, 2010 and 2019 were 0.82, 0.78, 0.82 and 0.77, respectively, suggesting that the SDI was notably clustered in space rather than randomly distributed. The expansion of unused land due to land use change, water withdrawal from the Amu Darya River and the discharge of salt downstream all contributed to land degradation in the ADD. This study provides several valuable insights into the land degradation monitoring and management of this salinized delta and similar settings worldwide.

Keywords: land degradation; salinization; remote sensing index; salinized land degradation index (SDI); Amu Darya delta (ADD)

1. Introduction

Land degradation can lead to reduced land productivity, population displacement, food insecurity and the destruction of ecosystems [1]. The report from the National Forestry and Grassland Administration of China (http://www.forestry.gov.cn/, accessed on 24 June 2021) shows that 197 countries have signed the United Nations Convention to Combat Desertification (UNCCD) as of January 2019; the problem has not been alleviated in recent decades and has instead progressively worsened [2,3]. Monitoring land degradation and revealing its characteristics is essential for the management and restoration of land quality.

Salinization induces land degradation, and this ecological problem is more prevalent in drylands [4]. In particular, in most irrigated areas of Central Asia, high-salinity water is used for irrigation, resulting in secondary salinization, which exacerbates land degradation [5]. Moreover, a severe ecological disaster was initiated with the gradual retreat of the Aral Sea due to the massive expansion of agricultural practices [6]. Consequently, the ecosystem around the Aral Sea has been almost destroyed, especially in the Amu Darya delta (ADD) [7–9], which has become one of the most severely degraded areas worldwide due to salinization [10]. The land degradation of the ADD caused by high levels of soil salinization has led to ecological and socio-economic problems, such as the withering of vegetation [5] and reduced agricultural yields [8]. Moreover, the ecology of the ADD is vulnerable to hydrological changes due to its dry climate. With the expansion of agricultural land, the structure of the water systems in the ADD has changed dramatically [8]. Numerous natural lakes and wetlands have disappeared and transformed into sparse vegetation or bare land [10,11]. The increasing land degradation is threatening the stability of ADD ecosystems [10]. However, the characteristics of land degradation in such a saline region remain unclear.

In recent decades, land degradation has attracted considerable research attention worldwide. Different indicators (e.g., vegetation index [12], desertification index [13], etc.) and methods (e.g., Analytic Hierarchy Process (AHP) [14], Entropy Weighting and Delphi [15], etc.) have been used to monitor land degradation. These studies have facilitated the understanding of the mechanism of land degradation at the regional and global scales. However, the characteristics of land degradation are different for each region (e.g., rocky desertification [16], sandy [17], salinization [18], etc.) These studies do not take into account the main characteristics of regional land degradation when establishing a land degradation assessment framework or index, which may affect the accuracy of the monitoring results. Previous studies have confirmed that the factors affecting land degradation vary region-wise [10,19]. Therefore, selecting indicators that are representative of the ecological characteristics of the region during the assessment can increase the rationality of the land degradation assessment. Then, in saline areas, the salinization index (SI) [20], which reflects information on soil salinity, should be considered when monitoring land degradation. In addition, indicators such as the normalized difference vegetation index (NDVI) [21], albedo [22,23] and soil moisture [13,24], extracted from remote sensing data, have been widely used to monitor regional land degradation. The NDVI is one of the most widely used indicators to monitor the land degradation, as it can accurately reflect the vegetation greenness and biomass information [25,26]. The surface albedo is closely related to the soil exposure. The increase in albedo can be used as an indirect indicator to detect the soil degradation in drylands [12,27]. Moreover, the land surface soil moisture index (LSM) can reflect the soil water content and is a key indicator to monitor the land degradation in drylands [28,29].

The combination of the aforementioned (NDVI, LSM, SI and albedo) indicators can provide a comprehensive understanding of land degradation in salinized areas for the reference of regional land management [10,12,30]. Thereby, a salinized land degradation index (SDI), including information on the salinity, vegetation, soil moisture and bareness, needs to be constructed to reflect the land degradation characteristics of salinized areas. Recently a method based on the principal component analysis (PCA) was developed to assess the regional ecological conditions [31–33]. The PCA method is a multidimensional data compression technique. This method allows the characteristics of the indicators to be coupled, and the weights of each factor are automatically and objectively assigned according to the contribution of each factor to the principal component [31,34]. In contrast to weighting methods such as AHP and Delphi, PCA prevents variations or errors in the definitions of weights caused by individual subjective experience [35,36]. Therefore, in this study, we attempted to (1) construct the SDI based on the PCA, (2) assess the reliability of the SDI in monitoring land degradation in salinized areas and (3) explore the spatial and temporal patterns of land degradation. Finally, the potential driving factors related to land degradation were discussed. Understanding the spatiotemporal characteristics of land degradation can likely contribute to the management of ADD land and sustainability of the ecosystem and provide guidance for future studies.

2. Materials and Methods

2.1. Study Area

The ADD is located south of the Aral Sea, downstream of the Amu Darya River (Figure 1). The region runs through Turkmenistan and Uzbekistan and covers an area of 6.3×10^4 km^2. The runoff from large permanent glaciers in the mountains and melting snow are the main water sources of the Amu Darya River [37]. The ADD has a typical continental climate characterised by extreme dryness throughout the year. The average annual temperature is approximately 13 °C [37]. The potential annual evapotranspiration can be as high as 1600 mm, and the average annual precipitation is less than 100 mm [37]. Such a dry climate makes the ADD one of the most ecologically fragile regions worldwide [9].

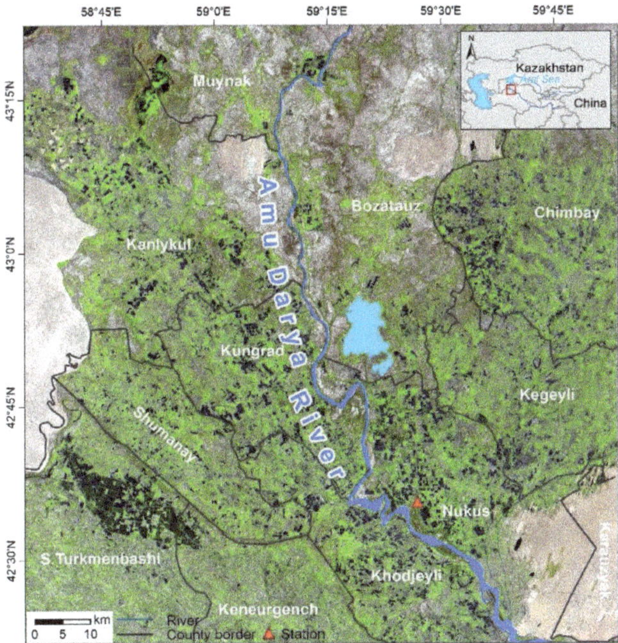

Figure 1. Location of the Amu Darya delta (ADD). Colour composite map of Landsat-8 OLI images in the ADD for 2019 in a colour combination of shortwave-infrared band 1, near-infrared and red band. Green represents vegetation, and brown represents bare soil.

However, as the main grain-producing region of the Aral Sea basin, dry climate and land use changes due to agricultural expansion have further exacerbated the salinization of the ADD [38]. Moreover, the changes in the political system after the disintegration of the Soviet Union have led to intensified conflicts in the use of water resources among different countries in the region, resulting in land degradation and a decline in the stability of the ecological system [39]. In this regard, a series of ecological conservation and restoration projects have been implemented or are about to be launched to mitigate the land degradation caused by salinization of the ADD [40]. Within this context, it is essential to investigate the spatial and temporal characteristics of the land degradation in the ADD, a typical salinized region, to provide reference for the ecosystem management of the delta.

2.2. Data and Pre-Processing

The datasets used in this study included satellite images, field soil salinity, temperature and precipitation, land use and water withdrawal and salt discharge data. The satellite images from the United States Geological Survey (http://earthexplorer.usgs.gov/, accessed on 5 June 2020), acquired on 19 July and 26 July 1990, July 21 and 30 July 2000, 10 July and 17 July 2010 (Landsat-5 TM) and 4 August and 11 August 2019 (Landsat-8 OLI/TIRS) were used. The selected images were in the growing season, and the time phase was similar, and there was basically no cloud coverage, which ensured the accuracy of the remote sensing index calculations [41,42]. The images were first radiometrically corrected with ENVI 5.1 software to convert the digital numbers to irradiance values and later atmospherically corrected using the Fast Line-of-sight Atmospheric Analysis of Spectral Hypercubes (FLAASH) module to eliminate the effects of noise generated during the imaging process [43,44]. The images acquired at different times were geometrically corrected using the two polynomials, and the root mean square error was controlled within 0.5 pixels [33]. Finally, the images were clipped based on the ADD boundary, and the water bodies were masked using the modified normalised difference water index (MNDWI) [45].

The salinity data sampled in the field on 18 March 2019 (see Figure S1 and Table S1 for details) were used to explore the feasibility of the SDI in assessing the land degradation in salinized areas. The average annual precipitation and temperature records during 1980–2016 were derived from Nukus Station in the ADD. The annual statistical data on water withdrawal and salt discharge data for 1990–2015 were obtained from the Amu Darya River basin database of the Inter-State Commission for Water Coordination of Central Asia (ICWC, http://www.cawaterinfo.net/, accessed on 10 June 2020). The land use data were obtained from the Xinjiang Institute of Ecology and Geography, Chinese Academy of Sciences and interpreted based on the Landsat images from 1990, 2000, 2010 and 2019 that have been applied in related studies in the ADD [10,46]. Following this dataset, the land use types were divided into five categories: cropland, forest, grassland, built-up land and bare soil.

2.3. Construction of the SDI

2.3.1. SI

Salinization is the major factor influencing the land degradation of the ADD. The SI extracted from the remote sensing image has been shown to be able to assess the characteristics of regional salinization [47–49]. In this work, an inversion model that has been proved to be applicable in the ADD [10,50] was selected to construct the SI. The model was derived using the following equation [20]:

$$\mathrm{SI} = \sqrt{\rho_{\mathrm{Blue}} \times \rho_{\mathrm{Red}}} \tag{1}$$

where $\rho\mathrm{Blue}$ and $\rho\mathrm{Red}$ are the blue and red bands of the Landsat TM and OLI imagery, respectively.

2.3.2. Albedo

The albedo is a key physical parameter related to the soil exposure [51]. In general, the albedo is higher in desert areas due to the sparse vegetation and soil exposure, and areas with high vegetation cover exhibit a lower albedo [23,52]. Therefore, the albedo was chosen to represent the surface exposure, and it was calculated as follows [51]:

$$\mathrm{Albedo} = 0.356\,\rho_{\mathrm{Blue}} + 0.130\,\rho_{\mathrm{Red}} + 0.373\,\rho_{\mathrm{NIR}} + 0.085\,\rho_{\mathrm{SWIR1}} \\ + 0.072\,\rho_{\mathrm{SWIR2}} - 0.018 \tag{2}$$

where $\rho\mathrm{Blue}$, $\rho\mathrm{Red}$, $\rho\mathrm{NIR}$, $\rho\mathrm{SWIR1}$ and $\rho\mathrm{SWIR2}$ denote the blue, red, near-infrared and two shortwave-infrared bands of the Landsat TM and OLI imagery, respectively. These bands are the same as those referred to in the following text.

2.3.3. NDVI

The NDVI is based on the structure absorbed by the plant leaf surface that reflects the parameters of the plant biomass and vegetation coverage [53]. This parameter has been successfully used to monitor land degradation at different scales [54], and its expression is as follows:

$$NDVI = (\rho_{NIR} - \rho_{Red}) / (\rho_{NIR} + \rho_{Red}) \tag{3}$$

2.3.4. LSM

The LSM is crucial to regulate the vegetation productivity, and it directly affects the regional land degradation [55]. The LSM can be calculated using the tasselled cap transformation through the following formulas [56,57]:

$$LSM_{TM} = 0.0315\,\rho_{Blue} + 0.2021\,\rho_{Green} + 0.3102\,\rho_{Red} + 0.1594\,\rho_{NIR} \\ -0.6806\,\rho_{SWIR1} - 0.6109\,\rho_{SWIR2} \tag{4}$$

$$LSM_{OLI} = 0.1511\,\rho_{Blue} + 0.1972\,\rho_{Green} + 0.3283\,\rho_{Red} + 0.3407\,\rho_{NIR} \\ -0.7117\,\rho_{SWIR1} - 0.4559\,\rho_{SWIR2} \tag{5}$$

where ρ denotes the corresponding bands of the Landsat TM and OLI imagery.

2.3.5. Constructing SDI Based on PCA

In this study, based on previous studies [33,58,59], the PCA method was used to synthesize the four selected indicators (SI, albedo, NDVI and LSM) to construct the SDI. Before using the PCA method to couple the SDI for 1990, 2000, 2010 and 2019, respectively, it is necessary to normalise the four indicators in the range of 0 to 1 [58] by Equation (6). The SDI obtained using Equation (7) and higher values of the SDI revealed more severe land degradation. Figure 2 illustrates the processing for SDI.

$$I_{normal} = (I - I_{min}) / (I_{max} - I_{min}) \tag{6}$$

$$SDI = PC1[f(SI, Albedo, NDVI, LSM)] \tag{7}$$

where I_{noraml} is the index value after standardisation, I is the numerical value of this index, and I_{max} and I_{min} are the maximum and minimum values of the relevant index, respectively.

2.4. Spatial Autocorrelation Analysis

A spatial autocorrelation analysis is an effective way to test whether the values of adjacent samples of a spatial variable are correlated [60]. In this study, the global Moran's I index and the local indicator of spatial association (LISA) were used to analyse the spatial correlation of the SDI.

Moran's I generates a global assessment for spatial autocorrelation, with Moran's I values ranging from −1 to 1 [61,62]. Moran's I value > 0 means that the SDI has a positive spatial autocorrelation, while Moran's I value < 0 means that the SDI has a negative spatial autocorrelation. The closer the value to 1, the stronger the positive spatial autocorrelation, and the closer the value to −1, the stronger the negative spatial autocorrelation. Moran's I = 0 means that there is no spatial autocorrelation, and the SDI has a random spatial distribution. The global Moran's I index was calculated using the following equations:

$$I = \frac{\sum_i^n \sum_{j \neq i}^n W_{ij}(x_i - \bar{x})(x_j - \bar{x})}{S^2 \sum_i^n \sum_{j \neq i}^n W_{ij}} \tag{8}$$

$$S^2 = \frac{1}{n}\sum_i^n (x_i - \bar{x})^2 \tag{9}$$

$$\bar{x} = \frac{1}{n}\sum_{i(j)}^n x_{i(j)} \tag{10}$$

$$Z_{Score} = \frac{1 - E(I)}{\sqrt{Var(I)}} \qquad (11)$$

where x_i and x_j are the values of the SDI at spatial locations i and j, respectively, x is the mean value of the SDI, S^2 is the mean squared deviation of the SDI, W_{ij} is the spatial weight value, which is expressed by the n-dimensional matrix W(n×n), var(I) is the variance of Moran's I, and E(I) is the expected value of Moran's I.

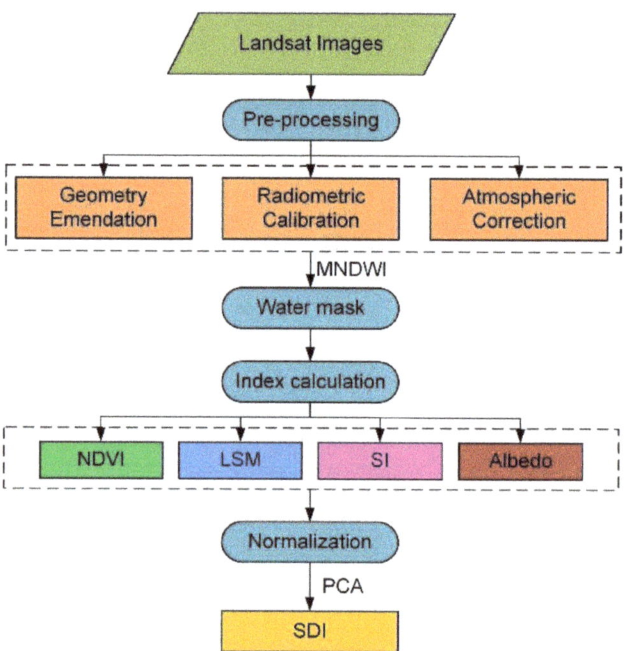

Figure 2. Flowchart of construction salinized land degradation index (SDI). MNDWI: modified normalised difference water index; NDVI: normalised difference vegetation index; LSM: land surface soil moisture index; SI, salinization index; PCA: principal component analysis; SDI: salinized land degradation index.

LISA is a local statistical method for spatial variables that reveals the spatial clustering characteristics of observations within spatially adjacent regions [33,60]. A positive LISA value indicates that the SDI is similar to the adjacent value and reveals a spatial pattern of high–high clustering (H–H, high values are near other high values) or low–low clustering (L–L, low values are near other low values). A negative LISA value indicates that the SDI is a spatial outlier and can include a high–low outlier value (H–L, a high value is near a low value) and a low–high outlier value (L–H, a low value is near a high value).

3. Results

3.1. Integration of the Remote Sensing Indexes Based on PCA

The SDI of the ADD was calculated by the PC1 of the four indicators. The PC1 results shown in Table 1 indicated that, during the studied years, the percent eigenvalues of PC1 were higher than 78%, revealing that PC1 integrated most of the information of the four indictors. Therefore, PC1 was chosen to construct the SDI in this study. The loading values of the four variables in PC1 were divided into two types according to their signs. The albedo and SI comprised one type with loading values that were positive, and the NDVI and LSM comprised the second type with loading values that were negative. The opposite signs of the two variables indicated that the corresponding contributions to the SDI value were opposing.

Table 1. Loadings of the four selected variables on the first principal component (PC1) and associated contributions in different study years.

	1990	2000	2010	2019
Loading of the SI	0.58	0.64	0.57	0.61
Loading of albedo	0.23	0.43	0.23	0.26
Loading of the NDVI	−0.32	−0.16	−0.46	−0.24
Loading of LSM	−0.71	−0.62	−0.64	−0.71
Eigenvalue contribution percentage (%)	78.61	83.24	86.35	87.67

SI: salinization index (SI); NDVI: normalised difference vegetation index; LSM: land surface soil moisture index.

The descriptive statistics of the PC1 indicated that the average PC1 value increased from 0.42 in 2010 to 0.43 in 2000, and the medium value of PC1 also increased from 0.38 to 0.40 (Table 2). From 2010 to 2019, on the contrary, the average PC1 value decreased from 0.41 to 0.30, and the medium value of PC1 also decreased from 0.38 to 0.28. The positive skewness in 1990, 2000, 2010 and 2019 indicated that the tail on the right side of the probability density function was longer or fatter than the left side. Overall, the statistical results of PC1 showed that land degradation accelerated from 1990 to 2000 and weakened from 2010 to 2019.

Table 2. Description statistics of PC1.

	Minimum	Maximum	Mean	Median	Skewness	Kurtosis	Standard Deviation
1990	0.04	1.15	0.42	0.38	0.48	−0.33	0.19
2000	0.04	1.21	0.43	0.40	0.31	0.14	0.16
2010	0.03	1.12	0.41	0.38	0.49	−0.71	0.18
2019	0.04	1.12	0.30	0.28	0.42	−0.80	0.18

Furthermore, the correlation coefficient between each indicator and the SDI and that among the indicators are shown in Figure 3 (at the 0.01 level of significance). For four years, the SDI exhibited a high correlation with each single indicator. In general, the SDI exhibited a positive correlation with the SI and albedo and a negative correlation with the NDVI and LSM (Figure 3). The SI exhibited the highest correlation with the SDI, and the positive correlations were 0.971, 0.983, 0.973 and 0.972 in 1990, 2000, 2010 and 2019, respectively. The albedo exhibited the highest negative correlation with the SDI in 2000, and the correlation was 0.915. The NDVI exhibited the highest correlation with the SDI in 2019 (−0.885). The correlation coefficients between the LSM and SDI were greater than −0.72, and the highest correlation with the SDI was observed in 2000 (−0.911).

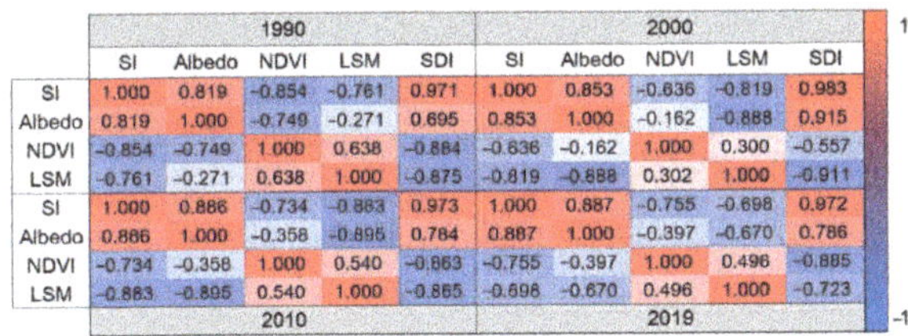

Figure 3. Correlations between pairs of the four selected indicators and their correlations with the SDI in different study years. SI: salinization index; NDVI: normalised difference vegetation index; LSM: land surface soil moisture index; SDI: salinized land degradation index. The blue and red colours represent negative and positive correlations, respectively (the darker the colour, the stronger the correlation).

3.2. Spatiotemporal Changes in the Land Degradation

To analyse the spatiotemporal characteristics of the land degradation during the different periods in the ADD, the SDI values were normalised by Equation (6) (range of 0 to 1). As the SDI approximates a normal distribution, we divided it into five categories by equal intervals to indicate the different land degradation levels [36,58]—namely, no degradation (0–0.2), slight degradation (0.20–0.4), moderate degradation (0.4–0.6), strong degradation (0.6–0.8) and extreme degradation (0.8–1). In summary, the land degradation level distribution was not uniform in space and varied over space and time. As shown in Figure 4, in terms of the land degradation level distribution, the extreme and strong degradation areas were clustered in the west and north of the ADD during the studied years. Areas with moderate degradation corresponded to a sporadic distribution in the middle and south of the ADD. Most areas in the middle of the ADD exhibited a slight degradation. The spatial distribution of the not-degraded areas in the study area showed a difference in the 4 years: in 1990 and 2000, the not-degraded areas were mainly distributed along the Amu Darya River, and a small portion appeared in the northwest and northeast corners of the ADD; in contrast, the not-degraded areas were mainly distributed in the middle of the study area in 2010 and 2019 and formed a "V" shape.

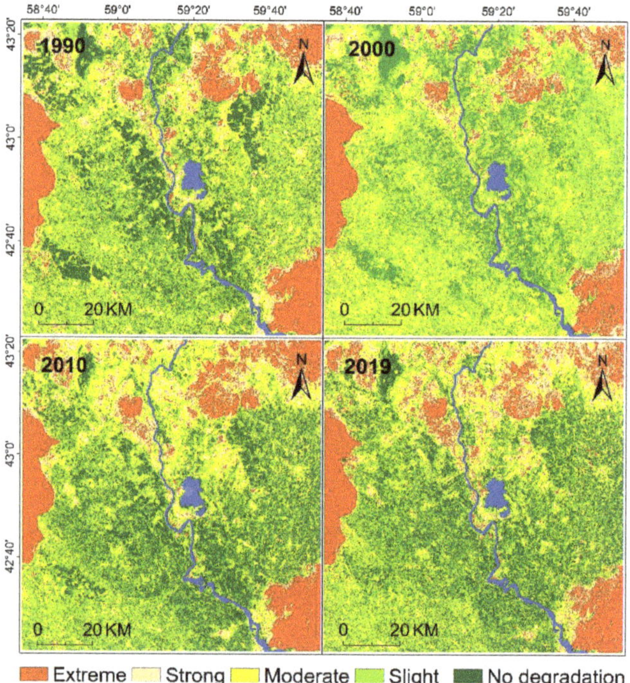

Figure 4. Spatial distribution of the land degradation levels in the ADD in each study year. Extreme: extreme degradation; Strong: strong degradation; Moderate: moderate degradation; Slight: slight degradation.

Figure 5 shows the percentage of the study area occupied by the five land degradation levels in 1990, 2000, 2010 and 2019. In general, the largest areas in the ADD were slight degradations in the years 1990–2019, which accounted for more than 26% of the total area covered by the Landsat images. From 1990 to 2000, the areas with extreme, strong and no degradation decreased, whereas the areas with slight and moderate degradations increased. From 2000 to 2010, the areas with extreme, strong and no degradation expanded; among which, the expansion of the no degradation regions was significant. In contrast, the areas with slight and moderate degradations exhibited a decreasing trend; in particular, for the areas with a slight degradation, the dynamic degree was 7%. From 2010 to 2019, a small increase from 13.30% to 15.56% of the total area was observed in the area with moderate degradation, and the areas of the other four levels did not change considerably.

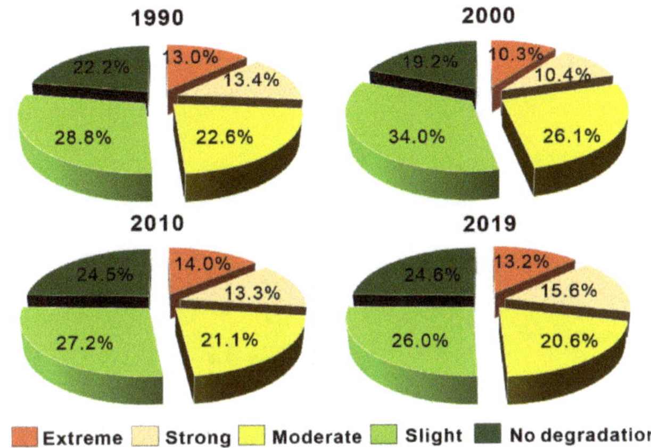

Figure 5. Proportion of land degradation levels in different study years. Extreme: extreme degradation; Strong: strong degradation; Moderate: moderate degradation; Slight: slight degradation.

Using a spatial analysis, the land degradation spatial distribution changes from 1990 to 2000, 2000 to 2010, 2010 to 2019 and 1990 to 2019 were mapped (Figure 6). We defined the figure elements as follows: development of land degradation across levels 1 or 2 corresponded to "Developed" (e.g., a change from "No degradation" to "Slight degradation" or "Moderate degradation"), a development across levels 3 or 4 corresponded to "Seriously developed" (e.g., a change from "No degradation" to "Strong degradation" or "Extreme degradation"), an improvement of the land degradation across levels 1 or 2 corresponded to "Improvement" (e.g., a change from "Extreme degradation" to "Strong degradation" or "Moderate degradation"), an improvement across levels 3 or 4 corresponded to "Significant improvement" (e.g., a change from "Extreme degradation" to "Slight degradation" or "No degradation"), and no change during the study periods corresponded to "Stable". The areas of land degradation dynamics for the four periods are shown in Table 3.

Table 3. Area and proportion of land degradation developed or an improvement of the ADD during different time periods.

Type	1990–2000		2000–2010		2010–2019		1990–2019	
	km^2	%	km^2	%	km^2	%	km^2	%
Seriously developed	130.0	1.1	110.1	0.9	40.3	0.3	217.8	1.8
Developed	2687.5	22.5	3728.8	31.3	2438.0	20.5	2901.7	24.3
Stable	5291.8	44.4	5094.0	42.7	7061.7	59.2	5385.8	45.2
Improvement	3653.8	30.6	2886.0	24.2	2354.3	19.7	3314.7	27.8
Significant improvement	160.5	1.4	104.8	0.8	29.8	0.3	103.4	0.9

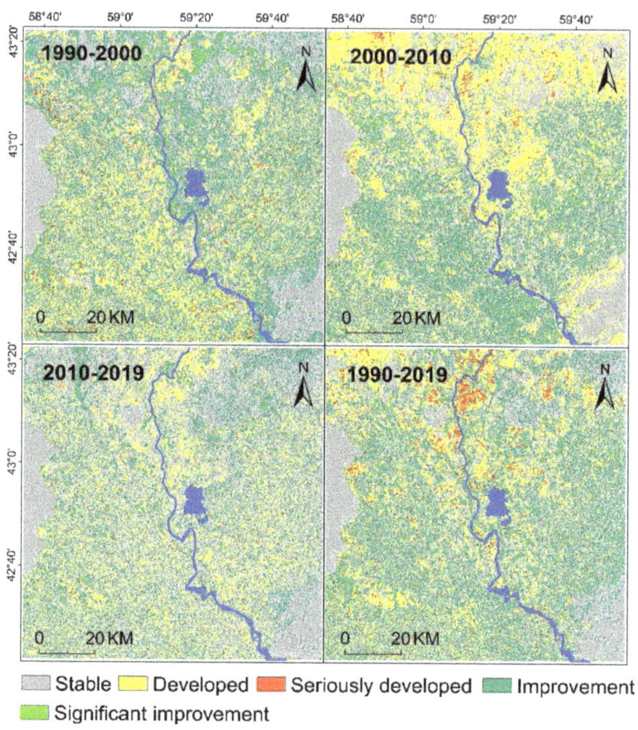

Figure 6. Spatial distribution of land degradation developed or improvement in the ADD in different time periods.

Overall, the stable areas accounted for a large proportion (more than 42%) during the four study periods (Table 3), and these areas were mainly located west and southeast of the ADD. From 1990 to 2000, the improvement areas covered 3653.78 km^2 (30.6%), and these areas were mainly clustered north of the ADD (Figure 6). Additionally, the developed areas covered 2687.53 km^2 (22.5%), and these areas mainly occupied the south and middle regions of the ADD. A smaller proportion (1.4%) of significant improvement areas was observed in the northern part of the ADD. In comparison, the proportion of the seriously developed areas was smaller (1.1%), and these areas were mainly located in the northwest corner of the ADD. From 2000 to 2010, the developed areas covered 3728.76 km^2 (31.3%), and the areas were mainly concentrated in the northern part of the study area. The improvement areas covered 2886.02 km^2 (24.2%) from 2000 to 2010 and were mainly located in the eastern and central parts of the ADD. The areas of seriously developed and significant improvement exhibited smaller proportions—0.92% and 0.88%, respectively—and the seriously developed areas were mainly observed downstream of the ADD. Compared with those in 2000–2010, during 2010–2019, the areas with developed, and the improvements exhibited decreasing trends and occupied 20.5% and 19.7% of the total area, respectively. Moreover, these regions were mainly clustered in the downstream of the ADD. The seriously developed and significant improvement areas occupied less than 0.4% of the total area. From 1990 to 2019, the developed areas covered 2901.65 km^2 (24.3%) and were mainly concentrated north of the ADD and in the downstream region of the Amu Darya River. The improvement regions, which occupied 27.8% of the total area, were mainly observed in the west and east of the ADD. The seriously developed regions, with an area of approximately 217.75 km^2, were mainly concentrated in the north of the ADD, and the areas with significant improvement were smaller, accounting for only 0.9% of the total area.

3.3. Spatial Autocorrelation Analysis of the SDI

To further clarify the spatial and temporal variabilities in the land degradation, the spatial autocorrelation of the SDI was examined.

The mapping of the global spatial autocorrelation of the SDI is shown in Figure 7. Most of the SDI values are distributed in the first and third quadrants, with H–H and L–L clustered in the first and third quadrants, respectively, indicating a strong positive spatial correlation between the spatial units in these two quadrants. The Moran's I values in 1990, 2000, 2010 and 2019 were 0.888, 0.856, 0.891 and 0.851, respectively, which were high values greater than zero. The results showed that the SDI of the ADD exhibited significant spatial clustering, which indicated a strong positive spatial correlation. The Moran's I values of the SDI decreased, increased and later decreased again in 1990, 2000, 2010 and 2019, exhibiting an overall decreasing trend during the study period.

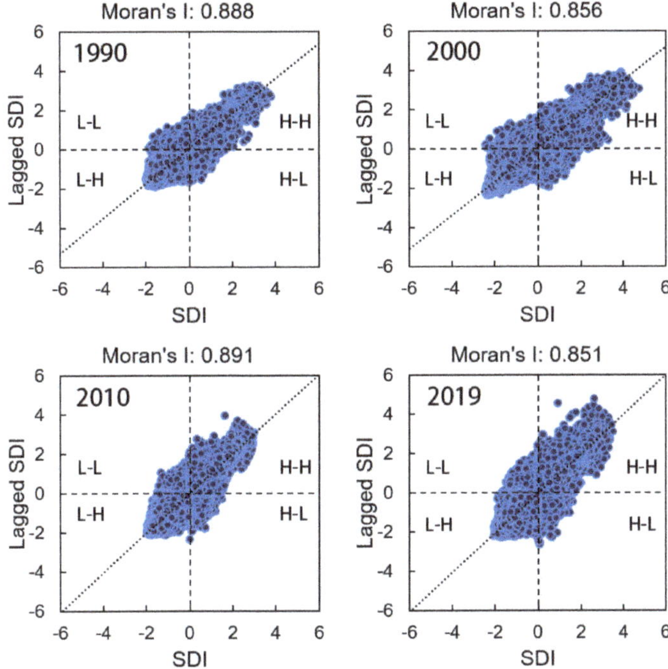

Figure 7. Moran scatter plot of the SDI in the ADD for each study year.

3.4. Spatial and Temporal Changes in Land Use and Salinization

Figure 8 displays the land use maps for 1990, 2000, 2010 and 2019. These land use maps indicated that croplands were the dominant land use type in the ADD. Grassland and forest were distributed in the north of the ADD and bare soil mainly in the edge and north of the ADD. Figure 8b displays the spatial variations in the land use types. The combinations with no land use type transformations, smaller conversion areas and built-up land transformations were merged into "Stable and others". From 1990 to 2010, the conversion of land use categories was mainly between grasslands and croplands. The conversion from grassland to cropland (GL to CL) was prominent in the northern part of the ADD, where the area of grassland decreased by 158.74 km² and 338.93 km² from 1990 to 2000 and 2000 to 2010, respectively (Table 4). However, the conversion of land use categories from 2010 to 2019 was mainly cropland to grassland (CL to GL), which was distributed in the northern and western parts of the ADD. The area of cropland decreased by 602.26 km², and the area of grassland increased by 492.37 km² during this period. In addition, there was partial degradation of grassland to bare soil (GL to BS) in the northern part of the ADD from 2010 to 2019. The conversion of land use categories throughout the study period was mainly from grassland to cropland and cropland to grassland.

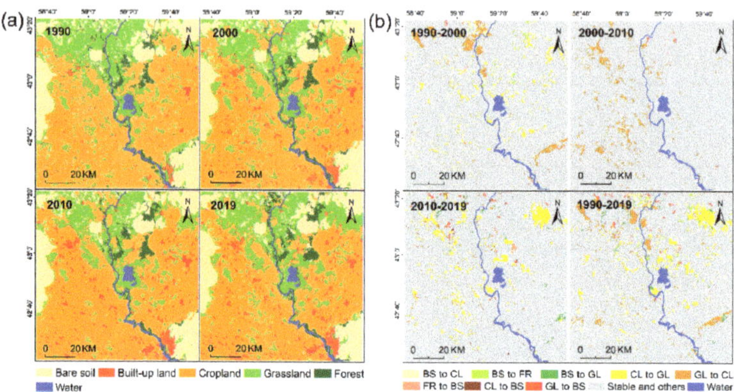

Figure 8. Land use change maps from 1990 to 2019. CL: cropland; GL: grassland; FR: forest; BS: bare soil. (**a**) land use maps for 1990, 2000, 2010 and 2019; (**b**) changes between different periods.

Table 4. Areas and percentages of different land use types from 1990 to 2019 in the ADD.

| | 1990 | | 2000 | | 2010 | | 2019 | |
Type	km²	%	km²	%	km²	%	km²	%
Bare soil	2165.71	18.42	1968.62	16.74	1946.80	16.56	1989.45	16.92
built-up land	205.52	1.75	528.64	4.50	570.81	4.85	588.47	5.00
Cropland	6593.86	56.08	6624.91	56.34	6952.42	59.13	6350.16	54.01
Grassland	2412.67	20.52	2253.93	19.17	1915.46	16.29	2407.83	20.48
Forest	380.66	3.24	382.32	3.25	372.93	3.17	422.51	3.59

Figure 9 reveals the spatial distribution (Figure 9a) and variation (Figure 9b) of the SI over the study time period. The values of the SI were larger at the edges and north of the ADD and smaller in the centre. The SI in the northern part of the ADD decreased from 1990 to 2000, while it increased significantly from 2000 to 2010. From 2010 to 2019, the SI increased slightly in the central part of the ADD and decreased in the northeast. Throughout the study period (1990–2019), the SI increased significantly in the northern part of the ADD and decreased at the edges. To reveal the relationship between land use change and salinization, the mean SI values for each land use type (built-up land and water were excluded) were calculated in Figure 10. The mean value of the SI for the land use type was the largest in 2000 compared to the other three years, indicating higher soil salinization. The mean value of the SI for the land use type was the largest in 2000 compared to the other three years, indicating higher soil salinization, with a decreasing trend in the SI from 2000 to 2019. Within each year, bare soil had the largest mean value of the SI, followed by grassland and forest.

The dynamic characteristics of salinization during land use change were illustrated in Figure 11. The ΔSI represents the difference in the SI over the study period (the following year minus the previous year). We counted a percentage of areas with ΔSI < 0 and ΔSI < 0 over the course of each land use type transfer. The SI increased during all land use type changes from 1990 to 2000 (more than 50% of the area with ΔSI > 0), which was related to the maximum SI value in 2000 mentioned earlier. The 2000–2010 SI values decreased for BS-CL and GL-CL (ΔSI < 0 over 80% of the area). The area with ΔSI < 0 dominated the land use change process from 2010 to 2019. During the conversion of bare soil to cropland (BS-CL), forest (BS-FR) and grassland (BS-GL) from 1990 to 2019, the area with ΔSI < 0 exceeded 50%, indicating a decrease in the SI.

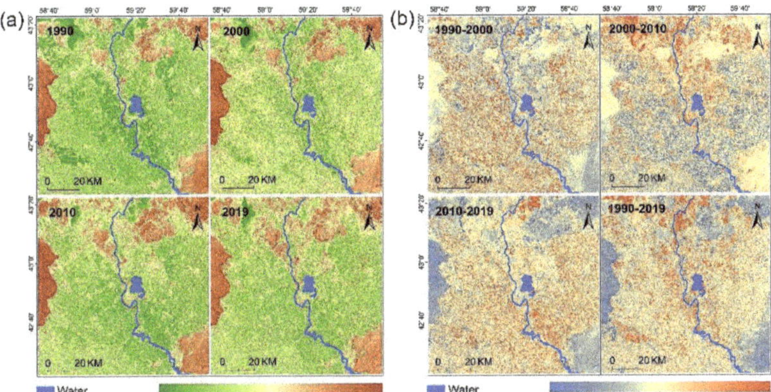

Figure 9. Spatial distribution of soil salinization from 1990 to 2019 (**a**) and spatial distribution of the salinity index (SI) changes between different periods (**b**).

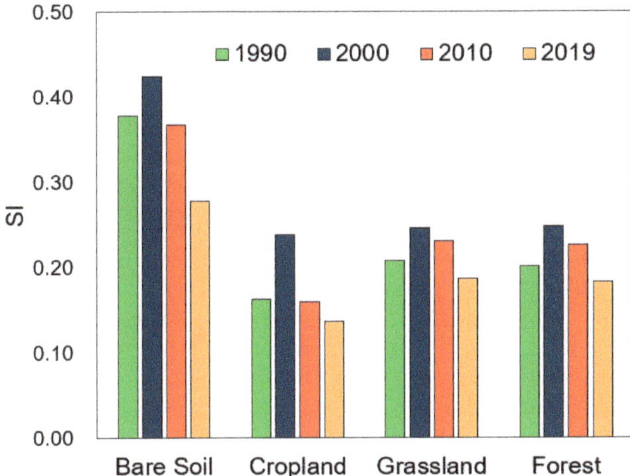

Figure 10. Mean values of the salinity index (SI) for land use types over the study period.

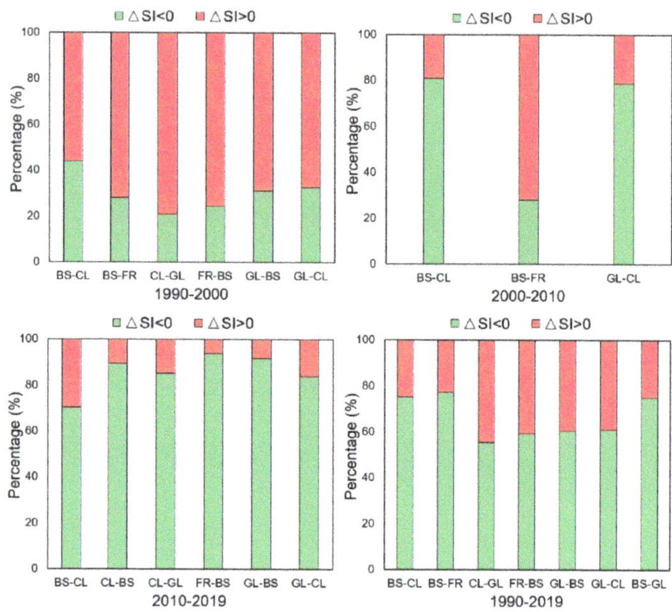

Figure 11. The percentage of ΔSI area for the land use categories that changed during the study period. ΔSI: the salinity index (SI) of the next year minus the SI of the previous year. BS–CL: bare soil to cropland; BS–FR: bare soil to forest; CL–GL: cropland to grassland; FR–BS: forest to bare soil; GL–BS: grassland to bare soil; GL–CL: grassland to cropland; BS–GL: bare soil to grassland.

4. Discussion

4.1. Effectiveness of the Proposed SDI

In this study, we established a new SDI by integrating the SI, albedo, NDVI and LSM indices based on the PCA method. The SDI was used to explore the land degradation characteristics for a typical salinized area, i.e., the ADD. We found that the regions with extreme land degradation were mainly distributed downstream and at the periphery of the ADD (Figure 4). The SDI-based results supported previous findings that the ecological risks and vulnerabilities are higher in the downstream and peripheral regions of the ADD [10,46], thereby demonstrating that the SDI can reflect the land degradation conditions of the ADD. To further evaluate the effectiveness of the proposed SDI, we evaluated its effectiveness by field survey data (Figure S1 and Table S1). The relationship between the SDI and soil salt content shown in Figure 12 indicates that the soil salt content was significantly positively correlated with the SDI ($R^2 = 0.89$, $p < 0.001$). The relationships between the four indices derived from Landsat imagery in 2019 and field-measured soil salinity are also presented in Figure S2. There was a positive correlation between the SI, albedo and measured soil salinity ($R^2 = 0.41$ and 0.43, respectively) and a negative correlation between the NDVI, LSM and measured soil salinity ($R^2 = 0.29$ and 0.19, respectively). In contrast to the single indices, the SDI showed a stronger agreement with the measured soil salinity. In general, although the land degradation is influenced by multiple aspects of the environment, this positive correlation suggests that the SDI can capture the salinity features pertaining to the land degradation, which provides potential evidence for the effectiveness of the index in monitoring the land degradation in salinized areas. In addition, the SI extracted from the remote sensing data exhibits a positive correlation with the SDI (Figure 3). The reliability of the SDI is also reflected in the other three indicators. High vegetation cover and sufficient soil moisture reduce the risk of land degradation, and this finding is supported by the negative correlation between the SDI and NDVI and LSM in our study (Figure 3). An increase in the albedo values leads to a higher SDI, which is related to the exposure information represented by the albedo and may be attributed to a strong coupling between the soil salinity information and albedo (Table 1 and Figure 3). These characteristics have also been reported previously [33]. These results indicate that the SDI can help reliably and efficiently monitor the land degradation of salinized areas. In addition, the SDI is composed of accessible remote sensing indicators and can thus be extended to other similar ecological environments [35,59,63].

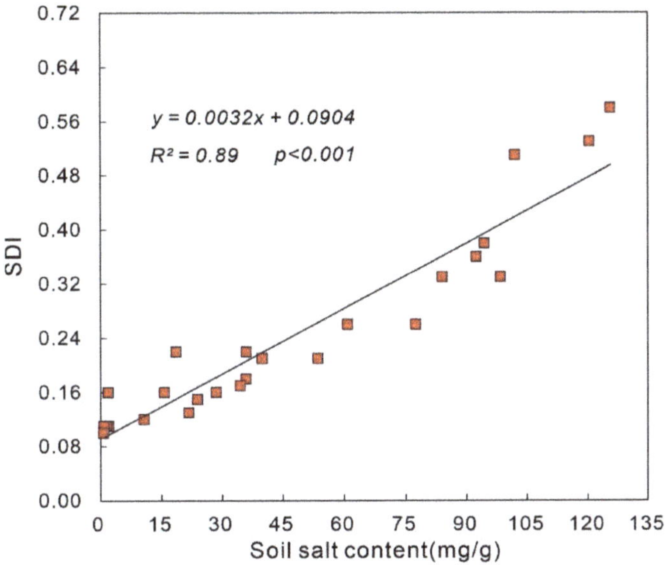

Figure 12. Relationship between the SDI and field-measured soil salt content.

4.2. Factors Influencing the Land Degradation in the ADD

Climate variables mainly affect the land degradation through changes in the precipitation and temperature [10,64]. The annual mean precipitation (AMP) and annual mean temperature (AMT) of Nurkus Weather Station are shown in Figure 13. It can be noted that the AMP decreased and AMT increased in the ADD in the past 40 years. From 1990 to 2019, land degradation developed over an area of more than 3000 km^2 (Table 3). Previous studies have raised concerns regarding the withering of grasslands and sparse vegetation caused by warming and dry climates in the ADD, warning that these aspects could accelerate land degradation [10,65].

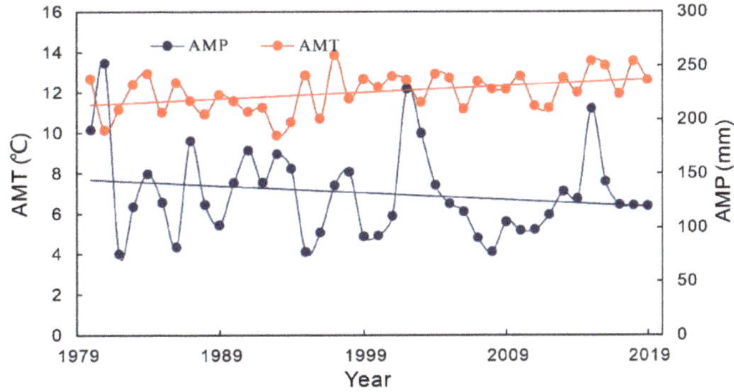

Figure 13. Changes in the annual mean precipitation and temperature measured at Nurkus Station from 1980 to 2016.

Amu Darya River is the main source of water for ADD living and irrigation. Over-watering has disrupted the water system of the ADD, leading to the disappearance of an amount of lakes, followed by local climate change, which has reduced the ecosystem stability of the ADD, particularly in the downstream region [7,10]. The widespread use of diffuse irrigation has caused individuals to compensate for inefficient irrigation by collecting large amounts of water, resulting in a reduction in the ecological water that sustains the ecosystems and the gradual withering of vegetation without sufficient water to support growth [9,66,67], exacerbating the land degradation. The extent of the land degradation followed the same trend as that of the water withdrawal from Amu Darya River. With the decline in the water withdrawal in 1990–2000 (Figure 14), the improvement in ADD land degradation was most pronounced, while, in 2000–2019, as the water withdrawal increased, the land degradation developed in a larger area than the improvement area. A more critical situation is that the reservoir built in the upper reaches of the Amu Darya River intercepted a large amount of the water [66], resulting in a decrease in the supply of ecological water downstream. The ecological effects caused by these factors were confirmed by our research: The land degradation downstream of the ADD was significantly degraded compared with the region in the study period (Figures 4 and 6). In addition, the higher levels of land degradation in the outer delta, farther from Amu Darya River, were likely caused by the lack of water supply to the ecosystem and the difficulties in the land management in the transboundary area [46,68]. This finding indicates that, to alleviate the ADD land degradation and the ecological crisis in the Aral Sea basin, further effort and cooperation is necessary in the rational allocation of water resources.

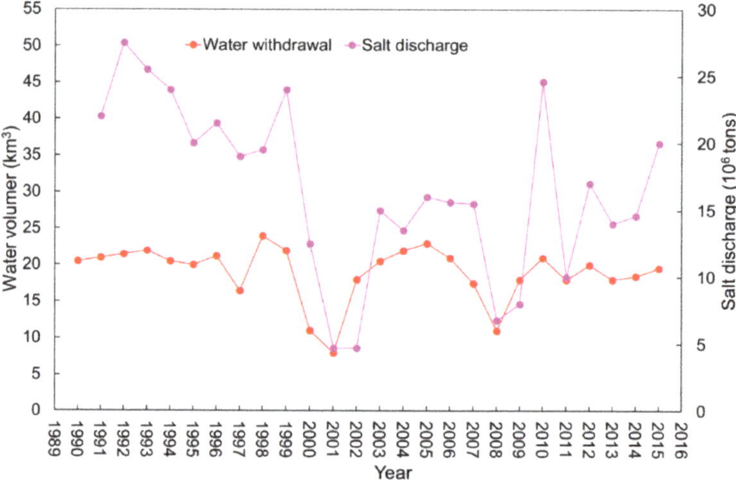

Figure 14. Annual changes in the water withdrawal and salt discharge in the ADD from 1990 to 2015.

In addition, the impacts of land-use changes on the land degradation cannot be ignored [10,69,70]. The consolidation and management of croplands can contribute to the mitigation of land degradation, and our research supports this perspective. We demonstrated that the area with no degradation occupied a larger proportion of cropland during most of the study period, while the area with extreme land degradation was mainly distributed on bare soil (Figure 15). In general, the risk of land degradation was reduced when land with sparse vegetation and bare soil was reclaimed as cropland, as crops contribute to higher ecosystem productivity and stability. Land degradation is more severe in the northern part of the ADD, where part of the cropland has been abandoned and converted into grassland or bare soil (Figure 8). Previous studies have indicated that the ADD is facing an ecological threat posed by the degradation of grasslands and croplands to bare soil [10,46]. Compared to the other land use categories, bare soil has a higher soil salinity

(Figure 10), and the conversion of land use types to bare soil not only reduces the biomass but also increases the risk of soil salinization.

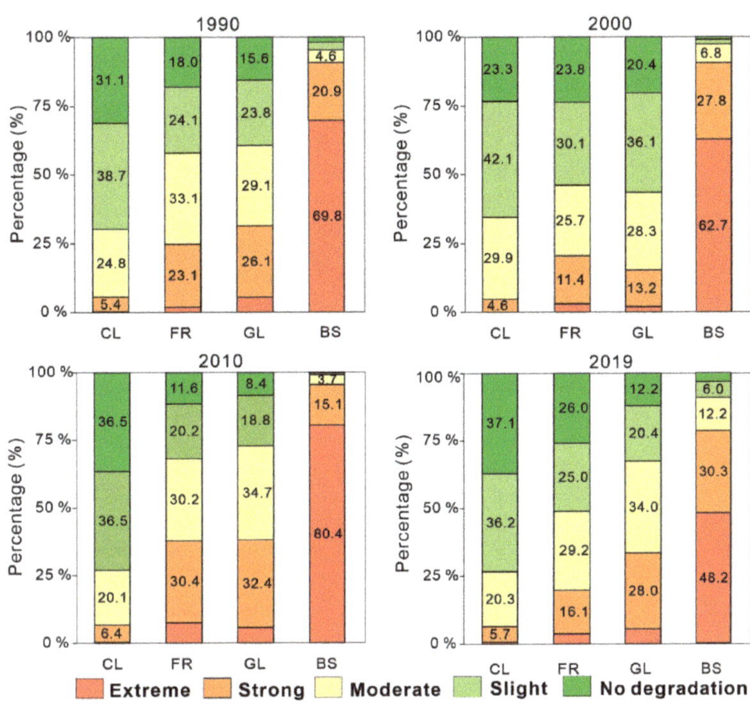

Figure 15. Percentage graphs showing the proportion of each land degradation level in different land use types. Extreme: extreme degradation; Strong: strong degradation; Moderate: moderate degradation; Slight: slight degradation.CL: cropland; FR: forest; GL: grassland; BS: bare soil.

However, to reduce the salt content of croplands, a large amount of water is acquired from Amu Darya River to rinse the cropland soil, which further aggravates the water deficit of other ecosystems. In addition, the excess salinity from croplands is discharged by widely distributed channels to the Amu Darya River, as well as to the lakes downstream of the delta, resulting in increased salinity in the river water and a significant accumulation of salinity downstream (Figure 9) [10]. The increase in the discharge was the most significant after 2000 (Figure 14). The extreme land degradation distribution patterns were noted to be clustered downstream of the ADD (Figures 4 and 6). These ecological effects caused by the large accumulation of salt in the downstream region were confirmed by our research. In addition, with the disintegration of the Soviet Union, the socialist economy turned into a market economy, and the gradual influx of the rural population into cities led to certain croplands eventually transforming into unused lands with a low biodiversity [71], thereby accelerating land degradation.

The 15th initiative of the Sustainable Development Goals (SDGs) aims to achieve land degradation neutrality by 2030. To effectively alleviate land degradation of the ADD and promote the further realisation of SDGs, based on the abovementioned factors influencing the SDI, the following corresponding measures and countermeasures are proposed. First, salinization treatment technology should be implemented to alleviate the promotion of land degradation caused by salinity, and reservoirs for storing salt can be built to reduce the ecological pressure caused by the transportation of salt from the alkali drainage canal that discharges into the downstream area of the ADD. Second, a drip irrigation system can be promoted to achieve precision irrigation and enhance the irrigation efficiency to relieve the pressure of water resources required to maintain the stability of the land degradation. Furthermore, farmers are encouraged to maintain the stability and biodiversity of croplands through agricultural subsidy policies. In addition, ecological conservation projects can be considered to mitigate the impacts of climate change on the land degradation.

5. Conclusions

We coupled multiple remote sensing indices (SI, NDVI, albedo and LSM) to construct a new SDI by using the PCA method. The proposed index integrated the soil salinity, soil bareness, soil moisture and vegetation coverage and made it possible to identify the characteristics of the regional land degradation, especially in salinized areas. To test the reliability of the SDI, the index was applied to the typical ADD region to monitor the spatial and temporal dynamics of the land degradation.

The results indicated that the NDVI and LSM adversely influenced the land degradation, while the SI and albedo had positive effects. The SI was strongly positive correlated with the SDI, with an average correlation coefficient of 0.97. Regions with extreme and strong land degradation were mostly clustered west and north of the ADD. The temporal and spatial dynamics of the SDI indicated that the land degradation in the ADD developed by approximately 26% (including seriously developed and developed areas) from 1990 to 2019, and the degradation was mainly concentrated in the downstream region of the ADD. The areas exhibiting improvement accounted for approximately 28% of the total area of the ADD and were mainly centred in the eastern and central parts. Among them, the area of land degradation developed from 2000 to 2010 was the largest (approximately 32%), while the improvement area was 25%. The results of spatial autocorrelation analysis showed that the SDI values of Moran's I in 1990, 2000, 2010 and 2019 were 0.89, 0.86, 0.89 and 0.85, respectively, which showed that the SDI was clearly clustered in space rather than randomly distributed.

The drying climate and excessive water withdrawal from the Amu Darya River exacerbated the land degradation in the ADD, especially in 2000–2019; as the water withdrawal increased, the land degradation developed into a larger area than the improvement area. The expansion of unused land increases the risk of land degradation, with higher levels of land degradation on unused land than on other types during the study period. In addition, a large amount of salt discharged from croplands downstream of the ADD results in the downstream being the most degraded area of land.

Clarifying the characteristics of land degradation of salinized areas is conducive to the restoration and promotion of sustainable terrestrial ecosystems. Our study revealed land degradation characteristics at the interannual scale of the ADD based on the SDI, which provided an efficient decision-making basis for regional land management. Nevertheless, some limitations still exist in this research. For example, the seasonal and continuous dynamics of land degradation have not been taken well into consideration due to the limited temporal resolution of the Landsat satellites. Constructing a SDI with the high temporal resolution MODerate resolution Imaging Spectroradiometer (MODIS) has the potential to enable the seasonal and continuous temporal monitoring of land degradation on a large scale, which is further work that deserves to be advanced.

Supplementary Materials: The following are available online at https://www.mdpi.com/article/10.3390/rs13152851/s1: Figure S1: Soil sampling sites of the ADD. Table S1: Field sampling data of soil salt.

Author Contributions: T.Y. and G.J. designed the research. T.Y. processed the data and wrote the manuscript. A.B. and G.Z. revised the manuscript. L.J., Y.Y. and X.H. provided the analysis tools and technical assistance. All authors contributed to the final version of the manuscript by proofreading and offering constructive comments. All authors have read and agreed to the published version of the manuscript.

Funding: This research was supported by the Strategic Priority Research Program of Chinese Academy of Sciences (Grant No. XDA19030301) and the Open Foundation of State Key Laboratory of Desert and Oasis Ecology, Xinjiang Institute of Ecology and Geography, Chinese Academy of Sciences (G2019-02-03).

Data Availability Statement: The data is available upon request.

Acknowledgments: We thank the journal's editors and reviewers for their kind comments and valuable suggestions to improve the quality of this paper.

Conflicts of Interest: The authors declare no conflict of interest.

References

1. Gisladottir, G.; Stocking, M. Land degradation control and its global environmental benefits. *Land Degrad. Dev.* **2005**, *16*, 99–112. [CrossRef]
2. Gibbs, H.K.; Salmon, J.M. Mapping the world's degraded lands. *Appl. Geogr.* **2015**, *57*, 12–21. [CrossRef]
3. Huang, J.; Yu, H.; Guan, X.; Wang, G.; Guo, R. Accelerated dryland expansion under climate change. *Nat. Clim. Chang.* **2015**, *6*, 166–171. [CrossRef]
4. D'Odorico, P.; Bhattachan, A.; Davis, K.F.; Ravi, S.; Runyan, C.W. Global desertification: Drivers and feedbacks. *Adv. Water Resour.* **2013**, *51*, 326–344. [CrossRef]
5. Ivushkin, K.; Bartholomeus, H.; Bregt, A.K.; Pulatov, A. Satellite Thermography for Soil Salinity Assessment of Cropped Areas in Uzbekistan. *Land Degrad. Dev.* **2017**, *28*, 870–877. [CrossRef]
6. MICKLIN, P.P. Desiccation of the Aral Sea: A Water Management Disaster in the Soviet Union. *Science* **1988**, *241*, 1170–1176. [CrossRef]
7. Dubovyk, O.; Menz, G.; Khamzina, A. Land Suitability Assessment for Afforestation with *Elaeagnus AngustifoliaL.* in Degraded Agricultural Areas of the Lower Amudarya River Basin. *Land Degrad. Dev.* **2014**, *27*, 1831–1839. [CrossRef]
8. Khamzina, A.; Lamers, J.P.A.; Vlek, P.L.G. Tree establishment under deficit irrigation on degraded agricultural land in the lower Amu Darya River region, Aral Sea Basin. *For. Ecol. Manag.* **2008**, *255*, 168–178. [CrossRef]
9. Schlüter, M.; Khasankhanova, G.; Talskikh, V.; Taryannikova, R.; Agaltseva, N.; Joldasova, I.; Ibragimov, R.; Abdullaev, U. Enhancing resilience to water flow uncertainty by integrating environmental flows into water management in the Amudarya River, Central Asia. *Glob. Planet. Chang.* **2013**, *110*, 114–129. [CrossRef]
10. Jiang, L.; Jiapaer, G.; Bao, A.; Li, Y.; Guo, H.; Zheng, G.; Chen, T.; De Maeyer, P. Assessing land degradation and quantifying its drivers in the Amudarya River delta. *Ecol. Indic.* **2019**, *107*, 105595. [CrossRef]
11. Guo, H.; Bao, A.; Ndayisaba, F.; Liu, T.; Jiapaer, G.; El-Tantawi, A.M.; De Maeyer, P. Space-time characterization of drought events and their impacts on vegetation in Central Asia. *J. Hydrol.* **2018**, *564*, 1165–1178. [CrossRef]
12. Mariano, D.A.; Santos, C.A.C.D.; Wardlow, B.D.; Anderson, M.C.; Schiltmeyer, A.V.; Tadesse, T.; Svoboda, M.D. Use of remote sensing indicators to assess effects of drought and human-induced land degradation on ecosystem health in Northeastern Brazil. *Remote. Sens. Environ.* **2018**, *213*, 129–143. [CrossRef]
13. Jiang, L.; Bao, A.; Jiapaer, G.; Guo, H.; Zheng, G.; Gafforov, K.; Kurban, A.; De Maeyer, P. Monitoring land sensitivity to desertification in Central Asia: Convergence or divergence? *Sci. Total Environ.* **2019**, *658*, 669–683. [CrossRef]
14. Jiang, L.; Jiapaer, G.; Bao, A.; Kurban, A.; Guo, H.; Zheng, G.; De Maeyer, P. Monitoring the long-term desertification process and assessing the relative roles of its drivers in Central Asia. *Ecol. Indic.* **2019**, *104*, 195–208. [CrossRef]
15. Cheng, W.; Xi, H.; Sindikubwabo, C.; Si, J.; Zhao, C.; Yu, T.; Li, A.; Wu, T. Ecosystem health assessment of desert nature reserve with entropy weight and fuzzy mathematics methods: A case study of Badain Jaran Desert. *Ecol. Indic.* **2020**, *119*, 106843. [CrossRef]
16. Pei, J.; Wang, L.; Wang, X.; Niu, Z.; Kelly, M.; Song, X.-P.; Huang, N.; Geng, J.; Tian, H.; Yu, Y.; et al. Time Series of Landsat Imagery Shows Vegetation Recovery in Two Fragile Karst Watersheds in Southwest China from 1988 to 2016. *Remote Sens.* **2019**, *11*, 2044. [CrossRef]
17. Li, J.; Xu, B.; Yang, X.; Qin, Z.; Zhao, L.; Jin, Y.; Zhao, F.; Guo, J. Historical grassland desertification changes in the Horqin Sandy Land, Northern China (1985–2013). *Sci. Rep.* **2017**, *7*. [CrossRef]
18. Jin, Z.; Guo, L.; Wang, Y.; Yu, Y.; Lin, H.; Chen, Y.; Chu, G.; Zhang, J.; Zhang, N. Valley reshaping and damming induce water table rise and soil salinization on the Chinese Loess Plateau. *Geoderma* **2019**, *339*, 115–125. [CrossRef]
19. Yang, M.; Nelson, F.E.; Shiklomanov, N.I.; Guo, D.; Wan, G. Permafrost degradation and its environmental effects on the Tibetan Plateau: A review of recent research. *Earth-Sci. Rev.* **2010**, *103*, 31–44. [CrossRef]
20. Khan, N.M.; Sato, Y. Environmental land degradation assessment in semi-arid Indus basin area using IRS-1B LISS-hII data. In Proceedings of the Igarss 2001: Scanning the Present and Resolving the Future, Sydney, Australia, 9–13 July 2001; Volume 5, pp. 2100–2102.
21. Bai, Z.G.; Dent, D.L.; Olsson, L.; Schaepman, M.E. Proxy global assessment of land degradation. *Soil Use Manag.* **2008**, *24*, 223–234. [CrossRef]
22. Zhao, Y.; Wang, X.; Novillo, C.J.; Arrogante-Funes, P.; Vázquez-Jiménez, R.; Berdugo, M.; Maestre, F.T. Remotely sensed albedo allows the identification of two ecosystem states along aridity gradients in Africa. *Land Degrad. Dev.* **2019**, *30*, 1502–1515. [CrossRef]

23. Houspanossian, J.; Giménez, R.; Jobbágy, E.; Nosetto, M. Surface albedo raise in the South American Chaco: Combined effects of deforestation and agricultural changes. *Agric. For. Meteorol.* **2017**, *232*, 118–127. [CrossRef]

24. Kumar, S.; Singh, A.K.; Singh, R.; Ghosh, A.; Chaudhary, M.; Shukla, A.K.; Kumar, S.; Singh, H.V.; Ahmed, A.; Kumar, R.V. Degraded land restoration ecological way through horti-pasture systems and soil moisture conservation to sustain productive economic viability. *Land Degrad. Dev.* **2019**, *30*, 1516–1529. [CrossRef]

25. Holm, A. The use of time-integrated NOAA NDVI data and rainfall to assess landscape degradation in the arid shrubland of Western Australia. *Remote. Sens. Environ.* **2003**, *85*, 145–158. [CrossRef]

26. Symeonakis, E.; Karathanasis, N.; Koukoulas, S.; Panagopoulos, G. Monitoring Sensitivity to Land Degradation and Desertification with the Environmentally Sensitive Area Index: The Case of Lesvos Island. *Land Degrad. Dev.* **2016**, *27*, 1562–1573. [CrossRef]

27. Liu, F.; Chen, Y.; Lu, H.; Shao, H. Albedo indicating land degradation around the Badain Jaran Desert for better land resources utilization. *Sci. Total Environ.* **2017**, *578*, 67–73. [CrossRef] [PubMed]

28. Ibrahim, Y.; Balzter, H.; Kaduk, J.; Tucker, C. Land Degradation Assessment Using Residual Trend Analysis of GIMMS NDVI3g, Soil Moisture and Rainfall in Sub-Saharan West Africa from 1982 to 2012. *Remote Sens.* **2015**, *7*, 5471–5494. [CrossRef]

29. Yang, C.; Li, Q.; Chen, J.; Wang, J.; Shi, T.; Hu, Z.; Ding, K.; Wang, G.; Wu, G. Spatiotemporal characteristics of land degradation in the Fuxian Lake Basin, China: Past and future. *Land Degrad. Dev.* **2020**, *31*, 2446–2460. [CrossRef]

30. Sommer, S.; Zucca, C.; Grainger, A.; Cherlet, M.; Zougmore, R.; Sokona, Y.; Hill, J.; Della Peruta, R.; Roehrig, J.; Wang, G. Application of indicator systems for monitoring and assessment of desertification from national to global scales. *Land Degrad. Dev.* **2011**, *22*, 184–197. [CrossRef]

31. Xu, H.; Wang, M.; Shi, T.; Guan, H.; Fang, C.; Lin, Z. Prediction of ecological effects of potential population and impervious surface increases using a remote sensing based ecological index (RSEI). *Ecol. Indic.* **2018**, *93*, 730–740. [CrossRef]

32. Guo, B.; Fang, Y.; Jin, X.; Zhou, Y. Monitoring the effects of land consolidation on the ecological environmental quality based on remote sensing: A case study of Chaohu Lake Basin, China. *Land Use Policy* **2020**, *95*, 104569. [CrossRef]

33. Jing, Y.; Zhang, F.; He, Y.; Kung, H.-T.; Johnson, V.C.; Arikena, M. Assessment of spatial and temporal variation of ecological environment quality in Ebinur Lake Wetland National Nature Reserve, Xinjiang, China. *Ecol. Indic.* **2020**, *110*, 105874. [CrossRef]

34. Seddon, A.W.R.; Macias-Fauria, M.; Long, P.R.; Benz, D.; Willis, K.J. Sensitivity of global terrestrial ecosystems to climate variability. *Nature* **2016**, *531*, 229–232. [CrossRef]

35. Hu, X.S.; Xu, H.Q. A new remote sensing index based on the pressure-state-response framework to assess regional ecological change. *Sci. Pollut. Res.* **2019**, *26*, 5381–5393. [CrossRef]

36. Hu, X.S.; Xu, H.Q. A new remote sensing index for assessing the spatial heterogeneity in urban ecological quality: A case from Fuzhou City, China. *Ecol. Indic.* **2018**, *89*, 11–21. [CrossRef]

37. Lee, S.O.; Jung, Y. Efficiency of water use and its implications for a water-food nexus in the Aral Sea Basin. *Agric. Water Manag.* **2018**, *207*, 80–90. [CrossRef]

38. Kumar, N.; Khamzina, A.; Tischbein, B.; Knöfel, P.; Conrad, C.; Lamers, J.P.A. Spatio-temporal supply–demand of surface water for agroforestry planning in saline landscape of the lower Amudarya Basin. *J. Arid Environ.* **2019**, *162*, 53–61. [CrossRef]

39. Schettler, G.; Oberhänsli, H.; Stulina, G.; Mavlonov, A.A.; Naumann, R. Hydrochemical water evolution in the Aral Sea Basin. Part I: Unconfined groundwater of the Amu Darya Delta—Interactions with surface waters. *J. Hydrol.* **2013**, *495*, 267–284. [CrossRef]

40. Ablekim, A.; Ge, Y.; Wang, Y.; Hu, R. The Past, Present and Feature of the Aral Sea. *Arid Zone Res.* **2019**, *36*, 7–18.

41. Shen, H.; Abuduwaili, J.; Ma, L.; Samat, A. Remote sensing-based land surface change identification and prediction in the Aral Sea bed, Central Asia. *Int. J. Environ. Sci. Technol.* **2018**, *16*, 2031–2046. [CrossRef]

42. Li, J.; Yang, X.; Jin, Y.; Yang, Z.; Huang, W.; Zhao, L.; Gao, T.; Yu, H.; Ma, H.; Qin, Z.; et al. Monitoring and analysis of grassland desertification dynamics using Landsat images in Ningxia, China. *Remote. Sens. Environ.* **2013**, *138*, 19–26. [CrossRef]

43. Bi, S.; Li, Y.; Wang, Q.; Lyu, H.; Liu, G.; Zheng, Z.; Du, C.; Mu, M.; Xu, J.; Lei, S.; et al. Inland Water Atmospheric Correction Based on Turbidity Classification Using OLCI and SLSTR Synergistic Observations. *Remote Sens.* **2018**, *10*, 1002. [CrossRef]

44. Nazeer, M.; Nichol, J.E.; Yung, Y.-K. Evaluation of atmospheric correction models and Landsat surface reflectance product in an urban coastal environment. *Int. J. Remote Sens.* **2014**, *35*, 6271–6291. [CrossRef]

45. Xu, H. Modification of normalised difference water index (NDWI) to enhance open water features in remotely sensed imagery. *Int. J. Remote Sens.* **2007**, *27*, 3025–3033. [CrossRef]

46. Yu, T.; Bao, A.; Xu, W.; Guo, H.; Jiang, L.; Zheng, G.; Yuan, Y.; Nzabarinda, V. Exploring Variability in Landscape Ecological Risk and Quantifying Its Driving Factors in the Amu Darya Delta. *Int. J. Environ. Res. Public Health* **2019**, *17*, 79. [CrossRef] [PubMed]

47. Whitney, K.; Scudiero, E.; El-Askary, H.M.; Skaggs, T.H.; Allali, M.; Corwin, D.L. Validating the use of MODIS time series for salinity assessment over agricultural soils in California, USA. *Ecol. Indic.* **2018**, *93*, 889–898. [CrossRef]

48. Yu, R.; Liu, T.; Xu, Y.; Zhu, C.; Zhang, Q.; Qu, Z.; Liu, X.; Li, C. Analysis of salinization dynamics by remote sensing in Hetao Irrigation District of North China. *Agric. Water Manag.* **2010**, *97*, 1952–1960. [CrossRef]

49. Zhang, T.-T.; Zeng, S.-L.; Gao, Y.; Ouyang, Z.-T.; Li, B.; Fang, C.-M.; Zhao, B. Using hyperspectral vegetation indices as a proxy to monitor soil salinity. *Ecol. Indic.* **2011**, *11*, 1552–1562. [CrossRef]

50. Gorji, T.; Sertel, E.; Tanik, A. Monitoring soil salinity via remote sensing technology under data scarce conditions: A case study from Turkey. *Ecol. Indic.* **2017**, *74*, 384–391. [CrossRef]

51. Liang, S.L. Narrowband to broadband conversions of land surface albedo I Algorithms. *Remote Sens. Environ.* **2001**, *76*, 213–238. [CrossRef]

52. Kuusinen, N.; Stenberg, P.; Korhonen, L.; Rautiainen, M.; Tomppo, E. Structural factors driving boreal forest albedo in Finland. *Remote Sens. Environ.* **2016**, *175*, 43–51. [CrossRef]
53. Easdale, M.H.; Bruzzone, O.; Mapfumo, P.; Tittonell, P. Phases or regimes? Revisiting NDVI trends as proxies for land degradation. *Land Degrad. Dev.* **2018**, *29*, 433–445. [CrossRef]
54. Zhumanova, M.; Mönnig, C.; Hergarten, C.; Darr, D.; Wrage-Mönnig, N. Assessment of vegetation degradation in mountainous pastures of the Western Tien-Shan, Kyrgyzstan, using eMODIS NDVI. *Ecol. Indic.* **2018**, *95*, 527–543. [CrossRef]
55. Babaeian, E.; Sadeghi, M.; Jones, S.B.; Montzka, C.; Vereecken, H.; Tuller, M. Ground, Proximal, and Satellite Remote Sensing of Soil Moisture. *Rev. Geophys.* **2019**, *57*, 530–616. [CrossRef]
56. Baig, M.H.A.; Zhang, L.F.; Shuai, T.; Tong, Q.X. Derivation of a tasselled cap transformation based on Landsat 8 at-satellite reflectance. *Remote Sens. Lett.* **2014**, *5*, 423–431. [CrossRef]
57. Crist, E.P. A tm tasseled cap equivalent transformation for reflectance factor data. *Remote. Sens. Environ.* **1985**, *17*, 301–306. [CrossRef]
58. Shan, W.; Jin, X.; Ren, J.; Wang, Y.; Xu, Z.; Fan, Y.; Gu, Z.; Hong, C.; Lin, J.; Zhou, Y. Ecological environment quality assessment based on remote sensing data for land consolidation. *J. Clean. Prod.* **2019**, *239*, 118126. [CrossRef]
59. Xu, H.; Wang, Y.; Guan, H.; Shi, T.; Hu, X. Detecting Ecological Changes with a Remote Sensing Based Ecological Index (RSEI) Produced Time Series and Change Vector Analysis. *Remote Sens.* **2019**, *11*, 2345. [CrossRef]
60. Anselin, L. Local Indicators of Spatial Association—LISA. *Geogr. Anal.* **1995**, *27*, 93–115. [CrossRef]
61. Zhang, F.; Yushanjiang, A.; Jing, Y. Assessing and predicting changes of the ecosystem service values based on land use/cover change in Ebinur Lake Wetland National Nature Reserve, Xinjiang, China. *Sci. Total Environ.* **2019**, *656*, 1133–1144. [CrossRef] [PubMed]
62. Li, H.F.; Calder, C.A.; Cressie, N. Beyond Moran's I: Testing for spatial dependence based on the spatial autoregressive model. *Geogr. Anal.* **2007**, *39*, 357–375. [CrossRef]
63. He, C.Y.; Gao, B.; Huang, Q.X.; Ma, Q.; Dou, Y.Y. Environmental degradation in the urban areas of China: Evidence from multi-source remote sensing data. *Remote. Sens. Environ.* **2017**, *193*, 65–75. [CrossRef]
64. Turner, K.G.; Anderson, S.; Gonzales-Chang, M.; Costanza, R.; Courville, S.; Dalgaard, T.; Dominati, E.; Kubiszewski, I.; Ogilvy, S.; Porfirio, L.; et al. A review of methods, data, and models to assess changes in the value of ecosystem services from land degradation and restoration. *Ecol. Model.* **2016**, *319*, 190–207. [CrossRef]
65. Asarin, A.E.; Kravtsova, V.I.; Mikhailov, V.N. Amudarya and Syrdarya Rivers and Their Deltas. In *The Aral Sea Environment*; Kostianoy, A.G., Kosarev, A.N., Eds.; Springer: Berlin/Heidelberg, Germany, 2010; pp. 101–121.
66. Shi, H.; Luo, G.; Zheng, H.; Chen, C.; Hellwich, O.; Bai, J.; Liu, T.; Liu, S.; Xue, J.; Cai, P.; et al. A novel causal structure-based framework for comparing a basin-wide water–energy–food–ecology nexus applied to the data-limited Amu Darya and Syr Darya river basins. *Hydrol. Earth Syst. Sci.* **2021**, *25*, 901–925. [CrossRef]
67. Sun, J.; Li, Y.P.; Suo, C.; Liu, Y.R. Impacts of irrigation efficiency on agricultural water-land nexus system management under multiple uncertainties-A case study in Amu Darya River basin, Central Asia. *Agric. Water Manag.* **2019**, *216*, 76–88. [CrossRef]
68. Conrad, C.; Dech, S.W.; Hafeez, M.; Lamers, J.P.A.; Tischbein, B. Remote sensing and hydrological measurement based irrigation performance assessments in the upper Amu Darya Delta, Central Asia. *Phys. Chem. Earth Parts A B C* **2013**, *61–62*, 52–62. [CrossRef]
69. Zhang, K.; Yu, Z.; Li, X.; Zhou, W.; Zhang, D. Land use change and land degradation in China from 1991 to 2001. *Land Degrad. Dev.* **2007**, *18*, 209–219. [CrossRef]
70. Nababa, I.; Symeonakis, E.; Koukoulas, S.; Higginbottom, T.; Cavan, G.; Marsden, S. Land Cover Dynamics and Mangrove Degradation in the Niger Delta Region. *Remote Sens.* **2020**, *12*, 3619. [CrossRef]
71. Zhou, Y.; Zhang, L.; Xiao, J.; Williams, C.A.; Vitkovskaya, I.; Bao, A. Spatiotemporal transition of institutional and socioeconomic impacts on vegetation productivity in Central Asia over last three decades. *Sci. Total Environ.* **2019**, *658*, 922–935. [CrossRef] [PubMed]

 *remote sensing*

Article

Soil Salinity Assessment in Irrigated Paddy Fields of the Niger Valley Using a Four-Year Time Series of Sentinel-2 Satellite Images

Issaka Moussa [1,2], Christian Walter [1,*], Didier Michot [1], Issifou Adam Boukary [3], Hervé Nicolas [1], Pascal Pichelin [1] and Yadji Guéro [2]

[1]   SAS, INRAE, Institut Agro, 35000 Rennes, France; issakam1968@gmail.com (I.M.); didier.michot@agrocampus-ouest.fr (D.M.); herve.nicolas@agrocampus-ouest.fr (H.N.); pascal.pichelin@agrocampus-ouest.fr (P.P.)
[2]   Faculty of Agronomy, University Abdou Moumouni, Niamey PO Box 10896, Niger; yadjidjibril@yahoo.fr
[3]   Institut National de la Recherche Agronomique du Niger (INRAN), Niamey 8001, Niger; adamboukar@gmail.com
*   Correspondence: christian.walter@inrae.fr; Tel.: +33-223485439

Received: 31 August 2020; Accepted: 8 October 2020; Published: 16 October 2020

Abstract: Salinization is a major soil degradation threat in irrigated systems worldwide. Irrigated systems in the Niger River basin are also affected by salinity, but its spatial distribution and intensity are not currently known. The aim of this study was to develop a method to detect salt-affected soils in irrigated systems. Two complementary approaches were tested: salinity assessment of bare soils using a salinity index (SI) and monitoring of indirect effects of salinity on rice growth using temporal series of a vegetation index (NDVI). The study area was located south of Niamey (Niger) in two irrigated systems of rice paddy fields that cover 6.5 km^2. We used remote-sensing and ground-truth data to relate vegetation behavior and reflectance to soil characteristics. We explored all existing Sentinel-2 images from January 2016 to December 2019 and selected cloud-free images on 157 dates that covered eight successive rice-growing seasons. In the dry season of 2019, we also sampled 44 rice fields, collecting 147 biomass samples and 180 topsoil samples from January to June. For each field and growing season, time-integrated NDVI (TI-NDVI) was estimated, and the SI was calculated for dates on which bare soil conditions (NDVI < 0.21) prevailed. Results showed that since there were few periods of bare soil, SI could not differentiate salinity classes. In contrast, the high temporal resolution of Sentinel-2 images enabled us to describe rice-growing conditions over time. In 2019, TI-NDVI and crop yields were strongly correlated (r = 0.77 with total biomass yield and 0.82 with grain yield), while soil electrical conductivity was negatively correlated with both TI-NDVI (r = −0.38) and crop yield (r = −0.23 with total biomass and r = −0.29 with grain yield). Considering the TI-NDVI data from 2016–2019, principal component analysis followed by ascending hierarchical classification identified a typology of five clusters with different patterns of TI-NDVI during the eight growing seasons. When applied to the entire study area, this classification clearly identified the extreme classes (i.e., areas with high or no salinity). Other classes with low TI-NDVI (i.e., during dry seasons) may be related to areas with moderate or seasonal soil salinity. Finally, the high temporal resolution of Sentinel-2 images enabled us to detect stresses on vegetation that occurred repeatedly over the growing seasons, which may be good indicators of soil constraints due to salinity in the context of the irrigated paddy systems of Niger. Further research will validate the ability of the method developed to detect moderate soil salinity constraints over large areas.

Keywords: salinization; irrigated systems; Niger River basin; salinity index; vegetation index; TI-NDVI; Sentinel-2 images; high temporal resolution

---

1. Introduction

Irrigated agriculture covers 275 million ha worldwide (i.e., 20% of cultivated land) and accounts for 40% of world food production [1]. In the semi-arid area of the middle Niger valley (Niger) (Figure 1), irrigation techniques have been developed to respond to aridity and an increasing population [2]. Three factors have contributed to the rapid development of irrigation: (i) successive droughts from 1972 to 1973 and 1983 to 1984, which made people aware of serious risks to rainfed production; (ii) the high yields rapidly obtained in irrigated rice (*Oryza sativa*) and vegetable production; and (iii) commitment of the national government, farmers' organizations and several donors to irrigated agriculture. The first irrigated systems in Niger were constructed during the colonial period in the 1930s. From 1934 to 2011, 36 irrigated rice-growing systems were built along the middle Niger valley in Niger. Irrigation has improved the country's food security but has led to serious soil degradation by salinity. Irrigation facilities in many of these irrigated systems have aged and have not been renovated, and their drainage systems are not functioning. Excessive irrigation of less drained soil leads to waterlogging and soil salinization. The absence of a drainage network in clayey soils, which is characteristic of the study area [3], increases the concentration and precipitation of soluble salts at the soil surface but also in the subsoil and groundwater. Salt precipitates appear as white spots, mainly in the bright red oxidized clay horizons and in association with yellow iron spots. This salinization process threatens the sustainability of crop production in the study area by decreasing yields or reducing them to zero. The mechanisms by which soil salinity affects plant growth are generally known and have been summarized by many researchers [4–6]. In the middle Niger valley, rice is the main crop grown in irrigated fields. Rice can tolerate salinity without a reduction in yield up to a threshold of 3.0 dS/m of electrical conductivity of saturated paste (ECe) [7]. To preserve soil as a natural resource and maintain sustainable crop production in the study area, the spatial extent and level of salinity must be known. This can be done by performing field surveys to measure soil EC or electrical resistivity. Due to the large size of the Niger River basin (>85 km^2), however, doing so would require large amounts of time, labor and money. Using the potential of remote sensing along with other data sources is currently a promising method to map salinity at a large scale with high accuracy.

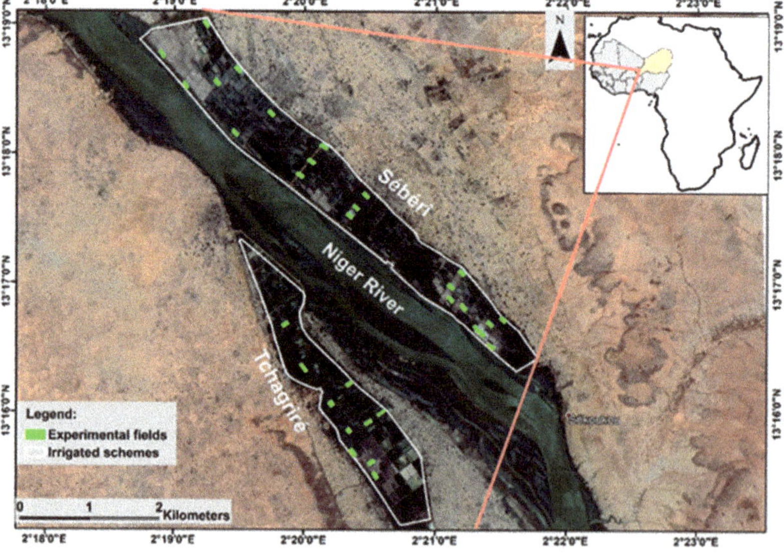

Figure 1. Location of the study area and aerial view of the experimental fields in the Sébéri and Tchagriré irrigation systems south of Niamey (Niger).

The use of remote sensing to monitor and map soil salinity dates to the 1990s [8]. Many researchers have used remote-sensing and ground-truth data (i.e., direct or indirect measurements of soil) to assess and monitor salinity [9,10]. A variety of remote-sensing data have been used to identify and monitor salt-affected areas: aerial photographs, visible and infrared multispectral images, video images, infrared thermography, microwave images and data collected by airborne geophysics and electromagnetic induction [11]. Two remote-sensing methods exist to identify soil salinity: (i) observing surface conditions of bare soil (i.e., salt efflorescence) and (ii) monitoring the behavior of vegetation affected by salinity over time [8]. Before the most recent generation of satellites was developed, many researchers used a classic single-date approach [12] because most older satellites passed over a given location too infrequently to enable monitoring. However, this single-date approach has some limits: although highly saline and non-saline soils can be detected easily, intermediate salinity classes are difficult to differentiate. Moreover, if vegetation is present, bare soil cannot be seen, and vegetation behavior cannot be monitored with a single-date approach.

The most recent generation of satellites (Sentinel-2, Landsat 8, SPOT 6 & 7) offers high spatial resolution and frequent passes over a given location, which makes identifying salinity variations and monitoring vegetation behavior appear possible. A variety of remote-sensing indices, such as the salinity index (SI) [13], brightness index, normalized difference salinity index and normalized difference vegetation index (NDVI) [14], have been developed to estimate the salinity of bare soil and monitor vegetation behavior in saline environments. These indices have been combined with ground-truth data. For instance, soil salinity was measured by electromagnetic induction and then combined with multi-year Landsat 7 reflectance data to map salt-affected soils in the western San Joaquin Valley, California, USA [9]. Electrical conductivity (EC) ground-truth measurements and various Sentinel-2 spectral parameters were used to create more reliable soil salinity maps in the Ebinur Lake region, Xinjiang, China [15]. To monitor salt-affected areas in Turkey, researchers performed multi-temporal monitoring of salinity using field EC measurements and spectral indices derived from multi-year Landsat 5 and 8 satellite images [16]. In practice, most studies have focused more on detecting severely salt-affected areas than on detecting and monitoring slightly or moderately affected areas.

This study aimed to develop a method to detect potential areas of soil salinity using multi-spectral and high-resolution Sentinel-2 satellite images combined with field data using two complementary approaches: (i) observing salinity of bare soil using the SI and (ii) monitoring vegetation behavior from 2016–2019 (eight growing seasons) in the arid zone of the Niger River basin. The study involved a four-year time series of Sentinel-2 remote-sensing images and field measurements of biomass and topsoil characteristics of cultivated rice fields.

2. Materials and Methods

2.1. Study Area

The study area corresponded to the irrigated systems of Sébéri and Tchagriré (6.5 km^2), located in the Niger River basin 50 km southeast of Niamey, the capital of Niger (13°16′35.32″ N, 2°21′31.81″ E) (Figure 1). The climate of the area corresponds to the dry tropical zone of the Sudano–Sahelian type. Annual precipitation has high spatial, temporal and interannual variability and a general trend towards a southward shift of isohyets over the past 30 years. Mean annual precipitation is ca. 510 mm/year (standard deviation (SD) = 100 mm/year) (National Meteorological Service of Niger). Precipitation is irregularly distributed in space and time: it peaks in August (150 mm) and is lowest in October (22 mm) and May (20 mm). Mean monthly temperature is 36 °C during the hottest period (April), with a maximum temperature of 47 °C. Minimum mean monthly temperatures are 25 °C and are observed from December to January. Evaporation varies from 1700 to 2100 mm/year. The water deficit is thus large during the dry season and is accentuated by the Harmattan, a dry continental trade wind from the Sahara. Consequently, the study area has two seasons: a dry season from October to May and a wet season from June to September [17].

Paddy fields in Sébéri and Tchagriré are located on the lowest alluvial terrace of the Niger Valley (Figure 1). The soils are Vertisols (60–74% clay in topsoil and 52–85% in subsoil) but also acidic, with a pH in water of 4.0–6.2 (SD = 0.5), an EC of 0.01–7.20 dS.m^{-1} (SD = 2.3 dS.m^{-1}), and they have low hydraulic conductivity at water saturation, which ranges from 2.8×10^{-8} in the topsoil horizon to 1.5×10^{-8} m s^{-1} at a depth of 50 cm [18,19]. In Sébéri, EC generally increases from the areas next to the sand dunes to the river's floodwater protection dike [19]. The types of salts found are hexahydrite, gypsum, epsomite and, secondarily, wattevilleite and sodium carbonate hydrate [19].

The observed long-term downward trend in rice yields can be explained by several factors: poor management of agricultural equipment, lack of technical supervision, soil degradation due to salinity and failure to respect the cropping calendar. Implementing a fixed and common cropping calendar for all rice-growing systems would create two growing seasons per year during optimal climatic conditions. It provides for two harvests, one in the dry cropping season (mid-November to mid-May) and the other in the wet cropping season (mid-June to mid-December) (Figure 2).

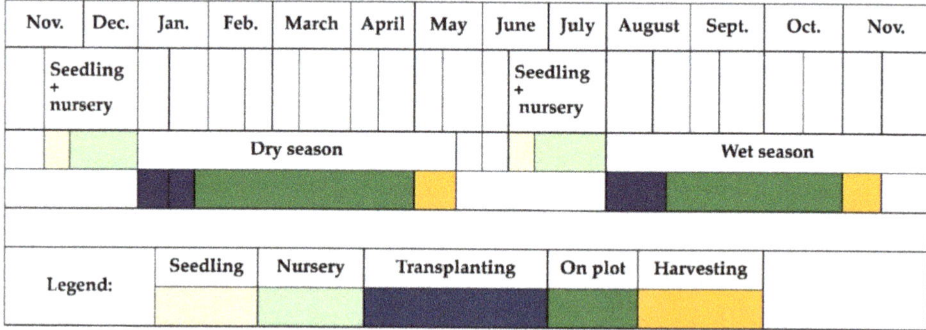

Figure 2. Double-cropping calendar for rice in the middle Niger valley (adapted from [20]).

2.2. Field Data Collection Strategy

In the Niger River basin, rice is grown in 0.25 ha (25 m × 100 m) fields. We selected 44 and 20 fields in the Sébéri and Tchagriré irrigated systems, respectively, for data collection on two dates in the 2019 dry season, each of which corresponded to a date when the Sentinel-2 satellite passed over the study area. We chose the 64 fields based on the level of salinity in the study area and whether they were cultivated with rice or bare. Three sampling plots of 1 m^2 each were set up on one diagonal of each of the 0.25 ha fields selected to collect soil and phenological parameters.

The first data collection campaign was performed on 8 February 2019, near the start of the growing season, when fields were already flooded and covered by a layer of irrigation water a few centimeters thick. The phenology of rice plants was assessed in a quarter (0.25 m^2) of each 1 m^2 plot. Then, all aerial biomass of this quarter was cut, weighed and dried under laboratory conditions to determine dry matter. The topsoil (0–30 cm) was sampled in each plot using an auger, and one composite sample per field was obtained by carefully mixing the three elementary plot samples. Given the rapidity of auger sampling and the very low hydraulic conductivity of these clay soils, sampling under a water layer did not significantly change the EC and pH values of the soil, as shown in previous studies performed in these irrigated systems [19]. Consequently, salt losses during soil sampling are very low, ensuring good reliability of collected samples and good representativeness of soil salinity measurements. The composite samples were air-dried and ground to pass through a 2-mm sieve. Then, pH in water (pH$_{1:5}$) and EC in water (EC$_{1:5}$) were analyzed in the laboratory following ISO 10390 and ISO 11265, respectively.

The second data collection campaign was performed on 4 May 2019, during the harvest period, when fields were not flooded. Phenological assessment and soil sampling were identical to those during the first campaign. Moreover, rice grain was collected from an undisturbed 0.25 m^2 quarter to estimate grain yield. Soil and phenological parameters were analyzed statistically for each field.

2.3. Remote-Sensing Data Collection

Remote-sensing data and in situ observations were processed in multiple steps (Figure 3). First, optical images were obtained from Sentinel-2 satellites. The Sentinel-2 mission is a constellation of two satellites that are 180° out-of-phase on the same orbit, which increases the frequency of passes over a given location (i.e., every five days). Each satellite records 13 bands, with three spatial resolutions: blue, green, red and near-infrared bands at 10 m resolution; red-edge and mid-infrared bands at 20 m resolution; and aerosol and water-vapor bands at 60 m resolution. Sentinel-2's utility thus lies in its high revisit frequency and high spatial resolution.

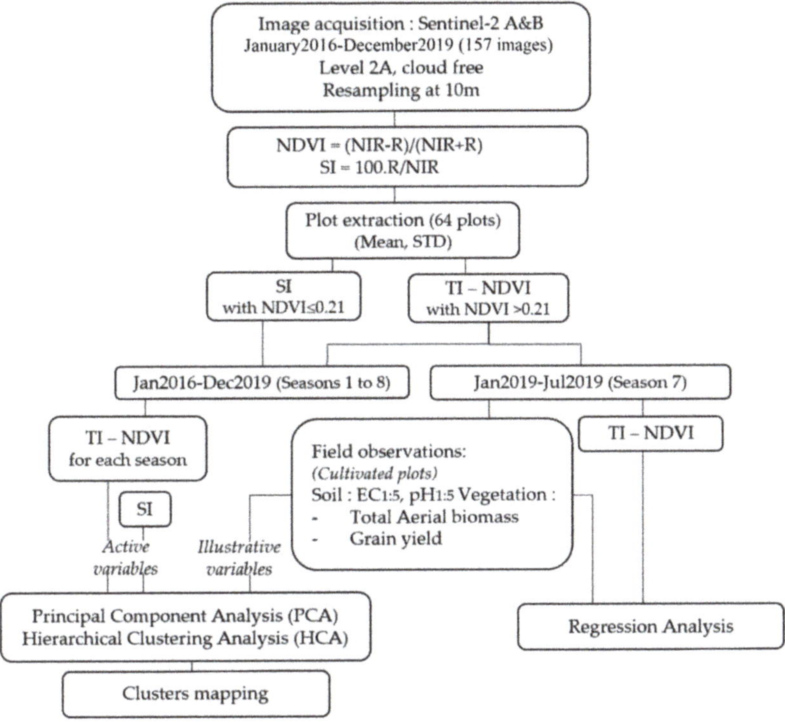

Figure 3. General data-processing flowchart.

All existing pre-processed Sentinel-2 images from January 2016 to December 2019 were downloaded. Pre-processing was performed by the MACCS (Multi-sensor Atmospheric Correction and Cloud Screening) processing chain, which consists of three successive steps: (i) cloud detection (using the satellite cirrus band), (ii) aerosol-thickness estimation and (iii) atmospheric correction. This processing chain, developed by CESBIO and CNES, provides ortho-rectified atmospheric-corrected surface reflectance images of 100 km × 100 km as a final product [21].

Overall, 157 preprocessed Sentinel-2 images were downloaded and then resampled at a spatial resolution of 10 m. Reflectance values that were missing due to cloud cover were estimated as the mean reflectance value of the two dates before and after the missing date. Finally, the study area of the irrigated systems of Sébéri and Tchagriré was extracted from the entire image.

A second step consisted of distinguishing the periods during which a plot was bare or covered with vegetation. To do this, the NDVI was first calculated per pixel (Table 1) [22], and then the mean and standard deviation of the NDVI of each plot's pixels was calculated. A pixel was considered to be part of a plot if its centroid fell within the plot's polygon. Since the plots measured 25 m × 100 m, each contained ca. 25 pixels. Based on the literature [23,24], plots with a mean NDVI ≤0.21 or >0.21 were considered as bare soil or vegetation, respectively.

Table 1. Characteristics and equations of the bare soil and vegetation indices selected. RED and NIR indicate red and near-infrared bands of the Sentinel-2 images.

Spectral Index	Equation	Characteristics
Salinity index (SI)	SI = (RED/NIR) × 100 in [25]	Created to detect saline soils.
Normalized difference vegetation index (NDVI)	NDVI = (NIR − RED)/(NIR + RED) [22] Varies from −1.0 to +1.0.For vegetation, NDVI varies from 0.2–0.8.	A standardized index for vegetation cover and chlorophyll activity. Used to monitor drought and monitor and predict agricultural production.

Means and standard deviations of NDVI were used to create a 4-year time series of NDVI for each plot. For the periods when the plots were estimated to be bare, we calculated an SI [25] (Table 1). Otherwise, for the periods when vegetation was dominant, we created a time series of NDVI over the four years of study.

A few days before rice planting and during the first part of the vegetation development cycle, a water layer covers the soil. This layer is thin, no more than a few centimeters thick, and has high turbidity. The NDVI of this water layer ranges from 0.10–0.21 before planting and then decreases as the rice is transplanted and the vegetation canopy develops. In comparison, NDVI values of the river near the plots are negative and close to −0.3.

The last step consisted of calculating the time-integrated NDVI (TI-NDVI) [26,27] for each growing season for fields with vegetation. TI-NDVI for a given growing season was calculated as follows:

$$TI - NDVI = \sum_{t=d1}^{de} \left[\frac{1}{n} \sum_{i=1}^{n} \left(\left(\frac{NDVI_{i,t} + NDVI_{i,t+1}}{2} \right) - 0.21 \right) \times (d(t+1) - d(t)) \right] \tag{1}$$

where $d1$ and de are the day of the year of the start and end (harvest) of the growing season (the same for all fields), respectively; $d(t)$ is the day of the year of date t for the set of days for which Sentinel-2 images are available during the growing season; n is the number of pixels within the field; and $NDVI_{i,t}$ is the NDVI of pixel i on date t. Units of TI-NDVI are expressed as NDVI.days.

Figure 4 illustrates the calculation of TI-NDVI for two successive growing seasons in 2019 derived from NDVI estimates based on existing Sentinel-2 images.

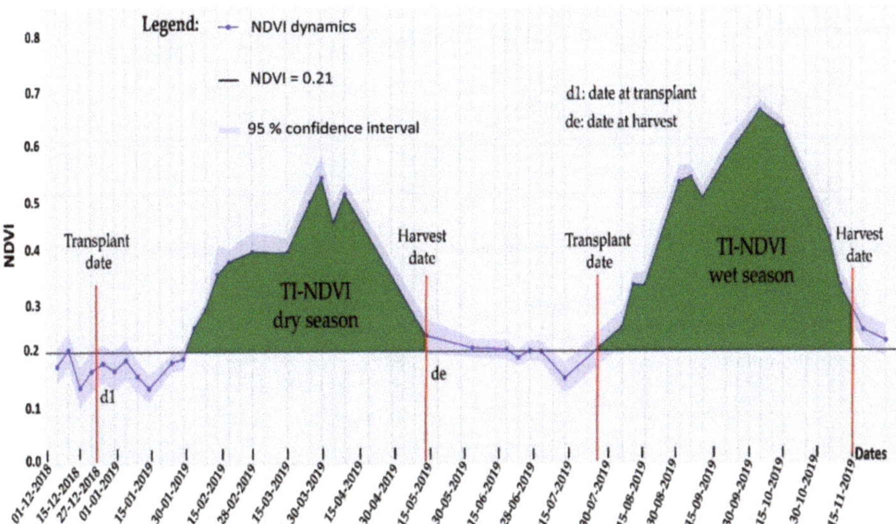

Figure 4. Example of normalized difference vegetation index (NDVI) dynamics for a given field for the dry and wet seasons of 2019 (derived from 42 cloud-free images among 60 existing Sentinel-2 images) and time-integrated NDVI (TI-NDVI) calculation for the two growing seasons.

2.4. Multidimensional Analysis of the Data

Principal component analysis (PCA) was performed using the FactoMiner package in R software [28] to analyze the multidimensional space of the remote-sensing and field data from 2019. Field data were pH and EC at the start and end of the growing seasons, aerial biomass and grain yields of the 64 fields. Qualitative variables were added: pH class, salinity class, land use and position in relation to the direction of river flow. Remote-sensing data were the mean SI of each field, TI-NDVI of each of the eight growing seasons and the number of seasons that each field was used to produce rice in dry and wet seasons. Remote-sensing data were considered as active variables and field data as illustrative variables.

PCA was followed by hierarchical clustering analysis (HCA) using the module HCPC (hierarchical clustering on principal components) [29] in R software [30] to create clusters of fields that behaved similarly. Finally, supervised classification for grids (SCG) was applied to 10-m resolution grids of the entire area that described the remote-sensing variables used in the PCA-HCA analysis (i.e., 8 grids with the TI-NDVI of each pixel for the 8 seasons and 1 grid with the SI of each pixel) to represent the spatial distribution of the clusters. Each pixel is assigned to the cluster with the shortest Mahalanobis distance, and the distance to the nearest cluster is analyzed to assess the quality of assignment to a cluster. SAGA-GIS 2.3.2 [31], implemented in QGIS 3.4.3, was used to perform the SCG with a maximum-likelihood algorithm.

3. Results

3.1. NDVI Dynamics over the Eight Growing Seasons

Figure 5 shows the dynamics of mean NDVI estimated from the 25 pixels within each of the three fields. The threshold of NDVI = 0.21 was selected to identify periods without live vegetation, and it enabled the identification of the short intercropping period after harvest when field irrigation was stopped, the soil plowed and irrigation started again before transplanting the next rice crop.

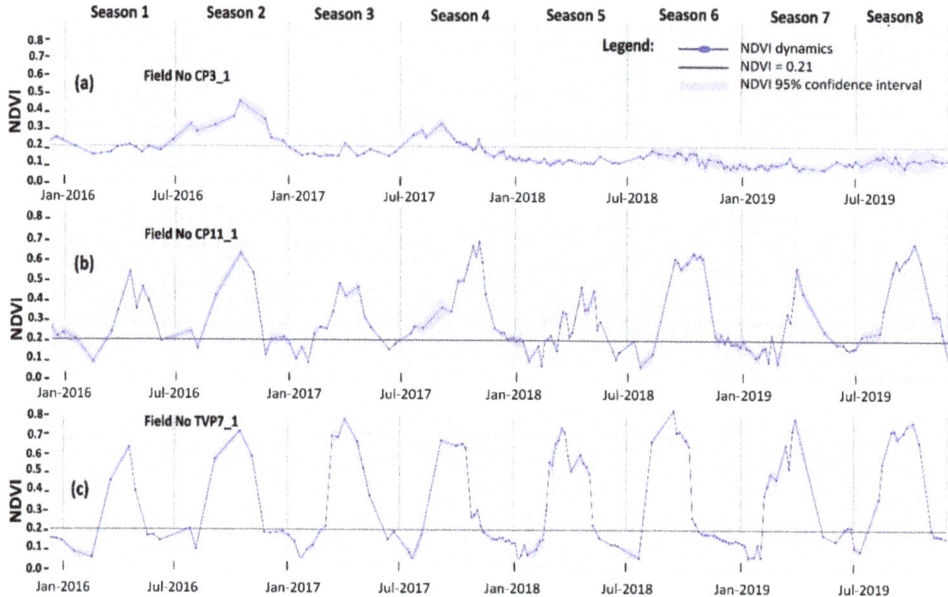

Figure 5. Example of dynamics of mean NDVI for three fields over eight growing seasons from January 2016 to December 2019: (**a**) Field CP3_1, cultivated in season 2 but not in the other seasons; (**b**) Field CP11-1, always cultivated but with lower NDVI in dry seasons than wet seasons; and (**c**) Field TVP7, always cultivated and with high NDVI in both dry and wet seasons. For each field, mean NDVI and its 95% confidence interval were estimated from the field's ca. 25 pixels.

Non-cultivated fields had NDVI dynamics below the threshold or slightly above due to weed growth (Figure 5a). When cultivated during a growing season, some fields had lower NDVI due to constrained plant growth, which could have been due in part to soil salinity. Among the fields cultivated during both dry and wet seasons, two groups of cultivated fields were identified: (i) those with lower NDVI during the dry seasons than wet seasons (Figure 5b) and (ii) those with smaller differences between dry and wet seasons and generally high NDVI (Figure 5c).

3.2. Spatial Variation in TI-NDVI over the Eight Growing Seasons

For the 8 seasons and 44 irrigated fields monitored in Sébéri, TI-NDVI was always significantly lower during dry seasons than wet seasons (Figure 6a): mean TI-NDVI per season ranged from 15 to 19 NDVI.days during the four dry seasons and 29 to 34 NDVI.days during the four wet seasons. TI-NDVI also varied greatly among fields for a given season and showed systematic trends, with values frequently lower in some fields in the south and northwest of the irrigation system and significantly higher in fields in the center (Figure 6b). This spatial variation was lower in some seasons (e.g., seasons 2 and 6) but higher in others, e.g., season 7 in 2019, when field data were collected.

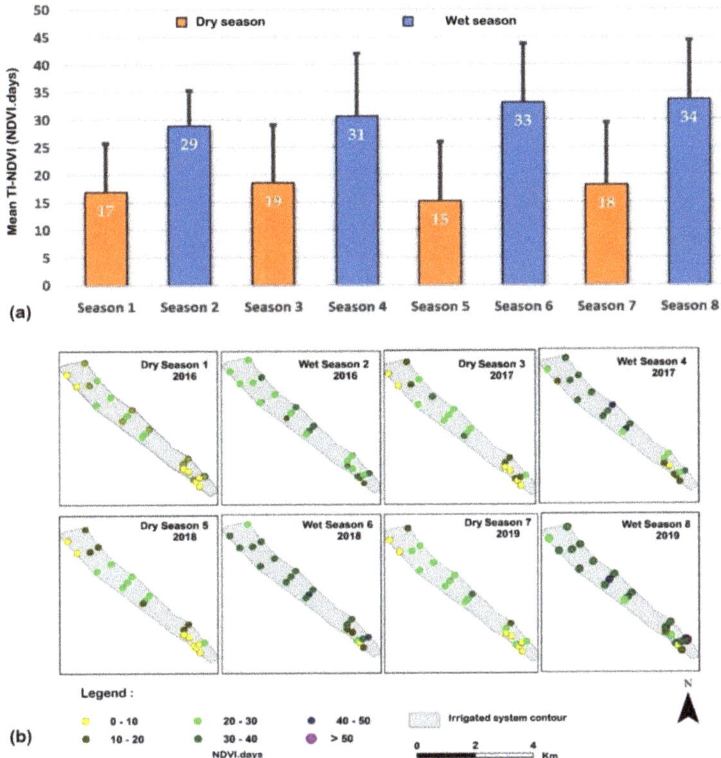

Figure 6. Temporal and spatial variations in the time-integrated normalized difference vegetation index (TI-NDVI) for the 44 fields of the Sébéri irrigation system: (**a**) mean and standard deviation of TI-NDVI for each season; (**b**) maps of TI-NDVI estimated for the eight growing seasons.

3.3. Salinity Measured in the Field in 2019

For the 64 fields monitored in 2019 during the dry season, $EC_{1:5}$ ranged from 0.01 to 5.36 dS/m at the start of the season and 0.44 to 6.19 dS/m at the end of the season (SD = 1.16 and 1.33 dS/m, respectively) (Table 2). The pH ranged from 5.52 to 6.68 in January and 5.29–6.25 in June (SD = 0.44 and 0.52, respectively). Differences between means and SDs for the two dates were not significant for $EC_{1:5}$ or pH. Total aerial biomass and grain yield at harvest varied greatly (Table 2), and 18 fields had no rice production.

Table 2. Statistics of field data for the dry growing season in 2019 (i.e., season 7) (n = 64, including 18 non-cultivated fields).

Statistic	Soil $EC_{1:5}$ (dS/m)		Soil $pH_{1:5}$		Total Aerial Biomass (g/m^2)	Grain Yield (g/m^2)
	Start of Season	At Harvest	Start of Season	At Harvest	At Harvest	At Harvest
Mean	0.40	0.44	5.52	5.29	914.2	389.5
Median	0.03	0.04	5.56	5.33	1048.5	420
SD	1.16	1.33	0.44	0.52	687.8	309.0
Min	0.01	0.01	4.45	4.00	0	0
Max	5.36	6.19	6.68	6.25	2269	1249

3.4. Correlation between Remote-Sensing Data and Field Data during the 2019 Dry Season

A strong and significant ($p < 0.05$) positive correlation was observed between TI-NDVI in season 7 and both total aerial biomass ($r = 0.77$, $p < 2 \times 10^{-14}$) and grain yield ($r = 0.82$, $p < 2.2 \times 10^{-16}$) (Table 3), which confirms that TI-NDVI is a good indicator of rice vegetation growth and its final yield. Soil $EC_{1:5}$ was negatively correlated with TI-NDVI at the start and end of the season ($r = -0.38$, $p = 0.002$ for the start, $p = 0.002$ for the end), with total biomass ($r = -0.23$, $p = 0.062$) and, significantly, with grain yield ($r = -0.29$, $p = 0.019$).

Table 3. Pearson correlations between the TI-NDVI indicator derived from Sentinel-2-images and soil ($EC_{1:5}$, $pH_{1:5}$) and crop indicators (total biomass, grain yield). Data were collected for 64 fields during the dry season of 2019 (i.e., season 7). SS and ES indicate the start and end of the growing season, respectively. Bold values indicate significant ($p < 0.05$) correlations.

	TI-NDVI	Soil $EC_{1:5}$ SS	Soil $EC_{1:5}$ ES	Soil pH SS	Soil pH ES	Total Aerial Biomass	Grain Yield
TI-NDVI	1.00						
Soil $EC_{1:5}$_SS	**−0.38**	1.00					
Soil $EC_{1:5}$_ES	**−0.38**	**0.99**	1.00				
Soil pH SS	**0.35**	**−0.63**	**−0.63**	1.00			
Soil pH_ES	0.16	**−0.62**	**−0.62**	**0.74**	1.00		
Total Aerial Biomass	**0.77**	−0.23	−0.23	**0.35**	0.1	1.00	
Grain Yield	**0.82**	**−0.29**	**−0.28**	**0.34**	0.07	**0.72**	1.00

3.5. PCA and HCA Analysis of Remote-Sensing Data over the Eight Growing Seasons

The first axis of the PCA was positively correlated with all of the TI-NDVI variables (Figure 7), which indicates that the spatial variability in TI-NDVI among fields was the main factor that influenced variations in the dimensional space. The second axis was correlated with SI but also distinguished the TI-NDVI estimates of the dry seasons (1, 3, 5 and 7) from those of the wet seasons (2, 4, 6 and 8).

While soil pH was little represented in the first PCA plane, soil $EC_{1:5}$ measured at the start and end of season 7 was negatively correlated with the first axis and thus with TI-NDVI variables. More surprisingly, neither soil $EC_{1:5}$ variable was correlated with SI. Following PCA, HCA identified five clusters of fields that behaved similarly during the eight growing seasons according to the remote-sensing data (Figure 8).

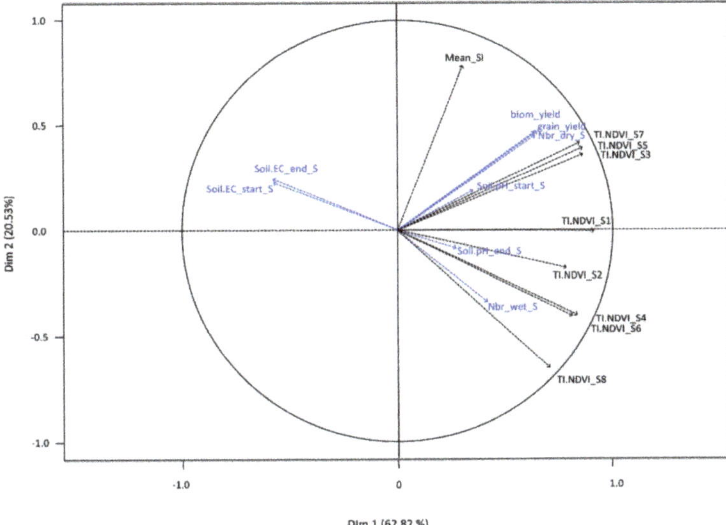

Figure 7. First plane of principal component analysis with variables derived from remote sensing as active variables (black) and soil and vegetation variables as illustrative variables (blue). Abbreviations: TI-NDVI_SN: time-integrated normalized difference vegetation index for season N; Mean_SI: mean salinity index; biom_yield: biomass yield at harvest; grain yield: grain yield at harvest; Soil EC_start_s: soil electrical conductivity at the start of the season; soil EC_end_s: soil electrical conductivity at the end of the season; soil pH_start_s: soil pH at the start of the season; Soil pH_end_season: soil pH at the end of the season; Nbr_dry_s: number of dry growing seasons during the time series; and Nbr_wt_s: number of wet growing seasons during the time series.

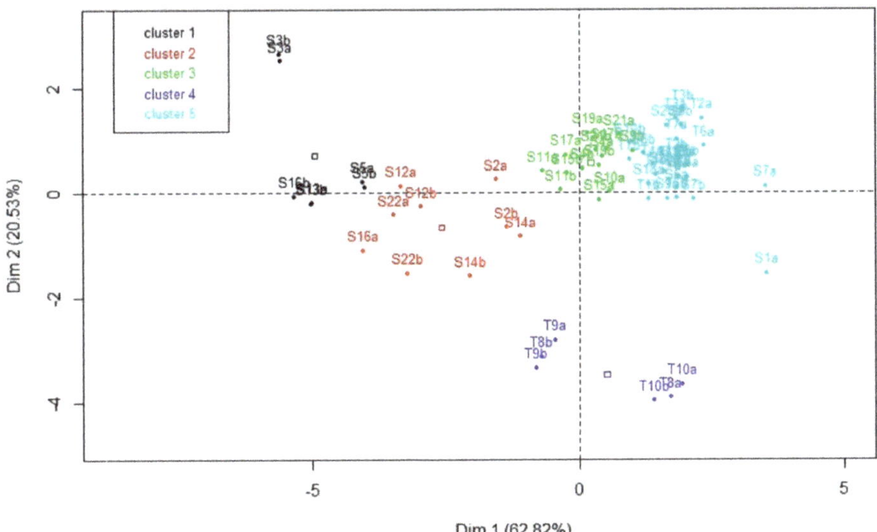

Figure 8. Representation on the first plane of principal component analysis (PCA) of the five clusters of fields defined by ascending hierarchical classification applied to the first three axes of the PCA. Points are labeled with field codes starting with S or T respectively for Seberi or Tchagrire irrigation system, while squares indicate barycenters of clusters.

3.6. Description of the Field Clusters

The five clusters differed in their remote-sensing and field data (Table 4):

- Cluster 1 had low TI-NDVI in both dry and wet seasons. In 2019, $EC_{1:5}$ was highest in this cluster, soil pH was very acidic and total biomass and grain yield were zero. The maximum $EC_{1:5}$ of some fields in this cluster (5.36 dS/m) indicates that this cluster had the highest salinity.
- Cluster 2 had low TI-NDVI in dry seasons, due to rare cultivation, and a higher TI-NDVI in wet seasons, but one that was still lower than those of clusters 3–5. In 2019, soil $EC_{1:5}$ was not significantly higher than those of clusters 3–5, but total biomass and grain yield were significantly lower.
- Cluster 3 had moderate TI-NDVI in dry and wet seasons and a significantly higher SI. In 2019, soil $EC_{1:5}$ was low, pH was relatively high, and mean total biomass and grain yield were the second highest among the clusters.
- Cluster 4 had low TI-NDVI in dry seasons due to frequent non-cultivation but high TI-NDVI in wet seasons. In 2019, soil $EC_{1:5}$ and pH were low, and the fields were not cultivated.
- Cluster 5 had extremely high TI-NDVI in both dry and wet seasons. In 2019, soil $EC_{1:5}$ was low, pH was relatively high and total biomass and grain yield were the highest.

Table 4. Mean (and standard deviation) per cluster of fields defined by ascending hierarchical clustering for the variables derived from Sentinel-2 images (TI-NDVI, SI) and from field data collected in 2019 ($EC_{1:5}$, pH, total biomass, grain yield). For each variable, different letters indicate significant differences in the mean among clusters according to a $p < 0.05$ Tukey test at a 95% confidence level.

		2016–2019			Dry Season 2019			
Cluster	No. of Fields	Dry Season TI-NDVI (NDVI.Days)	Wet Season TI-NDVI (NDVI.Days)	Mean SI	Soil $EC_{1:5}$ (dS/m)	Soil pH	Total Biomass (g/m^2)	Grain Yield (g/m^2)
1	7	1.0 (1.2) [a]	12.8 (4.4) [a]	72.9 (3.0) [ab]	2.6 (2.4) [b]	5.0 (0.6) [a]	0 (0) [a]	0 (0) [a]
2	9	5.2 (6.1) [b]	25.5 (3.1) [b]	72.5 (2.4) [a]	0.6 (1.1) [a]	5.5 (0.2) [bc]	314 (572) [a]	109 (217) [a]
3	14	19.7 (3.6) [d]	30.6 (1.8) [c]	74.8 (1.3) [b]	0.1 (0.0) [a]	5.7 (0.2) [c]	1308 (319) [b]	527 (179) [b]
4	6	10.1 (4.0) [c]	44.6 (5.0) [e]	70.1 (0.9) [a]	0.1 (0.0) [a]	5.2 (0.3) [ab]	0 (0) [a]	0 (0) [a]
5	28	25.4 (1.9) [e]	35.9 (3.1) [d]	75.0 (1.7) [b]	0.05 (0.1) [a]	5.6 (0.4) [bc]	1335 (414) [b]	592 (205) [b]

3.7. Mapping the Clusters over the Entire Study Area

Using the eight grids with the TI-NDVI of each pixel for the eight seasons and the grid with the SI of each pixel, SCG enabled each pixel to be attached to one of the five clusters defined from the 64 study fields considered as training areas. Figure 9 shows the distribution of the five clusters over the entire study area, with the associated distance to the nearest cluster, which indicates when the cluster represents a given pixel well (i.e., small distance) or poorly (i.e., large distance). Clusters 1 and 2, both of which had low TI-NDVI during the dry seasons since they were often not cultivated, were located mainly in the north and south of the Sébéri system (Figure 9). Cluster 5, with the highest TI-NDVI in both dry and wet seasons, predominated in the north of the Tchagriré system and the center of the Sébéri system. Cluster 3, with moderate TI-NDVI in both seasons, represented large areas in Sébéri in intermediate positions between clusters 1 or 2 and 5, but also in non-agricultural areas (e.g., paths, natural areas), which had large distances to the nearest cluster. Cluster 4 occupied small areas in Sébéri and the south of Tchagriré, often near fields of cluster 2.

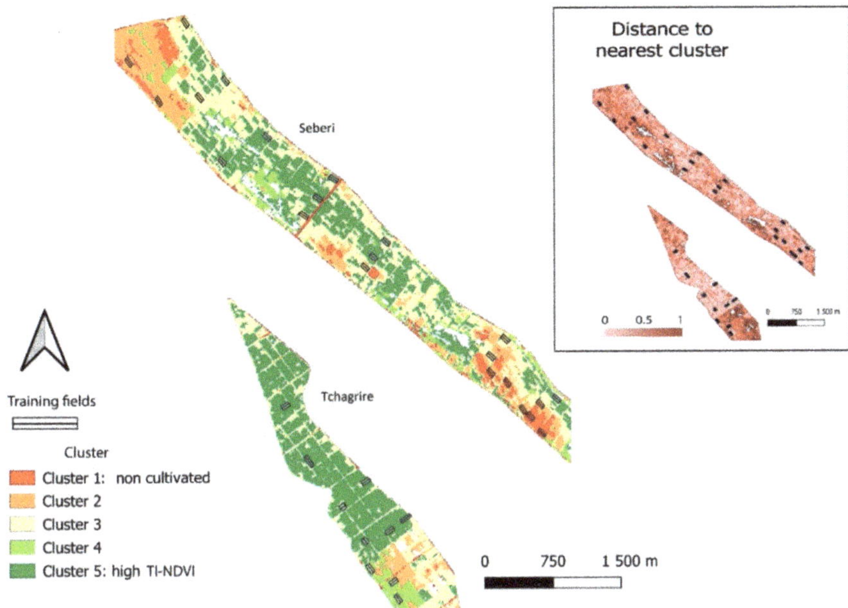

Figure 9. Mapping of the five clusters over the entire study area of the irrigated systems of Sébéri and Tchagriré using supervised classification that assigned each pixel to the nearest cluster.

4. Discussion

This study continuously monitored spatio-temporal dynamics of SI and NDVI in paddy fields using Sentinel-2 satellite images over eight growing seasons from 2016 to 2019. Based on field data collected (biomass, EC and pH) during the 2019 dry growing season, a relation between spectral indices (NDVI and SI) and field data was established to understand the behavior of rice crops and relate the spatio-temporal variation and pattern of spectral indices to salinity.

4.1. Variation in Spectral Indices among Crop Seasons

In the 2016–2019 time series, the spectral indices used (SI and NDVI) varied in different ways. First, SI was estimated at dates when bare soils prevailed according to an NDVI threshold (NDVI <0.21). SI averaged over the 4 years of study varied in a narrow range and was weakly correlated with soil $EC_{1:5}$; only highly saline areas, generally not cultivated, had higher SI values, due to the presence of salt crusts at the surface. These results may be explained by the short periods during which bare soil can be observed in irrigated paddy field systems; they generally last less than a month between successive crops, which corresponded to 3–5 dates when Sentinel-2 images were available. Thus, only a few dates when SI could be estimated were available. The soil may also be covered by crop residues or change drastically in water content due to the stopping or starting of irrigation after harvest or before preparatory work for planting rice. These factors may interfere with the estimation of SI and limit its ability to distinguish soil salinity.

In contrast, NDVI dynamics could be followed at a fine temporal resolution since cloud-free images were available at 133 dates during the eight growing seasons, with a mean of 21 dates per season since the end of 2017. We observed significant differences in TI-NDVI among the fields and eight growing seasons. TI-NDVI, which differed greatly between fields in a given season and over time, was used to differentiate areas with constraints to vegetation growth in the irrigated systems. In a given season, non-cultivated or cultivated areas with constraints had low TI-NDVI in dry and wet seasons (clusters 1 and 2) (Table 4). Areas cultivated throughout the year with few constraints had high TI-NDVI in wet seasons and moderate TI-NDVI in dry seasons (cluster 3), while zones cultivated in dry and wet seasons without any constraints had high TI-NDVI in both seasons (cluster 5) (Figure 8). Nonetheless, there were areas near the main drainage channel with salinity constraints and moderate TI-NDVI that were cultivated only in wet growing seasons. They may have occurred because they are always wet and wild grasses grow there in the dry seasons, which influences the NDVI. TI-NDVI also differed significantly between dry and wet seasons (Figure 5a,b). Wet growing seasons had higher mean TI-NDVI than dry ones. TI-NDVI varied from 29 to 34 NDVI.days in wet seasons and 15 to 19 NDVI.days in dry seasons. This result is due to the climatic conditions in the study area in dry seasons (hot, dry wind, high temperature) that limit crop growth. In addition, high temperature favors the rise of salt from lower horizons to the rooting zone of crops, which damages their roots. In wet seasons, conditions are more favorable for crop growth. Over the time series, wet and dry seasons varied among years, again due to climatic conditions (temperature in dry seasons and precipitation in wet seasons).

4.2. Temporal and Spatial Patterns of NDVI

Soil salinity influences vegetation density and crop growth, as explained by NDVI. TI-NDVI was used in this study to indicate constraints to crop growth and density, as in previous studies [32]. Five clusters were obtained to differentiate areas with constraints to rice growth in the two irrigated systems. Non-cultivated areas had the lowest TI-NDVI and were considered likely to be areas with high salinity constraints (cluster 1) (Figure 9). Areas cultivated only in wet seasons because of constraints had low TI-NDVI and may be considered to have moderate salinity constraints that limit crop growth (clusters 2 and 4) (Figure 9). Areas cultivated in both seasons with few constraints had high TI-NDVI and may be considered to have few salinity constraints (cluster 3) (Figure 9). Finally, areas cultivated without constraints had the highest TI-NDVI (cluster 5) (Figure 9) and were considered zones without salinity constraints. Our results show that the spatial pattern of TI-NDVI corresponds to the spatial distribution of problematic patches in the study area. The main constraint may be salinity, while other constraints could be explained by soil type, microtopography and farming practices. Mean ECe measured during ground truthing was calculated for each cluster and assigned to the respective TI-NDVI of the clusters to validate the four salinity classes obtained from the classification of TI-NDVI. Based on EC measurements, visual observations and knowledge of the terrain [33], mean TI-NDVI of the clusters were classified into four classes of soil salinity [34]: non-saline, slightly, moderately and very saline (i.e., ECe < 2, 2–4, 4–8 and 8–16 dS/m, respectively). Measured soil EC alone could not explain the level of salinity of the clusters (Table 4); however, integrated interpretation using the EC, mean TI-NDVI, total biomass and grain yield with visual observations of the study area could be used. Cluster 1, consisting of abandoned fields with low mean TI-NDVI in dry and wet seasons, no rice biomass or grain yield, maximum EC of 5.36 dS/m^2 and white efflorescence on the soil surface, can be considered saline soil. Clusters 2 and 4, consisting of fields cultivated only in wet seasons and rarely in dry seasons, despite having a lower level of surface salinity, have low TI-NDVI in dry seasons, low biomass and grain yield in the few fields cultivated in dry seasons and no yield in non-cultivated fields. Salt efflorescence is present in these fields. The maximum EC was 2.59 dS/m^2, which may not reflect the true EC of the clusters. These two clusters can be considered moderately saline. Clusters 3 and 5 consisted of cultivated fields in both dry and wet seasons. Based on measured EC, they were classified as non-saline, but since cluster 3 has lower TI-NDVI, biomass and grain yields than cluster 5, cluster 3 has few salinity constraints. Thus, cluster 3 can be considered slightly saline and cluster 5 non-saline.

4.3. Field EC Variation and Salinity

In the Séb́eri irrigated system, EC generally increased from the areas next to the sand dunes to the river's floodwater protection dike, perhaps because the dike has modified the functioning of the soils next to it [18]. In the Tchagriŕe irrigated system, EC generally increased from the areas next to the river to the main drainage ditch, likely due to topography. Areas at lower elevations had higher EC than those at higher elevations. Overall, the pattern of EC followed those of TI-NDVI, biomass and grain yield. EC measured in the two systems does not reflect the real situation in the field. Areas that showed signs of salinity (i.e., abandoned fields, low NDVI, low yields) had low measured EC, which indicates that a salt stock may lie in the lower horizons. To map salinity in this situation, measured EC should be compared to other maps (e.g., TI-NDVI, yield, soil, elevation) of the same site with similar sampling patterns or resolution. Doing so may provide useful insights into other parameters that could explain salinity.

5. Conclusions

This study developed a step-by-step method to estimate constraints on the growth of rice, the most important of which is salinity. Dense time series of Sentinel-2 images over eight growing seasons enabled us to describe the behavior of rice biomass and to differentiate fields where rice is subjected to stress during growing seasons. In irrigated systems, periods of bare soil were brief, and the SI derived from Sentinel-2 images could not differentiate soil salinity of fields. Monitoring vegetation behavior over four years by deriving the NDVI from Sentinel-2 images and calculating the TI-NDVI was able to differentiate fields based on constraints that limit rice growth. Several constraints can occur in areas subjected to stresses but can be related locally to soil salinity and verified by field sampling, which can be guided by TI-NDVI classification. This approach is particularly adapted to irrigated rice systems in which monoculture prevails and differences in TI-NDVI are not caused by different crops.

Author Contributions: I.M., C.W. and D.M. conceived the topic and developed the method. P.P. performed the technical work of downloading and processing images and calculating spectral indices. I.M., C.W. and D.M. wrote the manuscript, which was improved by H.N., Y.G. and I.A.B. participated actively in the fieldwork. All authors have read and agreed to the published version of the manuscript.

Funding: This study was made possible thanks to financial support of the EU-funded Agrocampus ERASMUS+ cooperation project in Rennes, France. This program financed a one-year's internship for M.I. and his supervision.

Acknowledgments: This study was performed with the administrative and technical support of the soil science laboratory from UMR SAS, INRAE and Institut Agro.

Conflicts of Interest: The authors declare no conflict of interest.

References

1.	World Water Assessment Programme (WWAP). Fact 24: Irrigated Land. United Nations Educational, Scientific and Cultural Organization (UNESCO). Available online: http://www.unesco.org/new/en/natural-sciences/environment/water/wwap/facts-and-figures/all-facts-wwdr3/fact-24-irrigated-land/ (accessed on 26 August 2020).

2.	Barbiéro, L.; Van Vliet-Lanoe, B. The alkali soils of the middle Niger valley: Origins, formation and present evolution. *Geoderma* **1998**, *84*, 323–343. [CrossRef]

3.	Guéro, Y. Organisation et Propriétés Fonctionnelles des sols de la Vallée du Moyen Niger. Ph.D. Thesis, Tunis University, Tunis, Tunisia, Niamey University, Niamey, Niger, 1987.

4.	Munns, R.; Cramer, G.R.; Ball, M.C. Interactions between Rising CO_2 Soil Salinity, and Plant Growth. In *Carbon Dioxide and Environmental Stress*; Luo, Y., Mooney, H.A., Eds.; Academic Press: Cambridge, MA, USA, 1999; pp. 139–167.

5.	Shrivastava, P.; Kumar, R. Soil salinity: A serious environmental issue and plant growth promoting bacteria as one of the tools for its alleviation. *Saudi J. Biol. Sci.* **2015**, *22*, 123–131. [CrossRef] [PubMed]

6.	Pessarakli, M.; Szabolcs, I. Soil Salinity and Sodicity as Particular Plant/Crop Stress Factors. In *Handbook of Plant and Crop Stress*, 2nd ed.; Pessarakli, M., Ed.; Marcel Dekker, Inc.: New York, NY, USA, 1999; ISBN 978-0-8247-1948-7.

7.	Maas, E.V.; Grattan, S.R. Crop Yields as Affected by Salinity. In *Agronomy Monographs*; Skaggs, R.W., van Schilfgaarde, J., Eds.; American Society of Agronomy: Madison, WI, USA, 2015; pp. 55–108. ISBN 978-0-89118-230-6.

8. Mougenot, B.; Pouget, M.; Epema, G.F. Remote sensing of salt affected soils. *Remote Sens. Rev.* **1993**, *7*, 241–259. [CrossRef]

9. Scudiero, E.; Skaggs, T.H.; Corwin, D.L. Regional scale soil salinity evaluation using Landsat 7, western San Joaquin Valley, California, USA. *Geoderma Reg.* **2014**, *2–3*, 82–90. [CrossRef]

10. Scudiero, E.; Skaggs, T.H.; Corwin, D.L. Comparative regional-scale soil salinity assessment with near-ground apparent electrical conductivity and remote sensing canopy reflectance. *Ecol. Indic.* **2016**, *70*, 276–284. [CrossRef]

11. Metternicht, G.I.; Zinck, J.A. Remote sensing of soil salinity: Potentials and constraints. *Remote Sens. Environ.* **2003**, *85*, 1–20. [CrossRef]

12. Nicolas, H.; Walter, C. Detecting salinity hazards within a semiarid contextby means of combining soil and remote-sensing data. *Geoderma* **2006**, *134*, 217–230.

13. Wang, J.; Ding, J.; Yu, D.; Ma, X.; Zhang, Z.; Ge, X.; Teng, D.; Li, X.; Liang, J.; Lizaga, I.; et al. Capability of Sentinel-2 MSI data for monitoring and mapping of soil salinity in dry and wet seasons in the Ebinur Lake region, Xinjiang, China. *Geoderma* **2019**, *353*, 172–187. [CrossRef]

14. Tripathi, N.K.; Rai, B.K.; Dwivedi, P. Spatial modeling of soil alkalinity in GIS environment using IRS data. In Proceedings of the 18th Asian Conference on Remote Sensing, Kuala Lumpur, Malaysia, 20–24 October 1997.

15. Khan, N.M.; Rastoskuev, V.V.; Sato, Y.; Shiozawa, S. Assessment of hydrosaline land degradation by using a simple approach of remote sensing indicators. *Agric. Water Manag.* **2005**, *77*, 96–109. [CrossRef]

16. Gorji, T.; Sertel, E.; Tanik, A. Monitoring soil salinity via remote sensing technology under data scarce conditions: A case study from Turkey. *Ecol. Indic.* **2017**, *74*, 384–391. [CrossRef]

17. Food and Agriculture Organization. *AQUASTAT Profil de Pays-Niger*; Organisation des Nations Unies pour l'Alimentation et l'Agriculture: Rome, Italie, 2015; p. 18.

18. Guéro, Y. Contribution à L'étude des Mécanismes de Dégration Physico-Chimique des sols Sous Climat Sahélien: Exemple pris Dans la Vallée du Moyen Niger. Ph.D. Thesis, Université de Niamey, Niamey, Niger, 2000.

19. Adam, I. Cartographie Fine et Suivi Détaillé de la Salinité des sols d'un Périmètre Irrigué au Niger en vue de leur Remédiation. Ph.D. Thesis, Uiversité de Bretagne Occidentale, Brest, France, Université Abdou Moumouni de Niamey, Niamey, Niger, 2011.

20. Ndanga Kouali, G.; Lévite, H.; Anid, M. *Diagnostic Participatif Rapide et Planification des Actions du Périmètre de Daïbéri (Département de Tillabéri-NIger)*; Technical Report; l'Association Nigérienne pour l'Irrigation et le Drainage (ANID): Niamey, Niger, 2010; p. 58.

21. Hagolle, O.; Huc, M.; Desjardins, C.; Auer, S.; Richter, R. *MAJA Algorithm Theoretical Basis Document*; Technical Report; CNES, CESBIO and DLR: Berlin, Germany, 2017.

22. Rouse, J.W.; Haas, R.H.; Schell, J.A.; Deering, D.W.; Harlan, J.C. *Monitoring the Vernal Advancement of Retrogradation of Natural Vegetation, Type III, Final Report*; NASA/GSFC: Greenbelt, MD, USA, 1974; pp. 1–137.

23. Shabou, M.; Mougenot, B.; Chabaane, Z.; Walter, C.; Boulet, G.; Aissa, N.; Zribi, M. Soil Clay Content Mapping Using a Time Series of Landsat TM Data in Semi-Arid Lands. *Remote Sens.* **2015**, *7*, 6059–6078. [CrossRef]

24. Roussillon, J. Développement de Méthodes Innovantes de Cartographie de L'occupation du sol à Partir de Séries Temporelles D'images Haute Résolution Visible (NDVI). Master's Thesis, Le Mans Université, Le Mans, France, 2015.

25. Allbed, A.; Kumar, L.; Aldakheel, Y.Y. Assessing soil salinity using soil salinity and vegetation indices derived from IKONOS high-spatial resolution imageries: Applications in a date palm dominated region. *Geoderma* **2014**, *230–231*, 1–8. [CrossRef]

26. Ma, C.; Guo, Z.; Zhang, X.; Han, R. Annual integral changes of time serial NDVI in mining subsidence area. *Trans. Nonferrous Met. Soc. China* **2011**, *21*, s583–s588. [CrossRef]

27. Reed, B.C.; Brown, J.F.; VanderZee, D.; Loveland, T.R.; Merchant, J.W.; Ohlen, D.O. Measuring phenological variability from satellite imagery. *J. Veg. Sci.* **1994**, *5*, 703–714. [CrossRef]

28. Husson, F.; Josse, J.; Pages, J. *Principal Component Methods—Hierarchical Clustering—Partitional Clustering: Why Would We Need to Choose for Visualizing Data?* Technical Report; Agrocampus: Angers, France, 2010.

29. Husson, F.; Josse, J.; Pagès, J. Analyse de données avec R-Complémentarité des méthodes d'analyse factorielle et de classification. In Proceedings of the 42èmes Journées de Statistique, Marseille, France, 25–29 May 2010.

30. R Core Team. European Environment Agency. Available online: https://www.eea.europa.eu/data-and-maps/indicators/oxygen-consuming-substances-in-rivers/r-development-core-team-2006 (accessed on 26 August 2020).

31. Conrad, O.; Bechtel, B.; Bock, M.; Dietrich, H.; Fischer, E.; Gerlitz, L.; Wehberg, J.; Wichmann, V.; Böhner, J. System for Automated Geoscientific Analyses (SAGA) v. 2.1.4. *Geosci. Model Dev.* **2015**, *8*, 1991–2007. [CrossRef]

32. Kundu, A.; Denis, D.; Patel, N.; Dutta, D. A Geo-spatial study for analysing temporal responses of NDVI to rainfall. *Singap. J. Trop. Geogr.* **2018**, *39*, 107–116. [CrossRef]

33. Abbas, A.; Khan, S.; Hussain, N.; Hanjra, M.A.; Akbar, S. Characterizing soil salinity in irrigated agriculture using a remote sensing approach. *Phys. Chem. Earth Parts A/B/C* **2013**, *55–57*, 43–52. [CrossRef]

34. Shahid, S.; Rahman, K. Soil Salinity Development, Classification, Assessment, and Management in Irrigated Agriculture. In *Handbook of Plant and Crop Stress*, 3rd Ed.; Pessarakli, M., Ed.; CRC Press: Boca Raton, FL, USA, 2011; Volume 20102370, pp. 23–39. ISBN 978-1-4398-1396-6.

Article

Framework for Accounting Reference Levels for REDD+ in Tropical Forests: Case Study from Xishuangbanna, China

Guifang Liu [1,2], Yafei Feng [1], Menglin Xia [1], Heli Lu [1,2,3,*], Ruimin Guan [1], Kazuhiro Harada [4] and Chuanrong Zhang [5]

[1] College of Environment and Planning/Key Research Institute of Yellow River Civilization and Sustainable Development & Collaborative Innovation Center on Yellow River Civilization of Henan Province, Henan University, Kaifeng 475004, China; guif@henu.edu.cn (G.L.); 104753190151@henu.edu.cn (Y.F.); xmllin@henu.edu.cn (M.X.); 490707753@henu.edu.cn (R.G.)
[2] Key Laboratory of Geospatial Technology for the Middle and Lower Yellow River Regions (Henan University), Ministry of Education/National Demonstration Center for Environment and Planning, Henan University, Kaifeng 475004, China
[3] Henan Key Laboratory of Earth System Observation and Modeling, Henan University, Kaifeng 475004, China
[4] Graduate School of Bioagricultural Sciences, Department of Forest and Environmental Resources Sciences, Nagoya University, Nagoya 4648601, Japan; harada@agr.nagoya-u.ac.jp
[5] Department of Geography & Center for Environmental Sciences and Engineering, University of Connecticut, Storrs, CT 06269-4148, USA; chuanrong.zhang@uconn.edu
[*] Correspondence: luheli@henu.edu.cn; Tel.: +86-037123881850

Abstract: The United Nations' expanded program for Reducing Emissions from Deforestation and Forest Degradation (REDD+) aims to mobilize capital from developed countries in order to reduce emissions from these sources while enhancing the removal of greenhouse gases (GHGs) by forests. To achieve this goal, an agreement between the Parties on reference levels (RLs) is critical. RLs have profound implications for the effectiveness of the program, its cost efficiency, and the distribution of REDD+ financing among countries. In this paper, we introduce a methodological framework for setting RLs for REDD+ applications in tropical forests in Xishuangbanna, China, by coupling the Good Practice Guidance on Land Use, Land Use Change, and Forestry of the Intergovernmental Panel on Climate Change and land use scenario modeling. We used two methods to verify the accuracy for the reliability of land classification. Firstly the accuracy reached 84.43%, 85.35%, and 82.68% in 1990, 2000, and 2010, respectively, based on high spatial resolution image by building a hybrid matrix. Then especially, the 2010 Globeland30 data was used as the standard to verify the forest land accuracy and the extraction accuracy reached 86.92% and 83.66% for area and location, respectively. Based on the historical land use maps, we identified that rubber plantations are the main contributor to forest loss in the region. Furthermore, in the business-as-usual scenario for the RLs, Xishuangbanna will lose 158,535 ha ($158,535 \times 10^4$ m^2) of forest area in next 20 years, resulting in approximately 0.23 million t (0.23×10^9 kg) CO$_2$e emissions per year. Our framework can potentially increase the effectiveness of the REDD+ program in Xishuangbanna by accounting for a wider range of forest-controlled GHGs.

Keywords: reference levels; REDD+; greenhouse gas emissions; Xishuangbanna; monitoring and reporting

Citation: Liu, G.; Feng, Y.; Xia, M.; Lu, H.; Guan, R.; Harada, K.; Zhang, C. Framework for Accounting Reference Levels for REDD+ in Tropical Forests: Case Study from Xishuangbanna, China. *Remote Sens.* **2021**, *13*, 416. https://doi.org/10.3390/rs13030416

Received: 30 December 2020
Accepted: 20 January 2021
Published: 26 January 2021

Publisher's Note: MDPI stays neutral with regard to jurisdictional claims in published maps and institutional affiliations.

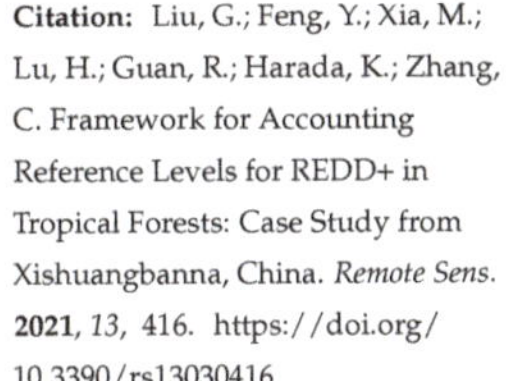

1. Introduction

Forests account for almost half of the global terrestrial carbon pool, and the vegetation within them alone (excluding soils) holds approximately 75% of all living carbon. The total carbon content in forest ecosystems is estimated to be 638 Gt [1–5]. Tropical forests play a particularly important role in the global carbon budget because they contain as much carbon in their vegetation and soils as all the temperate-zone and boreal forests combined [6–12]. Per unit area, tropical forests store, on average, approximately 50%

more carbon than their nontropical counterparts. Scientists agree that to achieve the goals of the United Nation's Framework Convention on Climate Change (UNFCCC), namely avoiding irreversible damage to the climate system, global warming must not exceed 2 °C [13–16]. However, concentrations of CO_2 in the atmosphere are already so high that global emissions will likely peak before they start to decline. Thus, in order to remain under the above-mentioned threshold, emissions from all major sources (i.e., from developed countries, major developing country emitters, and deforestation) must begin to decline within the next decade [17–19].

The Conference of the Parties (COP) agreed that Reducing Emissions from Deforestation and Forest Degradation (REDD+) with the enhancement of the removal of greenhouse gas (GHG) emissions by forests in developing countries could support the goals of the framework through the positive incentives provided by the UNFCCC. The general consensus at Doha in 2012 following last year's COP17 was that the results of financing, safeguards, measurements, and reporting and verification for REDD+ were mixed. In addition, significant progress has been widely recognized as having been made only within the technical arena relating to reference levels (RLs). The 19th Conference of the Parties to the UNFCCC (COP19) and the 9th Conference of the Parties to the Kyoto Protocol (CMP9) were jointly held on the topic of REDD+ funding in Warsaw, Poland, and in-depth discussions on the action points were conducted.

GHG-based compensation for REDD+ requires an agreement on emission RLs. Key elements for setting these RLs include the ability to measure changes throughout all forested areas, the use of consistent methodologies at repeated intervals to obtain accurate results, and the verification of results with ground-based or very high-resolution observations [20–24]. RLs have profound implications for the effectiveness of climate-related policies, cost efficiency, and distribution of REDD+ financing, and they involve a number of tradeoffs [25–30]. In this paper, referring to the business-as-usual scenario, we introduce a methodological framework for setting RLs for REDD+ applications in tropical forests in Xishuangbanna, China, by coupling the Good Practice Guidance (GPG) on Land Use, Land Use Change, and Forestry published by the Intergovernmental Panel on Climate Change (IPCC) and land use scenario modeling. This study contributes to the literature by highlighting key challenges for setting RLs as part of the REDD+ program.

2. Data and Methodology

2.1. Research Area

Not only is the forest in the Xishuangbanna region (Figure 1) the world's largest preserved area located in the northernmost part of the Earth, but it is home to the majority of tropical forest ecosystems in China as well. The geology, climate, and soil of Xishuangbanna are suitable for the growth and reproduction of various organisms. Moreover, 4500 species of higher plants have been recorded in Xishuangbanna, accounting for about one-seventh of the total number of higher plants in China. The native vegetation types include those found in tropical rain forests, montane rain forests, tropical monsoon forests, subtropical evergreen broad-leaved forests, deciduous broad-leaved forests, warm coniferous forests, and bamboo forests as well as shrubs and grasses [31–34]. In recent years, due to the increase in the population, intensification of anthropogenic activities, the enabling climate, and suitable terrain conditions in the area, the cultivation of rubber and other tropical and economically important crops has risen rapidly. Thus, the changes to the forest have been very dramatic.

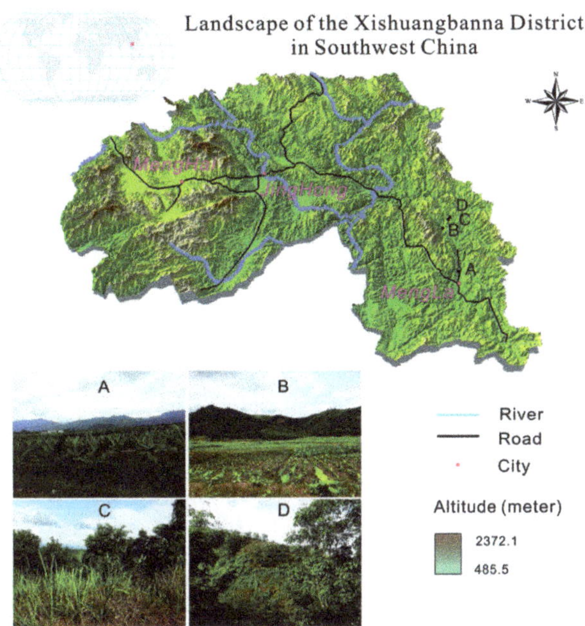

Figure 1. Landscape of the Xishuangbanna, China.

2.2. IPCC's Good Practice Guidance

The IPCC's existing GPG for Land Use, Land Use Change, and Forestry provides the recommended approach to account for fluctuations in carbon stocks resulting from changes in the use and management of forests. This framework has been accepted by all Parties in the Bali Action Plan of COP13 [35–37]. The IPCC's GPG framework refers to two basic inputs for forest carbon accounting, namely activity data and emission factors. Activity data in the REDD+ context refer to the areal extent of emissions. For example, in the context of deforestation, activity data refer to the area of deforestation, presented in hectares (10^4 m^2) over a known time period. Emission factors refer to the emission or removal of GHGs per unit activity. The emission or removal of GHGs resulting from land use conversion ultimately alters ecosystem carbon stocks.

2.2.1. Emissions Factors

To estimate emissions factors, the required number of sample plots was determined to the necessary accuracy using the size of the forest area and other available resources. Provisional surveys and/or existing data can be utilized to establish sample sizes, and tools also exist to calculate sample sizes based on fixed precision levels or given fixed inventory costs [38–41]. In the event carbon stocks and flows are to be monitored over the long term, permanent sites should be considered in order to reduce between-site variability and to capture actual trends as opposed to short-term fluctuations [42].

In the study, the aboveground biomass density map was sourced from Global Forest Watch (http://www.globalforestwatch.org/). This map is a global aboveground biomass density map produced in 2000 according to the method devised by Baccini [43]. Based on the improved methodology, the resolution can be increased to as much as 30 m. The aboveground biomass density map of Xishuangbanna region was extracted using the mask extraction method.

More recently, Maurizio Santoro [44] have proposed an integration methodology for estimation of aboveground biomass density for around the year 2010 by combining SAR, LiDAR, and optical observations together with other datasets such as auxiliary datasets from forest inventories, additional remote sensing observations, climate variables, and ecosystems classifications. We compared with the latest biomass map developed by Santoro against Global Forest Watch product.

2.2.2. Activity Data

Estimation of activities associated with national-level deforestation monitoring is practically possible only via remote sensing [45–49]. Since the early 1990s, changes in forest area have been monitored from space with confidence. Some countries have had well-established operational systems for over a decade.

Taking into account the availability of the data and their matching, this study used the Enhanced Thematic Mapper Plus/Thematic Mapper (ETM+/TM) remote sensing images related to Path number 130 and Row number 044, Path number 131 and Row number 045, Path number 130 and Row number 045, Path number 129 and Row number 045 in the study area in 1990, 2000, and 2010 to interpret land use changes(Table 1). The TM/ETM+ remote sensing images and normalized difference vegetation index (NDVI) data were sourced from the US Geological Survey (USGS, http://earthexplorer.usgs.gov/). We mosaiced the TM/ETM+ remote sensing images of the same year, used ENVI to perform geometric correction and radiometric correction, converted all the map data projections to WGS84/UTM Zone47N (EPSG: 32647), used the Xishuangbanna administrative vector map for mask extraction, and performed cropping to obtain the images of Xishuangbanna.

Table 1. Landsat imagery used in this study.

Satellite (Sensor)	Path Number and Row Number	Time	Resolution/m	Cloudiness/%
Landsat5 TM	130/044	1990-01-06	30/120	3.8
Landsat5 TM	131/045	2000-03-13	30/60/15	0.02
Landsat7 ETM+	129/045	2010-04-04	30/60/15	1.59

The vegetation in the study area changes obviously with the seasons, and the NDVI values at different times, thus, have a greater influence on the research results. Thus, this study synthesized the maximum value of the NDVI data in multiple phases of the same year and also eliminated the influence of cloud cover on the research results.

According to the characteristics of land use cover in the study area (Table 2), we divided the land use cover into eight types: forestland, shrubland, grassland, cultivated land, rubber forest, tea gardens, construction land, and water. The training samples were determined using QuickBird images in Google Earth. The terrain of the study area is relatively complex, and many "homogeneous spectrum" phenomena occur in the interpretation of remote sensing images. To avoid this phenomenon, it is necessary to select as many training samples as possible. Different band combinations of Landsat7 ETM images have different characteristics. The selection of the training samples was carried out according to these characteristics. The ETM541 band combination is helpful for distinguishing different vegetation types when supplemented by NDVI data. The training samples of natural forests, shrubs, rubber plantations, and tea gardens were selected. The ETM453 band combination was used to select the cultivated land and water, while the construction land was extracted through the ETM743 band combination, and the remainder was categorized as other land. Supervised classification was performed using the selected training samples to obtain the preliminary classification results, and the accuracy test was conducted. If the results did not agree, the training samples were reselected, and the supervised classification and accuracy tests were reperformed until they were ideal. Finally, the classification results were recoded, clustered, and eliminated, and the broken patches were merged into the adjacent largest classification to unify the smallest unit.

Table 2. Land use classification followed for this study.

Type Code	Land Use Type	Land Use Type Interpretation
1	Forestland	Primary and secondary forests
2	Shrubland	Forest coverage is less than 20%
3	Grassland	Less than 20% is covered by shrub, and grass is predominant
4	Cultivated land	Paddy fields and irrigated land
5	Rubber plantations	Man-made rubber plantations
6	Tea gardens	Man-made tea plantations
7	Construction land	Residential building land in urban and rural areas
8	Water	Water body

The Land Use Dynamic Index considers the transfer of land use types during the study period, and reflects the intensity of regional land use changes during this time. It is essential to find hot spots of land use changes at different spatial scales. It is one of the important parameters to analyze the dynamic changes in land use space [50]. Equation (1) was used to calculate the index.

$$K_i = \frac{U_{t_1} - U_{t_2}}{U_{t_1}} \times \frac{1}{t_2 - t_1} \times 100\% \tag{1}$$

K_i is the land use dynamic degree for land use type i in a certain period of time, U_{t_1} and U_{t_2} are the number of certain land use types at the start of the period t_1 and its end t_2, respectively, and $t_2 - t_1$ is the research duration.

$$S = \left[\sum_{i=1}^{n} \left(\frac{\Delta S_{i-j}}{S_i} \right) \right] \times 100 \times \frac{1}{t} \times 100\% \tag{2}$$

S is the comprehensive land use dynamic degree in the study area corresponding to t time period. ΔS_{i-j} is the area of land use type converted i converted to other land use types in the study period; S_i is the area of type i land use type at the beginning of the study; t is the time period of land use change.

2.3. Land Use Scenario Modeling for Reference Levels

Land use simulation is based on years of known land use changes. It predicts future land use changes. Most land use models used to simulate the process of land use change typically need to solve two problems: the quantity problem and the distribution problem. The quantity problem refers to how much of the land area has changed, while the distribution problem involves pinpointing where those land changes occurred. This study applied the Land Change Modeler (LCM) [51–53], which uses the Markov chain model to predict the number of future land use changes, and then calculates the distribution location of these changes according to the Multilayer Perceptron (MLP) model.

Markov chain is a kind of "no after-effect" random stored procedure, as it assumes that the state of the current variable is only related to its previous state, not to its states at other moments. Therefore, it has good operability and is used in the simulation of various land use changes. In Equation (3) of Markov chain, for any positive integer n and possible states $i_0, i_1, ..., i_n$ of the random variables,

$$P(X_n = i_n | X_{n-1} = i_{n-1}) = P(X_n = i_n | X_0 = i_0, X_1 = i_1, ..., X_{n-1} = i_{n-1}) \tag{3}$$

As the land use change conforms to the basic characteristics of the Markov process, it can be regarded as a Markov process. Therefore, the Markov chain analysis can describe the land use change process and predict the future land use change trend. It is an important transformation tool in land use change modeling. However, the following prerequisites must be fulfilled [54–56]: (1) In a certain area, different types of land use should be transformable into each other, (2) the conversion between different types of land use can include many events, which are difficult to describe with a specific formula, and (3) within the time limit of the study, the conversion status of the land use structure is relatively stable, which meets the requirements of the Markov chain. Moreover, the area ratio of the mutual conversion between the types of land uses equals the state transition probability.

MLP is a very widely used neural network in remote-sensing image processing, especially remote sensing image classification. MLP was used in the model primarily to calculate the land use change potential, that is, the future conversion probability between each land use type. The process involved analyzing future land use by establishing a land use driving force model and the quantitative relationship between each land use type to assess the probability of change. Based on the calculated potential distribution of soil use changes, the location of possible future land use changes can be determined. The back propagation algorithm used in MLP consists of two parts, namely the forward propagation of information and the backward propagation of errors. In the forward propagation process, the input information is calculated from the input layer through the hidden layer to the output layer, and each layer for the state of a neuron only affects the state of the next layer of neurons [57,58]. If the expected output is not obtained in the output layer, the error change value of the output layer is calculated, and then turned to reverse propagation, and the error signal is returned back along the original connection path through the network to modify the weights of neurons in each layer until the desired target is reached. During the forward propagation process, the state of the activated neuron is updated layer-by-layer from the input layer to the output layer, as shown in Equation (4):

$$x_j = \sum_i a_i w_{ji}. \tag{4}$$

x_j represents the total input received by neuron j, w_{ji} represents the weight between neurons j and i, a_i denotes neuron i once x_j is calculated. The most commonly used mapping function is the S (sigmoid) function, as shown in Equation (5).

$$a_j = f(x_j) = \cfrac{1}{1 + \cfrac{1}{\exp\left(\frac{x_j}{T}\right)}}. \tag{5}$$

It is crucial to check the accuracy and effect of the model to determine whether the model needs to be adjusted. The Receiver Operating Characteristic (ROC) curve test evaluates the model by comparing the predicted land change probability distribution map with the actual changed 0–1 map (the changed land value is 1, and the unchanged land value is 0) [59–61]. This step converts the simulated and reference images into a 2×2 table, with each table corresponding to a different threshold. The number of pixels within the thresholds of A, B, C, and D create the statistical figure for each ROC curve threshold. The following data are produced; x and y form the point (x, y), where x is the ratio of classifications labeled as true−, namely D/(B + D), and y is the proportion of true+ classifications, that is, A/(A + C). In order to be expressed as a positive value on the x axis, the opposite part of true− is generally represented by B/(B + D). Thus, the ROC curve test provides the Area Under the ROC Curve (AUC), which is obtained using the following formula:

$$\text{AUC} = \sum_{i=1}^{n}(x_i - x_{i+1}) \times \left\{ y_i + \frac{y_{i+1} - y_i}{2} \right\} \tag{6}$$

where x_i refers to x for each threshold i, that is, B/(B + D), and y is calculated using D/(B + D).

3. Results and Discussion

3.1. Analysis of Historical Land Use

The land use maps in 1990, 2000, and 2010 and the accuracies are shown in Figure 2 and Appendix A. Firstly we randomly generated 2866, 2549, and 2481 sample points in 1990, 2000, and 2010 through hierarchical random sampling method. There were 1520 sample points, 1227 sample points, and 1008 sample points in forest area in 1990, 2000, and 2010, respectively. Then we evaluated the accuracy of classification for sample points based on Google earth.

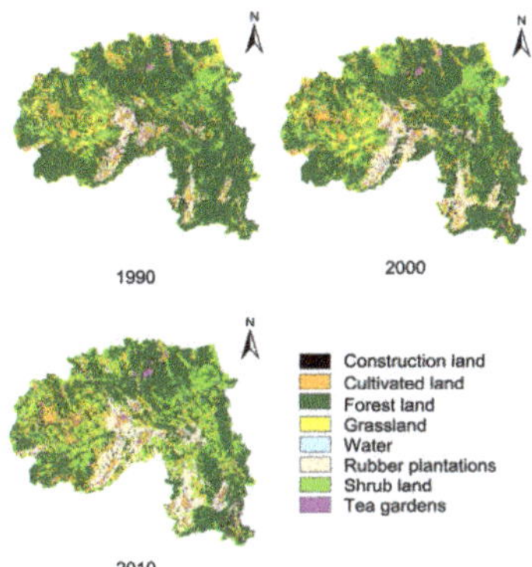

Figure 2. Land use maps and the accuracies in 1990, 2000, and 2010.

GlobeLand30, which was developed by the National Geomatics Center of China, is an open-access 30m resolution global land cover data product with an overall classification accuracy of over 80% [62,63]. We compared the area and spatial location of the forest land in 2010 extracted by Globeland30 with those in this study (Figure 3). Firstly, about 600 sample points are randomly generated within Xishuangbanna administrative region. Then these sample points are overlapped with Globeland30 and land use map respectively. Finally we evaluate the accuracy of land use map based on the consistency of forest land and nonforest land in Globeland30.

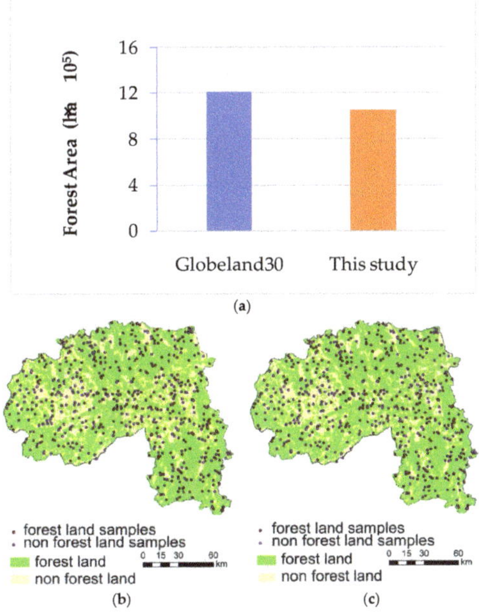

Figure 3. (**a**) Forest area in 2010 from Globeland30 and this study; (**b**)spatial distribution of the validation samples in Globeland30; (**c**) spatial distribution of the validation samples in this study.

In terms of the forest land area, the forest land area extracted from Globeland30 was 1.21×10^6 ha, and that from this study is 1.05×10^6 ha, with the accuracy 86.92%. In terms of spatial location of the forest land, among 600 randomly generated sample points, 376 were the forest land and 230 were nonforest land in Globeland30; in comparison, 331 sample points were the forest land and 275 sample points were nonforest land in this study. The overall accuracy is 83.66%, and the kappa coefficient is 0.657.

The areas, changes, and dynamics of the three types of land use in 1990, 2000, and 2010 are shown in Figure 4.

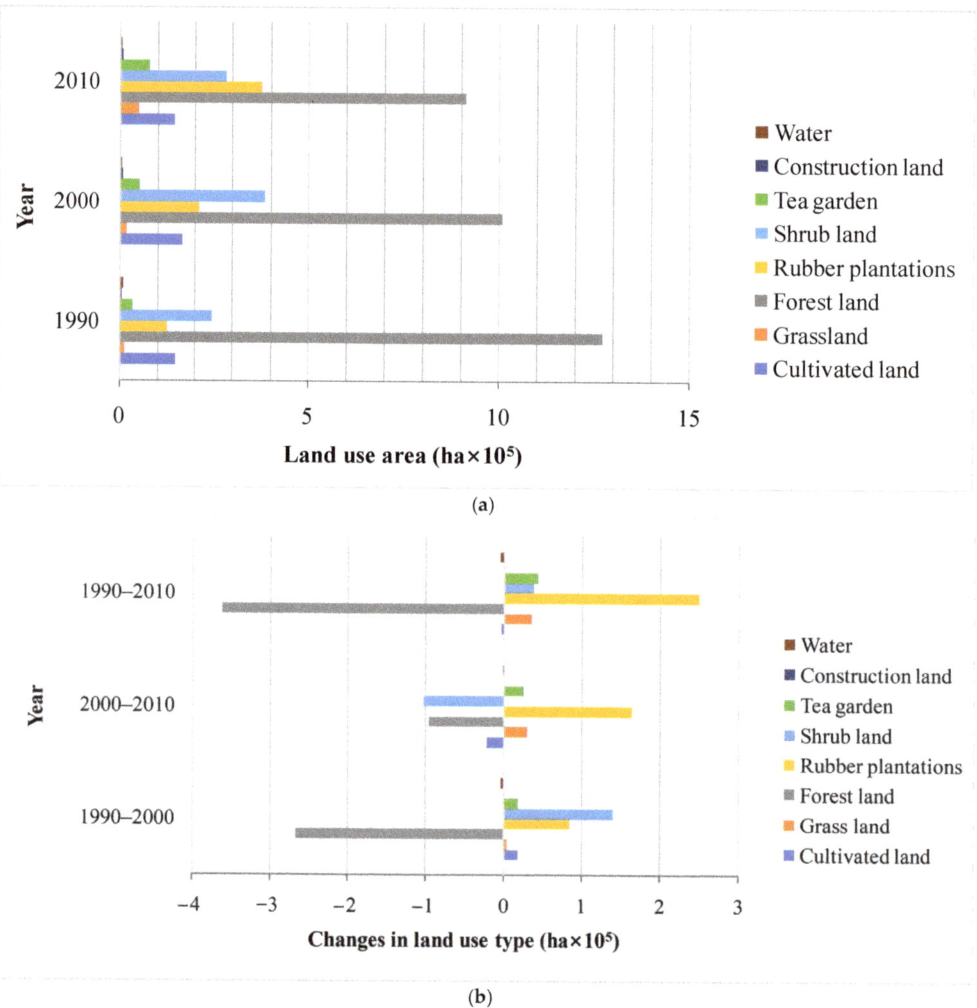

Figure 4. *Cont.*

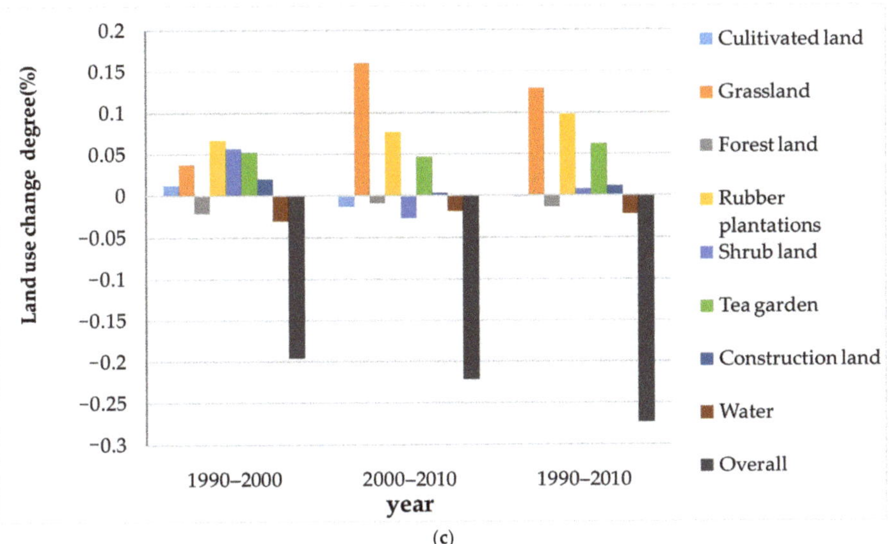

Figure 4. (**a**)Land use areas in 1990, 2000, and 2010, (**b**) changes in land use type, and (**c**) land use change degree in Xishuangbanna between 1990 to 2000, 2000 to 2010, and 1990 to 2010.

As shown in Figure 4, the areas under cultivated land, forested land, and water bodies in Xishuangbanna showed a downward trend from 1990 to 2010. Among them, the decrease in forest area is the most obvious, with a total reduction of 360,819 ha ($360{,}819 \times 10^4$ m^2) over the past 20 years, a dynamic land use change degree of -1.42%, a decrease of 265,491 ha ($265{,}491 \times 10^4$ m^2) from 1990 to 2000, and a reduction of 95,328 ha ($95{,}328 \times 10^4$ m^2) from 2000 to 2010. The area of cultivated land showed an increasing trend in the previous 10 years, marked by a rise of 18,153 ha ($18{,}153 \times 10^4$ m^2) and a dynamic land use change degree of 1.24%. The area of cultivated land decreased by a total of 21,456 ha ($21{,}456 \times 10^4$ m^2) in the latter 10 years, with a dynamic land use change degree of -1.31%. The area under water bodies declined continuously for the two decades, with a total reduction of 3996 ha (3996×10^4 m^2) and a dynamic degree of -2.17%. Grasslands, rubber plantations, shrubland, tea gardens, and construction land in Xishuangbanna region showed increasing trends from 1990 to 2010. Among them, the area of rubber plantations showed the most obvious growth, with a total increase of 249,948 ha ($249{,}948 \times 10^4$ m^2) in 20 years, and a dynamic land use change degree of 9.87%. Moreover, the area under tea gardens increased by 43,686 ha ($43{,}686 \times 10^4$ m^2) in the past 20 years, the dynamic land use change degree being 6.26%. Although the areas under grassland, shrubland, and construction land increased, the changes were relatively insignificant.

In summary, the economic development of the Xishuangbanna region and the improvement in people's quality of life led to a rise in the cultivation of cash crops such as rubber and tea in the region in the past 20 years, resulting in a large number of forests being felled.

During the period 1990–2000, carbon emissions for Global Forest Watch and Santoro datasets were 7.85 million t CO$_2$e and 5.63 million t CO$_2$e, respectively, with a difference of 28.30%. During the period 2000–2010, carbon emissions for Global Forest watch and Santoro datasets were 2.82 million t CO$_2$e and 2.00 million t CO$_2$e, respectively, with a difference of 28.81%. Carbon emissions for the period 1990–2000 were about 2.8 times as much as those for the period 2000–2010 (Figure 5).

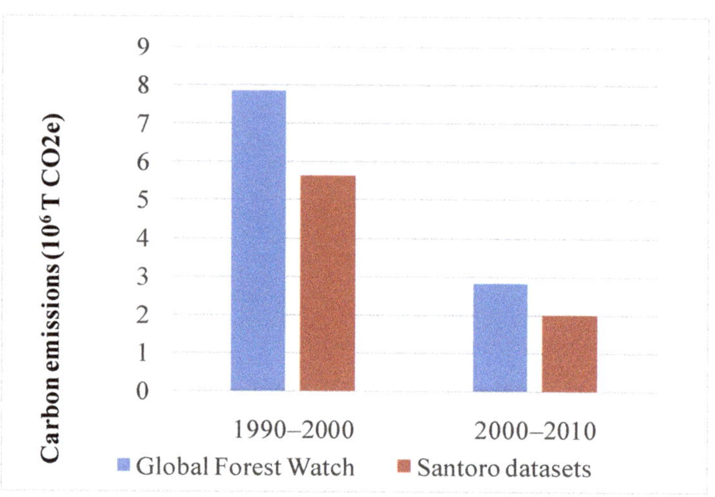

Figure 5. Carbon emissions for the period 1990–2000 and the period 2000–2010.

3.2. Influencing Factors of Land Use Change

There are many drivers that lead to deforestation and forest degradation within REDD+. Direct drivers are human activities or immediate actions that directly impact forest cover and loss of carbon such as agriculture expansion (both commercial and subsistence), infrastructure extension, and wood extraction. Indirect drivers are complex interactions of social, economic, political, cultural, and technological processes to cause deforestation or forest degradation. They act at multiple scales: international (markets, commodity prices), national (population growth, domestic markets, national policies, governance), and local circumstances (subsistence, poverty) [64–67]. Since RLs refer to the business-as-usual scenario, which means without any change in REDD+ drivers (situation, government, socio-economic forces, etc. that occur over time), this study only considered seven factors influencing land use change, namely distance to a road, distance to a river, elevation, slope, aspect, distance to an administrative center, and nature reserves (Table 3 and Figure 6).

Table 3. Factors influencing land use change and data acquisition methods.

No.	Influencing Factor	Data Acquisition Method
1	Elevation	Using the Shuttle Radar Topography Mission (SRTM) data, the topographic data of Xishuangbanna region were extracted through the mask
2	Slope	A slope map was generated from the extracted elevation data
3	Aspect	An aspect map was generated through the Digital Elevation Model (DEM)
4	Distance to a road	Using the traffic map, roads classified as level 3 and above in Xishuangbanna region were vectorized, and distance analysis was used to obtain the distribution map of the roads nearest to the studied areas in the region
5	Distance to a river	The main rivers in Xishuangbanna region were vectorized, and distance analysis was used to obtain the distribution map of the rivers closest to the studied areas in Xishuangbanna region
6	Distance to an administrative center	Distance analysis was conducted for all such centers in Xishuangbanna region
7	Nature reserves (limiting factors)	The distribution map showing Xishuangbanna's nature reserves was analyzed as land transfer within reserves is restricted

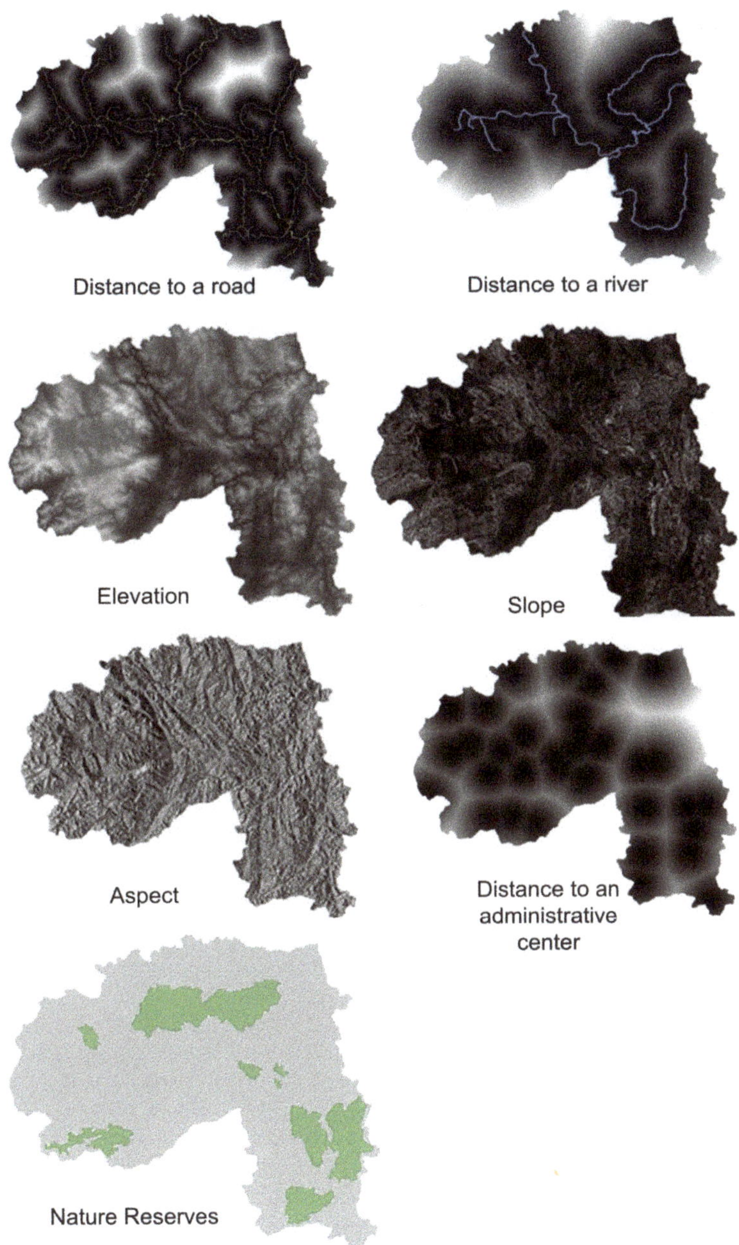

Figure 6. Influencing factors of land use change in Xishuangbanna region.

Cramer's V coefficients (Table 4) were calculated to measure the correlation between the above-mentioned factors impacting land use change and land distribution. The larger the value, the stronger the correlation.

Table 4. Cramer's V coefficients indicating correlations between the influencing factors of land use change and land distribution.

	Distance to a Road	Distance to a River	Elevation	Slope	Aspect	Distance to an Administrative Center
Overall	0.1334	0.0905	0.2539	0.1608	0.0431	0.1252
Woodland	0.0001	0.0001	0.3482	0.0001	0.0001	0.0001
Shrubland	0.2735	0.1809	0.1192	0.2131	0.0729	0.1871
Grass	0.0795	0.0815	0.1289	0.0406	0.0560	0.0933
Cultivated land	0.0462	0.0347	0.1355	0.0210	0.0216	0.0455
Rubber plantations	0.1550	0.0708	0.5297	0.1984	0.0570	0.1542
Tea gardens	0.2148	0.1277	0.1230	0.1665	0.0308	0.1516
Construction land	0.1411	0.0893	0.0451	0.1076	0.0294	0.1025
Other land	0.0492	0.0217	0.2130	0.0256	0.0199	0.0361

3.2.1. Distance to a Road

Besides playing a very important role in the economic and social development of a region, traffic conditions impact the land use status of a region. The overall correlation between the land type and distance from a road is 0.1334. Firstly, compared with the overall value, Cramer's V coefficient for shrubland and tea gardens is 0.2735 and 0.2148, respectively, which is much higher than the overall value. Thus, the distance from a road is a relative important factor affecting shrubland and tea gardens. Secondly, Cramer's V coefficient of the impact of the distance from a road on rubber plantations and construction land is 0.1550 and 0.1411, respectively, quite similar to the overall value. Thus, the affected land types dominated by road traffic in the Xishuangbanna region are shrubland, tea gardens, rubber plantations, and construction. It is evident that these land types are affected by anthropogenic activity. The reason of highest correlation between the shrubland and road is that it is very common in Xishuanbbanna to have roads built across shrubland rather other areas.

3.2.2. Distance to a River

The precipitation in Xishuangbanna region is abundant and evenly distributed. The dependence of most land use types on rivers is not obvious, except for shrubland and tea gardens. Among them, the influencing factor, namely the overall correlation value of the distance from a river to the land type is 0.0905, and the Cramer's V coefficients for tea gardens (0.1277) are higher than this overall value. This result indicates that the distance from a river is the main factor affecting tea gardens.

3.2.3. Terrain-Related Factors

Topographic factors play a very important limiting role in various production activities. The study area is mainly mountainous, and, thus, the topographic factors of elevation, slope, and aspect cannot be ignored. Firstly, the overall value of the correlation is 0.2539, and woodland and rubber plantations alone show higher correlation coefficients than this overall value (the corresponding Cramer's V coefficients are 0.3482 and 0.5297). During the period 1990–2010, the rubber plantation in Xishuangbanna continuously expanded from low-altitude flat valleys to mountainous areas in high altitudes due to high rubber price from the international market, population pressure, and economic development. This is the reason for the highest correlation between the elevation and rubber plantation. Cramer's V coefficients of elevation for shrubland, grassland, cultivated land, tea gardens, construction land, and other land are 0.1192, 0.1289, 0.1355, 0.1230, 0.0451, and 0.2130, respectively, indicating that their correlation coefficients are lower than the overall value.

Secondly, the slope affects the water distribution, wind speed, and soil texture required for crop growth. The overall value of the correlation for the slope is 0.1608, while Cramer's V coefficients for shrubland, rubber plantations, and tea gardens are 0.2131, 0.1984, and 0.1665, respectively, higher than the overall value. Thus, this factor can be regarded as the main factor impacting these land uses. However, in overall terms, Cramer's V coefficient is less than the corresponding values for grassland, cultivated land, construction land, and other land (0.0406, 0.0210, 0.1076, and 0.0256, respectively).

Finally, the aspect primarily affects the length of time and temperature for the growth and final yield of crops. The overall value in this case is 0.0431, while Cramer's V coefficients for shrubland, grassland, and rubber plantations are all greater than the overall value (0.0729, 0.0560, and 0.0570, respectively).

3.2.4. Distance to an Administrative Center

Governmental administrative organizations are typically located in townships. Given the increasingly strict forest protection policies being applied to Xishuangbanna region, areas closer to governmental administrative organizations can be conveniently supervised and regulated, resulting in a certain deterrent effect on forest destruction and illegal mining of local resources. The overall value of the distance from a township is 0.1252. The corresponding Cramer's V coefficients for rubber plantations, and tea gardens (0.1542, and 0.1516, respectively) are higher than the overall value. However, the coefficients for grassland, cultivated land, construction land, and other land (namely, 0.0933, 0.0455, 0.1025, and 0.0361, respectively) are less than the overall value. Therefore, rubber plantations and tea gardens are clearly (and expectedly) impacted by distance to a township, whereas this is not so for the remaining land use types.

3.2.5. Limiting Factor (Nature Reserve)

Xishuangbanna Nature Reserve is a national nature reserve consisting of five small sub-reserves, namely the Mengyang, Menglun, Mengla, Shangyong, and Manzhang Reserves. These sub-reserves are not geographically connected to each other and cover a total area of 242,500 ha (242,500 $\times$ 10^4 m^2). Notably, 12.68% of the total area of the Prefecture is allocated to nature conservation, namely the protection of the tropical forest ecosystem and its rare wildlife. Relatively little land change has been observed in the protected area, and man-made damage has also been effectively contained. In this study, the conversion rate of certain land use types, such as forestland, in the protected area was set to 0; in other words, anthropogenic activities in these areas are completely restricted.

3.3. Future Land Use Simulation Results and Inspection

The expansion of rubber and other cash crops has caused massive forest loss and fragmentation in Xishuangbanna. The region experienced the most severe forest losses and degradation particularly for the period 1990 to 2010. Therefore, we chosen the period 1990 to 2010 for REDD+ in Xishuangbanna as the baseline, which is crucial to measure the emission reduction performance and consequently to negotiate meaningful deforestation emission reduction targets. As a result, the land use change data for 1990 and 2000 were used as inputs to the model of the Markov chain and MLP, and the 2010 land use change data were used as the verification values to simulate future land use. The validation of AUC value from the ROC curve method is 0.8, indicating that the results provided by the model are ideal. The land use prediction results for the Xishuangbanna region in the next 20 years of 2016–2035 are shown in Figure 7.

Area under forestland shows a downward trend and is the largest change over the 20 years, with the areal reduction amounting to 158,535 ha (158,535 $\times$ 10^4 m^2). Conversely, the areas under rubber plantations, tea gardens, and cultivated land increase, with rubber plantations showing the highest increase (by 108,450 ha (108,450 $\times$ 10^4 m^2)). The areas under tea gardens and cultivated land also increase, but only slightly (by 39,204 ha (39,204 $\times$ 10^4 m^2) and 31,707 ha (31,707 $\times$ 10^4 m^2), respectively). The areas under shrubland, grassland, construction land, and water bodies remained stable. Thus, in the next 20 years, the Xishuangbanna region will undergo further deforestation; simultaneously, given its improved economic development and the rising human demand for resources, the cultivation of cash crops such as rubber and tea will continue to increase, which will add pressure on the region's forests.

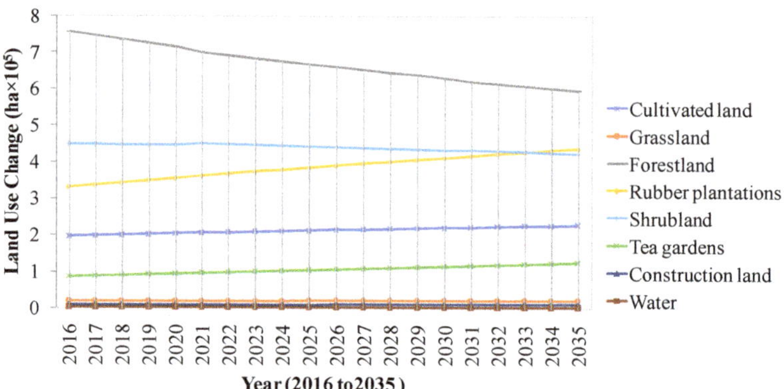

Figure 7. Land use forecast for Xishuangbanna region for the next 20 years (unit: ha or 10^4 m^2).

3.4. Reference Levels in Xishuangbanna

According to IPCC's Good Practice Guidance, the source/or sink estimates were determined by multiplying the activity data by a carbon stock coefficient (i.e., emission factor) at two points in time. In this study, the combination of the IPCC method and the land use change model showed that the carbon emissions from the study region obviously increased year by year over the 20 years of this study (Figure 8); the simulated growth trend provides an estimate of 0.35 million t CO$_2$e of annual carbon emissions on average. Simultaneously, the large increase in rubber plantations facilitated a rise in carbon absorption, resulting in average annual carbon sequestration of 0.13 million t CO$_2$e. Although the total amount of carbon sequestration attributable to cultivated land, grassland, shrubland, and tea gardens changed, the overall increase was not large. In general, the total carbon emissions in Xishuangbanna rose year by year during the past two decades. The average annual carbon emissions in the past two decades were estimated to be 0.23 million t CO$_2$e, while the total carbon emissions in the same time period amounted to 4.6866 million t CO$_2$e, indicating an obvious increase.

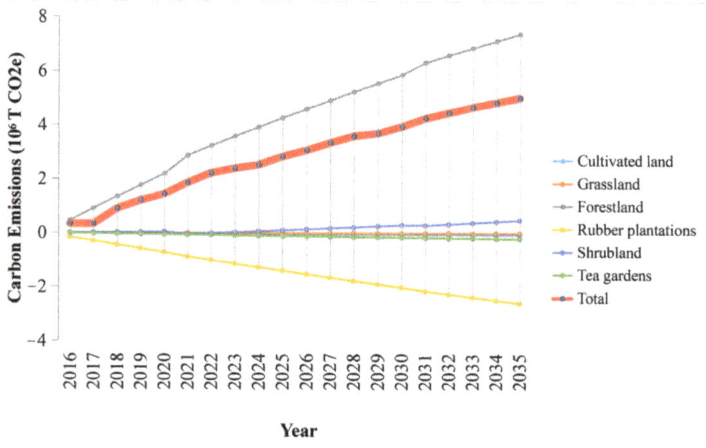

Figure 8. Reference levels in the Xishuangbanna region.

4. Conclusions

A careful assessment of RLs for REDD+ in Xishuangbanna, China provides significant insights to REDD+ project. The implications that emerge from this study are as follows.

1. We developed a methodological framework to estimate carbon emissions for the REDD+ program in the tropical forests of Xishuangbanna, China. By coupling IPCC's GPG and land use scenario modeling, we could successfully estimate the RLs. Within the framework, the Enhanced Thematic Mapper Plus/Thematic Mapper(ETM+/TM) remote sensing images in the study area were used to interpret land use changes in 1990, 2000, and 2010. The Land Use Dynamic Index was used for the transfer of land use types during the study period to identify that rubber plantations were the main contributor to forest loss in this region. The Markov chain model was used to predict the number of future land use changes and the Multilayer Perceptron model was applied to calculate the distribution location of these changes.

2. According to Paragraph 71 of Decision 1/CP.16, forests RLs are one of the elements to implement REDD+ activities for developing country parties. Moreover, the COP recognizes the importance and necessity of adequate and predictable financial and technology support for developing such RLs. Identifying these RLs is, therefore, a critical step in the provision of financial incentives and/or creation of carbon markets. Furthermore, they guide the design of the REDD+ strategy. In this study for the business-as-usual scenario of the RLs, Xishuangbanna will lose 158,535 ha ($158,535 \times 10^4$ m^2) of forest area in next 20 years, resulting in approximately 0.23 million t (0.23×10^9 kg) CO_2e emissions per year. This is due to the improved economic development and the rising human demand for resources, such as the cultivation of rubber and tea.

5. Future Scope

Estimating carbon emissions based on RLs is a multidisciplinary task. It requires expertise in forestry science, ecological modeling, statistics, remote sensing, and field techniques. Undertaking this exercise is demanding given global geographical diversity, and, thus, building technical capacity to this end is essential. Modeling future emissions based on historical trend rates and understanding the relationships between deforestation patterns and the drivers of deforestation are essential for RL estimation [68–70].

Remote sensing technology using optical sensors is capable of measuring the carbon content of different forest types when supported by field information from, for example, sample plots used to calibrate the technology. Using this methodology, a multitemporal set of remotely sensed data can be used to detect forest changes over time [71–73]. Thus, freely available Landsat images can provide reliable measurements of forest change, especially when complemented with high-resolution satellite imagery from sensors such as QuickBird, which provide data for image analysis training and validation.

Author Contributions: Conceptualization, H.L.; Data curation, G.L., C.Z. and K.H.; Formal analysis, K.H., Y.F. and M.X.; Investigation, Y.F., M.X. and R.G.; Methodology, G.L. and C.Z.; Project administration, H.L.; Resources, G.L.; Writing—original draft, H.L. and G.L.; Writing—review and editing, H.L. All authors have reviewed and agreed to the published version of the manuscript.

Funding: This study is under the auspices of NSFC42071267, NSFC41371525, Program for Innovative Research Team (in Science and Technology) in University of Henan Province (21IRT-STHN008), Dabieshan National Observation and Research Field Station of Forest Ecosystem at Henan andSYL20060111.

Acknowledgments: We are grateful to Quntao Yang's contribution to the research. We are also indebted to Liangyun Liu and Bowen Song, our colleagues at the Aerospace Information Research Institute (AIR) under the Chinese Academy of Sciences (CAS), for their kind help on China's GlobeLand30.

Conflicts of Interest: The authors declare no conflict of interest.

Appendix A

Table A1. The accuracies in 1990.

Reference Date (1990)										
Classification	Forest Land	Shrub Land	Cultivated Land	Rubber Plantations	Water	GrassLand	Constructionland	Tea Gardens	Total	UA
Forest land	1350	67	0	45	0	30	0	28	1520	88.82%
Shrub land	55	320	11	25	0	13	0	0	424	75.47%
Cultivated land	0	0	245	0	15	0	12	0	272	90.07%
Rubber plantations	24	12	0	277	0	10	0	0	323	85.76%
Water	0	12	0	0	40	13	0	0	65	61.54%
Grassland	15	0	15	0	0	65	0	0	95	68.42%
Construction land	0	0	10	0	0	0	45	0	55	81.82%
Tea gardens	0	12	15	11	0	0	0	74	112	66.07%
Total	1444	423	296	358	55	131	57	102	2866	
PA	93.49%	75.65%	82.77%	77.37%	72.73%	49.62%	78.95%	72.55%		OA = 84.30%
Omissionerror	0.065	0.243	0.172	0.226	0.273	0.504	0.211	0.275		Kappa = 0.770
Commission error	0.120	0.043	0.012	0.018	0.009	0.011	0.004	0.014		

Table A2. The accuracies in 2000.

Reference Date (2000)										
Classification	Forest Land	Shrub Land	Cultivated Land	Rubber Plantations	Water	Grassland	Constructionland	Tea Gardens	Total	UA
Forest land	1100	54	20	31	0	0	0	22	1227	89.65%
Shrub land	27	385	0	29	0	15	0	0	456	84.43%
Cultivated land	0	0	260	0	15	0	11	0	289	90.91%
Rubber plantations	15	12	0	163	0	10	0	0	200	81.50%
Water	0	12	0	0	30	13	0	0	55	54.55%
Grassland	0	0	15	0	0	57	0	0	72	79.17%
Construction land	0	0	14	0	0	0	45	0	59	76.27%
Tea garden	0	11	0	13	0	0	0	80	104	76.92%
Total	1142	474	309	236	45	95	56	102	2459	
PA	96.32%	81.22%	84.14%	69.07%	66.67%	60.00%	80.36%	78.43%		OA = 86.21%
Omission error	0.037	0.188	0.159	0.309	0.333	0.400	0.196	0.216		Kappa = 0.805
Commission error	0.096	0.036	0.014	0.017	0.010	0.006	0.006	0.010		

Table A3. The accuracies in 2010.

Reference Date (2010)										
Classification	Forest Land	Shrub Land	Cultivated Land	Rubber Plantations	Water	Grassland	Constructionland	Tea Gardens	Total	UA
Forest land	856	54	0	45	0	25	0	28	1008	84.92%
Shrub land	16	275	15	25	0	0	0	0	331	83.08%
Cultivated land	0	0	245	0	15	0	12	0	272	90.07%
Rubber plantations	38	12	0	390	0	10	0	0	450	86.67%
Water	0	14	0	0	40	0	0	0	54	74.07%
Grassland	20	0	25	0	0	80	0	0	125	64.00%
Construction land	0	0	14	0	0	0	45	0	59	76.72%
Tea gardens	0	20	17	13	0	0	0	132	182	72.53 %
Total	930	375	316	473	55	115	57	160	2481	
PA	92.04%	77.33%	77.53%	82.45%	72.73%	69.57%	78.95%	82.50%		OA = 83.15%
Omission error	0.080	0.267	0.225	0.175	0.273	0.304	0.211	0.175		Kappa = 0.781
Commission error	0.098	0.027	0.014	0.030	0.006	0.019	0.006	0.022		

References

1. FAO. *Global Forest Resource Assessment 2005*; FAO: Rome, Italy, 2006.
2. Baccini, A.; Walker, W.; Carvalho, L.; Farina, M.; Sulla-Menashe, D.; Houghton, R.A. Tropical forests are a net carbon source based on aboveground measurements of gain and loss. *Science* **2017**, *358*, 230–234. [CrossRef] [PubMed]
3. Hubau, W.; Lewis, S.; Phillips, O.; Affum-Baffoe, K.; Beeckman, H.; Cuní-Sanchez, A.; Daniels, A.K.; Ewango, C.E.N.; Fauset, S.; Mukinzi, J.M.; et al. Asynchronous Carbon Sink Saturation in African and Amazonian Tropical Forests. *Nature* **2020**, *579*, 80–87. [CrossRef] [PubMed]
4. Sagar, R.; Li, G.; Singh, J.S.; Wan, S. Carbon fluxes and species diversity in grazed and fenced typical steppe grassland of Inner Mongolia, China. *J. Plant Ecol.* **2019**, *12*, 10–22. [CrossRef]
5. Lu, H.; Liu, G.; Zhang, C.; Okuda, T. Approaches to quantifying carbon emissions from degradation in pan-tropic forests—Implications for effective REDD monitoring. *Land Degrad. Dev.* **2020**, *31*, 1890–1905. [CrossRef]

6. Barbier, E.B.; Lozano, R.; Rodríguez, C.M.; Troëng, S. Adopt a carbon tax to protect tropical forests. *Nature* **2020**, *578*, 213–216. [CrossRef]

7. Tagesson, T.; Schurgers, G.; Horion, S.; Ciais, P.; Tian, F.; Brandt, M.; Ahlström, A.; Wigneron, J.P.; Ardö, J.; Olin, S.; et al. Recent divergence in the contributions of tropical and boreal forests to the terrestrial carbon sink. *Nat. Ecol. Evol.* **2020**, *4*, 202–209. [CrossRef] [PubMed]

8. Hansen, M.C.; Potapov, P.; Tyukavina, A. Comment on "Tropical forests are a net carbon source based on aboveground measurements of gain and loss". *Science* **2019**, *363*, eaar3629. [CrossRef]

9. Melillo, J.M.; McGuire, A.D.; Kicklighter, D.W.; Moore, B.; Vorosmarty, C.J.; Schloss, A.L. Global climate change and terrestrial net primary production. *Nature* **1993**, *363*, 234–240. [CrossRef]

10. Liu, G.; Liu, Q.; Song, M.; Chen, J.; Zhang, C.; Meng, X.; Zhao, J.; Lu, H. Costs and Carbon Sequestration Assessment for REDD+ in Indonesia. *Forests* **2020**, *11*, 770. [CrossRef]

11. Lin, D.; Xia, J.; Wan, S. Climate warming and biomass accumulation of terrestrial plants: A meta-analysis. *New Phytol.* **2010**, *188*, 187–198. [CrossRef]

12. Liang, J.; Xia, J.; Liu, L.; Wan, S. Global patterns of the responses of leaf-level photosynthesis and respiration in terrestrial plants to experimental warming. *J. Plant Ecol.* **2013**, *6*, 437–447. [CrossRef]

13. Wang, H.; Chen, W. Modeling of energy transformation pathways under current policies, NDCs and enhanced NDCs to achieve 2-degree target. *Appl. Energy* **2019**, *250*, 549–557. [CrossRef]

14. Marx, A.; Kumar, R.; Thober, S.; Zink, M.; Wanders, N.; Wood, E.; Pan, M.; Sheffield, J.; Samaniego, L. Climate change alters low flows in Europe under a 1.5, 2, and 3 degree global warming. *Hydrol. Earth Syst. Sci. Discuss.* **2017**, *22*, 1017–1032. [CrossRef]

15. Ha, Y.; Teng, F. Midway toward the 2 degree target: Adequacy and fairness of the Cancun pledges. *Appl. Energy* **2013**, *112*, 856–865. [CrossRef]

16. Meinshausen, M.; Meinshausen, N.; Hare, W.; Raper, S.C.B.; Frieler, K.; Knutti, R.; Frame, D.J.; Allen, M.R. Greenhouse-gas emission targets for limiting global warming to 2 C. *Nature* **2009**, *458*, 1158–1162. [CrossRef]

17. Roopsind, A.; Sohngen, B.; Brandt, J. Evidence that a national REDD+ program reduces tree cover loss and carbon emissions in a high forest cover, low deforestation country. *Proc. Natl. Acad. Sci. USA* **2019**, *116*, 24492–24499. [CrossRef]

18. Sheng, J.; Qiu, H. Governmentality within REDD+: Optimizing incentives and efforts to reduce emissions from deforestation and degradation. *Land Use Policy* **2018**, *78*, 611–622. [CrossRef]

19. Seymour, F.; Boyd, W.; Stickler, C.; Duchelle, A.E.; Nepstad, D.; Bahar, N.H.A.; Rodriguez-Ward, D. *Jurisdictional Approaches to REDD+ and Low Emissions Development: Progress and Prospects*; World Resources Institute: Washington, WA, USA, 2018.

20. Johnson, B.A.; Scheyvens, H.; Samejima, H. Quantitative Assessment of the Earth Observation Data and Methods Used to Generate Reference Emission Levels for REDD+. In *Satellite Earth Observations and Their Impact on Society and Policy*; Onoda, M., Young, O.R., Eds.; Springer: Singapore, 2017.

21. Johnson, B.A.; Scheyvens, H.; Samejima, H.; Onoda, M. Characteristics of the remote sensing data used in the proposed UNFCCC REDD+ forest reference emission levels (FRELs). *ISPRSInt. Arch. Photogramm. Remote Sens. Spat. Inf. Sci.* **2016**, *8*, 669–672. [CrossRef]

22. Reimer, F.; Asner, G.P.; Joseph, S. Advancing reference emission levels in subnational and national REDD+ initiatives: A CLASlite approach. *Carbon Balance Manag.* **2015**, *10*, 5. [CrossRef]

23. Herold, M.; Achard, F.; DeFries, R.; Skole, D.; Brown, S.; Townshend, J. Report of the workshop on monitoring tropical deforestation for compensated reductions. In Proceedings of the GOFC-GOLD Symposium on Forest and Land Cover Observations, Jena, Germany, 21–22 March 2006.

24. Lu, H.; Liu, G. Opportunity Costs of Carbon Emissions Stemming from Changes in Land Use. *Sustainability* **2015**, *7*, 3665–3682. [CrossRef]

25. Lu, H.; Liu, G.; Huang, Z.; Yang, Q. Carbon, soil, and ecological benefits of REDD+ policies in Southwest China. *Sci. Asia* **2016**, *42*, 1–11. [CrossRef]

26. Romijn, E.; Ainembabazi, J.H.; Wijaya, A.; Herold, M.; Angelsen, A.; Verchot, L.; Murdiyarso, D. Exploring different forest definitions and their impact on developing REDD+ reference emission levels: A case study for Indonesia. *Environ. Sci. Policy* **2013**, *33*, 246–259. [CrossRef]

27. Bond, I.; Gran, G.M.; Kanounnikoff, W.S.; Hazlewood, P.; Wunder, S.; Angelsen, A. *Incentives to Sustain Forest Ecosystem Services: A Review and Lessons for REDD. Natural Resource Issues No. 16*; International Institute for Environment &Development: London, UK, 2009.

28. Hoff, R.; Rajão, R.; Leroy, P. Can REDD+ still become a market? Ruptured dependencies and market logics for emission reductions in Brazil. *Ecol. Econ.* **2019**, *161*, 121–129. [CrossRef]

29. Sukma, M.F. LatarbelakangbantuanJepangterhadap Indonesia melaluimekanisme Reducing Emission from Deforestation and Forest Degradation (REDD+) Tahun. Bachelor's Thesis, FakultasIlmuSosial dan IlmuPolitik UIN SyarifHidayatullah, Jakarta, Indonesia, March 2018.

30. Hamdan, O. *Forest Reference Emission Level For. REDD+ In Pahang, Malaysia (Research Pamphlet No. 141)*; Forest Research Institute Malaysia: Kepong, Malaysia, 2018.

31. Ye, L.; Karunarathna, S.C.; Li, H.; Xu, J.; Hyde, K.D.; Mortimer, P.E. A Survey of Termitomyces (Lyophyllaceae, Agaricales), Including a New Species, from a Subtropical Forest in Xishuangbanna, China. *Mycobiology* **2019**, *47*, 391–400. [CrossRef] [PubMed]

32. Mani, S.; Cao, M. Nitrogen and Phosphorus Concentration in Leaf Litter and Soil in Xishuangbanna Tropical Forests: Does Precipitation Limitation Matter? *Forests* **2019**, *10*, 242. [CrossRef]
33. Li, S.H.; Wang, S.J.; Zhang, Z.; Chen, M.K.; Cao, R.; Cao, Q.B.; Zuo, Q.Q.; Wang, P. Effects of ant nesting on the spatiotemporal dynamics of soil easily oxidized organic carbon in Xishuangbanna tropical forests, China. *J. Appl. Ecol.* **2019**, *30*, 413–419.
34. Goldberg, S.D.; Zhao, Y.; Harrison, R.D.; Monkai, J.; Li, Y.; Chau, K.; Xu, J. Soil respiration in sloping rubber plantations and tropical natural forests in Xishuangbanna, China. *Agric. Ecosyst. Environ.* **2017**, *249*, 237–246. [CrossRef]
35. Marziliano, P.; Veltri, A.; Menguzzato, G.; Pellicone, G.; Coletta, V. A comparative study between "Default Method" And "Stock Change Method" of Good Practice Guidance for Land Use, Land-Use Change and Forestry (IPCC, 2003) to evaluate carbon stock changes in Forest. In Proceedings of the Second International Congress of Silviculture, Florence, Italy, 26–29 November 2014; pp. 551–557.
36. Grassi, G.; Ravindranath, N.H.; Böttcher, H.; Elhassan, N.; Elsiddig, E.; House, J.I.; Matsumoto, M.; Ometto, J.P.; Sanquetta, C.; Searson, M.J.; et al. Afforestation, Reforestation, Deforestation. In *IPCC 2013 Revised Supplementary Methods and Good Practice Guidance Arising from the Kyoto Protocol (KP Supplement)*; Hiraishi, T., Krug, T., Tanabe, K., Srivastava, N., Jamsranjav, B., Fukuda, M., Troxler, T., Eds.; Institute for Global Environmental Strategies: Kitakyushu, Japan, 2014.
37. Ohgita, T. Results and Impressions of Expert Meetings for the Preparation of the IPCC Good Practice Guidance. *J. Jpn. Soc. Saf. Eng.* **2000**, *39*, 334–342.
38. Thind, M.P.S.; Wilson, E.J.; Azevedo, I.M.L.; Marshall, J.D. Marginal Emissions Factors for Electricity Generation in the Midcontinent ISO. *Environ. Sci. Technol.* **2017**, *51*, 14445–14452. [CrossRef]
39. Wolf, J.; Asrar, G.R.; West, T.O. Revised methane emissions factors and spatially distributed annual carbon fluxes for global livestock. *Carbon BalanceManag.* **2017**, *12*, 16. [CrossRef]
40. Li, A.; Zhang, A.; Zhou, Y.; Yao, X. Decomposition analysis of factors affecting carbon dioxide emissions across provinces in China. *J. Clean. Prod.* **2017**, *141*, 1428–1444. [CrossRef]
41. MacDicken, K. *A Guide to Monitoring Carbon Storage in Forestry and Agroforestry Projects*; Winrock International Institute for Agricultural Development: Arlington, VA, USA, 1997.
42. Brown, S. Measuring carbon in forests: Current status and future challenges. *Environ. Pollut.* **2002**, *116*, 363–372. [CrossRef]
43. Baccini, A.; Goetz, S.J.; Walker, W.S.; Laporte, N.T.; Sun, M.; Sulla-Menashe, D.; Hackler, J.; Beck, P.S.A.; Dubayah, R.; Friedl, M.A.; et al. Estimated carbon dioxide emissions from tropical deforestation improved by carbon-density maps. *Nat. Clim. Chang.* **2012**, *2*, 182–185. [CrossRef]
44. Santoro, M.; Cartus, O.; Carvalhais, N.; Rozendaal, D.; Avitabilie, V.; Araza, A.; Veiga, P.R. The global forest above-ground biomass pool for 2010 estimated from high-resolution satellite observations. *Earth Syst. Sci. Data Discuss.* **2020**. [CrossRef]
45. Anaya, J.A.; Gutiérrez-Vélez, V.H.; Pacheco-Pascagaza, A.M.; Palomino-Ángel, S.; Han, N.; Balzter, H. Drivers of Forest Loss in a Megadiverse Hotspot on the Pacific Coast of Colombia. *Remote Sens.* **2020**, *12*, 1235. [CrossRef]
46. Sarzynski, T.; Giam, X.; Carrasco, L.; Lee, J.S.H. Combining Radar and Optical Imagery to Map Oil Palm Plantations in Sumatra, Indonesia, Using the Google Earth Engine. *Remote Sens.* **2020**, *12*, 1220. [CrossRef]
47. De Bem, P.P.; De Carvalho Junior, O.A.; Fontes Guimarães, R.; Trancoso Gomes, R.A. Change Detection of Deforestation in the Brazilian Amazon Using Landsat Data and Convolutional Neural Networks. *Remote Sens.* **2020**, *12*, 901. [CrossRef]
48. DeFries, R.; Achard, F.; Brown, S.; Herold, M.; Murdiyarso, D.; Schlamadinger, B.; Souza, C. *Reducing Greenhouse Gas Emissions from Deforestation in Developing Countries: Considerations for Monitoring and Measuring*; Global Terrestrial Observing System: Rome, Italy, 2006.
49. Achard, F.; DeFries, R.; Herold, M.; Mollicone, D.; Pandey, D.; de Souza, C. Guidance on monitoring of gross changes in forest area. In *Reducing Greenhouse Gas Emissions from Deforestation and Degradation in Developing Countries: A Sourcebook of Methods and Procedures for Monitoring, Measuring and Reporting*; Global Observation for Forest Cover and Land Dynamics: Alberta, AB, Canada, 2008.
50. Wang, X.; Bao, Y. Study on the methods of land use dynamic change research. *Prog. Geogr.* **1999**, *18*, 81–87.
51. Gupta, R.; Sharma, L.K. Efficacy of Spatial Land Change Modeler as a forecasting indicator for anthropogenic change dynamics over five decades: A case study of Shoolpaneshwar Wildlife Sanctuary, Gujarat, India. *Ecol. Indic.* **2020**, *112*, 106171. [CrossRef]
52. Anand, J.; Gosain, A.K.; Khosa, R. Prediction of land use changes based on Land Change Modeler and attribution of changes in the water balance of Ganga basin to land use change using the SWAT model. *Sci.Total Environ.* **2018**, *644*, 503–519. [CrossRef]
53. Jahanifar, K.; Amirnejad, H.; Mojaverian, M.; Azadi, H. Land Change Detection and Identification of Effective Factors on Forest Land Use Changes: Application of Land Change Modeler and Multiple Linear Regression. *Eur. Online J. Nat. Soc. Sci.* **2018**, *7*, 554–565.
54. Mohamed, A.; Worku, H. Simulating urban land use and cover dynamics using cellular automata and Markov chain approach in Addis Ababa and the surrounding. *Urban Clim.* **2020**, *31*, 100545. [CrossRef]
55. Satya, B.A.; Shashi, M.; Deva, P. Future land use land cover scenario simulation using open source GIS for the city of Warangal, Telangana, India. *Appl. Geomat.* **2020**, *12*, 281–290. [CrossRef]
56. Shen, L.; Li, J.; Wheate, R.; Yin, J.; Paul, S.S. Multi-Layer Perceptron Neural Network and Markov Chain Based Geospatial Analysis of Land Use and Land Cover Change. *J. Environ. Inform. Lett.* **2020**, *3*, 29–39. [CrossRef]
57. Kussul, N.; Lavreniuk, M.; Skakun, S.; Shelestov, A. Deep Learning Classification of Land Cover and Crop Types Using Remote Sensing Data. *IEEE Geosci. Remote Sens. Lett.* **2017**, *14*, 778–782. [CrossRef]

58. Ozturk, D. Urban Growth Simulation of Atakum (Samsun, Turkey) Using Cellular Automata-Markov Chain and Multi-Layer Perceptron-Markov Chain Models. *Remote Sens.* **2015**, *7*, 5918–5950. [CrossRef]
59. Cheong, Y.L.; Leitão, P.; Lakes, T. Assessment of land use factors associated with dengue cases in Malaysia using Boosted Regression Trees. *Spat. Spatio-Temporal Epidemiol.* **2014**, *10*, 75–84. [CrossRef]
60. Tien Bui, D.; Shahabi, H.; Shirzadi, A.; Chapi, K.; Pradhan, B.; Chen, W.; Khosravi, K.; Panahi, M.; Bin Ahmad, B.; Saro, L. Land Subsidence Susceptibility Mapping in South Korea Using Machine Learning Algorithms. *Sensors* **2018**, *18*, 2464. [CrossRef]
61. Ghorbanzadeh, O.; Blaschke, T.; Aryal, J.; Gholamnia, K. A new GIS-based technique using an adaptive neuro-fuzzy inference system for land subsidence susceptibility mapping. *J. Spat. Sci.* **2018**, *94*, 1–17. [CrossRef]
62. Chen, J.; Cao, X.; Peng, S.; Ren, H. Analysis and Applications of GlobeLand30: A Review. *ISPRS Int. J. Geo-Inf.* **2017**, *6*, 230. [CrossRef]
63. Xu, C.; Chen, J.; Wu, H.; Li, R.; Zhao, Y.J. An improved pixel counting method for arbitrary zonal statistics on globeland30. *ISPRS Int. Arch. Photogramm. Remote Sens. Spat. Inf. Sci.* **2019**, *XLII-4/W20*, 101–103. [CrossRef]
64. Siyum, Z.G. Tropical dry forest dynamics in the context of climate change: Syntheses of drivers, gaps, and management perspectives. *Ecol. Process.* **2020**, *9*, 25. [CrossRef]
65. Geist, H.; Lambin, E. What drives tropical deforestation? A meta-analysis of proximate and underlying causes of deforestation based on subnational case study evidence. Land-Use and Land-Cover Change Project, International Geosphere-Biosphere Programme. *LUCC Rep. Ser.* **2002**, *52*, 4.
66. Geist, H.; Lambin, E. Proximate causes and underlying driving forces of tropical deforestation. *BioScience* **2002**, *52*, 143–150. [CrossRef]
67. Obersteiner, M.; Huettner, M.M.; Kraxner, F.; McCallum, I.; Aoki, K.; Bottcher, H.; Fritz, S.; Gusti, M.; Havlik, P.; Reyers, B.; et al. On fair, effective and efficient REDD mechanism design. *Carbon Balance Manag.* **2009**, *4*, 11. [CrossRef]
68. Ellis, E.A.; Sierra-Huelsz, J.A.; Ceballos, G.C.O.; Binnqüist, C.L.; Cerdán, C.R. Mixed Effectiveness of REDD+ Subnational Initiatives after 10 Years of Interventions on the Yucatan Peninsula, Mexico. *Forests* **2020**, *11*, 1005. [CrossRef]
69. Maraseni, T.N.; Poudyal, B.H.; Rana, E.; Khanal, S.C.; Ghimire, P.L.; Subedi, B.P. Mapping national REDD+ initiatives in the Asia-Pacific region. *J. Environ. Manag.* **2020**, *269*, 110763. [CrossRef]
70. Gallo, P.; Brites, A.; Micheletti, T. *REDD+ Achievements and Challenges in Brazil: Perceptions over time (2015-2019)*; CIFOR: Bogor City, Indonesia, 2020.
71. Gebhardt, S.; Wehrmann, T.; Ruiz, M.A.M.; Maeda, P.; Bishop, J.; Schramm, M.; Ressl, R. MAD-MEX: Automatic Wall-to-Wall Land Cover Monitoring for the Mexican REDD-MRV Program Using All Landsat Data. *Remote Sens.* **2014**, *5*, 3923–3943. [CrossRef]
72. Mitchell, A.L.; Rosenqvist, A.; Mora, B. Current remote sensing approaches to monitoring forest degradation in support of countries measurement, reporting and verification systems for REDD+. *Carbon Balance Manag.* **2017**, *12*, 9. [CrossRef]
73. Fry, B.P. Community forest monitoring in REDD+: The 'M' in MRV? *Environ. Sci. Policy* **2011**, *2*, 181–187.

 *remote sensing*

Article

Assessment of Land Degradation in Semiarid Tanzania—Using Multiscale Remote Sensing Datasets to Support Sustainable Development Goal 15.3

Jonathan Reith [1,2,3,*], Gohar Ghazaryan [2,4,5], Francis Muthoni [3] and Olena Dubovyk [2,4]

[1] Federal Statistical Office of Germany, 53117 Bonn, Germany
[2] Center for Remote Sensing of Land Surfaces (ZFL), University of Bonn, 53113 Bonn, Germany;
     gohar.ghazaryan@zalf.de (G.G.); odubovyk@uni-bonn.de (O.D.)
[3] International Institute for Tropical Agriculture (IITA), Duluti, Arusha P.O. Box 10, Tanzania;
     f.muthoni@cgiar.org
[4] Remote Sensing Research Group (RSRG), Department of Geography, University of Bonn,
     53115 Bonn, Germany
[5] Leibniz Centre for Agricultural Landscape Research (ZALF), 15374 Müncheberg, Germany
*   Correspondence: Jonathanreith@gmail.com

Abstract: Monitoring land degradation (LD) to improve the measurement of the sustainable development goal (SDG) 15.3.1 indicator ("proportion of land that is degraded over a total land area") is key to ensure a more sustainable future. Current frameworks rely on default medium-resolution remote sensing datasets available to assess LD and cannot identify subtle changes at the sub-national scale. This study is the first to adapt local datasets in interplay with high-resolution imagery to monitor the extent of LD in the semiarid Kiteto and Kongwa (KK) districts of Tanzania from 2000–2019. It incorporates freely available datasets such as Landsat time series and customized land cover and uses open-source software and cloud-computing. Further, we compared our results of the LD assessment based on the adopted high-resolution data and methodology (AM) with the default medium-resolution data and methodology (DM) suggested by the United Nations Convention to Combat Desertification. According to AM, 16% of the area in KK districts was degraded during 2000–2015, whereas DM revealed total LD on 70% of the area. Furthermore, based on the AM, overall, 27% of the land was degraded from 2000–2019. To achieve LD neutrality until 2030, spatial planning should focus on hotspot areas and implement sustainable land management practices based on these fine resolution results.

Keywords: land degradation neutrality; SDG; land productivity; land cover; NDVI; Landsat; vegetation-precipitation relationship; soil organic carbon; Google Earth Engine

Citation: Reith, J.; Ghazaryan, G.; Muthoni, F.; Dubovyk, O. Assessment of Land Degradation in Semiarid Tanzania—Using Multiscale Remote Sensing Datasets to Support Sustainable Development Goal 15.3. *Remote Sens.* **2021**, *13*, 1754. https://doi.org/10.3390/rs13091754

Academic Editor: Elias Symeonakis

Received: 31 March 2021
Accepted: 26 April 2021
Published: 30 April 2021

1. Introduction

Land degradation (LD) is defined as the "continuous reduction or loss of the productivity of the land due to a combination of natural and anthropogenic causes" [1]. It is a global problem and affects people, their livelihoods and nature. Studies suggest that up to 3.2 billion people live and depend on degraded lands [2] and that approximately a quarter of the world's lands are affected by LD [3,4]. Poor people, who often rely on agriculture, are most vulnerable to LD [5,6]. Lost ecosystem services due to land use and land cover (LULC) change and LD account for up to USD 10.5 trillion loss per year, which is about a sixth of the world's gross domestic product (GDP) [7]. Furthermore, biodiversity is declining globally, with tremendous losses in sub-Saharan Africa because of LD [6]. Projections suggest that lower productivity in the face of climate change will drive LULC change globally. Moreover, the population growth, combined with a changing diet, will have an enormous influence on agriculture and thus LD [8]. It is for these reasons that the world community introduced the sustainable development goal (SDG) 15.3, which aims to "restore degraded land and strive to achieve an LD-neutral world", highlighting the global importance of this issue [9,10].

Tanzania is a hot spot of LD, with more than half its area showing signs of degradation [2,11]. It has the highest annual forest area net loss in East Africa and the fifth-highest worldwide [12]. The cost of LD has been summed up to USD 2.3 billion annually in the first decade of the new millennium [13]. Seventy-five percent of the total labor force, mostly rural people, work and depend on the agricultural sector, which is accountable for about 30% of the GDP [14]. Although the cultivated area increased in the last years, the output per hectare (ha) decreased, both in annual and perennial crops, even though fertilizer consumption quadrupled at the same time [15]. The number of undernourished people is growing and is currently more than 30% [16]. The population is increasing while agricultural productivity is stagnating, and the economic dependency on natural goods is still high. The consequences of this dilemma area persisting pressure on land and, thus, a probable conversion of natural into cultivated land in the coming years. The poor people's food security is also at risk, and in the coming years, in the face of climate change, new insecurities are likely to arise [17]. This holds especially true for the rural semiarid central districts of Kiteto and Kongwa (KK).

Agricultural intensification and sustainable land management (SLM) are keys to halt and reverse LD [18–20]. One major constraint that prevents action is the lack of spatial information on the extent and magnitude of LD [18]. In contrast to the laborious fieldwork, remote sensing offers the unique opportunity to consistently assess vast areas over a long period [2–4]. Unfortunately, the existing LD maps have a coarse spatial resolution and provide inconsistent estimates of the affected area [8]. For example, previous estimates of the extent of LD in Tanzania range from 41% to half of the country [2,3,11]. These variations emanate from differences in definitions of LD, monitoring methods and lack of appropriate data [6,21]. In the course of SDG 15.3 implementation, standard methods for assessing LD were introduced, making reports more comparable.

This new standard methodology, recommended by the United Nations Convention to Combat Desertification (UNCCD), includes the usage of three sub-indicators for the complimentary assessment of LD [22]. The first sub-indicator, land cover (LC), reports changes in vegetation cover. The second, land productivity (LP), captures changes in ecosystem functions. The last, soil organic carbon (SOC), indicates slower changes resulting from biomass alterations [20]. The three sub-indicators are aggregated to form the land degradation indicator. Improvements in one indicator cannot compensate losses in others, as they are complementary and not additive. Thus, the "one-out, all-out" approach is applied whereby even if one indicator shows signs of decline and the others are positive, the land is deemed to be degraded [23].

The recent Tanzanian national LD-neutrality (LDN) report follows these guidelines [24]. However, it only assesses LD for the first ten years of the 21st century and mainly uses global default data with a coarse spatial resolution. The 1 km coarseresolution is inadequate to monitor LD in small mountainous and highly fragmented landscapes, as it may miss out on smaller than pixel size LD areas [25].

Overall, only a few studies have been published on the subject of SDG 15.3.1 monitoring and assessment. Gichenje and Godinho [26], for example, conducted a baseline assessment of the SDG indicator 15.3.1 for the years 1992 to 2015 using the Advanced Very High Resolution Radiometer (8 km, AVHRR) Normalized Difference Vegetation Index (NDVI) time series and the European Space Agency (ESA) Climate Change Initiative (CCI) LC map in Kenya. In Mozambique, Frederique et al. [27] analyzed only the LD sub-indicator LP trend using the Moderate Resolution Imaging Spectroradiometer (250 m, MODIS) NDVI from 2001 to 2016.

However, these studies share the common disadvantage of applying only default methodology and global datasets for national and subnational LD assessments. Though Akinyemi et al. (2020) used a customized 30m resolution LC map to assess the LC sub-indicator of SDG 15.3.1 in Botswana, this study relied on AVHRR time-series assessment for the LP sub-indicator. Furthermore, no studies exist in Africa that used high-spatial-resolution datasets the assessment of more than one sub-indicator of SDG 15.3.1. Therefore, it is vital to overcome the existing research gaps and use high-resolution spatial data to provide improved information on the SDG 15.3 [28].

In this light, the main aim of our study was to assess the SDG 15.3.1 indicator based on the newly adopted approach based on the higher resolution (compared to default UNCCD datasets) 30m Landsat time series and 30m LC maps and compare our results to the estimates of the SDG 15.3.1 based on the default UNCCD data and methods.

Our study addressed the following research questions:

- How much land is degraded, and where are the hotspots of LD in KK?
- How do the individual sub-indicators affect LD?
- Does using higher resolution data (30 m) improve the delineation of LD compared to moderate-resolution data (250 m)?

2. Materials and Methods

2.1. Study Area

The study site is situated in Kiteto and Kongwa districts, located in Dodoma and Manyara regions of Central Tanzania, respectively (Figure 1). The elevation ranges between 850 and 2100 m above sea level. The study area has a hot arid steppe climate [29]. The average monthly temperature stays between 19 and 25 °C all year, and the precipitation is roughly 600 mm a year, with interannual differences of 500 to 800 mm. Large parts of northern Kiteto and more minor areas of the mountainous region in Kongwa are protected areas for nature and landscape conservation.

2.2. Materials

The SDG 15.3.1 indicator and its three LDN sub-indicators were computed using the recommended default method (DM) with Trends.Earth [30] and the adapted methods (AM) using high-resolution Landsat (and other) datasets (Table 1).

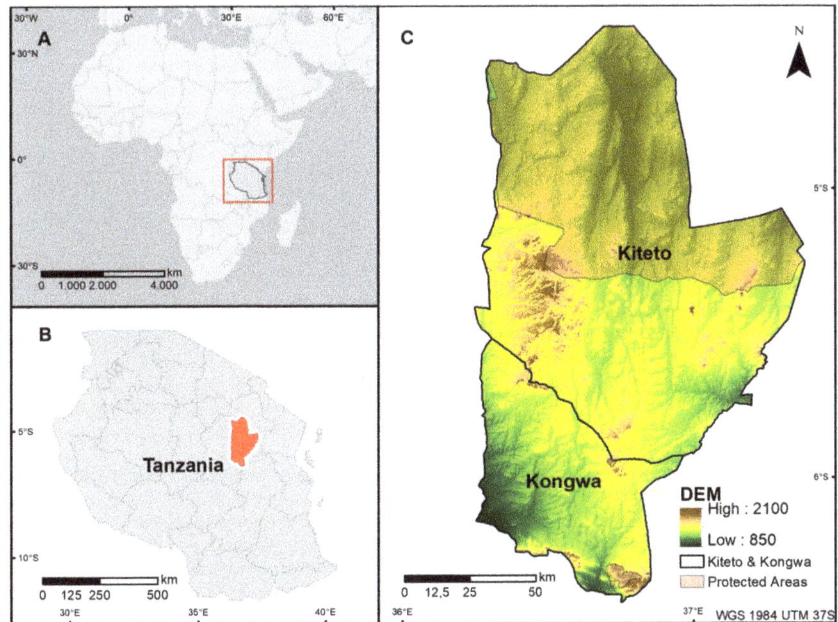

Figure 1. Location of the study area in Central Tanzania (**A,B**) and protected areas (**C**).

The DM (LC) map provided by the UNCCD is based on the 300 m ESA CCI LC map (Table 1). The AM utilized 30 m LC maps for 2000–2018 in the study area that the Regional Centre for Mapping of Resource for Development (RCMRD) developed. Both datasets were disaggregated into the six LC classes as defined by Intergovernmental Panel on Climate Change (IPCC), i.e., forestland, grassland, cropland, wetland, urban, and otherlands [31].

The recommended global default dataset uses the MOD-13Q1-coll6 (250 m) MODIS-NDVI products [30]. In contrast, the AM was calculated based on a 30 m resolution NDVI from a combination of Landsat 5, 7 and 8 (Table 1). The Landsat time series were accessed and analyzed using Google Earth Engine [32], based on atmospherically corrected surface reflectance collections (Table 1). The Landsat 5 and 7 data were spectrally harmonized with Landsat 8 series using linear transformation [33]. As a further step to improve the image quality, the fmask was adopted to mask out clouds and cloud shadows [34,35]. Generally, the images with cloud cover scores higher than 80% were removed. Finally, the NDVI was calculated for each image, and then the images of the same admission time were merged and clipped to the extent of the study area. As it is recommended to constrain the observation period to the growing season to reduce the number of irrelevant assets for the computation and enhance the quality of the time series [22], we used the imagery from November to June. When using Trends.Earth, there is no possibility to apply the computation to the growing season, so the DM uses the whole calendar year. In order to integrate the rainfall information, data from Climate Hazards Group InfraRed Precipitation with Station (CHIRPS) were used (Table 1).

Table 1. The datasets used for land degradation neutrality (LDN) reporting. For the land cover, the European Space Agency Climate Change Initiative (ESA-CCI) and the Regional Center for Mapping of Resources for Development (RCMDR) were used. Land productivity is based on the Moderate Resolution Imaging Spectroradiometer (MODIS) and Landsat. The precipitation is based on the Climate Hazards Group InfraRed Precipitation with Station data (CHIRPS). Lastly, the soil organic carbon content is derived from SoilGrids250m.

LDN Sub-Indicators	Method	Data	Resolution/Year	Reference
Land Cover	Default Method (DM)	ESA-CCI	300 m (2000–2015)	[36]
	Adapted Method (AM)	RCMRD	30 m (2000–2018)	[37]
Land Productivity	DM	MOD-13Q1-coll6	250 m (2000–2015)	[38]
	AM	Landsat 5	30 m (2000–2013)	[39]
		Landsat 7	30 m (2000–2019)	
		Landsat 8	30 m (2013–2019)	
	DM/AM	CHIRPS	0.05 arc° (2000–2019)	[40]
Soil Organic Carbon	DM/AM	SoilGrids250m	250 m	[41]

The SOC metrics were derived from the SoilGrids250m dataset [41] for the DM and the AM, as there is no national SOC database for Tanzania. SOC is measured at a depth of 30 cm and is stated as mass per area (e.g., tons per hectare (t/ha)) [22].

2.3. Methods

The calculation of the SDG 15.3.1 indicator is based on the "one out, all-out" approach (Ref. [23] and Figure 2). The three LD sub-indicators (LC change, LP decline and loss of SOC) are estimated, and if one indicator signals degradation, the LD indicator will reflect this as well. A baseline is needed to compare the progress of LDN. The baseline year (t_0) was set to be 2015 and is computed as the average of the period leading up to t_0 (2000–2015). The indicators are then remeasured in regular time intervals leading to 2030, and change is used to monitor the progress to accomplish LDN [20].

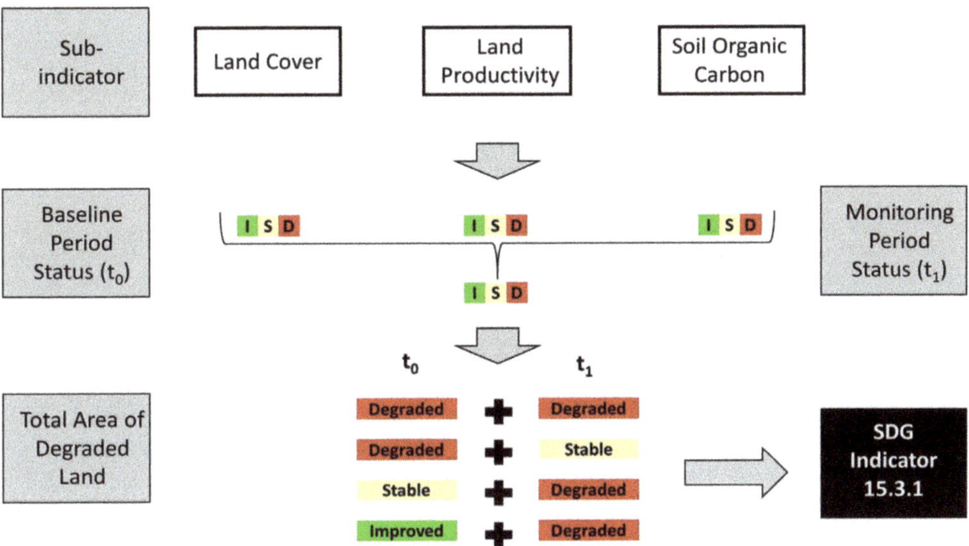

Figure 2. Steps to derive the sustainable development goal (SDG) indicator 15.3.1 from the sub-indicators. I represents Improvement, S represents Stable and D represents degraded (based on [31]).

To calculate the indicator for the reporting year 2019 (t_1), it is necessary first to assess the baseline util t_0 and then calculate the change from the baseline to t_1 (Figure 2). As a final step, combine both results. The details of the calculation of each indicator are explained in the following section. The three LD sub-indicators were created from satellite images using cloud-based geospatial computing. The indicators were calculated using Trends.Earth [30] and Google Earth Engine [32] for the DM and AM, respectively. As Trends.Earth currently only enables the computation for the baseline period (BP), the DM is only available from 2000 to 2015.

2.3.1. Sub-Indicator 1: Land Cover Transitions and Degradation

The first SDG 15.3.1 indicator is the LC change. To assess the LC degradation, the transitions between 2000–2015 and 2015–2018 were analyzed for the baseline and the first monitoring period (MP), respectively. To determine whether changes from one LC class to another are interpreted as degradation, a change matrix can help visualize the transitions (Table 2) based on the Good Practice Guidance by the UNCCD [31]. It is recommended to adopt this matrix for the national context. Therefore, transitions from grasslands to croplands were not considered LD for the AM to avoid tradeoff between ecosystems and food security and between nomadic and sedentary living.

2.3.2. Sub-Indicator 2: Loss of Land Productivity

LP is described as "the biological productive capacity of the land". It is closely associated with net primary productivity [42], which can be measured directly with earth observation methods [22]. NDVI is a widely used index detecting LP [26,43,44]. The LP sub-indicator consists of three distinct components, namely trend, state and performance.

The LP trend component measures the trajectory of change in productivity over time. It is calculated at the pixel level using linear regression and the Mann Kendall significance test [22,45,46]. Positive and negative changes in NDVI indicate increasing and decreasing productivity associated with vegetation recovery and degradation, respectively. The eight most recent years of data were used to create a new distinct and significant time series that is more responsive to present land conditions. Further, following [47], we accounted for the effect of rainfall variability on vegetation productivity trends by using the rain use efficiency (RUE) method.

The LP state component represents recent changes in LP compared to the BP. The yearly NDVI mean images of the shortened BP (2000–2012) were normalized and assigned to classes from 1 to 10 based on their percentiles. To avoid annual fluctuations, contemporary values of the three-year anteceding t_0 and t_1 were classified in this scheme. Areas with a reduction of two or more classes were classified as degraded, while the rise by two categories was interpreted as an improvement [31].

The LP performance component examines local productivity compared to similar ecoregions defined by the unique combination of SoilGrids [41], soil taxonomy great groups and LC classes (Table 1). The 90th percentile in each ecoregion was calculated as a proxy for the maximum productivity level. The LP performance was then calculated based on the ratio of the observed mean NDVI value per pixel and the $NDVI_{max\ (90th)}$. Values below 0.5 indicate regions where the LP is low and LD may prevail [31].

The overall LP sub-indicator is calculated based on the three components mentioned earlier. As the LP trend is based on a statistically significant test, it is most influential, and its status determines LP degradation. Only if both LP status and LP performance show negative results, does the LP indicator also show degradation [22]. If only the LP state component shows degradation, this could indicate "early signs of decline" because the other indicators may not have detected the most recent LD. Further, if only performance shows degradation, there is no temporal trend, and the land is classified as "stable but stressed" [22]. In contrast to the Good Practice Guidance by UNCCD, Trends.Earth (DM) also incorporates the "early signs of decline" state component into the LP degradation [30].

Table 2. Land cover transition matrix (2000–2015) based on the adapted methods (AM). Green, beige and brown colors indicate improving, stable and declining conditions of land cover categories, respectively. The area in km^2 and the possible cause of the land cover transition are indicated in the matrix. The change is based on the high-resolution land cover dataset.

		AM Land Cover Category in 2015 (km^2)						2000 Total (km^2)
		Forestland	Grassland	Cropland	Wetland	Urban	Otherland	
AM land cover category in 2000 (km^2)	Forestland	Stable 1969.4	Vegetation loss 226.8	Deforestation 237.5	Inundation 9.2	Deforestation 3	Vegetation loss 72.7	2519
	Grassland	Afforestation 36	Stable 6932.4	Agricultural expansion 806	Inundation 26.4	Urban expansion 40.4	Vegetation loss 253.5	8094.6
	Cropland	Afforestation 24.1	Withdrawal of agriculture 221.8	Stable 3622.3	Inundation 10.3	Urban expansion 14.7	Vegetation loss 76.2	3969.3
	Wetland	Woody encroachment 3.1	Waterbody drainage 53.4	Waterbody drainage 77.3	Stable 131.4	Waterbody drainage 3.4	Waterbody drainage 28.5	297.1
	Urban	Afforestation 0.4	Vegetation establishment 11.4	Agricultural expansion 32.8	Wetland establishment 0.5	Stable 141.5	Withdrawal of settlements 7.3	193.8
	Otherland	Afforestation 7.6	Vegetation establishment 118.2	Agricultural expansion 149.7	Wetland establishment 15.1	Urban expansion 8.1	Stable 1718.9	2017.5
2015 total (km^2)		2041	7563.9	4925.6	192.9	211	2157	17,091.4

2.3.3. Sub-Indicator 3: Degradation of Soil Organic Carbon

The Good Practice Guidance for the SOC sub-indicator is based on the maximum equilibrium SOC content at a location that is controlled by environmental factors such as rainfall, evaporation, solar radiation, and temperature [22]. The content can change based on three distinct change factors: First, the land-use factor represents SOC stock changes based on the type of land use. Second, the management factor reflects the management practice of the land use (e.g., grazing intensity on grasslands). Third, the input factor represents the different amounts of carbon input into the soil [22,48,49]. While the LULC change factor can be used with LC as a proxy, there are presently no sufficient datasets available to provide information about the management or the input for the other two indicators. Thus, the only indicator to assess SOC changes is the second LD indicator LC change [22].

3. Results

Three sub-indicators, namely LC transitions, LP decline, and SOC loss, were estimated to derive the SDG 15.3.1 indicator using the default and adapted methods. The patterns of each sub-indicator based on DM and AM are described in the following sections starting with the BP from 200 to 2015 for both DM and AM. The first monitoring period from 2015 to 2019 is only assessed using the AM, as the data necessary for this period are currently not available in Trends.Earth.

3.1. Sub-Indicator 1: Land Cover Transitions and Degradation

According to the DM based on the medium-resolution 300 m LC maps, over 99% of the study area remained stable in the BP (2000–2015) (Table A1). Urban areas covering less than 0.1% of the study area experienced the highest relative expansion (56%). The forestlands were the only other LC class that increased in area significantly (4.4%) in the BP.

In contrast to the DM, the AM with high-resolution (30 m) LC data revealed that 6.7% of the total area changed to a less desirable LC class, signifying LD, and only 2.3% of analyzed areas improved. The area of (semi)natural LC, such as forestlands (−19%), grasslands (−6.6%) and wetlands (0.1%), mostly declined, whereas the croplands recorded the highest spatial gain (24.2%) (Figure 3 and Table 2).

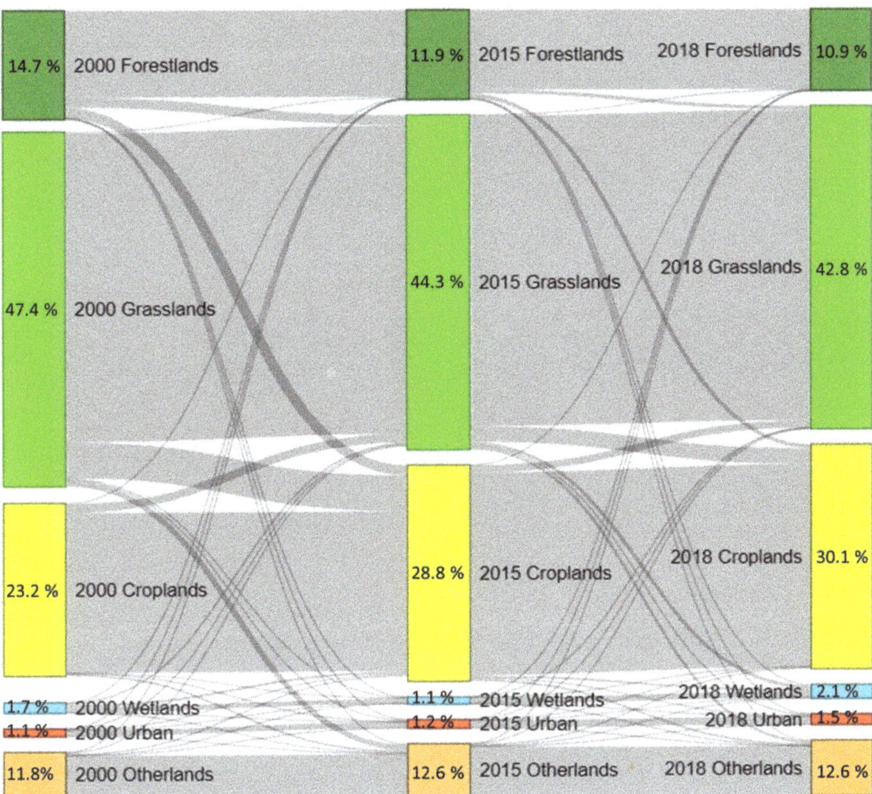

Figure 3. Sankey plot describing the land cover transitions between the years 2000, 2015 and 2018 using high-resolution land cover data. Bands represent the actual proportion of land that changed class over time.

The trend observed in the BP continued in the first years of the MP (Figure 3 and Table A2). Overall, from 2015 to 2018, 3.3% of the total area was degraded during the MP, while 1.2% of the area changed to a more desirable LC. Grass- and forestlands continued to decline by 3 to 9%, respectively, while anthropogenic(-influenced) covers such as cropland and urban areas expanded. Compared to about 3000 ha forests lost per year (a) in the BP, the rate doubled to 6000 ha/a in the MP. Similarly, the changes in croplands increased from 6000 ha/a in 2000–2015 to 7500 ha/a in 2015–2018.

3.2. Sub-Indicator 2: Loss of Land Productivity

The DM revealed that the LP sub-indicator showed degradation in 71.1% of the area during the BP from 2000 to 2015 (Table 3). The LP component trend showed "decline" in 26.8% of the area (Figure A1A). Another 44.3% of the study area showed "early signs of decline" (LP component state, Figure A2A), and the rest (28.9%) remained stable (Figure 4A). According to DM, croplands were most affected (48.4%) by LP decline in 2000–2015 (Figure 5). Forestlands with only about 11.7% marked as degraded were less affected compared to their actual LC share (Figure 5).

Table 3. The land productivity (LP) status in percent for the default (DM) and adapted methods (AM) for the baseline period from 2000 to 2015, as well as for the first monitoring period of 2015-2019. Furthermore, the land cover share of the degraded area in the target year is depicted.

		DM 2000–2015	AM 2000–2015	AM 2015–2019
	Degraded	71.1	8.2	12.2
LP Status (%)	Stable	28.9	91.3	87.7
	Improved	0	0.5	0.1

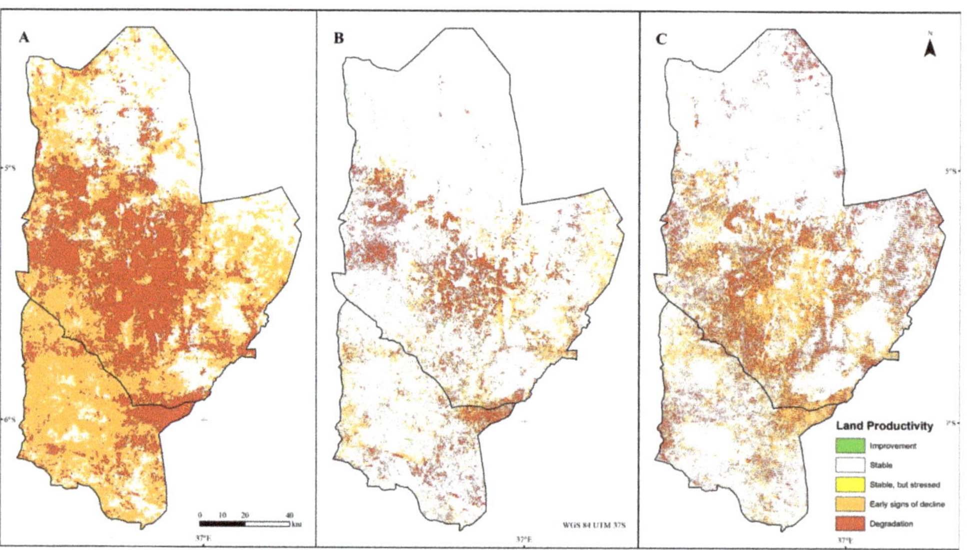

Figure 4. The land productivity sub-indicator generated using (**A**) the default approach with MODIS imagery, (**B**) the adapted approach with Landsat imagery for the baseline period, and (**C**) the adapted approach with Landsat imagery for the monitoring period 2015–2019.

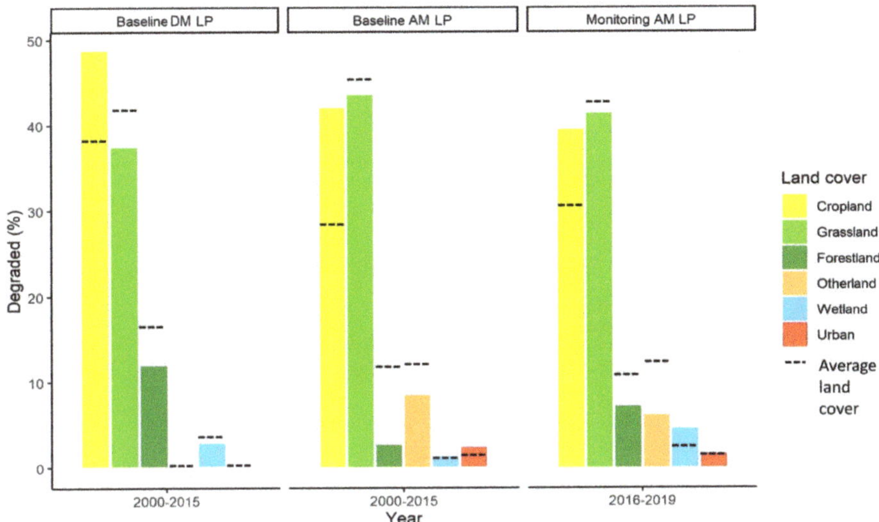

Figure 5. The bar chart showing the distribution of land productivity (LP) decline sub-indicators over the land cover classes using the default (DM) and adapted methods (AM). The dashed lines show the actual land cover share.

Based on the AM applied between 2000 and 2015, the final composite indicators of LP decline revealed that 8.2% of the study area was degraded between 2000 and 2015 (Table 3). This is nearly entirely based on the 8.2% "decline" of the LP trend component (Figure A1B). Further, 9.1% and 1.4% of the study area were marked as showing "early signs of decline" (Figure A2B) and "stable but stressed" areas (Figure A3B), respectively (Figure 4B). Grass- and croplands accounted for 43.5% and 42% of the degraded area (Figure 5). The decline in forestlands was, in turn, detected only on 2.6% of the total degraded area.

LP declined over 12.2% of the study area during the MP from 2015 to 2019 (Figure 4C). With an increase from 9.1% up to 17% of the area, the share of areas with "early signs of decline" (state component) was higher than during the BP (Figure A2). The area where LP was improving was reduced from 855 to 171 km^2 compared to the BP.

3.3. Sub-Indicator 3: Degradation of Soil Organic Carbon

Soil organic carbon was not directly computed but rather assessed through LC classes' alteration and the related change factors [49]. SOC did not change significantly with the DM during the BP from 2000 to 2015: 99.9% of the land did not change the in SOC content by more than 10% (Table 4). Changes in the individual LC classes were also neglectable.

In contrast to DM, the AM approach revealed that during the BP of 2000–2015, 8.4% of the land was degraded due to SOC diminishment, while 2.1% increased in SOC content (Figure 6). The average SOC stock declined from 51.2 to 50.2 t/ha in 2015, losing 1,592,423 t of carbon over 16 years (Table 4). Forestlands had significantly higher SOC stocks (62.2 t/ha) at t_0 than the other LC classes. Based on the transitions in LC, the amount of SOC in forests dropped by 19%, while SOC under agricultural use increased by 25.1%.

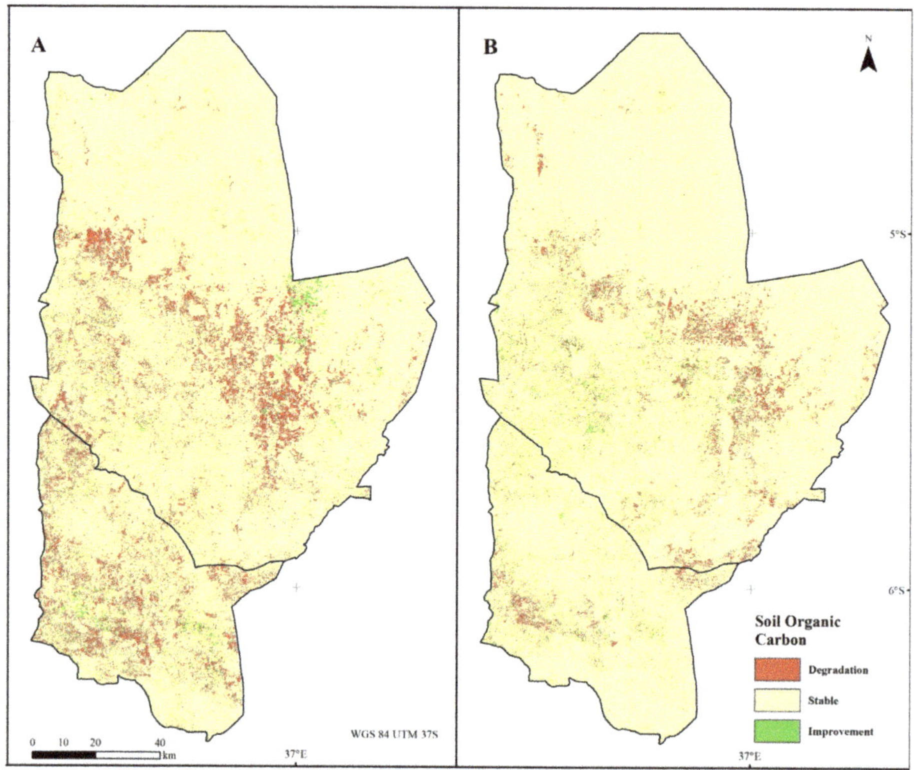

Figure 6. The soil organic carbon sub-indicator generated using the adapted approach with SoilGrids250m for the (**A**) baseline and (**B**) monitoring period.

In the MP, the SOC content experienced significant losses on 3.7% of the land. The same trend was observed in other LC classes (forest-, grass- and wetlands) that gradually lost SOC in the MP (Table 4).

Table 4. The soil organic carbon (SOC) content for the default (DM) and adapted methods (AM) for the baseline period from 2000 to 2015 as well as for the first monitoring period of 2015–2018.

		DM SOC		AM SOC		AM SOC
		2000	2015	2000	2015	2018
	Degraded	0.1		8.1		3.7
Status (%)	Stable	99.9		90		94.7
	Improved	0		2		1.7
	Study area	51.2	51.2	51.2	50.2	49.9
	Forestland	54.7	54.7	63.2	62.2	62
	Grassland	55	55	50.7	49.7	49.5
SOC (t/ha)	Cropland	46.2	46.2	46.5	46.9	46.9
	Wetland	45.1	45.1	49.2	47	46.7
	Urban	36.2	36.2	39.5	42.8	42.8
	Otherland	0	0	46.2	47.6	47.6

3.4. Combined Sustainable Development Indicator 15.3.1 for the Baseline and First Monitoring Period

During the BP from 2000 to 2015, the DM method identified 71.1% of KK's area as degraded and only as 0.5% improved (Figure 7A). This result is mainly caused by the sub-indicator LP, while the two other indicators LC and SOC showed nearly no degradation. The LP degradation was mainly driven by the state component of LP in 70.3% of the total area.

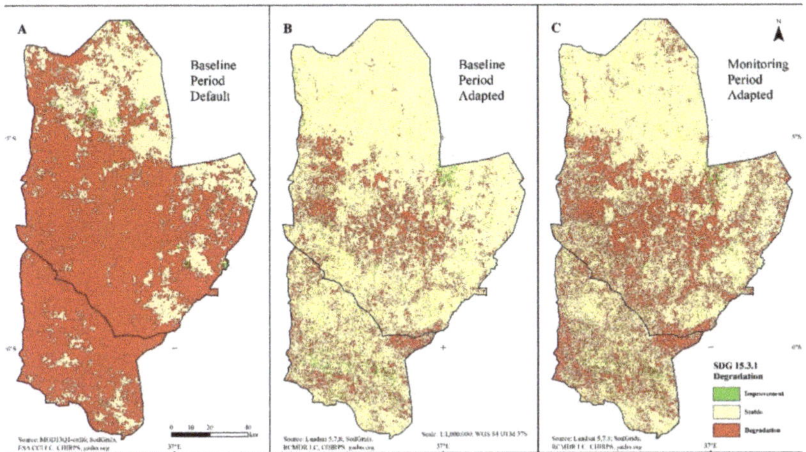

Figure 7. The sustainable development goal (SDG) 15.3.1 indicator "proportion of land that is degraded over total land area" for the baseline period with the (**A**) default and (**B**) adapted methods, and (**C**) for the first monitoring period using the adapted method.

On the contrary, during the BP, the AM showed that 16.4% of the area was degraded and 2.7% improved (Figure 7B). The distinct sub-indicators influenced the final indicator more evenly with 52.4%, 50% and 31.7% by SOC, LP and LC, respectively, compared to the DM.

The AM for the first MP (2015–2019) showed that 16% of the total area was degraded, 1.5% improved and more than 82% remained stable (Figure 7C). Forests and grasslands were the least affected among LC classes. Croplands (38%) and wetlands (7%) experienced the most degradation between 2015 and 2019. Over three-fourths of the degradation was driven by the LP sub-indicator, whereas LC and SOC only contributed 20% and 23% to LD, respectively.

3.5. Combined Sustainable Development Indicator 15.3.1 over 20 Years Using the AM

Over the whole period of 20 years (2000–2019), which results in the SDG 15.3.1 indicator at timestep t_1, 27.7% of KK was degraded, and 2.8% of KK improved (Figure 8A). Thus, the LD was widespread across the two studied districts and formed several LD clusters (Figure 7B,C). The degradation was not equally distributed over the study area: the biggest LD hotspots were Central and Western Kiteto, as well as Western Kongwa (Figure 8A). Even though the land covered by forests decreased and the land covered by crops increased from 2015 to 2018, the degraded proportion changed conversely as follows: The degraded area covered by forests increased to 3.9%, while the area covered by crops sank to 41.9%. While SOC's degraded area only changed slightly, the relative contribution sank from 50 to 30% (Figure 8B). The degraded area, which is solely influenced by LP, rose over 50% and interplayed with others over 70%.

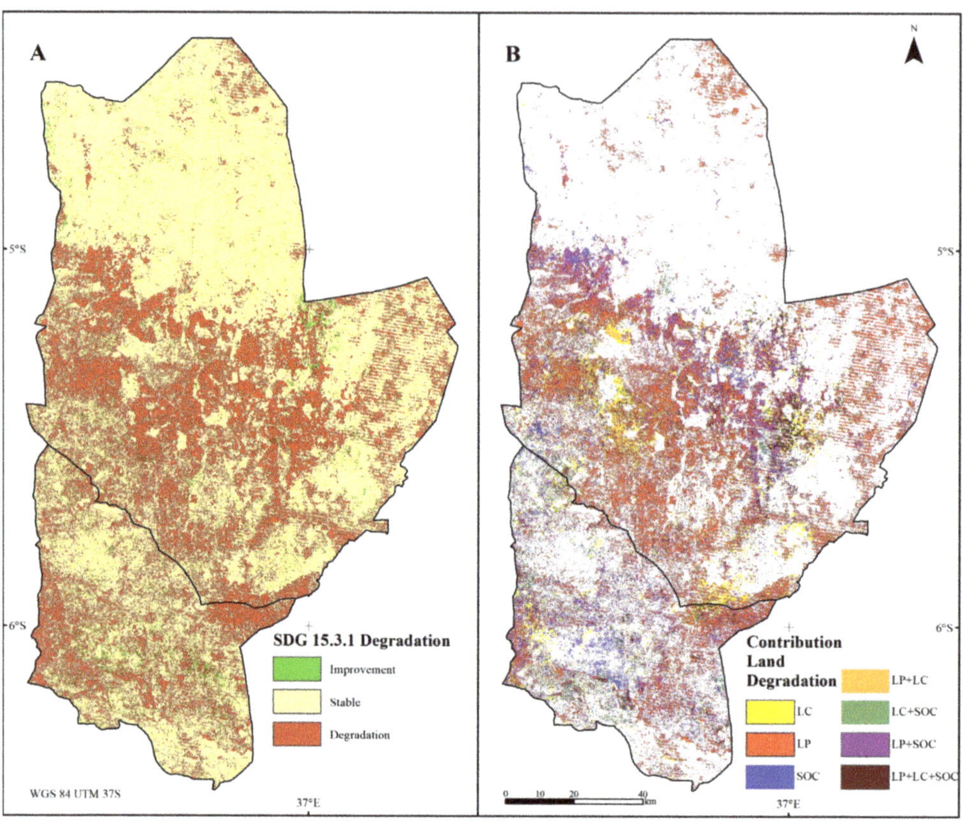

Figure 8. (**A**) The sustainable development goal (SDG) 15.3.1 indicator "proportion of land that is degraded over total land area" for the years 2000–2019 and (**B**) the contribution to the SDG 15.3.1 indicator by its three sub-indicators land cover (LC) change, land productivity (LP) decline and soil organic carbon (SOC) loss.

4. Discussion

The presented study is the first in Africa to support the monitoring of the SDG 15.3.1 indicator using fine-spatial-resolution (30 m) satellite time series data for LD assessment. This is a key contribution considering that previous studies used 250 m to 8 km resolution data [24,26,50] for LP sub-indicator monitoring, unlike our study that utilized long-term Landsat time series for SDG 15.3.1 monitoring. Furthermore, it is the first sub-national study that assesses the SDG 15.3.1 indicator in Tanzania for the BP and includes the MP until 2019. The first 4 out of 15 years of the SDG time frame are assessed and could help identify hotspot areas for targeting the appropriate measures to combat LD in the study area.

The presented LD assessment in KK districts confirmed that the LD problem is acute in Tanzania. The Tanzanian target is to achieve LDN by 2030 [24]. Both KK are part of declared LD hotspot regions, which need to improve 25% of the area based on the status at t_0. According to our analysis, only 2.7% of the land area has improved and 27.7% is degraded. Next to the (sub)national targets, there are also specific targets to avoid, minimize and reverse LD in Tanzania [24]. Among others, about half of the current national forest area should be restored, 50% of the national croplands should improve LP and the SOC content in croplands should rise to 54.5 t/ha [51]. Despite these more specific and ambitious targets, our results show a negative trend in all LD sub-indicators analyzed, suggesting that more efforts are needed to combat LD in the study area.

Precisely, instead of restoring forest areas, even more trees were cut over 19 years (14.7% to 10.9% tree cover). In croplands, LP degradation was above average, while the SOC content in croplands improved marginally. A possible explanation could be that restoration attempts using SLM practices had not yet shown effects, because it takes several years for the change to be monitored remotely [52,53]. Moreover, it takes decades for SOC to change [49,54]. Hence, it is of paramount importance to prioritize the detected LD hotspots for rehabilitation and SLM practices to reverse LD processes.

There are currently no sub-national studies for KK districts. With around 27% of the area in KK being degraded, it is less affected by LD compared to national assessments found in [2,3] or [11]. However, the comparison with these studies is difficult, as they used different monitoring periods (ending in the 2000s and 2016) and only a subset of the methodology (LP trend) and coarse resolution imagery (i.e., 8 km AVHRR data). This suggests that our study brought LD assessment in Tanzania one step further by assessing three components of LD according to the SDG 15.3.1 indicator. Further, using significantly higher spatial resolution, spatial datasets allowed us to reveal spatial patterns of LD beyond pixel sizes of 8 km [2,3,11] or 1 km [24].

Our study compared the results of the LD assessments based on default UNCCD-suggested datasets (250 m MODIS data used for LP sub-indicator and 300 m ESA CCI LC maps) and customized relatively high-resolution datasets (30 m Landsat data used for the LP sub-indicator and 30 m RCMRD LC maps). The resulting differences between LD estimates based on DM and AM were striking and could be primarily attributed to the difference in the pixel size of 6.25 ha (MODIS) versus 0.09 ha (Landsat), which could be critical in specific areas where fine LD patterns prevailed. This finding is confirmed by several studies highlighting the importance of using high-resolution imagery to detect LD, especially on heterogeneous landscapes, such as KK districts, dominated by heterogenous small-scale farms [50,55,56]. Recent studies that used ground-truth data for validation showed that using Landsat data for the LC sub-indicator captured LD better than using ESA-based 300 m datasets [50]. Nevertheless, certain factors could have impacted the AM, such as the scan-line failure in Landsat ETM+ data. To reduce the potential negative influence of this on our analysis, we applied several preprocessing steps confirmed to be effective in similar studies [56].

NDVI was applied in this study, although it was affected by soil brightness in areas with low vegetation cover. Other vegetation indices, such as MSAVI or MSAVI2, are less sensitive to soil optical properties in less vegetated areas and, therefore, can be used to detect a decline in vegetation productivity [57]. However, the alternative indices have significantly better results than NDVI only in areas where bare soils prevail. Further, Tüshaus et al. [58] compared NDVI with the Soil-Adjusted Vegetation Index (SAVI) and MERIS-based Terrestrial Chlorophyll Index (MTCI). The results indicated only little differences between the different vegetation indices. Nevertheless, the impact of different vegetation indices on the estimated LDN sub-indicators can be further tested.

Furthermore, our results pointed out that the ESA CCI LC did not reflect significant LC changes during the BP in KK districts. Other local estimates, such as the National Forest Resources Monitoring and Assessment of Tanzania Mainland [21] or Tanzanian Forest Reference Emission Level [59], suggest a change rate that is three to twenty times higher, respectively, for a similar period analyzed. Our result is in line with the study of Kimaro et al. [60], who investigated the LC change for the study area from 1987 to 2010. Their study indicated that the LC change was already in progress over 30 years ago with heavy declines in (semi)-natural landscapes. This suggests that our research offers advancement of sub-national assessment of LD in heterogeneous landscapes.

Our study revealed that the LP sub-indicator impacted LD in the study area the most (by 50%) using the AM. The remaining half is affected by SOC, LC, or by the combination of more than one sub-indicator. On the other hand, the LD indicator using the DM is nearly solely affected by the LP sub-indicator, which is primarily driven by the state component. This suggests two things: First, our AM is better suited to reflect the ongoing multidimensional degradation in KK districts. Second, even if the ongoing LULC change stops, the degradation will not halt because of the decline in LP.

This is well reflected in croplands, which were the worst affected land cover class, not only in LP decline but also in SOC loss. Due to the continuous cultivation of the agricultural lands combined with overgrazing and little fertilizer inputs, the crop yields in the study area are reportedly low, caused by the limited availability of soil nutrients and organic matter content [18]. Another study that assessed LD in Kenya in similar environmental and land use settings found that croplands experienced the highest decline in LP, indicating that unsustainable farming practices are widespread throughout Eastern Africa [26]. This has serious consequences, as already 30% of the Tanzanian population are undernourished [16], and the yield gap for the main crops needs to be closed for the population to sustain itself in the coming decades [61].

The soils in KK districts lost 1.6 million t of SOC due to LULC change from 2000–2018, according to our study. This is especially dire, as SOC is vital for soil quality and is a key ecosystem indicator [62]. The study by van der Esch et al. [63] suggests that due to LULC change, 27 Gt of SOC will be further lost globally by 2050, mainly in sub-Saharan Africa. Studies conducted in Tanzania found that higher SOC values on the farm level resulted in financial benefits for the farmers [64]. Thus, increasing SOC via SLM practices would not only improve farmers' living conditions but also allow slowing down ongoing SOC degradation.

In contrast to the LP and LC sub-indicators, which have a continuous basis with Landsat and Sentinel missions [65] and for which there are also further high-resolution maps available [66], the SOC sub-indicator still lacks good spatial and temporal coverage. Further, there are currently no sufficient datasets available to provide information about the management or the input for the SOC indicator. Thus, the SOC change is only approximated by the LC change sub-indicator, leading to a misbalance towards the LULC change in the overall SDG 15.3.1 indicator. At the moment of the analysis, the high spatial resolution SOC data by Innovative Solutions for Decision Agriculture (iSDA) based on [67] were not available. Further work should thus address this limitation and incorporate per availability high-resolution SOC data in the analysis, as well as conducting field validation of both approaches. At the beginning of 2021, the UNCCD updated the first version of the SDG 15.3.1 good practice guidance and innovated the methodology [68]. Future studies should therefore adopt this new approach in conjunction with newly available datasets.

The improvement of the subnational analysis with freely available data, the use of cloud computing platforms, and the source code's availability to perform LD assessment present an opportunity to upscale the analysis further and transfer the methods to other study areas.

5. Conclusions

The presented study demonstrates the potential of earth observation for LD monitoring with high spatial resolution data and uses cloud computing approaches with Google Earth Engine, and it improves the measurement of the SDG 15.3.1 indicator in the study area in Tanzania up until 2015 and 2019 at two different levels of spatial detail. Our study thus offers advancement of sub-national assessments of land degradation (LD) in heterogeneous landscapes. The improvement of the sub-national analysis with high-resolution data, the use of cloud computing platforms and the provision of the source code used here to perform LD assessment should encourage a transfer of the here presented approach to other study areas and/or the upscaling of the results of this study to the national level.

For this, we compared two approaches of assessing the SDG indicator 15.3.1 in Kiteto and Kongwa districts of Tanzania. The first method applied the global default (DM) medium resolution datasets proposed by the UNCCD for monitoring LD for the baseline period (BP, 2000–2015). The second method, the adapted method (AM), applied local land cover 30 m maps and 30 m Landsat to monitor LD for the baseline and the first monitoring period (MP, 2015–2019). The LD assessment for the BP reveals large differences between the DM and AM. Using the DM, nearly all degraded area stems from the LP sub-indicator based on 250 m MODIS imagery. In contrast, the degradation was less than 1% for the LC and SOC change sub-indicators, calculated based on ESA CCI LC (300 m) maps. The LD captured by the AM based on Landsat time series and 30 m LC data was evenly distributed between the three sub-indicators and revealed LD on 27.7% of the area. We, therefore, concluded that the results derived from medium-resolution datasets are likely to over- and underestimate the LD for different sub-indicators and, thus, might misinform policy- and decision-makers and land managers if used operationally. Further, our study concluded that the local datasets and high-resolution imagery are essential to capture subtle changes within the heterogeneous landscape in semiarid central Tanzania.

Our results confirmed that LD is currently ongoing in the study area. The LD did not halt after 2015 but spread further across the districts and formed several severe LD clusters. Therefore, to achieve the national LDN targets, it is crucial to address the most important LD causes, such as overgrazing and unsustainable farming in the study area. The application of SLM practices would enhance the low LP in croplands and prevent LULC change in KK districts.

Further work should incorporate high-resolution SOC data in the analysis and conduct field validation of LD assessments resulting from both approaches.

Author Contributions: Conceptualization, methodology, software and writing—original draft preparation, J.R.; writing—review and editing, J.R., G.G., F.M. and O.D.; visualization, J.R. and G.G.; supervision, F.M. and O.D.; funding acquisition, F.M. All authors have read and agreed to the published version of the manuscript.

Funding: This research was funded by the United States Agency for International Development, grant number AID-BFS-G-11-00002.

Data Availability Statement: The Google Earth Engine code can be found online on https://github.com/JAReith/SDG15.3.1.

Acknowledgments: The authors thank the German Academic Exchange Service for the generous funding by the PROMOS-program and the financial support by the International Institute for Tropical Agriculture for the fieldwork in Tanzania.

Appendix A

Table A1. Land cover transition matrix (2000–2015) based on the default methods (DM). Green, beige and brown colors indicate improving, stable and declining conditions of land cover categories, respectively. The area in km^2 and the possible cause of the land cover transition are indicated in the matrix. The change is based on the moderate-resolution land cover dataset.

		DM Land Cover Category in 2015 (km^2)						
		Forestland	Grassland	Cropland	Wetland	Urban	Otherland	2000 Total (km^2)
DM Land cover category in 2000 (km^2)	Forestland	Stable 2810.18	Vegetation loss 3,83	Deforestation 0.68	Inundation 0	Deforestation 0.49	Vegetation loss 0	2815.19
	Grassland	Afforestation 127.43	Stable 6976.03	Agricultural expansion 18.72	Inundation 0.74	Urban expansion 0	Vegetation loss 0	7122.92
	Cropland	Afforestation 3.09	Withdrawal of agriculture 4.2	Stable 6606.86	Inundation 0	Urban expansion 0.25	Vegetation loss 0	6614.40
	Wetland	Woody encroachment 0	Waterbody drainage 0	Waterbody drainage 0	Stable 537.22	Waterbody drainage 0.12	Waterbody drainage 0	537.34
	Urban	Afforestation 0	Vegetation establishment 0	Agricultural expansion 0	Wetland establishment 0	Stable 1.54	Withdrawal of settlements 0	1.54
	Otherland	Afforestation 0	Vegetation establishment 0	Agricultural expansion 0	Wetland establishment 0	Urban expansion 0	Stable 0	0
2015 total (km^2)		2940.71	6984.06	6626.25	537.96	2.41	0	17,091.40

Table A2. Land cover transition matrix in km^2 (2015–2018) based on the adapted methods (AM). Green, beige and brown colors indicate improvement, stable and decline of land cover category, respectively. The area and the possible cause of the land cover transition are indicated in the matrix. The change is based on the high-resolution land cover dataset.

		AM Land Cover Category in 2018 (km^2)						
		Forestland	Grassland	Cropland	Wetland	Urban	Otherland	2015 Total (km^2)
AM Land cover category in 2015 (km^2)	Forestland	Stable 1819.5	Vegetation loss 89.5	Deforestation 94.2	Inundation 19.3	Deforestation 5.4	Vegetation loss 13.1	2041
	Grassland	Afforestation 26.2	Stable 6993	Agricultural expansion 368.1	Inundation 77	Urban expansion 22.7	Vegetation loss 77	7563.9
	Cropland	Afforestation 9.3	Withdrawal of agriculture 175.4	Stable 4602.4	Inundation 50.9	Urban expansion 23	Vegetation loss 64.6	4925.6
	Wetland	Woody encroachment 0.7	Waterbody drainage 5.5	Waterbody drainage 7.4	Stable 173.7	Waterbody drainage 0.8	Waterbody drainage 4.7	192.9
	Urban	Afforestation 0.1	Vegetation establishment 2.2	Agricultural expansion 10	Wetland establishment 0.6	Stable 196.7	Withdrawal of settlements 1.4	211
	Otherland	Afforestation 2.7	Vegetation establishment 55.7	Agricultural expansion 69.7	Wetland establishment 30.6	Urban expansion 10.1	Stable 1988.3	2157
2018 total (km^2)		1858.4	7321.4	5151.7	352.2	258.7	2149.1	17,091.4

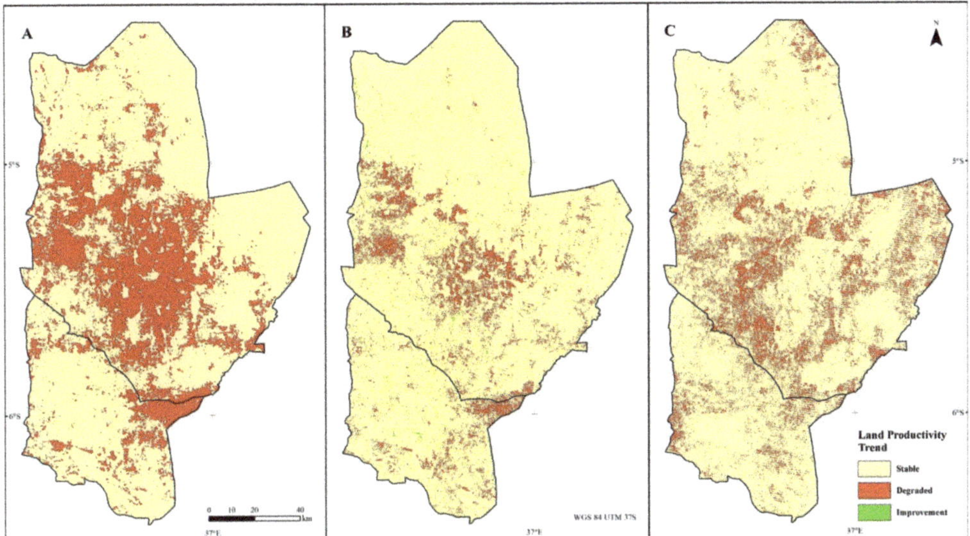

Figure A1. The land productivity component trend generated using (**A**) the default approach with MODIS imagery, (**B**) the adapted approach with Landsat imagery for the baseline period, and (**C**) the adopted approach with Landsat imagery for the monitoring period 2015–2019.

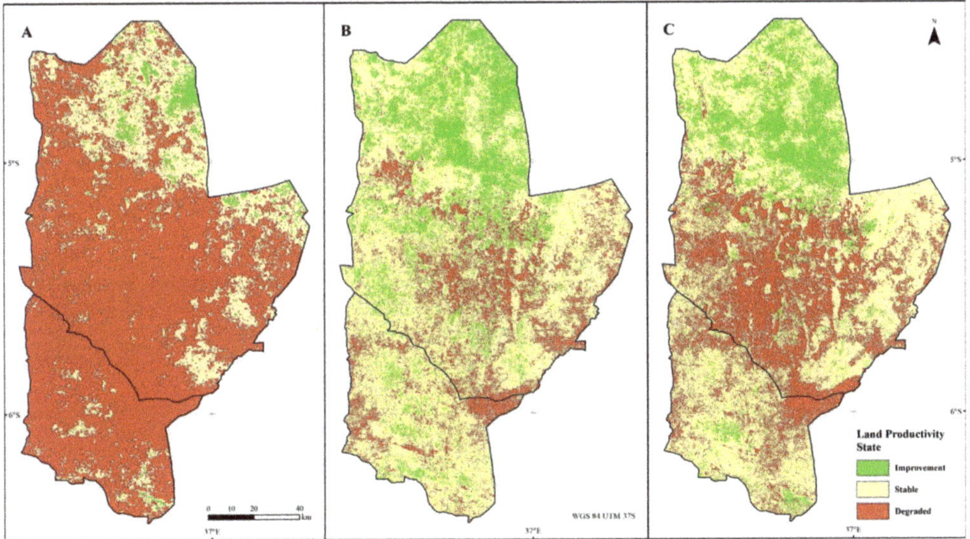

Figure A2. The land productivity component state generated using (**A**) the default approach with MODIS imagery, (**B**) the adapted approach with Landsat imagery for the baseline period, and (**C**) the adopted approach with Landsat imagery for the monitoring period 2015–2019.

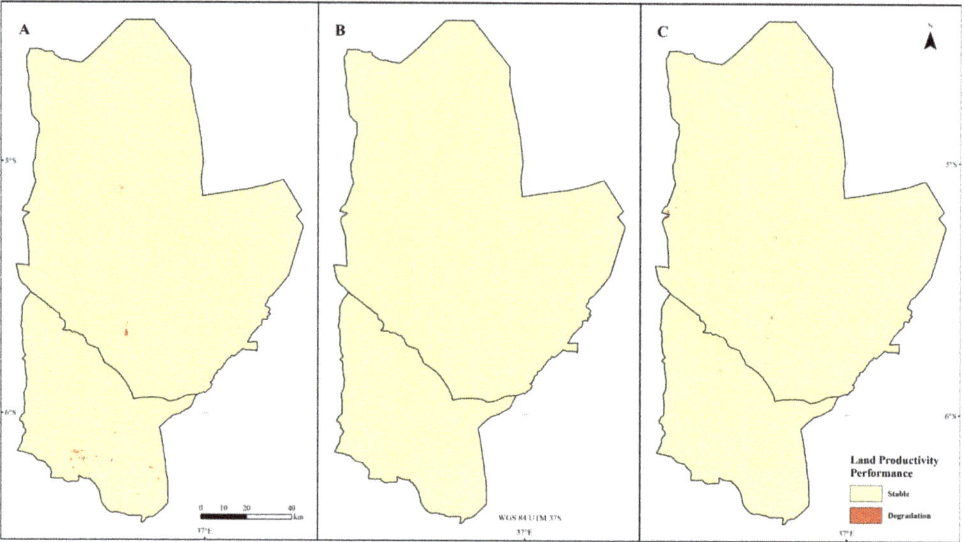

Figure A3. The land productivity component performance generated using (**A**) the default approach with MODIS imagery, (**B**) the adapted approach with Landsat imagery for the baseline period, and (**C**) the adopted approach with Landsat imagery for the monitoring period 2015–2019.

References

1. UNCCD. *Elaboration of an International Convention to Combat Desertification in Countries Experiencing Serious Drought and/or Desertification, Particularly in Africa*; UNCCD: Paris, France, 1994.
2. Le, Q.B.; Nkonya, E.; Mirzabaev, A. Biomass productivity-based mapping of global land degradation hotspots. In *Economics of Land Degradation and Improvement*; Nkonya, E., Mirzabaev, A., von Braun, J., Eds.; Springer Open: Cham, Switzerland, 2016; Volume 24, pp. 55–84. ISBN 978-3-319-19167-6.
3. Bai, Z.; Dent, D.L.; Olsson, L.; Schaepman, M.E. Proxy Global Assessment of Land Degradation. *Soil Use Manag.* **2008**, *24*, 223–234. [CrossRef]
4. Dubovyk, O. The Role of Remote Sensing in Land Degradation Assessments: Opportunities and Challenges. *Eur. J. Remote Sens.* **2017**, *50*, 601–613. [CrossRef]
5. Barbier, E.B.; Hochard, J.P. Land Degradation and Poverty. *Nat. Sustain.* **2018**, *1*, 623–631. [CrossRef]
6. IPBES. The IPBES Assessment Report on Land Degradation and Restoration; 2018. Bonn: Intergovernmental Science-Policy Platform on Biodiversity and Ecosystem Services (IPBES). Available online: https://www.ipbes.net/assessment-reports/ldr (accessed on 24 April 2021).
7. ELD Initiative. *The Value of Land: Prosperous Lands and Positive Rewards through Sustainable Land Management*; Stewart, N., Ed.; Economics of Land Degradation Initiative: Bonn, Germany, 2015; ISBN 978-92-808-6061-0.
8. IPCC Climate Change and Land. *IPCC Special Report on Climate Change, Desertification, Land Degradation, Sustainable Land Management, Food Security, and Greenhouse Gas Fluxes in Terrestrial Ecosystems*; IPCC: Geneve, Switzerland, 2019.
9. UNCCD Report of the Conference of the Parties on Its Seventeenth Session, Held in Durban from 28 November to 11 December 2011: Part Two: Action Taken by the Conference of the Parties at Its Eleventh Session: ICCD/COP(11)/23/Add.1. 2013. Available online: https://unfccc.int/resource/docs/2011/cop17/eng/09.pdf (accessed on 24 April 2021).
10. UN Transforming Our World: The 2030 Agenda for Sustainable Development; 2015. Available online: https://stg-wedocs.unep.org/bitstream/handle/20.500.11822/11125/unepswiosm1inf7sdg.pdf?sequence=1 (accessed on 24 April 2021).
11. Kirui, O.K.; Mirzabaev, A.; von Braun, J. Assessment of Land Degradation 'on the Ground' and from 'Above'. *SN Appl. Sci.* **2021**, *3*, 318. [CrossRef]
12. *FAO The Global Forest Resources Assessment 2015: Desk Reference*; Food and Agriculture Organization of the United Nations: Rome, Italy, 2015.
13. Kirui, O.K. Economics of land degradation and improvement in tanzania and malawi. In *Economics of Land Degradation and Improvement*; Nkonya, E., Mirzabaev, A., von Braun, J., Eds.; Springer Open: Cham, Switzerland, 2016; Volume 22, pp. 609–649, ISBN 978-3-319-19167-6.
14. FAO. *NBS Tanzania: CountrySTAT*; Food and Agriculture Organization of the United Nations: Rome, Italy; National Bureau of Statistics: Dar es Salam, Tanzania, 2020. Available online: http://tanzania.countrystat.org/home/en/ (accessed on 24 April 2021).

15. *NBS National Environment Statistics Report, 2017: Tanzania Mainland*; National Bureau of Statistics: Dar es Salam, Tanzania, 2018.
16. FAO. *FAO Stat Tanzania*; Food and Agriculture Organization of the United Nations: Rome, Italy, 2019; Available online: http://www.fao.org/faostat/en/#country/215 (accessed on 24 April 2021).
17. Wheeler, T.; von Braun, J. Climate Change Impacts on Global Food Security. *Science* **2013**, *341*, 508–513. [CrossRef]
18. Kimaro, A.A.; Mpanda, M.; Meliyo, J.L.; Ahazi, M.; Ermias, B.; Shepherd, K.D.; Coe, R.; Okori, P.; Mowo, J.G.; Bekunda, M. Soil Related Constraints for Sustainable Intensification of Cereal-Based Systems in Semi-Arid Central Tanzania. In *Proceedings of the Tropentag 2015: Management of Land Use Systems for Enhanced Food Security: Conflicts, Controversies and Resolutions, Berlin, Germany, 16–18 September 2015*; Tielkes, E., Ed.; Cuvillier Verlag: Göttingen, Germany, 2015; p. 675. ISBN 978-3-7369-9092-0.
19. Liniger, H.; Harari, N.; van Lynden, G.; Fleiner, R.; de Leeuw, J.; Bai, Z.; Critchley, W. Achieving Land Degradation Neutrality: The Role of SLM Knowledge in Evidence-Based Decision-Making. *Environ. Sci. Policy* **2019**, *94*, 123–134. [CrossRef]
20. Orr, B.J.; Cowie, A.L.; Castillo, V.M.; Chasek, P.; Crossman, N.D.; Erlewein, A.; Louwagie, G.; Maron, M.; Metternicht, G.I.; Minelli, S.; et al. *Scientific Conceptual Framework for Land Degradation Neutrality: A Report of the Science-Policy Interface*; United Nations Convention to Combat Desertification (UNCCD): Bonn, Germany, 2017; pp. 1–98.
21. Caspari, T.; van Lynden, G.; Bai, Z. *Land Degradation Neutrality: An Evaluation of Methods*; Ehlers, K., Ed.; Federal Ministry for the Environment, Nature Conservation, Building and Nuclear safety, Umweltbundesamt: Dessau-Roßlau, Germany, 2015.
22. URT. *URT Land Degradation Neutrality Target Setting Programme Report*; The United Republic of Tanzania (URT): Dar es Salam, Tanzania, 2018.
23. Giuliani, G.; Chatenoux, B.; Benvenuti, A.; Lacroix, P.; Santoro, M.; Mazzetti, P. Monitoring Land Degradation at National Level Using Satellite Earth Observation Time-Series Data to Support SDG15—Exploring the Potential of Data Cube. *Big Earth Data* **2020**, *5*, 1–20. [CrossRef]
24. Gichenje, H.; Godinho, S. Establishing a Land Degradation Neutrality National Baseline through Trend Analysis of GIMMS NDVI Time-Series. *Land Degrad. Dev.* **2018**, *29*, 2985–2997. [CrossRef]
25. Frederique, M.; Agnes, B.; Louise, L.; Clovis, G. Sensitivity Analysis of Land Productivity Change Calculation in Mozambique. In Proceedings of the IGARSS 2019—2019 IEEE International Geoscience and Remote Sensing Symposium, Yokohama, Japan, 28 July 2019; pp. 1633–1636.
26. Anderson, K.; Ryan, B.; Sonntag, W.; Kavvada, A.; Friedl, L. Earth Observation in Service of the 2030 Agenda for Sustainable Development. *Geo Spat. Inf. Sci.* **2017**, *20*, 77–96. [CrossRef]
27. Beck, H.E.; Zimmermann, N.E.; McVicar, T.R.; Vergopolan, N.; Berg, A.; Wood, E.F. Present and Future Köppen-Geiger Climate Classification Maps at 1-Km Resolution. *Sci. Data* **2018**, *5*, 180214. [CrossRef]
28. Conservation International TRENDS.EARTH. *Trends.Earth Documentation: Release 0.66*; TRENDS.EARTH: Arlington, GA, USA, 2019.
29. Sims, N.C.; England, J.R.; Newnham, G.J.; Alexander, S.; Green, C.; Minelli, S.; Held, A. Developing Good Practice Guidance for Estimating Land Degradation in the Context of the United Nations Sustainable Development Goals. *Environ. Sci. Policy* **2019**, *92*, 349–355. [CrossRef]
30. Gorelick, N.; Hancher, M.; Dixon, M.; Ilyushchenko, S.; Thau, D.; Moore, R. Google Earth Engine: Planetary-Scale Geospatial Analysis for Everyone. *Remote Sens. Environ.* **2017**, *202*, 18–27. [CrossRef]
31. Roy, D.P.; Kovalskyy, V.; Zhang, H.K.; Vermote, E.F.; Yan, L.; Kumar, S.S.; Egorov, A. Characterization of Landsat-7 to Landsat-8 Reflective Wavelength and Normalized Difference Vegetation Index Continuity. *Remote Sens. Environ.* **2016**, *185*, 57–70. [CrossRef]
32. Foga, S.; Scaramuzza, P.L.; Guo, S.; Zhu, Z.; Dilley, R.D.; Beckmann, T.; Schmidt, G.L.; Dwyer, J.L.; Joseph Hughes, M.; Laue, B. Cloud Detection Algorithm Comparison and Validation for Operational Landsat Data Products. *Remote Sens. Environ.* **2017**, *194*, 379–390. [CrossRef]
33. Zhu, Z.; Woodcock, C.E. Continuous Change Detection and Classification of Land Cover Using All Available Landsat Data. *Remote Sens. Environ.* **2014**, *144*, 152–171. [CrossRef]
34. Sims, N.C.; Green, C.; Newnham, G.J.; England, J.R.; Held, A.; Wulder, M.; Herold, M.; Cox, S.J.D.; Huete, A.R.; Kuma, L.; et al. *Good Practice Guidance: SDG Indicator 15.3.1: Proportion of Land That Is Degraded over Total Land Area, Version 1.0*; United Nations Convention to Combat Desertification (UNCCD): Bonn, Germany, 2017.
35. ESA. Land Cover CCI Product User Guide Version 2. Tech. Rep. 2017. Available online: http://www.esa-landcover-cci.org/?q=node/199 (accessed on 24 April 2021).
36. RCMRD Land Use Land Cover and Change Mapping Service. Available online: https://rcmrd.org/servir-land-use-land-cover-and-change-mapping-service (accessed on 30 March 2021).
37. USGS MOD13Q1 MODIS/Terra Vegetation Indices 16-Day L3 Global 250m SIN Grid V006. U.S. Geological Survey, Sioux Falls, United States of America. 2015. Available online: https://lpdaac.usgs.gov/products/mod13q1v006/ (accessed on 30 March 2021).
38. Wulder, M.A.; Loveland, T.R.; Roy, D.P.; Crawford, C.J.; Masek, J.G.; Woodcock, C.E.; Allen, R.G.; Anderson, M.C.; Belward, A.S.; Cohen, W.B.; et al. Current Status of Landsat Program, Science, and Applications. *Remote Sens. Environ.* **2019**, *225*, 127–147. [CrossRef]
39. Funk, C.; Peterson, P.; Landsfeld, M.; Pedreros, D.; Verdin, J.; Shukla, S.; Husak, G.; Rowland, J.; Harrison, L.; Hoell, A.; et al. The Climate Hazards Infrared Precipitation with Stations—A new Environmental Record for Monitoring Extremes. *Sci Data.* **2015**, *2*, 150066. [CrossRef] [PubMed]

40. Hengl, T.; Mendes de Jesus, J.; Heuvelink, G.B.M.; Ruiperez Gonzalez, M.; Kilibarda, M.; Blagotić, A.; Shangguan, W.; Wright, M.N.; Geng, X.; Bauer-Marschallinger, B.; et al. SoilGrids250m: Global Gridded Soil Information Based on Machine Learning. *PLoS ONE* **2017**, *12*, e0169748. [CrossRef] [PubMed]

41. Cowie, A.L.; Orr, B.J.; Castillo Sanchez, V.M.; Chasek, P.; Crossman, N.D.; Erlewein, A.; Louwagie, G.; Maron, M.; Metternicht, G.I.; Minelli, S.; et al. Land in Balance: The Scientific Conceptual Framework for Land Degradation Neutrality. *Environ. Sci. Policy* **2018**, *79*, 25–35. [CrossRef]

42. Clark, D.A.; Brown, S.; Kicklighter, D.W.; Chambers, J.Q.; Thomlinson, J.R.; Ni, J. Measuring Net Primary Production in Forests: Concepts and Field Methods. *Ecol. Appl.* **2001**, *11*, 356. [CrossRef]

43. Ivits, E.; Cherlet, M. *Land-Poductivity Dnamics Twards Itegrated Asessment of Land Degradation at Global Scales*; EUR, Scientific and Technical Research Series; Publications Office: Luxembourg, 2016; Volume 26052, ISBN 978-92-79-32354-6.

44. Landmann, T.; Dubovyk, O. Spatial Analysis of Human-Induced Vegetation Productivity Decline over Eastern Africa Using a Decade (2001–2011) of Medium Resolution MODIS Time-Series Data. *Int. J. Appl. Earth Obs. Geoinf.* **2014**, *33*, 76–82. [CrossRef]

45. Mann, H.B. Nonparametric Tests against Trend. *Econometrica* **1945**, *13*, 245. [CrossRef]

46. Kendall, M.G. *Rank Correlation Methods*; C. Griffin: Glasgow, UK, 1948.

47. Le Houerou, H.N. Rain Use Efficiency: A Unifying Concept in Arid-Land Ecology. *J. Arid Environ.* **1984**, *7*, 213–247.

48. *IPCC 2006 IPCC Guidelines for National Greenhouse Gas Inventories*; Eggleston, H.S.; Miwa, K.; Srivastava, N.; Tanabe, K. (Eds.) Institute for Global Environmental Strategies (IGES): Hayama, Japan, 2008.

49. Mattina, D.; Erdogan, H.E.; Wheeler, I.; Crossman, N.; Minelli, S.; Cumani, R. *Default Data: Methods and Interpretation: A Guidance Document for the 2018 UNCCD Reporting*; United Nations Convention to Combat Desertification (UNCCD): Bonn, Germany, 2018.

50. Akinyemi, F.O.; Ghazaryan, G.; Dubovyk, O. Assessing UN Indicators of Land Degradation Neutrality and Proportion of Degraded Land over Botswana Using Remote Sensing Based National Level Metrics. *Land Degrad. Dev.* **2020**. [CrossRef]

51. URT. *URT Voluntary Land Degradation Neutrality Targets and Associated Measures of the United Republic of Tanzania*; The United Republic of Tanzania (URT): Dar es Salam, Tanzania, 2018.

52. *GEF Value for Money Analysis for the Land Degradation projects of the GEF*; Independent Evaluation Office, Global Environment Facility: Washington, DC, USA, 2016.

53. Gonzalez-Roglich, M.; Zvoleff, A.; Noon, M.; Liniger, H.; Fleiner, R.; Harari, N.; Garcia, C. Synergizing Global Tools to Monitor ProgressTtowards Land Degradation Neutrality: Trends.Earth and the World Overview of Conservation Approaches and Technologies Sustainable Land Management Database. *Environ. Sci. Policy* **2019**, *93*, 34–42. [CrossRef]

54. Bernoux, M.; Feller, C.; Cerri, C.C.; Eschenbrenner, V.; Cerri, C.E.P. Soil carbon sequestration. In *Soil Erosion and Carbon Dynamics*; Advances in Soil Science; Roose, E., Lal, R., Feller, C., Barthes, B., Stewart, B., Eds.; CRC/Taylor & Francis: Boca Raton, FL, USA, 2006; pp. 13–22, ISBN 978-1-56670-688-9.

55. Fiorillo, E.; Maselli, F.; Tarchiani, V.; Vignaroli, P. Analysis of Land Degradation Processes on a Tiger Bush Plateau in South West Niger Using MODIS and LANDSAT TM/ETM+ Data. *Int. J. Appl. Earth Obs. Geoinf.* **2017**, *62*, 56–68. [CrossRef]

56. Venter, Z.S.; Scott, S.L.; Desmet, P.G.; Hoffman, M.T. Application of Landsat-Derived Vegetation Trends over South Africa: Potential for Monitoring Land Degradation and Restoration. *Ecol. Indic.* **2020**, *113*, 106206. [CrossRef]

57. Montandon, L.M.; Small, E.E. The Impact of Soil Reflectance on the Quantification of the Green Vegetation Fraction from NDVI. *Remote Sens. Environ.* **2008**, *112*, 1835–1845. [CrossRef]

58. Tüshaus, J.; Dubovyk, O.; Khamzina, A.; Menz, G. Comparison of Medium Spatial Resolution ENVISAT-MERIS and Terra-MODIS Time Series for Vegetation Decline Analysis: A Case Study in Central Asia. *Remote Sens.* **2014**, *6*, 5238–5256. [CrossRef]

59. URT. *URT Tanzania's Forest Reference Emission Level Submission to the UNFCCC*; The United Republic of Tanzania (URT): Dar es Salam, Tanzania, 2017.

60. Kimaro, A.A.; Weldesmayat, S.G.; Mpanda, M.; Swai, E.; Kayeye, H.; Nyoka, B.I.; Majule, A.E.; Perfect, J.; Kundhlade, G. *Final Technical Report for the Jumpstrat Projects: Evidence-Based Scaling-up of Evergreen Agriculture for Increasing Crop Productivity, Fodder Supply and Resilience of the Maize-Mixed and Agro-Pastoral Farming Systems in Tanzania and Malawi*; World Agroforestry Centre: Nairobi, Kenya, 2012.

61. van Ittersum, M.K.; van Bussel, L.G.J.; Wolf, J.; Grassini, P.; van Wart, J.; Guilpart, N.; Claessens, L.; de Groot, H.; Wiebe, K.; Mason-D'Croz, D.; et al. Can Sub-Saharan Africa Feed Itself? *Proc. Natl. Acad. Sci. USA* **2016**, *113*, 14964–14969. [CrossRef] [PubMed]

62. Chotte, J.-L.; Aynekulu, E.; Cowie, A.L.; Campbell, E.; Vlek, P.; Lal, R.; Kapovic-Solomun, M.; von Maltitz, G.P.; Kust, G.; Barger, N.; et al. *Realising the Carbon Benefits of Sustainable Land Management Practices: Guidelines for Estimation of Soil Organic Carbon in the Context of Land Degradation Neutrality Planning and Monitoring: A Report of the Science-Policy Interface*; United Nations Convention to Combat Desertification (UNCCD): Bonn, Germany, 2019.

63. van der Esch, S.; ten Brink, B.; Stehfest, E.; Bakkenes, M.; Sewell, A.; Bouwman, A.; Meijer, J.; Westhoek, H.; van den Berg, M.; van den Born, G.J.; et al. *Exploring Future Changes in Land Use and Land Condition and the Impacts on Food, Water, Climate Change and Biodiversity: Scenarios for the UNCCD Global Land Outlook*; Policy Report; PBL Netherlands Environmental Assessment Agency: The Hague, The Netherlands, 2017.

64. Bhargava, A.K.; Vågen, T.-G.; Gassner, A. Breaking Ground: Unearthing the Potential of High-Resolution, Remote-Sensing Soil Data in Understanding Agricultural Profits and Technology Use in Sub-Saharan Africa. *World Dev.* **2018**, *105*, 352–366. [CrossRef]

65. Malenovsky, Z.; Rott, H.; Cihlar, J.; Schaepman, M.E.; Garcia-Santos, G.; Fernandes, R.; Berger, M. Sentinels for Science: Potential of Sentinel-1, -2, and -3 Missions for Scientific Observations of Ocean, Cryosphere, and Land. *Remote Sens. Environ.* **2012**, *120*, 91–101. [CrossRef]
66. ESA CCI LAND COVER—S2 Prototype Land Cover 20 m Map of Africa. 2016. Available online: http://2016africalandcover20m.esrin.esa.int/ (accessed on 30 March 2021).
67. Hengl, T.; MacMillan, R.A. *Predictive Soil Mapping with R*; OpenGeoHub Foundation: Wageningen, The Netherlands, 2019; ISBN 978-0-359-30635-0.
68. Sims, N.C.; Newnham, G.J.; England, J.R.; Guerschman, J.; Cox, S.J.D.; Roxburgh, S.H.; Viscarra-Rossel, R.A.; Fritz, S. *Wheeler Good Practice Guidance. SDG Indicator 15.3.1, Proportion of Land That Is Degraded Over Total Land Area. Version 2.0*; United Nations Convention to Combat Desertification (UNCCD): Bonn, Germany, 2021.

MDPI
St. Alban-Anlage 66
4052 Basel
Switzerland
Tel. +41 61 683 77 34
Fax +41 61 302 89 18
www.mdpi.com

Remote Sensing Editorial Office
E-mail: remotesensing@mdpi.com
www.mdpi.com/journal/remotesensing